Advances in Cancer Therapy from Research to Clinical Practice—Surgical, Molecular or Systemic Management of Cancer

Advances in Cancer Therapy from Research to Clinical Practice—Surgical, Molecular or Systemic Management of Cancer

Editors

Nicolae Crisan
Călin Căinap

MDPI • Basel • Beijing • Wuhan • Barcelona • Belgrade • Novi Sad • Cluj • Manchester

Editors
Nicolae Crisan
Iuliu Hatieganu University of Medicine and Pharmacy
Cluj-Napoca
Romania

Călin Căinap
Iuliu Hatieganu University of Medicine and Pharmacy
Cluj-Napoca
Romania

Editorial Office
MDPI
St. Alban-Anlage 66
4052 Basel, Switzerland

This is a reprint of articles from the Special Issue published online in the open access journal *Medicina* (ISSN 1648-9144) (available at: https://www.mdpi.com/journal/medicina/special_issues/Cancer_Oncological_Disease).

For citation purposes, cite each article independently as indicated on the article page online and as indicated below:

Lastname, A.A.; Lastname, B.B. Article Title. *Journal Name* **Year**, *Volume Number*, Page Range.

ISBN 978-3-0365-8782-0 (Hbk)
ISBN 978-3-0365-8783-7 (PDF)
doi.org/10.3390/books978-3-0365-8783-7

Contents

About the Editors

Nicolae Crisan

Professor Nicolae Crisan is the Chief of the Urology Department and Manager of the Centre for Practical Skills and Simulation in Medicine within "Iuliu Hatieganu" University of Medicine and Pharmacy, Cluj-Napoca, Romania. He works within the Clinical Municipal Hospital and Regina Maria Hospital and he has extensive experience in robotic urologic surgery. He has completed multiple training stages: in oncologic and robotic surgery, within the Milano European Oncology Institute—Prof. Dr. Ottavio de Cobelli (2008); specialization in urologic robotic surgery at San Giuseppe Hospital in Milano—Prof. Dr Deliu Victor Matei (2008); and training at Da Vinci Surgical System Off-Site Training Programme for a Console Surgeon" IRCAD-EITS, Strasbourg, France. He has significant experience in 3D laparoscopy and robotics-assisted approaches for oncological urology. His current research concerns are as follows: robotic surgery, robotic and imaging (MRI) assistance of prostate biopsy, augmented-reality-guided surgery and applications of artificial intelligence in urology. He has published over 80 scientific papers, 32 book chapters, and 1 patent, and he has participated in 6 research national and international projects. He has participated as an invited speaker in more than 70 national and international medical conferences. Dr Nicolae Crisan is Vice-President of the Romanian Association of Urology and a member of the SIU Uro-Oncology—Kidney Working Group.

Călin Căinap

Călin Căinap, Professor in Medical Oncology, Habil, Md PhD.

Călin Căinap is the Head of the Oncology Department of Iuliu Hatieganu University of Medicine and Pharmacy Cluj Napoca (from 2015), and since 2022, has also been Head of the Medical Oncology Discipline. During his specialization, he had the opportunity to work in France; for 6 months he worked in Rouen at the Charle Nicolle Hospital in the unit of digestive cancerology, and for 1 year he worked at the Gustave Roussy Institute (Paris) in the hematology and autograft unit. After returning to Romania, he began his university career, starting as Assistant Professor (2004), then Senior Lecturer (2011), Associate Professor (2016), and finally Professor in Medical Oncology from 2021. He obtained his PhD Diploma in 2010 focusing on adjuvant chemotherapy in optimal resected gastric cancer, and achieved his Habilitate Diploma in 2020. He has published 79 scientific articles, 40 national conference speeches, and 5 national research grants.

In 2004, he became a Specialist in Medical Oncology and in 2008, Senior in Medical Oncology. Since 2017, he has been a member of ORDRE DES MEDECINS DU HAINAUT at the CONSEIL PROVINCIAL du HAINAUT (Belgium), and since 2018, he has been a member of ORDRE DES MEDECINS du Haute de Seine (France).

medicina MDPI

Editorial

Advances in Cancer Therapy from Research to Clinical Practice—Surgical, Molecular or Systemic Management of Cancer

Calin Cainap [1,*] and Nicolae Crisan [2]

1 Department of Medical Oncology, Iuliu Hatieganu University of Medicine and Pharmacy, 400012 Cluj-Napoca, Romania
2 Department of Surgical Specialities, Iuliu Hatieganu University of Medicine and Pharmacy, 400012 Cluj-Napoca, Romania; drnicolaecrisan@gmail.com
* Correspondence: calincainap2015@gmail.com

Cancer represents one of the most important general health problems of our day. The estimated incidence of new cases in 2020 was 19.3 million [1] versus 10.9 million in 2002 [2], which is an increase of 77%. The future is not very optimistic, as the number of new cases is expected to increase by at least 47% by 2040, reaching 28.4 million new cases [2]. The projections in the US are that, as of January 2022, the total number of patients who were diagnosed with cancer (curable or not, and alive at a definite point of analysis) represented 5.4% of the population [3].

Even when diagnosed at an early stage, cancer patients can experience a metastatic relapse. Gallicchio et al. estimated that, during a lifetime, the percentage of metastasisation can range from 30% for lung cancer to 72% for bladder cancer [4], even though the chance of survival increased throughout the analysed interval of time due to the increasing availability of new categories of drugs used in oncological treatments.

All of these improvements could not be reached without the translation of fundamental research into practical uses, ranging from the initial cellular level to the molecular level, and nowadays, we have a genetic understanding of cancerogenesis. Since the times of ancient Greece, multiple theories have been made about oncogenesis, and through 'step by step' discoveries, we have managed to learn how things work in the complex area of human biology, and new potentially valuable targets have emerged.

One of the most important dreams of healthcare personnel—from research, clinical, or laboratory specialities—is to solve a little or big part of the complicated oncology puzzle, which could contribute to saving lives. Every small discovery could represent a big step in curing more people, such as by making more effective treatments available, diagnosing cancer earlier, or reducing the risk for cancer.

This Special Issue, "Advances in Cancer Therapy from Research to Clinical Practice—Surgical, Molecular or Systemic Management of Cancer", was initially designed to allow the potential authors to share their work in the very complicated world of oncology using a large amount of big data, which can be practice changing.

The published articles covered a broad range of cancers, with the most important primary tumours being treated from a fundamental research or clinical practice point of view.

For breast cancer, for example, Lisencu et al. [5] tried to determine molecular predictive factors which are linked to the probability of obtaining a pathological response to neoadjuvant treatment. Selecting early those patients with the best chance of responding to a selective (and costly) therapy remains a challenging task. miRs could represent a good surrogate for the real-time evaluation of the tumour and its reaction to a specific treatment.

Moreover, Mkrantonakis et al. [6] tried to identify triple-negative breast cancer (TNBC) patients who have a better or worse response to immunomodulator agents, which are

Citation: Cainap, C.; Crisan, N. Advances in Cancer Therapy from Research to Clinical Practice—Surgical, Molecular or Systemic Management of Cancer. *Medicina* **2023**, *59*, 1309. https://doi.org/10.3390/medicina59071309

Received: 28 June 2023
Accepted: 12 July 2023
Published: 14 July 2023

newcomers in the breast cancer armamentarium. PD-L1 gene polymorphisms rs822336 (G > C) and rs822337 (T > A) seem to be differentially expressed in TNBC patients and could represent potential markers of unfavourable prognosis and diminished survival. Ortega et al. [7] showed that the expression of insulin receptor substrate 4 (IRS-4), cyclin D1, and cyclooxygenase 2 (COX-2) are overexpressed in invasive lobular breast cancer compared to invasive ductal cancer, which could explain the difference between these two types of breast cancer in terms of prognosis and survival.

The molecular types of cancer are not a fundamental discovery without any clinical relevance. As shown by Szep et al. [8], multiparametric breast MRI (magnetic resonance imaging), which provides morphological and functional images of breast tumours, is nowadays a part of the recommended workup for breast cancer. The molecular types can change the MRI aspect of breast cancer, whereas MRI can be used to predict the molecular types (but not replace the biopsy).

For lung cancer patients, Ahn et al. [9] searched for the expression of BRAF mutations in real-life unselected patients with non-small cell lung cancer. Despite reduced incidence in the analysed cohort, adenocarcinoma with micropapillary components seemed to have a privileged histology which harboured this mutation; this is an interesting finding that underlines the necessity of prioritising BRAF mutation testing.

In melanoma cancer, rechallenging with BRAF inhibitors becomes an attractive option for multiple-relapse patients. The polyclonal theory of cancer remains an important element to be taken into consideration when we are in a difficult situation after standard therapy failure. Ksomidis et al. [10] showed some interesting approaches to this clinical issue.

Inflammation in cancer could be a target for immunomodulatory treatment, but could also represent an unfavourable factor for patients, being responsible for cancer progression. The combined peripheral neutrophil–platelet index seems to be an unfavourable predictor factor for overall survival in resectable oesophageal squamous cell carcinoma (ESCC) patients, as shown by Peng et al. [11].

An absolute iron deficiency in colorectal cancer can impair not only the cardiac and respiratory functions, among others, but also the immune system's defence. Deficient patients could be affected more and could present with more advanced disease–lymphatic invasion, as shown by Fagarasan et al. [12].

For hepatocarcinoma, Chen et al. [13] showed that recurrence with a surgery option can be alternatively treated with microwave ablation (MWA) that is either associated or not associated with transarterial chemoembolisation (TACE). Whether TACE adds something or not in terms of survival or diminishes the incidence of relapse remains a controversial issue. Most likely, the previous treatment could more appropriately select the best candidates. For those with progression of hepatocarcinoma after TACE, an option was proposed by Wu et al. [14] using the association of oxaliplatin with raltitrexed (HAICROX).

A new chemotherapy association or a new modality of drug delivery? Chioreanu et al. [15] tested a compound of 215 nm with an improved capacity of encapsulation, a good rate of diffusion of the drug, and a better tolerance potential.

Body mass index can be a simple prognostic factor for prostate cancer, as shown in Popovici et al.'s article [16], due to its relative resistance to insulin and possible proliferative stimulation, which could be responsible for a more advanced stage of disease at diagnosis.

Al-Gharaibeh et al. [17] analysed the tridimensional changes in the earliest cellular events before the start of renal cancer. In testicular cancer, Rohozneanu et al. [18] analysed if 'less is more' for stage I to avoid unnecessary toxicities.

For gynaecologic cancers, Obradovic et al. [19] investigated the role of a protein from the immunophilin family with antiangiogenic properties, which have a significantly decreased expression in the case of endometrioid endometrial carcinoma compared with benign endometrial hyperplasia. Havasi et al. [20] demonstrated the molecular mechanism of platin hypersensitivity and the strategies to overpass it, assuring the continuation of the essential regimen with platin derivates in ovarian cancer. As shown by Boeriu et al. [21], this class of medication could induce even a benign pathology such as lipomatosis in children.

Multiple neoplasia represents a difficult pathology for an oncologist, as mentioned and developed in an article published in *Medicina* by Simescu et al. [22]. Parathyroid carcinoma and differentiated thyroid carcinoma are exceptionally rare conditions. Less than 100 cases of adrenal neuroblastoma in adult patients have been reported, with one of these being described by Telecan et al [23]. A rare clinical condition represented by Maffucci syndrome and astrocytoma with a rare IDH mutation was described by Ashirov et al. [24].

Cotorogea-Simion et al. [25] developed the mechanisms presumed to be implicated in the pathogenesis of respiratory insufficiency, which frequently occurs in hematologic diseases.

We hope that the readers will find answers to their questions or read about a finding related to their scientific interest in our Special Issue.

Acknowledgments: I would like to thank the authors who shared their important work in this Special Issue. I am grateful to the Editorial staff for their sustained activity and support, and for their confidence in me to organise this Special Issue. Many thanks to the Editorial team of this Special Issue for their devotion and hard work.

Conflicts of Interest: The author declares no conflict of interest.

References

1. Sung, H.; Ferlay, J.; Siegel, R.L.; Laversanne, M.; Soerjomataram, I.; Jemal, A.; Bray, F. Global Cancer Statistics 2020: GLOBOCAN Estimates of Incidence and Mortality Worldwide for 36 Cancers in 185 Countries. *CA Cancer J. Clin.* **2021**, *71*, 209–249. [CrossRef] [PubMed]
2. Parkin, D.M.; Bray, F.; Ferlay, J.; Pisani, P. Global cancer statistics, 2002. *CA Cancer J. Clin.* **2005**, *55*, 74–108. [CrossRef] [PubMed]
3. American Cancer Society. *Cancer Treatment & Survivorship Facts & Figures 2022–2024*; American Cancer Society: Atlanta, GA, USA, 2022.
4. Gallicchio, L.; Devasia, T.P.; Tonorezos, E.; Mollica, M.A.; Mariotto, A. Estimation of the Number of Individuals Living With Metastatic Cancer in the United States. *J. Natl. Cancer Inst.* **2022**, *114*, 1476–1483. [CrossRef]
5. Lisencu, L.A.; Roman, A.; Visan, S.; Bonci, E.-A.; Pașca, A.; Grigorescu, E.; Mustea, E.; Cismaru, A.; Irimie, A.; Lisencu, C.; et al. The Role of miR-375-3p, miR-210-3p and Let-7e-5p in the Pathological Response of Breast Cancer Patients to Neoadjuvant Therapy. *Medicina* **2022**, *58*, 1494. [CrossRef]
6. Makrantonakis, A.-E.; Zografos, E.; Gazouli, M.; Dimitrakakis, K.; Toutouzas, K.G.; Zografos, C.G.; Kalapanida, D.; Tsiakou, A.; Samelis, G.; Zagouri, F. PD-L1 Gene Polymorphisms rs822336 G>C and rs822337 T>A: Promising Prognostic Markers in Triple Negative Breast Cancer Patients. *Medicina* **2022**, *58*, 1399. [CrossRef]
7. Ortega, M.A.; Fraile-Martinez, O.; García-Montero, C.; Borja-Vergel, S.; Torres-Carranza, D.; Pekarek, L.; Arribas, C.B.; De León-Luis, J.A.; Sánchez-Rojo, C.; Alvarez-Mon, M.A.; et al. Patients with Invasive Lobular Carcinoma Show a Significant Increase in IRS-4 Expression Compared to Infiltrative Ductal Carcinoma—A Histopathological Study. *Medicina* **2022**, *58*, 722. [CrossRef]
8. Szep, M.; Pintican, R.; Boca, B.; Perja, A.; Duma, M.; Feier, D.; Fetica, B.; Eniu, D.; Dudea, S.M.; Chiorean, A. Multiparametric MRI Features of Breast Cancer Molecular Subtypes. *Medicina* **2022**, *58*, 1716. [CrossRef]
9. Ahn, H.Y.; Lee, C.H.; Lee, M.K.; Eom, J.S.; Jeong, Y.J.; Kim, Y.D.; Cho, J.S.; Lee, J.; Lee, S.J.; Shin, D.H.; et al. BRAF V600E Mutation of Non-Small Cell Lung Cancer in Korean Patients. *Medicina* **2023**, *59*, 1085. [CrossRef]
10. Kosmidis, C.S.; Papadopoulou, K.; Mystakidou, C.M.; Papadopoulou, E.; Mantalovas, S.; Varsamis, N.; Koulouris, C.; Theodorou, V.; Papadopoulos, K.; Sevva, C.; et al. Melanoma: BRAFi Rechallenge. *Medicina* **2023**, *59*, 975. [CrossRef]
11. Peng, M.-Y.; Zuo, Z.-G.; Cao, F.-J.; Yu, Y.-D.; Cai, X.-J.; Wan, G.-X. Elevated Preoperative NMPR Predicts an Unfavorable Chance of Survival in Resectable Esophageal Squamous Cell Carcinoma. *Medicina* **2022**, *58*, 1808. [CrossRef]
12. Făgărășan, V.; Andraș, D.; Amarinei, G.; Seicean, R.I.; Bințințan, V.V.; Dindelegan, G.C.; Căinap, C.I. Absolute and Functional Iron Deficiency in Colon Cancer: A Cohort Study. *Medicina* **2022**, *58*, 1202. [CrossRef] [PubMed]
13. Chen, G.; Chen, H.; Huang, X.; Cheng, S.; Zheng, S.; Wu, Y.; Zheng, T.; Chen, X.; Guo, X.; Zhang, Z.; et al. Recurrence Outcome in Hepatocellular Carcinoma within Milan Criteria Undergoing Microwave Ablation with or without Transarterial Chemoembolization. *Medicina* **2022**, *58*, 1016. [CrossRef] [PubMed]
14. Wu, Y.; Zheng, S.; Zhang, Z.; Chen, G.; Chen, X.; Zheng, T.; Guo, X.; Chen, H.; Wang, M.; Xie, X.; et al. Hepatic Arterial Infusion Chemotherapy with Oxaliplatin Plus Raltitrexed as an Alternative Option in Advanced Hepatocellular Carcinoma Patients with Failure of, or Unsuitability for, Transarterial Chemoembolization. *Medicina* **2022**, *58*, 1343. [CrossRef] [PubMed]
15. Chioreanu, A.; Mot, I.C.; Horhat, D.I.; Balica, N.C.; Sarau, C.A.; Morar, R.; Domuta, E.M.; Dumitru, C.; Negrean, R.A.; Bumbu, B.A.; et al. Development and Preliminary Characterization of Polyester-Urethane Microparticles Used in Curcumin Drug Delivery System for Oropharyngeal Cancer. *Medicina* **2022**, *58*, 1689. [CrossRef]
16. Popovici, D.; Stanisav, C.; Pricop, M.; Dragomir, R.; Saftescu, S.; Ciurescu, D. Associations between Body Mass Index and Prostate Cancer: The Impact on Progression-Free Survival. *Medicina* **2023**, *59*, 289. [CrossRef] [PubMed]
17. Al-Gharaibeh, N.S.; Shively, S.B.; Vortmeyer, A.O. Characterization of Microscopic Multicellular Foci in Grossly Normal Renal Parenchyma of Von Hippel-Lindau Kidney. *Medicina* **2022**, *58*, 1725. [CrossRef] [PubMed]

18. Rohozneanu, E.F.; Deac, C.; Căinap, C.I. A Systematic Review Investigating the Difference between 1 Cycle versus 2 Cycles of Adjuvant Chemotherapy in Stage I Testicular Germ Cell Cancers. *Medicina* **2023**, *59*, 916. [CrossRef]
19. Obradović, D.D.; Milić, N.M.; Miladinović, N.; McClements, L.; Oprić, D.M. Loss of Expression of Antiangiogenic Protein FKBPL in Endometrioid Endometrial Carcinoma: Implications for Clinical Practice. *Medicina* **2022**, *58*, 1330. [CrossRef]
20. Havasi, A.; Cainap, S.S.; Havasi, A.T.; Cainap, C. Ovarian Cancer—Insights into Platinum Resistance and Overcoming It. *Medicina* **2023**, *59*, 544. [CrossRef]
21. Boeriu, E.; Boc, A.G.; Borda, A.; Negrean, R.A.; Feciche, B.; Boeriu, A.I.; Horhat, F.G.; Mot, I.C.; Horhat, I.D.; Ravulapalli, M.; et al. Insights on Lipomatosis after Platinum-Based Chemotherapy Use in Pediatric Oncology: A Case Report. *Medicina* **2022**, *58*, 1715. [CrossRef]
22. Simescu, R.; Pop, M.; Piciu, A.; Muntean, V.; Piciu, D. Association of Parathyroid and Differentiated Thyroid Carcinomas: A Narrative Up-To-Date Review of the Literature. *Medicina* **2022**, *58*, 1184. [CrossRef]
23. Telecan, T.; Andras, I.; Bungardean, M.R.; Muntean, D.; Militaru, C.; Perciuleac, I.; Maga, A.; Crisan, N. Adrenal Gland Primary Neuroblastoma in an Adult Patient: A Case Report and Literature Review. *Medicina* **2023**, *59*, 33. [CrossRef] [PubMed]
24. Ashirov, N.; Mammadinova, I.; Moldabekov, A.; Zhetpisbaev, B.; Teltayev, D.; Ryskeldiyev, N.; Akshulakov, S. A Rare Co-Occurrence of Maffucci Syndrome and Astrocytoma with IDH1 R132H Mutation: A Case Report. *Medicina* **2023**, *59*, 1056. [CrossRef] [PubMed]
25. Cotorogea-Simion, M.; Pavel, B.; Isac, S.; Telecan, T.; Matache, I.-M.; Bobirca, A.; Bobirca, F.-T.; Rababoc, R.; Droc, G. What Is Different in Acute Hematologic Malignancy-Associated ARDS? An Overview of the Literature. *Medicina* **2022**, *58*, 1215. [CrossRef] [PubMed]

medicina MDPI

Article

Patients with Invasive Lobular Carcinoma Show a Significant Increase in IRS-4 Expression Compared to Infiltrative Ductal Carcinoma—A Histopathological Study

Miguel A. Ortega [1,2,3,*], Oscar Fraile-Martinez [1,2], Cielo García-Montero [1,2], Sandra Borja-Vergel [1], Diego Torres-Carranza [1], Leonel Pekarek [1,2,4], Coral Bravo Arribas [5,6], Juan A. De León-Luis [5,6], Cristina Sánchez-Rojo [7], Miguel Angel Alvarez-Mon [1,2], Natalio García-Honduvilla [1,2], Julia Buján [1,2], Santiago Coca [1,2], Melchor Alvarez-Mon [1,2,8], Miguel A. Saez [1,2,9,*,†] and Luis G. Guijarro [2,6,10,†]

1 Department of Medicine and Medical Specialities, Faculty of Medicine and Health Sciences, University of Alcalá, 28801 Alcala de Henares, Spain; oscarfra.7@hotmail.com (O.F.-M.); cielo.gmontero@gmail.com (C.G.-M.); sandraborjavergel@gmail.com (S.B.-V.); diegotc90@gmail.com (D.T.-C.); leonel.pekarek@gmail.com (L.P.); maalvarezdemon@icloud.com (M.A.A.-M.); natalio.garcia@uah.es (N.G.-H.); mjulia.bujan@uah.es (J.B.); s.coca@uah.es (S.C.); mademons@gmail.com (M.A.-M.)
2 Ramón y Cajal Institute of Sanitary Research (IRYCIS), 28034 Madrid, Spain; luis.gonzalez@uah.es
3 Cancer Registry and Pathology Department, Principe de Asturias University Hospital, 28806 Alcala de Henares, Spain
4 Oncology Service, Guadalajara University Hospital, 19002 Guadalajara, Spain
5 Department of Obstetrics and Gynecology, University Hospital Gregorio Marañón, 28009 Madrid, Spain; cbravoarribas@gmail.com (C.B.A.); jaleon@ucm.es (J.A.D.L.-L.)
6 Health Research Institute Gregorio Marañón, 28009 Madrid, Spain
7 Department of Obstetrics and Gynecology, Central University Hospital of Defence-UAH Madrid, 28801 Alcala de Henares, Spain; cristina.sanchez.rojo@gmail.com
8 Immune System Diseases-Rheumatology, Oncology Service an Internal Medicine, University Hospital Príncipe de Asturias, 28806 Alcala de Henares, Spain
9 Pathological Anatomy Service, Central University Hospital of Defence-UAH Madrid, 28801 Alcala de Henares, Spain
10 Unit of Biochemistry and Molecular Biology, Department of System Biology, University of Alcalá, 28801 Alcala de Henares, Spain
* Correspondence: miguel.angel.ortega92@gmail.com (M.A.O.); msaegal@oc.mde.es (M.A.S.)
† These authors contributed equally to this work.

Citation: Ortega, M.A.; Fraile-Martinez, O.; García-Montero, C.; Borja-Vergel, S.; Torres-Carranza, D.; Pekarek, L.; Arribas, C.B.; De León-Luis, J.A.; Sánchez-Rojo, C.; Alvarez-Mon, M.A.; et al. Patients with Invasive Lobular Carcinoma Show a Significant Increase in IRS-4 Expression Compared to Infiltrative Ductal Carcinoma—A Histopathological Study. *Medicina* **2022**, *58*, 722. https://doi.org/10.3390/medicina58060722

Academic Editors: Nicolae Crisan and Călin Căinap

Received: 25 April 2022
Accepted: 26 May 2022
Published: 28 May 2022

Abstract: *Background and Objectives:* Breast cancer (BC) is the first diagnosed type of cancer and the second leading cause of cancer-related mortality in women. In addition, despite the improvement in treatment and survival in these patients, the global prevalence and incidence of this cancer are rising, and its mortality may be different according to the histological subtype. Invasive lobular carcinoma (ILC) is less common but entails a poorer prognosis than infiltrative ductal carcinoma (IDC), exhibiting a different clinical and histopathological profile. Deepening study on the molecular profile of both types of cancer may be of great aid to understand the carcinogenesis and progression of BC. In this sense, the aim of the present study was to explore the histological expression of Insulin receptor substrate 4 (IRS-4), cyclooxygenase 2 (COX-2), Cyclin D1 and retinoblastoma protein 1 (Rb1) in patients with ILC and IDC. *Patients and Methods:* Thus, breast tissue samples from 45 patients with ILC and from 45 subjects with IDC were analyzed in our study. *Results:* Interestingly, we observed that IRS-4, COX-2, Rb1 and Cyclin D1 were overexpressed in patients with ILC in comparison to IDC. *Conclusions:* These results may indicate a differential molecular profile between both types of tumors, which may explain the clinical differences among ILC and IDC. Further studies are warranted in order to shed light onto the molecular and translational implications of these components, also aiding to develop a possible targeted therapy to improve the clinical management of these patients.

Keywords: breast cancer (BC); invasive lobular carcinoma (ILC); insulin receptor substrate 4 (IRS-4); cyclooxygenase 2 (COX-2); cyclin D1; retinoblastoma protein 1 (Rb1)

1. Introduction

Breast cancer (BC) is the most common type of diagnosed cancer in women and the second leading cause of cancer-related mortality in this group [1,2]. Following the American Cancer Society, 1 in 8 women will suffer from BC in their lives, and it has been projected that the global incidence of these tumors will reach 3.2 million new cases per year by 2050 [3]. Moreover, men can also suffer from BC, accounting for less than 1% of all cancers in men and less than 1% of all breast cancers for them [4]. Lifestyle habits like smoking, sedentarism, alcohol consumption and diet are some of the most important risk factors for suffering from BC, along with obesity, aging, race, hormonal/reproductive factors and history of familiar BC [5]. An early diagnosis and regular screening are crucial for a good prognosis and survival rate of BC patients. Indeed, thanks to these approaches, the 5-year relative survival rate can be over 80% in some developed countries [6]. However, the overall survival of these patients will depend mainly on two factors: the tumoral stage (invasiveness, metastasis), and the tumoral subtype [7]. According to its histological classification, the most common type of invasive BC is the infiltrative ductal carcinoma (IDC) of no special subtype [8]. Approximately 1 in 4 tumors are defined as histological 'special types', including at least 17 discrete pathological entities such as invasive lobular carcinoma (ILC) [9]. Furthermore, ILC is the second most common subtype of BC, and differs from IDC in a set of clinical and histopathological features. For instance, ILCs present more difficulties in their detection, often exhibit a poorer prognosis, more advanced stages, frequent late recurrences and lower responses to therapy [10]. In addition, there are a plethora of biological differences between these types of tumors, including in their molecular profiles, immune response, metastasis, metabolism and other hallmarks of cancer [11,12]. Thus, it is necessary to deepen examinations on the molecular and biological basis of ILC in order to develop further approaches and strategies to aid in the clinical management of this relevant malignancy.

Insulin receptor substrate 4 (IRS-4) is a relevant molecule altered in different types of malignancies like BC. It seems that this molecule promotes tumoral proliferation and therapy resistance due to the alteration of different molecular pathways [13–15]. Furthermore, previous studies have found that an overexpression of this component is frequently related to a poorer prognosis [16]. A similar role has been found with cyclooxygenase 2 (COX-2) in patients with BC, being frequently related to a poorer prognosis and a set of carcinogenic mechanisms [17]. In the same line, Cyclin D1 levels are often disrupted in different types of tumors, especially in BC, where approximately 50% of mammary carcinomas present an overexpression of this component [18]. Likewise, the retinoblastoma protein 1 (Rb1) is also involved in the tumorigenesis and impaired cell features, having been proposed as a promising therapeutic target of different types of tumors, including BC [18,19]. Despite the relevance and demonstrated role of these markers in the development and progression of BC, there is little evidence collected regarding the differential expression of each component in ILC vs. IDC.

Hence, the aim of the present review is to explore the histopathological detection of IRS-4, COX-2, Rb1 and Cyclin D1 in 45 patients with ILC in comparison to 45 IDC in order to establish potential biological differences between both types of tumors. With that purpose, we have conducted immunohistochemical studies of n patients with ILC and compared these with n patients with IDC.

2. Patients and Methods

2.1. Collection of Samples

For our study, we used paraffin-embedded sections of breast tissue from 45 patients diagnosed with invasive lobular carcinoma (ILC) and 45 patients diagnosed with infiltrative ductal carcinoma (IDC). The diagnosis followed the principles of Lakhani et al. [20]. The present study was designed as an observational, analytical, retrospective cohort study with longitudinal follow-up. Paraffin blocks and all details with extensive clinical information on patients and follow-up data were retrospectively reviewed.

The study was carried out in accordance with the basic ethical principles of autonomy, beneficence, non-maleficence and distributive justice, and its development followed the rules of Good Clinical Practice, the principles contained in the most recent Declaration of Helsinki (2013) and the Convention of Oviedo (1997). The data and information collected complied with current legislation on data protection (Organic Law 3/2018 of 5 December on the Protection of Personal Data and Guarantee of Digital Rights and Regulation (EU) 2016/679).

2.2. Histopathological and Immunohistochemical Studies

Immunohistochemical studies were performed on paraffin-embedded breast tissue samples. The antibody recovery step was described in the protocol specifications (Table 1). Antigen/antibody reactions were detected by the avidin-biotin complex (ABC) method, with avidin-peroxidase, following the protocols of Ortega et al. [21]. After incubation with the primary antibody (1 h and 30 min), samples were incubated with 3% BSA blocker (catalog #37525; Thermo Fisher Scientific, Inc., Waltham, MA, USA) and PBS overnight at 4 °C. The samples were then incubated with biotin-conjugated secondary antibody and diluted in PBS for 90 min at room temperature (RT; Rabbit IgG (RG-96, 1:1000, Sigma-Aldrich/Mouse IgG (F2012/045K6072) 1:300, Sigma-Aldrich, St. Louis, MI, USA). Avidin-peroxidase conjugate ExtrAvidin®-Peroxidase (Sigma-Aldrich; Merck KGaA, Darmstadt, Germany) was used for 60 min at RT (1:200 dilution with PBS), and then the level of protein expression was determined using a Chromogenic Diaminobenzidine (DAB) Substrate Kit (cat. no. SK-4100; Maravai LifeSciences, San Diego, CA, USA), which was prepared immediately prior to exposure (5 mL distilled water, two drops of buffer, four drops of DAB and two drops of hydrogen peroxide drops). The signal was developed with the chromogenic peroxidase substrate for 15 min at RT; this technique allows the detection of a brown staining. For the detection of each protein, sections of the same tissue were assigned as negative controls, substituting the incubation with the primary antibody for a blocking solution (PBS). In all tissues, contrast was performed with Carazzi hematoxylin for 15 min at RT.

2.3. Histopathological Evaluation

Tissue sections were viewed using a Zeiss Axiophot light microscope (Carl Zeiss, Oberkochen, Germany) equipped with an AxioCam HRc digital camera (Carl Zeiss, Oberkochen, Germany). Given the important role of the proteins studied, the evaluation of the histological results was carried out according to the intensity of expression for the immunohistochemical staining with Score. Therefore, the histological samples of patients diagnosed with breast cancer were classified as negative (0) or low/medium (1) and high (3) expression using the IRS-Score method [22]; the samples were evaluated by two independent pathologists (MAO, MAS), and in case of discrepancies, a third pathologist intervened (SC). For each established subject group, seven randomly selected microscopy fields were examined in each of the five sections. Subjects were classified as positive when the mean proportion of the labeled sample was greater than or equal to 5% of the total sample. This was completed by calculating the total percentage of the labeled tissue in each microscopy field to obtain a mean for the study sample as described [23]. The observation and quantification of the samples were carried out independently by two researchers.

2.4. Statistic Analysis

For the statistical analysis, the statistical package GraphPad Prism® 5.1 was used for the Mann–Whitney U test as appropriate. The data are provided as the mean ± standard deviation (SD). The error bars in the figures indicate the SD. Different levels of significance are distinguished as * $p < 0.05$, ** $p < 0.005$ and *** $p < 0.001$.

Table 1. Primary antibodies used, together with the dilutions and protocol specifications.

Antigen	Dilution	Provider	Protocol Specifications
IRS-4	1:250	Thermo Fisher Scientific—PA5-117329	Preincubation with Tris-EDTA buffer pH 9 and incubation with 0.1% TTX (Triton ×100 in TBS) for 5 min.
COX-2	1:750	Vitro, MAD-000335QD-3/V	————
Rb1	1:500	Vitro, MAD-000900QD-3/V	————
Cyclin D1	1:500	Vitro, MAD-000630QD-3/V	Preincubation with Tris-EDTA buffer pH 9 and incubation with 0.1% TTX (Triton ×100 in TBS) for 5 min.

3. Results

3.1. Clinical and Sociodemographic Characteristics of the Study Population

The present study was designed as an observational, analytical, retrospective cohort study with longitudinal follow-up. A total of 45 patients with ILC and 45 patients with IDC were analyzed, with a median age of 69.167 ± 13.663 years for ILC and 67.571 ± 10.717 years for IDC. All patients had a score greater than pT1. The percentage of expression of estrogen receptors was 70.000 ± 22.887% for ILC and 66.538 ± 28.091% for IDC. The percentage of expression of progesterone receptors was 54.333 ± 26.245% for ILC and 55.833 ± 33.086% for IDC. The Ki67 expression percentage was 12.609 ± 6.373% for ILC and 16.190 ± 7.731% for IDC.

3.2. Patients with Invasive Lobular Carcinoma Show a Significant Increase in the Expression of IRS-4

Our results demonstrate how patients with ILC show an increased expression of IRS-4 in the tissue compared to IDC patients. We observed how the IRS-Score expression score was 2.522 ± 0.574 in ILC and 1.322 ± 0.479 in IDC, *** $p < 0.001$ (Figure 1A–C).

3.3. Patients with Invasive Lobular Carcinoma Show a Significant Increase in the Expression of COX-2, Rb1 and Cyclin D1

In addition, we have observed how patients diagnosed with ILC show an increased expression of COX-2, Rb1 and Cyclin D1. In the case of COX-2, we observed how the IRS-Score expression score was 1.806 ± 0.605 in ILC and 1.533 ± 0.537 in IDC, * $p = 0.0267$ (Figure 2A–C).

For Rb1, we observed how the IRS-Score expression score was 2.000 ± 0.674 in ILC and 1.544 ± 0.620 in IDC, ** $p = 0.0022$ (Figure 3A–C). In the case of Cyclin D1, we observed how the IRS-Score expression score was 2.117 ± 0.527 in ILC and 1.639 ± 0.793 in IDC, ** $p = 0.0018$ (Figure 4A–C).

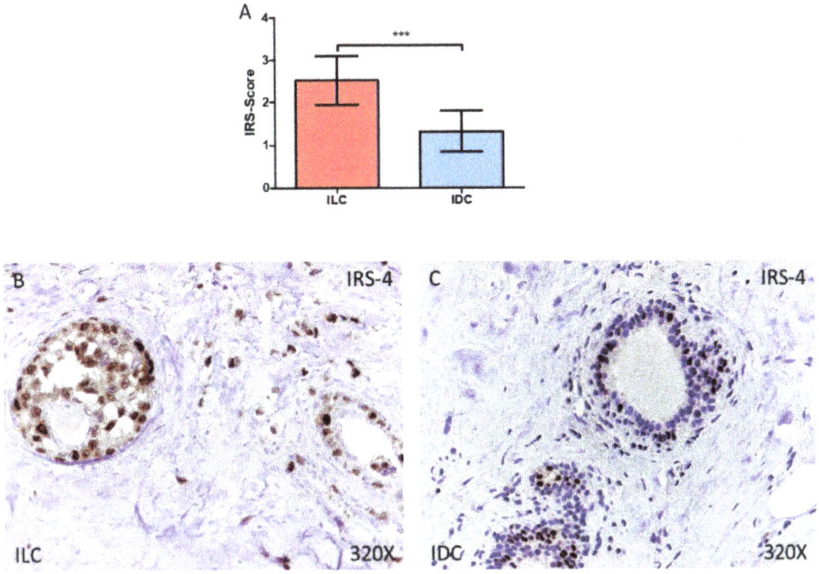

Figure 1. (**A**) IRS-Score for IRS-4 in patients diagnosed with invasive lobular carcinoma (ILC) and infiltrative ductal carcinoma (IDC). (**B**,**C**) Histological images of IRS-4 expression using immunohistochemical techniques in the breast tissue of patients diagnosed with ILC (**B**) and IDC (**C**). *** $p < 0.001$.

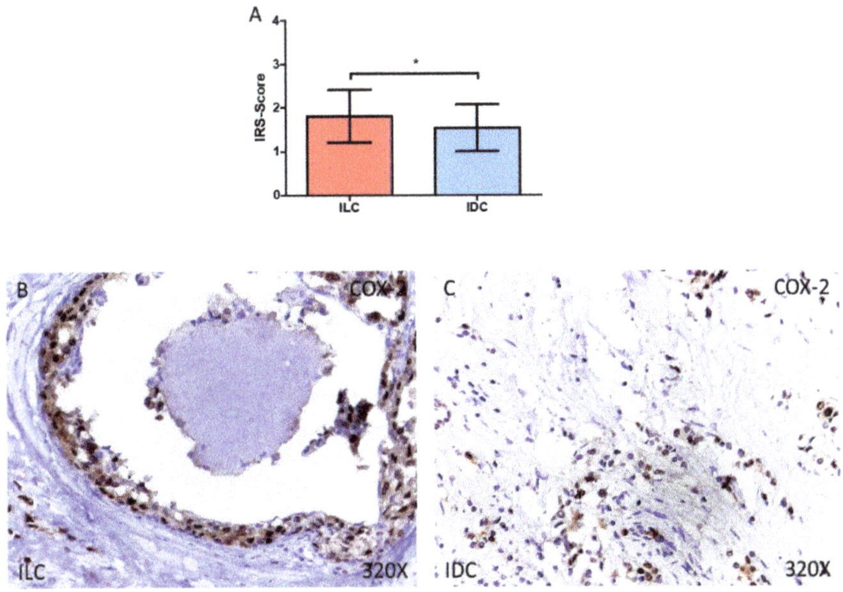

Figure 2. (**A**) IRS-Score for COX-2 in patients diagnosed with invasive lobular carcinoma (ILC) and infiltrative ductal carcinoma (IDC). (**B**,**C**) Histological images of COX-2 expression using immunohistochemical techniques in the breast tissue of patients diagnosed with ILC (**B**) and IDC (**C**). * $p < 0.05$.

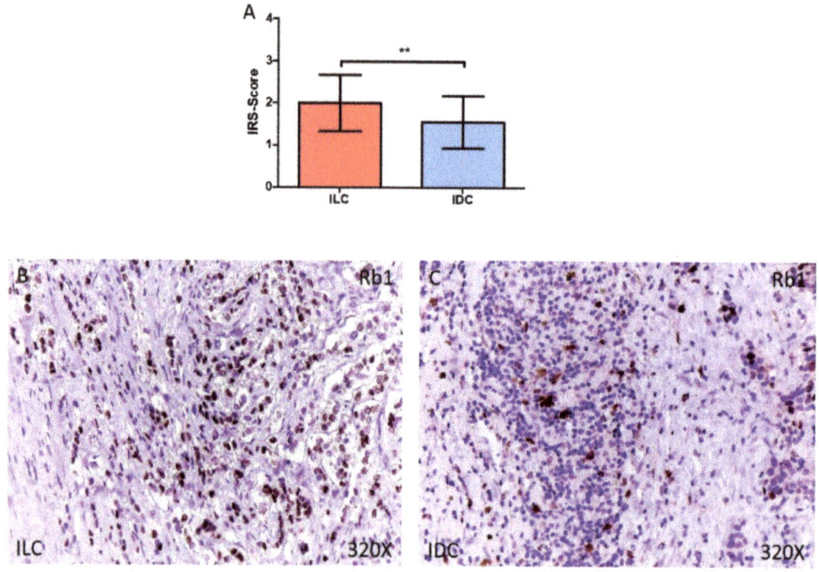

Figure 3. (**A**) IRS-Score for Rb1 in patients diagnosed with invasive lobular carcinoma (ILC) and infiltrative ductal carcinoma (IDC). (**B**,**C**) Histological images of Rb1 expression using immunohistochemical techniques in the breast tissue of patients diagnosed with ILC (**B**) and IDC (**C**). ** $p < 0.005$.

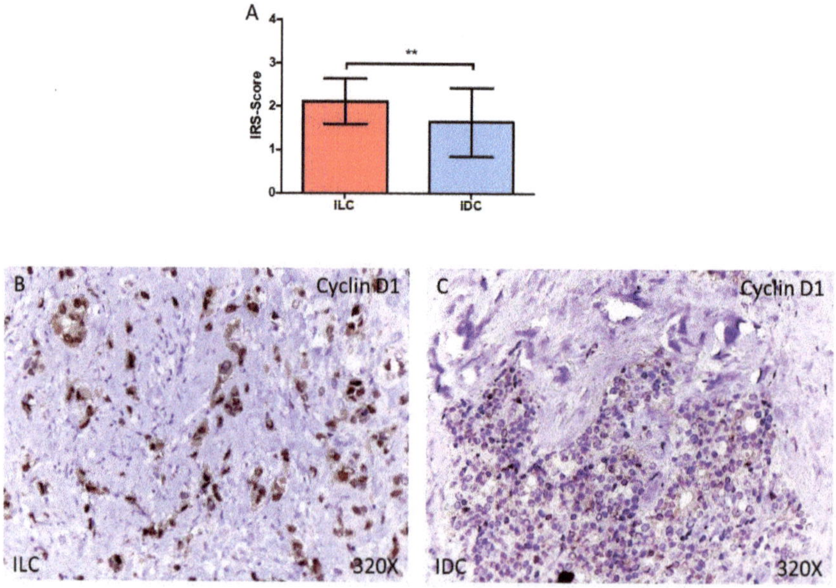

Figure 4. (**A**) IRS-Score for Cyclin D1 in patients diagnosed with invasive lobular carcinoma (ILC) and infiltrative ductal carcinoma (IDC). (**B**,**C**) Histological images of Cyclin D1 expression using immunohistochemical techniques in the breast tissue of patients diagnosed with ILC (**B**) and IDC (**C**). ** $p < 0.005$.

4. Discussion

The annual incidence and prevalence of BC are rising worldwide in the younger and elderly populations [24]. Despite the improvements achieved in early diagnosis, screening and survival, more knowledge is required to understand the biology of this cancer and their specific types in order to develop better approaches for the clinical management of these patients. In this sense, our study gains further insights into the biology of ILC in comparison to IDC, aiding to explain some of the differences in the clinical presentation and histopathological features of these types of tumors.

ILCs are frequently associated with a poorer prognosis than IDCs, and the biological differences between both types of tumors have received significant attention. For instance, differences in the expression of genes and specific proteins/receptors have been reported among both groups [25]. IRS-4 was a pivotal marker overexpressed in ILC in comparison to IDC. To our knowledge, this is the first study demonstrating a possible pathophysiological role of this marker in these tumors. IRS-4 is an adaptor protein acting as a constitutive activator of critical cell transduction pathways in cancer, leading to the activation of the PI3K/Akt pathway collaborating with the actions of the human epidermal growth factor receptor 2 (HER2) [13]. HER2 is a central receptor involved in BC development and stratification. According to the absence or presence of HER2, estrogen receptor (ER) and progesterone receptor (PR), the molecular classification of breast tumors can be luminal A, luminal B, HER2+ enriched cells and triple negative BC. Luminal subtypes are ER/PR+, being luminal A subtype negative for HER2 and with low levels of ki67, whereas luminal B presents the HER2+ receptor and high levels of ki67. HER2+ enriched cells only present this marker, and triple negative BC lacks all of these receptors [26,27]. It is of note, that patients with ILC were HER2+, whereas IDC subjects did not show HER2+ expression. HER2+ ILC has distinct clinical characteristics and immune landscapes compared to IDC, and a poorer prognosis [28]. Hence, the augmentation of both IRS-4 and HER2+ may act synergically with the progression of ILC. Moreover, compelling evidence has established the central role of PI3K in the development of BC, being closely related to cell growth, proliferation, survival, motility, metastasis, metabolism and immune modulation [29]. In addition, previous studies have also found an association between the activation of PI3K/Akt and other pathways such as Ras-MAPK by IRS-4 with therapeutic resistance and tumor progression in lung cancer [13]. Thus, IRS-4 could mediate many of these processes in ILCs due to its hyperactivation of PI3K/Akt and other signaling pathways, aiding to explain the worse prognosis of these patients in comparison to IDC.

Likewise, the expression of IRS-4 was also related to the activation of Cyclin D1 and Rb1 in colorectal cancer, two markers also altered in our study [30]. Because of the increased expression of these components, it is likely that IRS-4 may collaborate with the overexpression of Cyclin D1 and Rb1. Cyclin D1 and Rb1 are major regulators of the cell cycle. In the case of Cyclin D1, prior research has detected a substantial dysregulation of this marker in several types of cancer [31]. In the field of BC, the overexpression of Cyclin D1 is associated with abnormalities in the cell cycle and a set of carcinogenic mechanisms in the breast, also mediating the effects of estrogen in this tissue [32]. Soslow et al. [33] claimed that 82% (23 out of 28) of ILCs exhibited a high expression of Cyclin D1 in comparison to the 54% of IDCs (18 out of 34). In agreement with previous studies, we have observed an increased expression of Cyclin D1 in ILCs in comparison to IDCs. In the case of Rb1, it may act as either a tumor suppressor or as promoting tumor growth [34]. It seems that the high expression of Rb1 is associated with a high proliferation of different invasive breast tumors [35]. Interestingly, the levels of Rb1 appear to be correlated with those of glucose transporter 1 (GLUT-1) in BC, suggesting a promising therapeutic approach using GLUT-1 inhibitors in patients with high Rb1 expression [36]. However, more studies are required to better understand the role of Cyclin D1 and Rb1 in lobular carcinoma before drawing any conclusions.

Last but not least, COX-2 is an enzyme ubiquitously involved in the mammary carcinogenesis. Its expression appears to be directly correlated with the stage, cancer progression, angiogenesis and metastasis [37]. Furthermore, COX-2-driven prostaglandin E2 (PGE-2) biosynthesis is related to a plethora of aggressive carcinogenic mechanisms in BC, having been proposed as a promising therapeutic target and also being associated with a poorer prognosis [17,38]. COX-2 can be released by cancer-associated fibroblasts, M2 macrophages and cancer cells to the tumor microenvironment, inducing cancer stem cell-like activity and cell proliferation, angiogenesis, invasion, inflammation, metastasis and apoptotic resistance [39]. In consonance with our results, Holmes et al. [40] found that higher COX-2 expression was more observed in ILC than IDC, aiding to explain its worse prognosis. Other studies, however, found that COX-2 was highly expressed in both ILC and IDC, and the proportion of total COX-2 positive tumors range between studies [37,41]. These differences could be attributed to the different scoring systems and cut-off of immunohistochemistry, as these do not allow for the extraction of quantitative results, which may be a limitation of our study [42]. Similarly, COX-2 was also highly detected in lobular and ductal in situ carcinoma, being associated with an increased risk for developing subsequent invasive carcinomas [42,43], hence supporting the role of COX-2 in the early carcinogenesis as well.

5. Conclusions

Overall, we have found a differential molecular profile between ILC and IDC, showing an increased expression of IRS-4, COX-2, Cyclin D1 and Rb1 in favor of the former. Further studies are warranted in order to deepen exploration on the molecular and translational implications of these components, as well as to analyze a possible targeted therapy to improve the clinical management of these patients.

Author Contributions: Conceptualization, M.A.O., M.A.S. and L.G.G.; methodology, M.A.O. and M.A.S.; validation, J.B.; investigation, M.A.O., O.F.-M., C.G.-M., S.B.-V., D.T.-C., L.P., C.B.A., J.A.D.L.-L., C.S.-R., M.A.A.-M., N.G.-H., J.B., S.C., M.A.-M., M.A.S. and L.G.G.; data curation, M.A.S.; writing—original draft preparation, M.A.O., O.F.-M., C.G.-M., S.B.-V., D.T.-C., L.P., C.B.A., J.A.D.L.-L., C.S.-R., M.A.A.-M., N.G.-H., J.B., M.A.-M., M.A.S. and L.G.G.; writing—review and editing, M.A.O., O.F.-M., C.G.-M., S.B.-V., D.T.-C., L.P., C.B.A., J.A.D.L.-L., C.S.-R., M.A.A.-M., N.G.-H., J.B., M.A.-M., M.A.S. and L.G.G.; visualization, M.A.O., S.C. and M.A.S.; supervision, M.A.O. and L.G.G.; project administration, M.A.O. and M.A.-M.; funding acquisition, M.A.-M. All authors have read and agreed to the published version of the manuscript.

Funding: The study was supported by the Comunidad de Madrid (B2017/BMD-3804 MITIC-CM, B2020/MITICAD-CM) and HALEKULANI, S.L.

Institutional Review Board Statement: The study was approved by the Hospital Universitario Gregorio Marañon (SAR18/2020)-UCM Madrid.

Informed Consent Statement: Not applicable.

Data Availability Statement: The datasets used and/or analyzed during the present study are available from the corresponding author on reasonable request.

Conflicts of Interest: The authors declare no conflict of interest.

References

1. Simon, A.; Robb, K. Breast Cancer. In *Cambridge Handbook of Psychology, Health and Medicine*, 2nd ed.; Cambridge University Press: Cambridge, UK, 2021; pp. 577–580. [CrossRef]
2. Bober, P.; Tomková, Z.; Alexovič, M.; Ropovik, I.; Sabo, J. The unfolded protein response controls endoplasmic reticulum stress-induced apoptosis of MCF-7 cells via a high dose of vitamin C treatment. *Mol. Biol. Rep.* **2019**, *46*, 1275–1284. [CrossRef] [PubMed]
3. Tao, Z.Q.; Shi, A.; Lu, C.; Song, T.; Zhang, Z.; Zhao, J. Breast Cancer: Epidemiology and Etiology. *Cell Biochem. Biophys.* **2015**, *72*, 333–338. [CrossRef] [PubMed]
4. Yousef, A.J.A. Male Breast Cancer: Epidemiology and Risk Factors. *Semin. Oncol.* **2017**, *44*, 267–272. [CrossRef] [PubMed]
5. Ataollahi, M.R.; Sharifi, J.; Paknahad, M.R.; Paknahad, A. Breast cancer and associated factors: A review. *J. Med. Life* **2015**, *8*, 6.
6. Sun, Y.S.; Zhao, Z.; Yang, Z.N.; Xu, F.; Lu, H.J.; Zhu, Z.Y.; Shi, W.; Jiang, J.; Yao, P.P.; Zhu, H.P. Risk Factors and Preventions of Breast Cancer. *Int. J. Biol. Sci.* **2017**, *13*, 1387. [CrossRef]
7. Łukasiewicz, S.; Czeczelewski, M.; Forma, A.; Baj, J.; Sitarz, R.; Stanisławek, A. Breast Cancer-Epidemiology, Risk Factors, Classification, Prognostic Markers, and Current Treatment Strategies-An Updated Review. *Cancers* **2021**, *13*, 4287. [CrossRef]
8. Oluogun, W.A.; Adedokun, K.A.; Oyenike, M.A.; Adeyeba, O.A. Histological Classification, Grading, Staging, and Prognostic Indexing of Female Breast Cancer in an African Population: A 10-Year Retrospective Study. *Int. J. Health Sci.* **2019**, *13*, 3.

9. Weigelt, B.; Horlings, H.M.; Kreike, B.; Hayes, M.M.; Hauptmann, M.; Wessels, L.F.A.; de Jong, D.; van de Vijver, M.J.; Van't Veer, L.J.; Peterse, J.L. Refinement of Breast Cancer Classification by Molecular Characterization of Histological Special Types. *J. Pathol.* **2008**, *216*, 141–150. [CrossRef]
10. Adachi, Y.; Ishiguro, J.; Kotani, H.; Hisada, T.; Ichikawa, M.; Gondo, N.; Yoshimura, A.; Kondo, N.; Hattori, M.; Sawaki, M.; et al. Comparison of Clinical Outcomes between Luminal Invasive Ductal Carcinoma and Luminal Invasive Lobular Carcinoma. *BMC Cancer* **2016**, *16*, 248. [CrossRef]
11. Du, T.; Zhu, L.; Levine, K.M.; Tasdemir, N.; Lee, A.V.; Vignali, D.A.A.; van Houten, B.; Tseng, G.C.; Oesterreich, S. Invasive Lobular and Ductal Breast Carcinoma Differ in Immune Response, Protein Translation Efficiency and Metabolism. *Sci. Rep.* **2018**, *8*, 7205. [CrossRef]
12. Ortega, M.A.; Fraile-Martínez, O.; Guijarro, L.G.; Casanova, C.; Coca, S.; Álvarez-Mon, M.; Buján, J.; García-Honduvilla, N.; Asúnsolo, Á. Á. The Regulatory Role of Mitochondrial MicroRNAs (MitomiRs) in Breast Cancer: Translational Implications Present and Future. *Cancers* **2020**, *12*, 2443. [CrossRef] [PubMed]
13. Hao, P.; Huang, Y.; Peng, J.; Yu, J.; Guo, X.; Bao, F.; Dian, Z.; An, S.; Xu, T.R. IRS4 Promotes the Progression of Non-Small Cell Lung Cancer and Confers Resistance to EGFR-TKI through the Activation of PI3K/Akt and Ras-MAPK Pathways. *Exp. Cell Res.* **2021**, *403*, 112615. [CrossRef] [PubMed]
14. Ikink, G.J.; Hilkens, J. Insulin Receptor Substrate 4 (IRS4) Is a Constitutive Active Oncogenic Driver Collaborating with HER2 and Causing Therapeutic Resistance. *Mol. Cell. Oncol.* **2017**, *4*, e1279722. [CrossRef] [PubMed]
15. Guijarro, L.G.; Sanmartin-Salinas, P.; Pérez-Cuevas, E.; Toledo-Lobo, M.V.; Monserrat, J.; Zoullas, S.; Sáez, M.A.; Álvarez-Mon, M.A.; Bujan, J.; Noguerales-Fraguas, F.; et al. Possible Role of IRS-4 in the Origin of Multifocal Hepatocellular Carcinoma. *Cancers* **2021**, *13*, 2560. [CrossRef] [PubMed]
16. Ortega, M.A.; Pekarek, L.; Garcia-Montero, C.; Fraile-Martinez, O.; Saez, M.A.; Asúnsolo, A.; Alvarez-Mon, M.A.; Monserrat, J.; Coca, S.; Toledo-Lobo, M.V.; et al. Prognostic Role of IRS-4 in the Survival of Patients with Pancreatic Cancer. *Histol. Histopathol.* **2022**, *9*, 18432. [CrossRef]
17. Jana, D.; Sarkar, D.K.; Ganguly, S.; Saha, S.; Sa, G.; Manna, A.K.; Banerjee, A.; Mandal, S. Role of Cyclooxygenase 2 (COX-2) in Prognosis of Breast Cancer. *Indian J. Surg. Oncol.* **2014**, *5*, 59. [CrossRef]
18. Barnes, D.M.; Gillett, C.E. Cyclin D1 in Breast Cancer. *Breast Cancer Res. Treat.* **1998**, *52*, 1–15. [CrossRef]
19. Robinson, T.J.W.; Liu, J.C.; Vizeacoumar, F.; Sun, T.; Maclean, N.; Egan, S.E.; Schimmer, A.D.; Datti, A.; Zacksenhaus, E. RB1 Status in Triple Negative Breast Cancer Cells Dictates Response to Radiation Treatment and Selective Therapeutic Drugs. *PLoS ONE* **2013**, *8*, e78641. [CrossRef]
20. Lakhani, S.R. *WHO Classification of Breast Tumours: WHO Classification of Tumours*; International Agency for Research on Cancer: Lion, France, 2019; Volume 2.
21. Ortega, M.A.; Fraile-Martínez, O.; Pekarek, L.; Alvarez-Mon, M.A.; Asúnsolo, Á.; Sanchez-Trujillo, L.; Coca, S.; Buján, J.; Álvarez-Mon, M.; García-Honduvilla, N.; et al. Defective expression of the peroxisome regulators PPARα receptors and lysogenesis with increased cellular senescence in the venous wall of chronic venous disorder. *Histol. Histopathol.* **2021**, *36*, 547–558. [CrossRef]
22. Ortega, M.A.; Pekarek, L.; Fraile-Martinez, O.; Garcia-Montero, C.; Saez, M.A.; Asúnsolo, A.; Alvarez-Mon, M.A.; Monserrat, J.; Ruiz-Llorente, L.; García-Honduvilla, N.; et al. Implication of ERBB2 as a Predictive Tool for Survival in Patients with Pancreatic Cancer in Histological Studies. *Curr. Oncol.* **2022**, *29*, 2442–2453. [CrossRef]
23. Ortega, M.A.; Romero, B.; Asúnsolo, Á.; Sainz, F.; Martinez-Vivero, C.; Álvarez-Mon, M.; Buján, J.; García-Honduvilla, N. Behavior of Smooth Muscle Cells under Hypoxic Conditions: Possible Implications on the Varicose Vein Endothelium. *BioMed Res. Int.* **2018**, *2018*, 7156150. [CrossRef] [PubMed]
24. Huang, J.; Chan, P.S.; Lok, V.; Chen, X.; Ding, H.; Jin, Y.; Yuan, J.; Lao, X.Q.; Zheng, Z.J.; Wong, M.C. Global Incidence and Mortality of Breast Cancer: A Trend Analysis. *Aging* **2021**, *13*, 5748. [CrossRef] [PubMed]
25. Bertucci, F.; Orsetti, B.; Nègre, V.; Finetti, P.; Rougé, C.; Ahomadegbe, J.C.; Bibeau, F.; Mathieu, M.C.; Treilleux, I.; Jacquemier, J.; et al. Lobular and Ductal Carcinomas of the Breast Have Distinct Genomic and Expression Profiles. *Oncogene* **2008**, *27*, 5359–5372. [CrossRef] [PubMed]
26. Schnitt, S.J. Will Molecular Classification Replace Traditional Breast Pathology? *Int. J. Surg. Pathol.* **2010**, *18*, 162–166. [CrossRef] [PubMed]
27. Sørlie, T.; Perou, C.M.; Tibshirani, R.; Aas, T.; Geisler, S.; Johnsen, H.; Hastie, T.; Eisen, M.B.; van de Rijn, M.; Jeffrey, S.S.; et al. Gene Expression Patterns of Breast Carcinomas Distinguish Tumor Subclasses with Clinical Implications. *Proc. Natl. Acad. Sci. USA* **2001**, *98*, 10869–10874. [CrossRef] [PubMed]
28. Chumsri, S.; Li, Z.; Shachner, T.; Advani, P.; Sideras, K.; Moreno-Aspitia, A.; Colon-Otero, G.; Knutson, K.L.; Nassar, A.; Perez, E.A.; et al. Outcome and Immune Landscape of HER2-Positive Invasive Lobular Carcinoma in the North Central Cancer Treatment Group (NCCTG) N9831 (Alliance) Trial. *J. Clin. Oncol.* **2021**, *39*, 535. [CrossRef]
29. Ortega, M.A.; Fraile-Martínez, O.; Asúnsolo, Á.; Buján, J.; García-Honduvilla, N.; Coca, S. Signal Transduction Pathways in Breast Cancer: The Important Role of PI3K/Akt/MTOR. *J. Oncol.* **2020**, *2020*, 9258396. [CrossRef]
30. Sanmartín-Salinas, P.; del Val Toledo Lobo, M.; Noguerales-Fraguas, F.; Londoño, M.T.; Jiménez-Ruiz, A.; Guijarro, L.G. Insulin Receptor Substrate-4 Is Overexpressed in Colorectal Cancer and Promotes Retinoblastoma-Cyclin-Dependent Kinase Activation. *J. Gastroenterol.* **2018**, *53*, 932–944. [CrossRef]

31. Qie, S.; Diehl, J.A. Cyclin D1, Cancer Progression, and Opportunities in Cancer Treatment. *J. Mol. Med.* **2016**, *94*, 1313–1326. [CrossRef]
32. Roy, P.G.; Thompson, A.M. Cyclin D1 and Breast Cancer. *Breast* **2006**, *15*, 718–727. [CrossRef]
33. Soslow, R.A.; Carlson, D.L.; Horenstein, M.G.; Osborne, M.P. A Comparison of Cell Cycle Markers in Well-Differentiated Lobular and Ductal Carcinomas. *Breast Cancer Res. Treat.* **2000**, *61*, 161–170. [CrossRef] [PubMed]
34. Pietiläinen, T.; Lipponen, P.; Aaltomaa, S.; Eskelinen, M.; Kosma, V.M.; Syrjänen, K. Expression of Retinoblastoma Gene Protein (Rb) in Breast Cancer as Related to Established Prognostic Factors and Survival. *Eur. J. Cancer* **1995**, *31A*, 329–333. [CrossRef]
35. Ceccarelli, C.; Santini, D.; Chieco, P.; Taffurelli, M.; Gamberini, M.; Pileri, S.A.; Marrano, D. Retinoblastoma (RB1) Gene Product Expression in Breast Carcinoma. Correlation with Ki-67 Growth Fraction and Biopathological Profile. *J. Clin. Pathol.* **1998**, *51*, 818. [CrossRef] [PubMed]
36. Zhang, X.; Pang, X.; Zhang, Z.; Liu, Q.; Zhang, H.; Xiang, Q.; Cui, Y. Co-Expression and Prognosis Analyses of GLUT1-4 and RB1 in Breast Cancer. *BMC Cancer* **2021**, *21*, 1026. [CrossRef]
37. Costa, C.; Soares, R.; Reis-Filho, J.S.; Leitão, D.; Amendoeira, I.; Schmitt, F.C. Cyclo-Oxygenase 2 Expression Is Associated with Angiogenesis and Lymph Node Metastasis in Human Breast Cancer. *J. Clin. Pathol.* **2002**, *55*, 429. [CrossRef]
38. Harris, R.E.; Casto, B.C.; Harris, Z.M. Cyclooxygenase-2 and the Inflammogenesis of Breast Cancer. *World J. Clin. Oncol.* **2014**, *5*, 677. [CrossRef]
39. Hashemi Goradel, N.; Najafi, M.; Salehi, E.; Farhood, B.; Mortezaee, K. Cyclooxygenase-2 in Cancer: A Review. *J. Cell. Physiol.* **2019**, *234*, 5683–5699. [CrossRef]
40. Holmes, M.D.; Chen, W.Y.; Schnitt, S.J.; Collins, L.; Colditz, G.A.; Hankinson, S.E.; Tamimi, R.M. COX-2 Expression Predicts Worse Breast Cancer Prognosis and Does Not Modify the Association with Aspirin. *Breast Cancer Res. Treat.* **2011**, *130*, 657–662. [CrossRef]
41. Singh-Ranger, G.; Salhab, M.; Mokbel, K. The Role of Cyclooxygenase-2 in Breast Cancer: Review. *Breast Cancer Res. Treat.* **2008**, *109*, 189–198. [CrossRef]
42. Hoellen, F.; Kelling, K.; Dittmer, C.; Diedrich, K.; Friedrich, M.; Thill, M. Impact of Cyclooxygenase-2 in Breast Cancer | Anticancer Research. *Anticancer Res.* **2011**, *31*, 4359–4367.
43. Perrone, G.; Zagami, M.; Santini, D.; Vincenzi, B.; Gullotta, G.; Morini, S.; Battista, C.; Guerriero, G.; Altomare, V.; Tonini, G.; et al. COX-2 Expression in Lobular in Situ Neoplasia of the Breast: Correlation with Histopathological Grading System According to the Tavassoli Classification. *Histopathology* **2007**, *51*, 33–39. [CrossRef] [PubMed]

medicina

MDPI

Article

Recurrence Outcome in Hepatocellular Carcinoma within Milan Criteria Undergoing Microwave Ablation with or without Transarterial Chemoembolization

Guobin Chen [1,2,†], Hong Chen [1,2,†], Xing Huang [1], Sisi Cheng [1], Susu Zheng [1,2], Yanfang Wu [1,2], Tanghui Zheng [1,2], Xiaochun Chen [1,2], Xinkun Guo [1,2], Zhenzhen Zhang [1,2], Xiaoying Xie [1,2,3,*] and Boheng Zhang [1,2,3,*]

1 Department of Hepatic Oncology, Zhongshan Hospital (Xiamen), Fudan University, Xiamen 361015, China; chen.guobin@zsxmhospital.com (G.C.); chen.hong@zsxmhospital.com (H.C.); huang.xing@zsxmhospital.com (X.H.); cheng.sisi@zsxmhospital.com (S.C.); zheng.susu@zsxmhospital.com (S.Z.); wu.yanfang@zsxmhospital.com (Y.W.); zheng.tanghui@zsxmhospital.com (T.Z.); chen.xiaochun@zsxmhospital.com (X.C.); guo.xinkun@zsxmhospital.com (X.G.); zhang.zhenzhen@zsxmhospital.com (Z.Z.)

2 Xiamen Clinical Research Center for Cancer Therapy, Xiamen 361015, China

3 Department of Hepatic Oncology, Zhongshan Hospital, Fudan University, Shanghai 200032, China

* Correspondence: xie.xiaoying@zs-hospital.sh.cn (X.X.); zhang.boheng@zs-hospital.sh.cn (B.Z.)

† These authors contributed equally to this work.

Abstract: *Background and Objectives*: The recurrence outcome in patients who underwent microwave ablation (MWA) with or without transarterial chemoembolization (TACE) for hepatocellular carcinoma (HCC) within Milan criteria remains unclear. The aim of this retrospective study was to identify the predictive factors of recurrence in these patients. *Materials and Methods*: From May 2018 to April 2021, 66 patients with HCC within Milan criteria were enrolled. Local tumor progression (LTP) and recurrence-free survival (RFS) were evaluated. Univariate and multivariate analyses were used to evaluate the risk factors of recurrence. The propensity score analysis was conducted to reduce potential confounding bias. *Results*: During the median follow-up of 25.07 months (95% confidence interval [CI], 21.85, 28.28), the median time to LTP and RFS were 20.10 (95%CI, 14.67, 25.53) and 13.03 (95%CI, 6.36, 19.70) months. No group difference (MWA vs. MWA + TACE) was found in 1-year cumulative LTP ($p = 0.575$) and RFS ($p = 0.515$), but meaningful significant differences were found in two-year recurrence (LTP, $p = 0.007$ and RFS, $p = 0.037$). Univariate and multivariate analyses revealed that treatment received before ablation was an independent risk factor of LTP (hazard ratio [HR] 4.37, 95%CI, 1.44, 13.32) and RFS (HR 3.41, 95%CI, 1.49, 7.81). *Conclusions*: The LTP and RFS in the MWA group were similar to that in the MWA combined with TACE. For HCC within Milan criteria, both groups preferentially selected MWA. More endeavor and rigorous surveillance should be taken to relapse prevention, in patients who have received previous treatment.

Keywords: hepatocellular carcinoma; microwave thermotherapy; chemoembolization; recurrence

Citation: Chen, G.; Chen, H.; Huang, X.; Cheng, S.; Zheng, S.; Wu, Y.; Zheng, T.; Chen, X.; Guo, X.; Zhang, Z.; et al. Recurrence Outcome in Hepatocellular Carcinoma within Milan Criteria Undergoing Microwave Ablation with or without Transarterial Chemoembolization. *Medicina* **2022**, *58*, 1016. https://doi.org/10.3390/medicina58081016

Academic Editors: Nicolae Crisan, Călin Căinap and Konstantinos Dimas

Received: 23 June 2022
Accepted: 25 July 2022
Published: 29 July 2022

1. Introduction

Primary liver cancer is the third leading cause of cancer death. Globalcan2020 estimated 906,000 new cases in 2020, with approximately 75–85% cases being hepatocellular carcinoma (HCC) [1]. Thermal ablation, transplantation and resection were curative therapies for patients with HCC within the Milan criterion and are the available options in the China liver cancer (CNLC) staging system [2], the European Association for the Study of the Liver (EASL) [3], and the American Association for the Study of Liver Diseases (AASLD) guidelines [4]. However, approximately 20% of patients with hepatocellular carcinoma may experience a survival benefit from resection and liver transplantation [5]. Moreover, some patients have missed out on surgery owing to poor liver function (such as severe liver cirrhosis), location of the tumor nodules or rejection of surgery [6]. Liver transplantation is also limited due to the lack of availability of liver transplants and the high cost. Therefore, there is a need for a less invasive and effective treatment method.

Microwave ablation (MWA) is a relatively new ablation technique in thermal ablation and has the demonstrated benefits of safety, as well as being effective and minimally invasive [7]. When compared to radiofrequency ablation (RFA), MWA has the advantage of being less susceptible to the heat sink effect and provides a larger ablation zone [8–10]. There has been no demonstrated difference between MWA and RFA in efficacy or local tumor progression [11]. A meta-analysis depicted an analogical efficacy and safety between MWA and RFA. However, MWA displayed a preponderance in reducing the rate of five-year recurrences [12]. Some retrospective studies have shown that MWA achieved long-term oncologic outcomes for ≤4 cm HCC and equivalent metastasis and recurrence rates for ≤5 cm HCC when compared with surgery [7,13]. Unfortunately, to improve the clinical outcomes, it is not enough for HCC with a single treatment.

In addition to being the first-line treatment for intermediate-stage HCC, transarterial chemoembolization (TACE) can also be used for early-stage HCC [14]. TACE combined with MWA has been shown to be an effective treatment with a mean overall survival (mOS) rate of 54.9 months in early HCC [15]. The recurrence pattern of HCC was shown to be relative to post-recurrence survival [16]. Therefore, the purpose of this study was to clarify the risk factors for recurrence in patients with HCC within the Milan criteria, receiving MWA with or without TACE.

2. Materials and Methods

2.1. Study Population

This retrospective study was approved by the ethics committees of Xiamen Branch, Zhongshan Hospital, Fudan University (authorization number B2019-010). All patients were diagnosed with HCC according to the EASL guidelines [3]. The study enrolled 66 patients with HCC who received MWA with or without TACE from May 2018 to April 2021 in our institution, as shown in Figure 1. Patients were selected according to the following criteria:

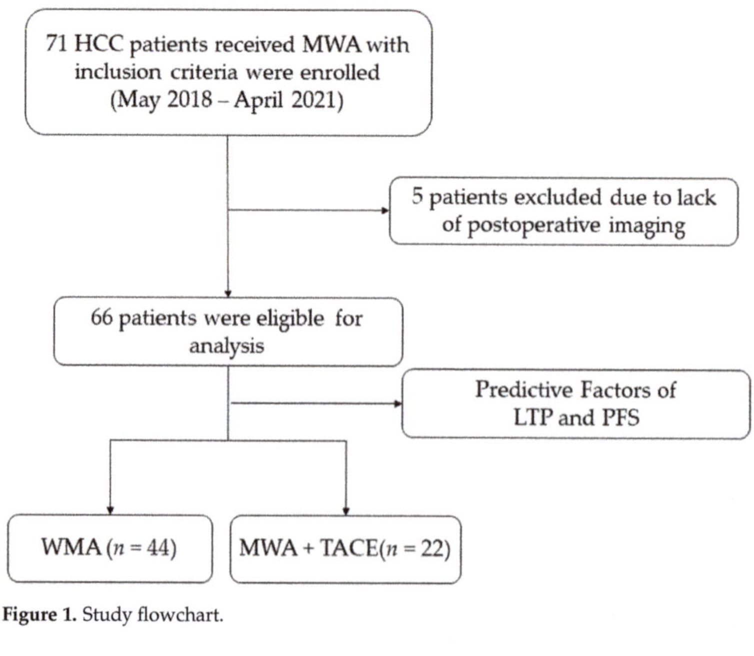

Figure 1. Study flowchart.

(1) HCC within the Milan criteria (single tumor $\leq$ 5 cm or two to three tumors $\leq$ 3 cm without vascular invasion) [17] who had received MWA with or without TACE as the treatment; (2) Child–Pugh score $\leq$ 7; (3) Eastern Cooperative Oncology Group performance status 0 or 1; and (4) refusal or unfitness for surgery and liver transplantation.

The exclusion criteria were (1) missing examination imaging from one to three-month postoperative period; (2) a lack of complete follow-up information; (3) other malignant tumors; and (4) other anti-cancer therapy received less than one month prior to intervention.

Demographics, oncology characteristics and some serological markers from within seven days before the operation were collected for analysis. The albumin–bilirubin (ALBI) score was calculated as follows: ($\log_{10}$ total bilirubin (μmol/L) $\times$ 0.66) + (serum albumin (g/L) $\times$ -0.085) [18]. The tumor location was recorded, including the tumor nodule position such as hepatic subcapsular, near large vessels, diaphragm or gallbladder. Treatment before ablation was defined as treatment up to one month prior to the intervention, such as TACE, conventional surgery, system treatment, radiotherapy or other anti-tumor treatment.

2.2. MWA Procedure

The tumor location and size were assessed for all patients by contrast enhanced ultrasound (CEUS) and contrast enhanced magnetic resonance imaging (CE-MRI) before the operation. Under the guidance of real-time ultrasonic imaging (US), MWA was performed with commercially available MWA systems (Covidien Medical Devices Technology Co., Ltd., Mansfield, MA, USA) after local or general anesthesia. An ablation needle with antennae was inserted into the tumor. During the MWA procedure, ablation session, time (range 1–10 min each session) and energy power (range 40–100 W each session) were determined depending on the tumor size, shape and location. The melting range was to achieve an ablated margin of at least 5 mm around the tumor. Finally, needle-path ablation was used to prevent post-operative bleeding and needle-path metastases.

2.3. TACE Combined with MWA Procedure

TACE and MWA procedures performed in the same session similar to Roberto Iezzi et.al commenting [19]. In brief, angiography was performed prior to ablation. All TACE procedures were performed by an experienced physician starting with a routine Seldinger puncture in the arteria cruralis after local anesthesia. A 5 French catheter (Progreat Lambda, Terumo, Japan) was selected to perform arteriography of the celiac and common hepatic artery to identify the tumor and feeders. Afterwards, MWA was performed as described above. CEUS (with sonoview as the enhancing agent) was used to evaluate the completeness of the ablation. If there was a residual tumor, an additional ablation was performed. Subsequently, a microcatheter was used to superselect the tumor-feeding branch and embolization was performed with 1–3 mL of lipiodol, with or without 10 mg of epirubicin (water-in-oil technique was used to mix chemotherapeutic agents and iodized oil which was described in previous study [20]). Finally, angiography was repeated to evaluate the extent of the lipiodol deposits. The brief procedure was shown in Supplemental Figure S1.

2.4. Follow-Up

All patients received regular follow-up after the operation. A CE-MRI/CT was performed one to three months after MWA with or without TACE. If the tumor was completely ablated, the patient was followed-up every two to three months. Combination therapy such as immunotherapy, antiangiogenic therapy, surgery etc. would then be conducted based on tumor recurrence. The study was censored on 2 February 2022.

2.5. Evaluation of Therapeutic Outcomes

The treatment response was evaluated by the modified response evaluation criteria in solid tumors (mRECIST) [21]. A non-specifically trained radiologist may magnify variability in the evaluation of treatment outcomes [22]. The assessment of treatment response was calculated by two different experts, one of whom has over 10 years of experience in radiology and another who has 10 years of experience in interventional radiology. A one-month postoperative CE-MRI/CT was used for assessing the initial therapeutic effect. Local tumor progression time (LTPt) was defined as the time between the operation date and the date when any residual or new-onset tumor around the ablation zone was discovered in the same liver lobe. Recurrence-free survival (RFS) was defined as the interval from the date of MWA with or without TACE to the date of HCC recurrence. Overall survival (OS) was defined as the time from the operation time until the time of death or the last follow-up recorded. Adverse events that occurred within one week of the interventional operation were recorded at follow-up, as were complications that were considered likely to be MWA/TACE-related.

2.6. Statistical Analysis

All analyses were performed using SPSS 21 software (IBM Corp., Armonk, NY, USA). Data were presented as mean ± standard deviation for continuous variables, which were analyzed by an independent *t*-test similar to the previous study [23]. Categorical data were defined as frequency (as a percentage) and calculated by applying a chi-square test or Fisher's exact test. LTP, RFS and OS were compared with a log-rank test using the Kaplan–Meier method. The risk factors for recurrence were analyzed by a univariate analysis. Multivariate Cox regression models were built to include all variables found to be $p \leq 0.1$ in the univariate analyses. A 1:1 propensity score with logistic regression was performed for balancing variables [24], with a caliper distance of 0.1. All comparisons were two-sided, and p values less than 0.05 were considered statistically significant.

3. Results

3.1. Patient Characteristics

The baseline patient characteristics of the two groups are summarized in Table 1. Of the 66 patients, 87.9% patients suffered from chronic viral hepatitis B (HBV), and 75.8% suffered from hepatic cirrhosis. The majority of patients presented with BCLC Stage A (84.8%) and received other treatment before ablation (63.4%). The 44 patients who underwent MWA had a significantly smaller tumor size compared to the 22 patients who received MWA combined with TACE (18.36 ± 8.47 vs. 25.09 ± 10.31 mm, p = 0.006). In addition, the MWA group had a smaller number of tumor nodules (11.4% multiple nodules) according to MWA plus the TACE group (multiple nodules 11.4% vs. 36.4%, p = 0.038). The lymphocyte counts (1.70 ± 0.61 vs. 1.37 ± 0.43) and ALBI (−3.07 ± 0.31 vs. −3.25 ± 0.27, p = 0.027) of the MWA group were significantly higher than MWA plus TACE.

Table 1. Baseline patient characteristics stratified by therapy [mean SD/N (%)].

Characteristics	Overall (N = 66)	MWA (N = 44)	MWA + TACE (N = 22)	*p* Value
age	61.14 ± 11.18	61.36 ± 10.89	60.68 ± 11.98	0.817
Sex				
male	51 (77.3%)	31 (70.5%)	20 (90.9%)	0.062
female	15 (22.7%)	13 (29.5%)	2 (9.1%)	
Hepatic Cirrhosis				
yes	50 (75.8%)	34 (77.3%)	16 (72.7%)	0.685
no	16 (24.2%)	10 (22.7%)	6 (27.3%)	
Diabetes				
yes	12 (18.2%)	8 (18.2%)	4 (18.2%)	>0.999
no	54 (81.8%)	36 (81.8%)	18 (81.8%)	
Hypertension				
yes	17 (25.8%)	13 (29.5%)	4 (18.2%)	0.320
no	49 (74.2%)	31 (70.5%)	18 (81.8%)	

Table 1. *Cont.*

Characteristics	Overall (N = 66)	MWA (N = 44)	MWA + TACE (N = 22)	*p* Value
HBV				
yes	58 (87.9%)	39 (88.6%)	19 (86.4%)	>0.999
no	8 (12.1%)	5 (11.4%)	3 (13.6%)	
Tumor diameter (mm)	20.61 ± 9.59	18.36 ± 8.47	25.09 ± 10.31	0.006
BCLC				
A	56 (84.8%)	39 (88.6%)	17 (77.3%)	0.396
B	10 (15.2%)	5 (11.4%)	5 (22.7%)	
Tumor location				
special location	48 (72.7%)	33 (75%)	15 (68.2%)	0.716
traditional location	18 (27.3%)	11 (25%)	7 (31.8%)	
Tumor number				
single	53 (80.3%)	39 (88.6%)	14 (63.6%)	0.038
multiple	13 (19.7%)	5 (11.4%)	8 (36.4%)	
Treatment before ablation				
yes	24 (36.4%)	18 (40.9%)	6(27.3%)	0.278
no	42 (63.6%)	26 (59.1%)	16(72.7%)	
Baseline AFP(ng/mL)	433.81 ± 1666.24	285.50 ± 1219.54	730.44 ± 2325.90	0.310
Baseline Lymphocytes (×109/L)	1.59 ± 0.58	1.70 ± 0.61	1.37 ± 0.43	0.034
Baseline Monocytes(×109/L)	0.50 ± 0.17	0.51 ± 0.17	0.47 ± 0.18	0.452
Baseline ALB (g/L)	44.51 ± 3.21	43.86 ± 3.18	45.77 ± 2.94	0.022
Baseline ALT (U/L)	28.49 ± 14.91	30.61 ± 16.70	24.36 ± 9.64	0.061
Baseline AST (U/L)	30.06 ± 13.79	31.56 ± 15.90	27.14 ± 7.77	0.137
Baseline rGT (U/L)	60.56 ± 53.02	64.48 ± 56.12	52.73 ± 46.44	0.400
Baseline LDH(U/L)	197.89 ± 44.45	192.88 ± 43.41	207.68 ± 45.83	0.207
CD4/CD8	1.98 ± 1.21	2.04 ± 1.38	1.87 ± 0.85	0.626
IL-6 (pg/mL)	6.61 ± 7.11	7.11 ± 8.57	5.65 ± 2.71	0.474
TNF-α (pg/mL)	12.07 ± 12.21	10.43 ± 11.14	15.09 ± 13.77	0.203
Baseline ALBI	−3.13 ± 0.31	−3.07 ± 0.31	−3.25 ± 0.27	0.027
Baseline CRP(mg/L)	3.10 ± 8.28	6.85 ± 16.84	2.13 ± 2.38	0.627

Median with standard deviation are shown for quantitative variables and counts with proportions are shown for categorical variables. Tumor special location including tumor nodule in the position such as hepatic subcapsular, near large vessels, diaphragm and gallbladder. Abbreviations: Ref-Reference; HBV-Hepatitis B Virus infection; AFP-alpha-fetoprotein; ALB-albumin; ALT-alanine transaminase; AST-aspartate aminotransferase; γGT-γ-glutamyltranspeptidas; LDH-lactate dehydrogenase; ALBI-albumin–bilirubin; CRP-c-reactive protein.

3.2. Treatment Outcomes and Complications

The rate of technical success of the ablations was 100%. During the median follow-up period of 25.07 months (95% CI, 21.85, 28.28), 6.1% of patients died. There was no significant difference between the two groups ($p = 0.396$). The estimated OS rates of the MWA group were 100% and 97% at one year and two years, respectively, and 100% and 85%, respectively, in the MWA plus TACE group (Figure 2a). A total of 37/66 (56.1%) ablations demonstrated LTP. The median LTPt (mLTPt) was 20.10 months (95% CI, 14.67, 25.53). The 1-year and 2-year cumulative LTP incidence was 40% and 53%, respectively, in the MWA group and 38% and 88%, respectively, in the MWA plus TACE group. RFS occurred in 41/66 (62.1%) patients. The median RFS (mRFS) time was 13.03 months (95% CI, 6.36, 19.70). The cumulative RFS was estimated to be 45% and 38% in the MWA group, respectively, and 61% and 15% in the MWA plus TACE group, respectively, at one year and two years. There were no significant differences found in mLTPt ($p = 0.575$, Figure 2b) and mRFS ($p = 0.515$, Figure 2c), but statistically significant differences were observed in 2-year cumulative LTP ($p = 0.007$) and RFS ($p = 0.037$) incidence.

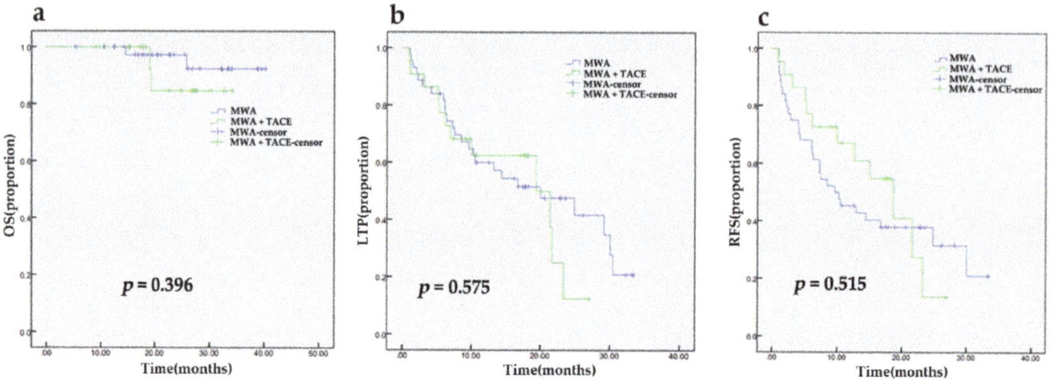

Figure 2. The Kaplan–Meier analysis of overall survival (**a**), time to local tumor recurrence (**b**) and recurrence-free survival (**c**) after MWA with or without TACE.

The adverse events were evaluated by Common Terminology Criteria for Adverse Events (CTCAE) Version 5.0 (Department of Health and Humen services, USA). As shown in Table 2, 80.3% of patients experienced an adverse reaction, of which the majority were Grade ½ adverse events. Neither the Grade ½ complication rates or the Grade ¾ level adverse events were significantly different between the MWA group and the MWA plus TACE group, ½ ($p = 0.595$) and ¾ ($p = 0.735$), respectively.

Table 2. Adverse Events and Complications.

Categories	Overall (N = 66)			MWA (N = 44)		MWA + TACE (N = 22)	
Grade	**1–4 level**	**½ level**	**¾ level**	**½ level**	**¾ level**	**½ level**	**¾ level**
Adverse events	53 (80.3%)	52 (78.8%)	12 (18.2%)	36 (81.8%)	7 (15.9%)	16 (72.7%)	5 (22.7%)
Fever	2 (3.0%)	2 (3.0%)	0 (0%)	1 (2.3%)	0 (0%)	1 (4.5%)	0 (0%)
Nausea or vomiting	13 (19.7%)	13 (19.7%)	0 (0%)	8 (18.2%)	0 (0%)	5 (22.7%)	0 (0%)
Fatigue	5 (7.6%)	5 (7.6%)	0 (0%)	2 (4.5%)	0 (0%)	3 (13.6%)	0 (0%)
Abdominal pain/distension	4 (6.1%)	4 (6.1%)	0 (0%)	3 (6.8%)	0 (0%)	1 (2.3%)	0 (0%)
Total bilirubin elevation, transient	1 (1.5%)	1 (1.5%)	0 (0%)	1 (2.3%)	0 (0%)	0 (0%)	0 (0%)
ALT elevation	45 (68.2%)	42 (63.6%)	3 (4.5%)	28 (63.6%)	2 (4.5%)	14 (63.6%)	1 (2.3%)
AST elevation	53 (80.3%)	43 (65.2%)	10 (15.2%)	30 (68.2%)	7 (15.9%)	13 (59.1%)	3 (13.6%)

3.3. Univariable and Multivariable Analyses for LTP and RFS

In the univariate analysis, Barcelona Clinic Liver Cancer (BCLC) stage ($p = 0.097$), tumor location ($p = 0.086$), treatment before ablation ($p < 0.001$), baseline lymphocyte counts ($p = 0.024$), baseline monocyte count ($p = 0.034$), TNF-α ($p = 0.050$) and ALBI ($p = 0.064$) were taken into multivariate Cox models for LTP. As shown in Table 3, receiving treatment before ablation (HR 4.37, 95% CI, 1.44, 13.32) and baseline ALBI (HR 4.31, 95% CI, 1.17, 15.92) independently predicted the LTPt. As shown in Table 4, treatment before ablation was associated with RFS ($p = 0.003$) and was verified as an independent predictor for RFS by multivariate analysis (HR 3.41, 95% CI, 1.49, 7.81).

Table 3. Univariable and multivariable predictors of LTP.

Characteristics	Univariate Analysis HR (95% CI)	Univariate Analysis p Value	Multivariate Analysis HR (95% CI)	Multivariate Analysis p Value
age	0.99 (0.96, 1.02)	0.571		
Sex				
male	1.00 (Ref)	0.146		
female	0.50 (0.19, 1.28)			
Hepatic Cirrhosis				
yes	1.00 (Ref)	0.166		
no	1.68 (0.81, 3.52)			
Diabetes				
yes	1.00 (Ref)	0.455		
no	0.75 (0.35, 1.60)			
Hypertension				
yes	1.00 (Ref)	0.638		
no	1.20 (0.56, 2.54)			
HBV				
yes	1.00 (Ref)	0.845		
no	0.92 (0.38, 2.22)			
Tumor diameter (mm)	1.01 (0.97, 1.04)	0.768		
BCLC				
A	1.00 (Ref)	0.097	1.00 (Ref)	0.845
B	1.96 (0.89, 4.33)		1.11 (0.40, 3.04)	
Tumor location				
special location	1.00 (Ref)	0.086	1.00 (Ref)	0.181
traditional location	0.53 (0.25, 1.10)		0.55 (0.23, 1.32)	
Tumor number				
Single	1.00 (Ref)			
Multiple	1.46 (0.68, 3.14)	0.332		
Treatment before ablation				
no	1.00 (Ref)	<0.001	1.00 (Ref)	0.009
yes	4.77 (2.05, 11.07)		4.37 (1.44, 13.32)	
Baseline AFP (ng/mL)	1.00 (1.00, 1.00)	0.564		
Baseline Lymphocytes (×109/L)	0.47 (0.24, 0.90)	0.024	1.32 (0.52, 3.39)	0.561
Baseline Monocytes (×109/L)	0.10 (0.01, 0.84)	0.034	0.07 (0.00, 1.45)	0.086
Baseline ALB (g/L)	0.93 (0.85, 1.03)	0.167		
Baseline ALT (U/L)	0.99 (0.97, 1.02)	0.489		
Baseline AST (U/L)	1.00 (0.97, 1.02)	0.732		
Baseline rGT (U/L)	1.00 (0.99, 1.01)	0.798		
Baseline LDH (U/L)	1.01 (0.99, 1.01)	0.19		
CD4/CD8	0.89 (0.66, 1.20)	0.444		
IL-6 (pg/mL)	0.98 (0.92, 1.04)	0.448		
TNF-α (pg/mL)	1.03 (1.00, 1.06)	0.05	1.03 (0.99, 1.05)	0.071
Baseline ALBI	2.57 (0.95, 6.95)	0.064	4.31 (1.17, 15.92)	0.028
Baseline CRP (mg/L)	0.98 (0.94, 1.03)	0.984		

Median with standard deviation are shown for quantitative variables and counts with proportions shown for categorical variables. Tumor special location including tumor nodule in the position such as hepatic subcapsular and near large vessels, diaphragm or gallbladder. Abbreviations: Ref-Reference; HBV-Hepatitis B Virus infection; AFP-alpha-fetoprotein; ALB-albumin; ALT-alanine transaminase; AST-aspartate aminotransferase; γGT-γ-glutamyltranspeptidase; LDH-lactate dehydrogenase; ALBI-albumin–bilirubin; CRP-c-reactive protein.

Table 4. Univariable and multivariable predictors of RFS.

Characteristics	Univariate Analysis HR (95% CI)	p Value	Multivariate Analysis HR (95% CI)	p Value
age	0.98 (0.95, 1.01)	0.139		
Sex				
male	1.00 (Ref)	0.121		
female	1.99 (0.83, 4.76)			
Hepatic Cirrhosis				
yes	1.00 (Ref)	0.755		
no	1.12 (0.55, 2.30)			
Diabetes				
yes	1.00 (Ref)	0.455		
no	0.75 (0.36, 1.59)			
Hypertension				
yes	1.00 (Ref)	0.162		
no	1.70 (0.81, 3.59)			
HBV				
yes	1.00 (Ref)	0.734		
no	1.16 (0.49, 2.78)			
Tumor diameter (mm)	0.98 (0.95, 1.02)	0.38		
BCLC				
A	1.00 (Ref)	0.373		
B	1.42 (0.66, 3.09)			
Tumor location				
special location	1.00 (Ref)	0.705		
traditional location	0.87 (0.44, 1.76)			
Tumor number				
single	1.00 (Ref)	0.802		
multiple	1.10 (0.52, 2.31)			
Treatment before ablation				
no	1.00 (Ref)	0.003	1.00 (Ref)	0.004
yes	3.10 (1.47, 6.54)		3.41 (1.49, 7.81)	
Baseline AFP (ng/mL)	1.00 (1.00, 1.00)	0.425		
Baseline Lymphocytes (×109/L)	0.57 (0.31, 1.05)	0.07	0.95 (0.43, 2.11)	0.896
Baseline Monocytes (×109/L)	0.16 (0.02, 1.13)	0.067	0.21 (0.02, 2.51)	0.219
Baseline ALB (g/L)	0.97 (0.88, 1.06)	0.447		
Baseline ALT (U/L)	1.00 (0.98, 1.03)	0.809		
Baseline AST (U/L)	1.00 (0.98, 1.02)	0.904		
Baseline rGT (U/L)	1.00 (0.99, 1.01)	0.430		
Baseline LDH (U/L)	1.00 (0.99, 1.01)	0.984		
CD4/CD8	0.86 (0.59, 1.26)	0.447		
IL-6 (pg/mL)	0.98 (0.92, 1.04)	0.491		
TNF-α (pg/mL)	1.02 (0.99, 1.04)	0.295		
Baseline ALBI	1.86 (0.73, 4.79)	0.196		
Baseline CRP (mg/L)	0.98 (0.93, 1.03)	0.435		

Median with standard deviation are shown for quantitative variables and counts with proportions shown for categorical variables. Tumor special location including tumor nodule in the position such as hepatic subcapsular, near large vessels, diaphragm and gallbladder. Abbreviations: Ref-Reference; HBV-Hepatitis B Virus infection; AFP-alpha-fetoprotein; ALB-albumin; ALT-alanine transaminase; AST-aspartate aminotransferase; γGT-γ-glutamyltranspeptidase; LDH-lactate dehydrogenase; ALBI-albumin–bilirubin; CRP-c-reactive protein.

3.4. Subgroup Analysis

In the subgroup analysis, 29/66 (43.9%) and 24/66 (36.4%) of patients had been treated with TACE or traditional surgery, respectively, at least one month before ablation. The univariate analysis showed that TACE ($p < 0.003$ and $p = 0.001$, respectively) and radiotherapy ($p < 0.001$) were associated with LTP and RFS. In the multivariate analysis, TACE (HR 3.92, 95% CI, 1.72, 8.93) and radiotherapy (HR 17.95 95% CI, 4.10, 78.71) were independent predictors of LTP. The risk factors for RFS were antiangiogenic therapy (HR 2.54, 95% CI, 0.99, 8.93) and radiotherapy (HR 8.41, 95% CI, 2.29, 30.89) (Table 5).

Table 5. Subgroup analysis of the treatment before ablation [N/(%)].

Method		LTP		PFS	
	Number	HR (95% CI)	*p* Value	HR (95% CI)	*p* Value
Univariate Analysis					
TACE	29 (43.9%)	4.56 (2.14, 9.75)	<0.001	2.27 (1.18, 4.35)	0.014
Thermal ablation	17 (25.8%)	1.43 (0.98, 2.08	0.063	1.73 (0.86, 3.48)	0.124
Antiangiogenic therapy	7 (10.6%)	2.19 (1.38, 3.46)	0.001	3.54 (1.45, 8.62)	0.005
Surgery	24 (36.4%)	1.05 (0.75, 1.47)	0.775	1.10 (0.58, 2.09)	0.765
Radiotherapy	3 (4.5%)	4.08 (1.98, 8.41)	<0.001	8.33 (2.27, 30.54)	0.001
Multivariate Analysis					
TACE	29 (43.9%)	3.92 (1.72, 8.93)	0.001	1.96 (1.00, 3.88)	0.053
Thermal ablation	17 (25.8%)	1.05 (0.43, 2.58)	0.912	——	——
Antiangiogenic therapy	7 (10.6%)	2.59 (0.93, 7.23)	0.068	2.54 (0.99, 6.44)	0.049
Radiotherapy	3 (4.5%)	17.95 (4.10, 78.71)	<0.001	8.41 (2.29, 30.89)	0.001

Only the variables found to be $p \leq 0.1$ in the univariate analyses were taken into Multivariate Cox regression models.

The rate of LTP and RFS was compared between the MWA and MWA plus TACE group after propensity score matching. As depicted in Supplemental Table S1, for mLTPt ($p = 0.945$) and mRFS ($p = 0.28$), there were still no significant differences between the two groups after balancing the variables.

4. Discussion

In the present study, we demonstrated that recurrence and major complication rates of the MWA group were similar to the MWA combined with TACE group, meeting the Milan criteria in terms of LTP and RFS. Furthermore, treatment before ablation in particular with TACE, antiangiogenic therapy and radiotherapy were independent risk factors for tumor recurrence.

MWA is a promising thermal technique because of its efficacy and safety. A meta-analysis proved that MWA had a relative risk of 0.93 (95% CI, 0.78, 1.14) compared to RFA for 1-year LTP in early HCC [25]. In previous research, 729 patients with HCC within the Milan criteria undergoing MWA or surgical resection (SR) were analyzed retrospectively. They identified that MWA achieved comparable long-term oncologic outcomes such as LTP or disease-free survival (DFS) with SR for ≤4 cm HCC [7]. A randomized controlled trial that screened 278 patients with 3–5 cm HCC reported that the one-year recurrence rates in the TACE combined with MWA group was significantly lower than the MWA or TACE groups [26]. However, TACE was followed by MWA after 15 days in this study. TACE combined with simultaneous DynaCT-guided MWA was reported as an outstanding method for the treatment of <5 cm HCC in contrast to TACE [27]. With this method, the mean PFS was 28.22 months longer than the TACE group. Furthermore, a single tumor less than 3 cm showed a prolonged PFS and OS when performed by the TACE combined with MWA [15]. To our knowledge, the recurrence outcome of MWA with or without simultaneous TACE has yet to be researched in depth. Moreover, few studies have focused on all HCCs meeting the Milan criteria.

In the current study, the mLTPt was greater than in another recent study (20.10 vs. 9.60 months) [28], which is of note particularly because we did not exclude patients who relapsed after receiving other antitumor therapies before ablation. Univariate and multivariate analyses revealed that treatment before ablation was a significant risk factor for LTP and RFS. As shown in the subgroup analysis, patients with TACE, antiangiogenic therapy or radiotherapy demonstrated earlier relapse. As described in a previous study, TACE was an independent risk factor for worse PFS when used to treat HCC 5 cm or smaller [29,30]. Moreover, Salas et al. found that radiotherapy was a key factor influencing the incidence of local recurrence in solitary fibrous tumors [31]. To the best of our knowledge, the present study is the first to explore occurrence outcomes of HCC treated with MWA with or without TACE. Patients who received treatment before ablation tended to possess poor tumor characteristics, and treatment such as radiotherapy may increase VEGF/plt level, which is associated with poor outcomes [32]. This may explain the recurrence of LTP and RFS.

The treatment of MWA combined with TACE seemed to have a similar result to MWA with regard to LTP and RFS in our results. For LTP and RFS at two years, the late relapse rate of the combined therapy group was significantly worse than the MWA group. However, there were differences in the baseline variables relating to tumor size, nodule number, lymphocyte counts and ALBI. Tumor size is one of the critical factors for complete ablation and survival. For a single tumor nodule ≤3 cm, priority should be given to MWA over treatment with TACE [29]. However, for lesions in the range 3.1–5 cm, this study demonstrated that MWA had similar effects as TACE in OS. A multicenter observational study demonstrated that tumor size and number were crucial prognostic factors for HCC with TACE [33]. Another study has also shown that lymphocyte counts and ALBI play an important role in the carcinogenesis and progression of HCC. In our research, the combination therapy group had larger and more numerous tumor nodules; moreover, lymphocyte counts and ALBI were higher than in the MWA group [34,35]. This appears to suggest that the combination treatment was superior for the prevention of tumor recurrence. However, after balancing the variables, this was not validated. This may have been due to the small sample size. Further studies are required to explore the possibility. Howsoever, MWA will be effective in HCC within Milan criteria and the single procedure doesn't meet an early relapse.

No serious adverse events (Grade 4) were recorded for the entire follow-up period. However, 18.2% patients suffered Grade 3 events; all of them recovered with symptomatic treatment. This indicated that both treatment methods were safe for patients.

Some limitations were identified in the study. First, it was a single center retrospective study with a small sample, and the two groups differed significantly on some variables. Selection bias is inescapable; however, we used the propensity matching score to lessen the effect. Second, 87.9% of our study cohort had an HBV infection; different etiologies may influence tumor characteristics. Thirdly, ultrasound was used to ensure complete ablation in the study. However, post-ablation ultrasound images may be affected by gas or inflammatory edema around the ablation site. Three dimensional digital subtraction and angiography technology may eliminate the influence as previous study have reported [15,27]. This deserves further exploration. Finally, the difference of intraoperative medications and procedures in MWA and TACE may have affected the outcome of treatment. To minimize the differences, all procedures were performed by the same team. Therefore, further research with a large stratified multicenter patient cohort is necessary to validate our results.

5. Conclusions

In short, the LTP and RFS in the MWA group were comparable to that in the group treated with MWA combined with TACE. In our results, for HCC meeting the Milan criteria, priority should be given to MWA when making treatment choices between MWA and MWA combined with TACE. The approach possesses further relevant advantages such as the decrease of patient discomfort and cost savings due to unnecessary TACE is not required. Moreover, receiving treatment before ablation was an independent risk factor for recurrence. When patients receive prior treatment, more rigorous surveillance should be taken to closely observe them for recurrence. Besides, further prospective studies with larger samples are required to clarify the distinction between various treatments for HCC.

Supplementary Materials: The following supporting information can be downloaded at: https://www.mdpi.com/article/10.3390/medicina58081016/s1, Table S1. The analysis of mean LTPt and RFS around propensity score matching. Figure S1. The procedures of TACE combined with MWA for a 54-years-old male with HCC. (a): Preoperative CE-MRI found a tumor nodule in the subscapular of right lobe. (b): Angiography demonstrated the location of the tumor and tumor-supplying arteries. (c): The puncture was inserted through the tumor and performed the MWA. (d): The embolization of tumor-feeding branch and angiography shown that tumor staining disappeared. (e) After 1 month of treatment, CE-MRI displayed tumor necrosis.

Author Contributions: Conceptualization, B.Z. and X.X.; methodology, B.Z., G.C. and X.H.; software, G.C., X.H. and H.C.; validation, G.C. and S.C.; Operation, B.Z., X.X., H.C. and G.C.; formal analysis, G.C. and H.C.; investigation, S.Z., Y.W., T.Z., X.C., H.C., Z.Z. and X.G.; resources, S.C.; data curation, B.Z. and X.X.; writing—original draft preparation, G.C. and X.H.; writing—review and editing, G.C., X.X. and B.Z.; visualization, H.C. and X.H.; supervision, X.X. and B.Z.; project administration, G.C.; funding acquisition, B.Z. and G.C.; All authors have read and agreed to the published version of the manuscript.

Funding: This study was supported by grants from Key Program of Xiamen Medical and Health (grant number 3502Z20191105) and Health Youth Scientific research project in Fujian (grant number 2019-2-64).

Institutional Review Board Statement: The study was conducted in accordance with the Declaration of Helsinki, and approved by the ethics committees of Xiamen Branch, Zhongshan hospital, Fudan university (approval number B2019-010 and 14 January 2019).

Informed Consent Statement: Informed consent was not required from patients because retrospectively collected data were anonymized and waived for the analysis.

Data Availability Statement: The data presented in this study are available from the corresponding authors. The data are not publicly available due to privacy restrictions.

Conflicts of Interest: The authors declare that they have no conflict of interest.

References

1. Sung, H.; Ferlay, J.; Siegel, R.L.; Laversanne, M.; Soerjomataram, I.; Jemal, A.; Bray, F. Global Cancer Statistics 2020: GLOBOCAN Estimates of Incidence and Mortality Worldwide for 36 Cancers in 185 Countries. *CA Cancer J. Clin.* **2021**, *71*, 209–249. [CrossRef]
2. Xie, D.Y.; Ren, Z.G.; Zhou, J.; Fan, J.; Gao, Q. 2019 Chinese clinical guidelines for the management of hepatocellular carcinoma: Updates and insights. *Hepatobiliary Surg. Nutr.* **2020**, *9*, 452–463. [CrossRef] [PubMed]
3. European Association for the Study of the Liver. EASL Clinical Practice Guidelines: Management of hepatocellular carcinoma. *J. Hepatol.* **2018**, *69*, 182–236. [CrossRef] [PubMed]
4. Heimbach, J.K.; Kulik, L.M.; Finn, R.S.; Sirlin, C.B.; Abecassis, M.M.; Roberts, L.R.; Zhu, A.X.; Murad, M.H.; Marrero, J.A. AASLD guidelines for the treatment of hepatocellular carcinoma. *Hepatology* **2018**, *67*, 358–380. [CrossRef] [PubMed]
5. Finn, R.S.; Zhu, A.X. Evolution of Systemic Therapy for Hepatocellular Carcinoma. *Hepatology* **2021**, *73*, 150–157. [CrossRef]
6. El-Serag, H.B.; Mason, A.C. Rising incidence of hepatocellular carcinoma in the United States. *N. Engl. J. Med.* **1999**, *340*, 745–750. [CrossRef] [PubMed]
7. Dou, J.; Cheng, Z.; Han, Z.; Liu, F.; Wang, Z.; Yu, X.; Yu, J.; Liang, P. Microwave ablation vs. surgical resection for treatment naive hepatocellular carcinoma within the Milan criteria: A follow-up of at least 5 years. *Cancer. Biol. Med.* **2021**. [CrossRef]
8. Llovet, J.M.; De Baere, T.; Kulik, L.; Haber, P.K.; Greten, T.F.; Meyer, T.; Lencioni, R. Locoregional therapies in the era of molecular and immune treatments for hepatocellular carcinoma. *Nat. Rev. Gastroenterol. Hepatol.* **2021**, *18*, 293–313. [CrossRef] [PubMed]
9. Andreano, A.; Huang, Y.; Meloni, M.F.; Lee, F.T., Jr.; Brace, C. Microwaves create larger ablations than radiofrequency when controlled for power in ex vivo tissue. *Med. Phys.* **2010**, *37*, 2967–2973. [CrossRef] [PubMed]
10. Smolock, A.R.; Shaw, C. Hepatic Microwave Ablation in Challenging Locations. *Semin. Interv. Radiol.* **2019**, *36*, 392–397. [CrossRef]
11. Yu, J.; Yu, X.L.; Han, Z.Y.; Cheng, Z.G.; Liu, F.Y.; Zhai, H.Y.; Mu, M.J.; Liu, Y.M.; Liang, P. Percutaneous cooled-probe microwave versus radiofrequency ablation in early-stage hepatocellular carcinoma: A phase III randomised controlled trial. *Gut* **2017**, *66*, 1172–1173. [CrossRef] [PubMed]
12. Facciorusso, A.; Abd El Aziz, M.A.; Tartaglia, N.; Ramai, D.; Mohan, B.P.; Cotsoglou, C.; Pusceddu, S.; Giacomelli, L.; Ambrosi, A.; Sacco, R. Microwave Ablation Versus Radiofrequency Ablation for Treatment of Hepatocellular Carcinoma: A Meta-Analysis of Randomized Controlled Trials. *Cancers* **2020**, *12*, 3796. [CrossRef] [PubMed]
13. Wang, Z.L.; Liang, P.; Dong, B.W.; Yu, X.L.; Yu, D.J. Prognostic factors and recurrence of small hepatocellular carcinoma after hepatic resection or microwave ablation: A retrospective study. *J. Gastrointest. Surg.* **2008**, *12*, 327–337. [CrossRef]
14. Chang, Y.; Jeong, S.W.; Young Jang, J.; Jae Kim, Y. Recent Updates of Transarterial Chemoembolilzation in Hepatocellular Carcinoma. *Int. J. Mol. Sci.* **2020**, *21*, 8165. [CrossRef] [PubMed]
15. Li, Z.; Hao, D.; Jiao, D.; Zhang, W.; Han, X. Transcatheter Arterial Chemoembolization Combined with Simultaneous Cone-beam Computed Tomography-guided Microwave Ablation in the Treatment of Small Hepatocellular Carcinoma: Clinical Experiences From 50 Procedures. *Acad. Radiol.* **2021**, *28*, S64–S70. [CrossRef] [PubMed]
16. Huang, J.; Huang, W.; Guo, Y.; Cai, M.; Zhou, J.; Lin, L.; Zhu, K. Risk Factors, Patterns, and Long-Term Survival of Recurrence After Radiofrequency Ablation With or Without Transarterial Chemoembolization for Hepatocellular Carcinoma. *Front. Oncol.* **2021**, *11*, 638428. [CrossRef]

17. Yao, F.Y.; Ferrell, L.; Bass, N.M.; Bacchetti, P.; Ascher, N.L.; Roberts, J.P. Liver transplantation for hepatocellular carcinoma: Comparison of the proposed UCSF criteria with the Milan criteria and the Pittsburgh modified TNM criteria. *Liver. Transplant.* **2002**, *8*, 765–774. [CrossRef]
18. Johnson, P.J.; Berhane, S.; Kagebayashi, C.; Satomura, S.; Teng, M.; Reeves, H.L.; O'Beirne, J.; Fox, R.; Skowronska, A.; Palmer, D.; et al. Assessment of liver function in patients with hepatocellular carcinoma: A new evidence-based approach-the ALBI grade. *J. Clin. Oncol.* **2015**, *33*, 550–558. [CrossRef]
19. Iezzi, R.; Pompili, M.; Gasbarrini, A.; Bonomo, L. Sequential or combined treatment? That is the question. *Radiology* **2014**, *272*, 612–613. [CrossRef]
20. Renzulli, M.; Peta, G.; Vasuri, F.; Marasco, G.; Caretti, D.; Bartalena, L.; Spinelli, D.; Giampalma, E.; D'Errico, A.; Golfieri, R. Standardization of conventional chemoembolization for hepatocellular carcinoma. *Ann. Hepatol.* **2021**, *22*, 100278. [CrossRef]
21. Lencioni, R.; Llovet, J.M. Modified RECIST (mRECIST) assessment for hepatocellular carcinoma. *Semin. Liver Dis.* **2010**, *30*, 52–60. [CrossRef]
22. Tovoli, F.; Renzulli, M.; Negrini, G.; Brocchi, S.; Ferrarini, A.; Andreone, A.; Benevento, F.; Golfieri, R.; Morselli-Labate, A.M.; Mastroroberto, M.; et al. Inter-operator variability and source of errors in tumour response assessment for hepatocellular carcinoma treated with sorafenib. *Eur. Radiol.* **2018**, *28*, 3611–3620. [CrossRef]
23. Chen, G.; Xie, X.; Wang, M.; Guo, X.; Zhang, Z.; Zhang, L.; Zhang, B. Prognostic Significance of Tumor Growth Rate (TGR) in Patients with Huge Hepatocellular Carcinoma Undergoing Transcatheter Arterial Chemoembolization. *Curr. Oncol.* **2022**, *29*, 423–432. [CrossRef]
24. Rubin, D.B. Estimating causal effects from large data sets using propensity scores. *Ann. Intern. Med.* **1997**, *127*, 757–763. [CrossRef]
25. Gupta, P.; Maralakunte, M.; Kumar, M.P.; Chandel, K.; Chaluvashetty, S.B.; Bhujade, H.; Kalra, N.; Sandhu, M.S. Overall survival and local recurrence following RFA, MWA, and cryoablation of very early and early HCC: A systematic review and Bayesian network meta-analysis. *Eur. Radiol.* **2021**, *31*, 5400–5408. [CrossRef]
26. Zaitoun, M.M.A.; Elsayed, S.B.; Zaitoun, N.A.; Soliman, R.K.; Elmokadem, A.H.; Farag, A.A.; Amer, M.; Hendi, A.M.; Mahmoud, N.E.M.; Salah El Deen, D.; et al. Combined therapy with conventional trans-arterial chemoembolization (cTACE) and microwave ablation (MWA) for hepatocellular carcinoma >3–<5 cm. *Int. J. Hyperth.* **2021**, *38*, 248–256. [CrossRef]
27. Li, Z.; Jiao, D.; Han, X.; Si, G.; Li, Y.; Liu, J.; Xu, Y.; Zheng, B.; Zhang, X. Transcatheter arterial chemoembolization combined with simultaneous DynaCT-guided microwave ablation in the treatment of small hepatocellular carcinoma. *Cancer Imaging* **2020**, *20*, 13. [CrossRef]
28. Hermida, M.; Cassinotto, C.; Piron, L.; Aho-Glele, S.; Guillot, C.; Schembri, V.; Allimant, C.; Jaber, S.; Pageaux, G.P.; Assenat, E.; et al. Multimodal Percutaneous Thermal Ablation of Small Hepatocellular Carcinoma: Predictive Factors of Recurrence and Survival in Western Patients. *Cancers* **2020**, *12*, 313. [CrossRef]
29. Cui, R.; Wang, X.H.; Ma, C.; Liu, T.; Cheng, Z.G.; Han, Z.Y.; Liu, F.Y.; Yu, X.L.; Yu, J.; Liang, P. Comparison of Microwave Ablation and Transarterial Chemoembolization for Single-Nodule Hepatocellular Carcinoma Smaller Than 5cm: A Propensity Score Matching Analysis. *Cancer Manag. Res.* **2019**, *11*, 10695–10704. [CrossRef]
30. Lee, S.H.; Jin, Y.J.; Lee, J.W. Survival benefit of radiofrequency ablation for solitary (3-5 cm) hepatocellular carcinoma: An analysis for nationwide cancer registry. *Medicine* **2017**, *96*, e8486. [CrossRef] [PubMed]
31. Salas, S.; Resseguier, N.; Blay, J.Y.; Le Cesne, A.; Italiano, A.; Chevreau, C.; Rosset, P.; Isambert, N.; Soulie, P.; Cupissol, D.; et al. Prediction of local and metastatic recurrence in solitary fibrous tumor: Construction of a risk calculator in a multicenter cohort from the French Sarcoma Group (FSG) database. *Ann. Oncol.* **2017**, *28*, 1979–1987. [CrossRef]
32. Suh, Y.G.; Lee, E.J.; Cha, H.; Yang, S.H.; Seong, J. Prognostic values of vascular endothelial growth factor and matrix metalloproteinase-2 in hepatocellular carcinoma after radiotherapy. *Dig. Dis.* **2014**, *32*, 725–732. [CrossRef]
33. Wang, Q.; Xia, D.; Bai, W.; Wang, E.; Sun, J.; Huang, M.; Mu, W.; Yin, G.; Li, H.; Zhao, H.; et al. Development of a prognostic score for recommended TACE candidates with hepatocellular carcinoma: A multicentre observational study. *J. Hepatol.* **2019**, *70*, 893–903. [CrossRef] [PubMed]
34. Carr, B.I.; Bag, H.G.; Ince, V.; Akbulut, S.; Ersan, V.; Usta, S.; Isik, B.; Ogut, Z.; Tuncer, A.; Yilmaz, S. A Combination of Blood Lymphocytes and AST Levels Distinguishes Patients with Small Hepatocellular Carcinomas from Non-cancer Patients. *J Gastrointest. Cancer* **2021**, *52*, 1211–1216. [CrossRef] [PubMed]
35. Ni, J.Y.; Fang, Z.T.; An, C.; Sun, H.L.; Huang, Z.M.; Zhang, T.Q.; Jiang, X.Y.; Chen, Y.T.; Xu, L.F.; Huang, J.H. Comparison of albumin-bilirubin grade, platelet-albumin-bilirubin grade and Child-Turcotte-Pugh class for prediction of survival in patients with large hepatocellular carcinoma after transarterial chemoembolization combined with microwave ablation. *Int. J. Hyperth.* **2019**, *36*, 841–853. [CrossRef]

medicina MDPI

Review

Association of Parathyroid and Differentiated Thyroid Carcinomas: A Narrative Up-To-Date Review of the Literature

Razvan Simescu [1,2], Miana Pop [1,2], Andra Piciu [3,4,*], Valentin Muntean [1,2] and Doina Piciu [4,5]

1 Humanitas Hospital Cluj-Napoca, 400664 Cluj-Napoca, Romania
2 Department of Surgery, University of Medicine and Pharmacy Iuliu Hatieganu Cluj-Napoca, 400347 Cluj-Napoca, Romania
3 Department of Medical Oncology, University of Medicine and Pharmacy Iuliu Hatieganu Cluj–Napoca, 400347 Cluj-Napoca, Romania
4 Institute of Oncology Prof. Dr. I. Chiricuta Cluj-Napoca, 400015 Cluj-Napoca, Romania
5 Doctoral School, University of Medicine and Pharmacy Iuliu Hatieganu Cluj-Napoca, 400347 Cluj-Napoca, Romania
* Correspondence: piciuandra@gmail.com

Abstract: Aim: Parathyroid carcinoma (PC) is a rare endocrine malignancy that represents 0.005% of all malignant tumors. Associated PC and differentiated thyroid carcinoma (DTC) is an exceptionally rare condition, and the preoperative diagnostics and proper treatment are challenging. Almost all PCs and the majority of DTCs are diagnosed postoperatively, making correct surgical treatment questionable. Specific guidelines for parathyroid and thyroid carcinomas association treatment are lacking. The purposes of our study were to identify the association between parathyroid and thyroid carcinomas, to analyze the available published data, and to evaluate the possible relationship between preoperative diagnostic and surgical decision-making, and outcome-related issues. **Material and methods:** We performed a literature review of several databases from the earliest records to March 2022, using controlled vocabulary and keywords to search for records on the topic of PC and WDTC pathological association. The reference lists from the initially identified articles were analyzed to obtain more references. **Results:** We identified 25 cases of PC and DTC association, 14 more than the latest review from 2021. The mean age of patients was 55, with a female to male ratio of about 3:1. Exposure to external radiation was identified in only one patient, although it is considered a risk factor the development of both PC and DTC. The preoperative suspicion of PC was stated by the authors in only 25% of cases, but suspicion based on clinical, laboratory, ultrasound (US), and fine needle aspiration (FNA) criteria could have been justified in more than 50% of them. With neck ultrasound, 40% of patients presented suspicious features both for PC and thyroid carcinoma. Intra-operatory descriptions of the lesions revealed the highest suspicion (83.3%) of PC, but en bloc resection was recommended and probably performed in only about 50% of the cases. Histopathological examinations of the thyroid revealed different forms of papillary thyroid carcinoma (PTC) in most cases. Postoperative normocalcemia was achieved in 72% of patients, but follow-up data was missing in about 25% of cases. **Conclusion:** Associated PC and DTC is an exceptionally rare condition, and the preoperative diagnostic and treatment of the patients is a challenge. However, in most cases pre- and intraoperative suspicious features are present for identification by a highly specialized multidisciplinary endocrine team, who can thus perform the optimal treatment to achieve curability.

Keywords: parathyroid carcinoma; well differentiated thyroid cancers; tumor association; preoperatory suspicion; en bloc resection

Citation: Simescu, R.; Pop, M.; Piciu, A.; Muntean, V.; Piciu, D. Association of Parathyroid and Differentiated Thyroid Carcinomas: A Narrative Up-To-Date Review of the Literature. *Medicina* **2022**, *58*, 1184. https://doi.org/10.3390/medicina58091184

Academic Editors: Nicolae Crisan and Călin Căinap

Received: 11 August 2022
Accepted: 25 August 2022
Published: 30 August 2022

1. Background/Introduction

Parathyroid carcinoma is a cause of primary hyperparathyroidism (PHPT), PA (parathyroid adenoma), and PH (parathyroid hyperplasia), but less than 1% of PHPT cases are due to PC [1,2]. The peak incidence of PC occurs in the fifth decade of life, and there is no gender predilection [3]. DTCs arising from thyroid follicular epithelial cells are by far the most frequent thyroid cancers. PTC accounts for approximately 80–85% of DTC cases, while FTC accounts for approximately 12% [4]. In recent years, an increase in the incidence of thyroid carcinoma, in particular mPTC [5] has been observed, more marked among middle-aged women (aged 35–64 years) [6]. The incidence rates of DTC increased sharply in women during reproductive age, then declined and equalized with men by 85 years of age [7]. The reported female to male ratio was 2.9:1 [5,8].

Association of parathyroid and thyroid carcinomas is extremely rare [9–11]. When it occurs, PC is more likely to associate with DTC, mainly PTC [12]. Most PCs are sporadic, but they can also occur in the context of genetic syndromes. While some authors considered the combined occurrence of parathyroid and thyroid cancers a coincidence [13], others identified hypercalcemia or factors like epithelial and insulin growth factor as having a goitrogenic effect [14–16]. Sporadic PC and DTC have both been linked with exposure to external radiation [17,18].

In terms of treatment recommendations, specific guidelines for synchronous parathyroid and thyroid carcinomas treatment are absent. To date, the only accepted curative treatment for PC is en bloc resection of the affected parathyroid gland, with hemithyroidectomy of the ipsilateral lobe and surrounding adherent tissue including enlarged lymph nodes preoperatively diagnosed or identified during surgery [19–21]. DTC treatment is also surgical, and the extent of resection is determined by the extent of local and regional spread of the disease [22].

Almost all PCs and the majority of DTCs have been diagnosed postoperatively, making proper surgical treatment questionable. Preoperative suspicion of PC is difficult and rare, partly because clinical manifestations are often due to other causes of hyperparathyroidism [11,23,24]. DTC is often clinically silent, and half of all cases were found to present as incidental findings during physical examination or on neck US, or as previously unsuspected histological findings [25].

No specific imaging method can differentiate PC from benign parathyroid disease [21], but neck US has the greatest potential to raise both DTC and PC suspicion [26]. While FNA is useful in DTC preoperative diagnosis, it should be avoided in parathyroid lesions [27,28].

Intra-operative findings have also been described that raise the suspicion of PC [29,30]. However, a low percentage of patients were reported to have undergone en bloc resection, most likely because their PCs were not identified either preoperatively or even intraoperatively [31,32].

The extremely rare and problematic association between parathyroid and thyroid carcinoma prompted us to perform an up-to-date literature review to identify a possible relationship between preoperative diagnostic workup data, surgical decision-making, and outcome related issues. Recent reviews by Lam-Chung et al. [9] in 2020 and De Falco et al. in 2021 [10] documented confusing numbers of case reports (15 and 11 cases, respectively) with only 10 of them being superimposable. Moreover, the cases and some of their data were presented in the form of tables providing only partial data, for comparison with the author's case presentation.

Due to the rarity of parathyroid carcinoma and of the association with thyroid carcinoma, there are neither guidelines nor standard recommendations, and there have been no extensive randomized studies. Therefore, the aim of the study was to review the majority of publications in this field and provide an overview of the most common diagnostic and treatment strategies. This review can offer a comprehensive picture of the challenges faced during preoperative diagnosis, intraoperative decision-making, proper treatment, and case documentation, to enable improvement in future studies and patient treatments.

2. Methods

We performed a comprehensive search of databases including PubMed, Google Scholar, and Web of Science, from the earliest records in 1979 to August 2022. The search was carried out independently by two of the authors, using controlled vocabulary and keywords to search for records on the topic of PC and DTC pathological association. When the potentially eligible papers were retrieved, the full text publications were evaluated for eligibility.

Our focus was on information regarding demographics, relevant signs and symptoms, patients' relevant medical and family history, preoperatory laboratory and imaging findings, indications for surgery, intraoperative findings and descriptions, surgical treatment details, final diagnostics, outcome, and follow-up. Exclusion criteria were a language other than English, inaccessibility of the full text, no case presentation, and not enough data presented. The workup of selecting eligible papers for our review is schematically presented in Figure 1.

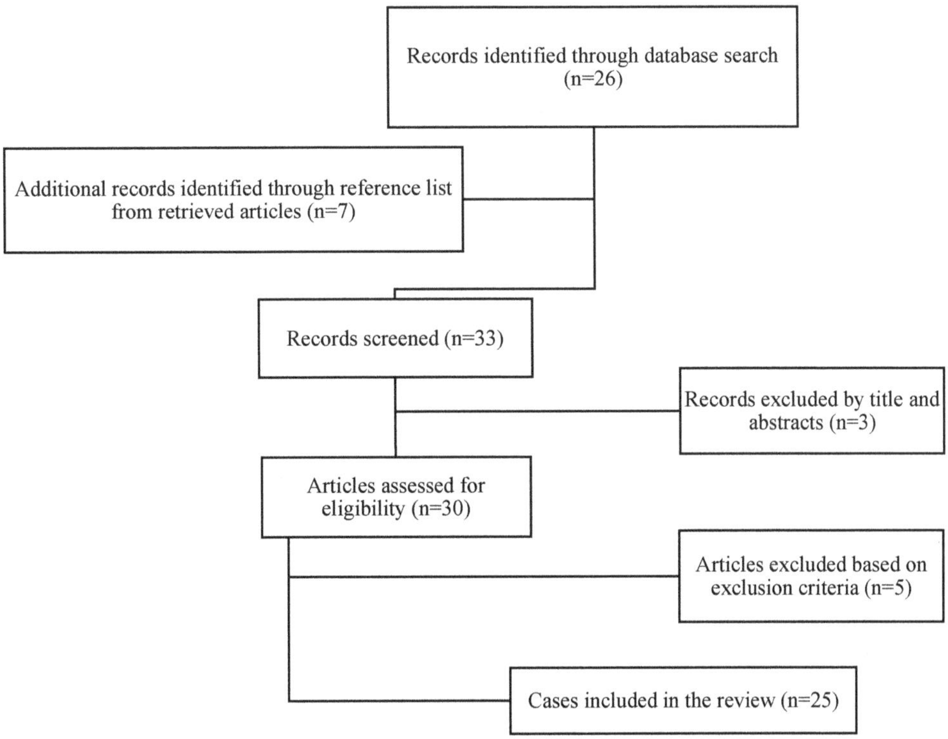

Figure 1. Workup for selecting relevant papers selected from the PubMed, Google Scholar, and Web of Science databases.

Relevant data were independently retrieved by two of the authors, who are certified endocrine surgeons, and then confronted and analyzed with additional input from two more experts, i.e., an endocrinologist and a nuclear medicine physician.

3. Results

According to our literature review, 25 cases of parathyroid and thyroid carcinoma association were identified and analyzed, see Table 1. Mean age of all the patients was 54.88 years (range: 21–89), with means of female and male patients at 56.37 (range: 29–89) and 51 (range: 21–68), respectively, and 60% of the cases were over 50 years old. There was a predominance of female cases (19 cases, 76%), compared with male (six cases, 24%).

Table 1. Features of patients with synchronous parathyroid and differentiated thyroid carcinomas.

No.	Authors Year (Reference)	Age Sex	Suspiciou Clinical Data EBRT+ Fh+ hPHCa+	Ca > 14 mg/dL	PTH > 5x NV	Suspicious US Parathyroid/FNA + Thyroid/FNA	^{99m}TC MIBI. SPECT/CT Uptake	Possible Preop. PC Suspicion	Stated Indication for Surgery	PC Location	Intraop. PC Suspicion	En Bloc Resection PC+Thyro	PC (cm) (g) IHC	NMTC (cm) Bilateral/Multifocal pTMN	Assoc. PA/PH	Outcom N/P/R	FLL-U (mo)
1.	**De Falco** et al., 2021[10]	63, M	No	No	No	No + Yes (FNA – benign)	^{99m}**TC MIBI.** Yes	NO	PA NG	Left inferior	No	Yes	1.2 No	2x mPTC (0.8, 0.6) pT1a(m)NxMx	No	N	84
2.	**Lam-Chung** et al., 2020 [9]	50, F	Yes **hPHCa+**	Yes	Yes	No + No	**SPECT/CT** Yes	Yes	PC suspicion –	Left superior	N/A	Yes	2.4 Yes	PTC (1.3) pT1bNxMx	No	N	1 ½
3.	**Kalthoum** et al., 2020 [33]	60, F	Yes	No	Yes	No + Yes	^{99m}**TC MIBI.** Yes	Yes	PHPT –	Left superior	Yes	Yes	4 No	2x mPTC (NS) pT1a(m)NxMx	No	N	N/A
4.	**D'cruz** et al., 2020 [12]	89, F	Yes	No	No	Yes + No	^{99m}**TC MIBI.** Yes	Yes	PA NG	Right inferior	Yes	Yes	1.7 Yes	FTC (3.5) pT2NxMx	PA	N	N/A
5.	**Edafe** et al., 2019 [34]	46, F	Yes **hPHCa+**	No	Yes	No + Yes (FNA – PTC)	^{99m}**TC MIBI.** Yes	Yes	PC suspicion PTC	Right ?	Yes	Yes	3.3 No	PTC + mPTC (>4, NS) pT4(m)NxMx	No	N	12
6.	**Kuzu** et al., 2017 [35]	52, F	Yes **hPHCa+**	No	No	No (FNA wash-out) + Yes (FNA – benign)	^{99m}**TC MIBI.** No	Yes	PA NG	Right inferior	Yes	Yes	1.8 No	PTC + mPTC (1, NS) pT1b(m)NxMx	No	N	12
7.	**Baek** et al., 2017 [36]	68, F	Yes **hPHCa+**	No	Yes	Yes + Yes (FNA – AUS/FLUS)	^{99m}**TC MIBI.** Yes	Yes	PA NG	Left inferior	Yes	N/S	4.2 No	mPTC (NS) pT1a(m)NxMx	No	N	6
8.	**Demir** et al., 2017 [37]	29, F	Yes **hPHCa+**	Yes	Yes	Yes + Yes	^{99m}**TC MIBI.** Yes	Yes	PA NG	Right ?	Yes	NS	2.8 No	PTC (1.6) pT1bNxMx	No	N	N/A
9.	**Aljabri** et al., 2017 [38]	72, F	Yes **hPHCa+**	No	Yes	∅ + Yes (FNA –benign)	^{99m}**TC MIBI.** No	Yes	– NG	Right Inferior	Yes	Yes	4.5 Yes	mPTC (0.2) pT1a(m)NxMx	No	N	1
10.	**Dikmen** et al., 2017 [21]	57, M	No	No	No	Yes+ ∅	^{99m}**TC MIBI.** Yes	Yes	Persistent elevated Ca + PTH after ePC removal	Mediastinal + left inferior	No	No	30 + 21 **(2xPC)** Yes	mPTC (0.2) pT1a(m)NxMx	ePC	N	N/A
11.	**Neslihan** et al., 2016 [39]	65, F	No	No	Yes	No + Yes (FNA – benign)	^{99m}**TC MIBI.** No	Yes	PC suspicion –	Left inferior	N/A	Yes	2.5 **Positive surgical margins** Yes	2x mPTC (0.5, 0.2) pT1a(m)NxMx	No	N	N/A

Table 1. *Cont.*

No.	Authors Year (Reference)	Age Sex	Suspiciou Clinical Data EBRT+ Fh+ hPHCa+	Ca > 14 mg/dL	PTH > 5x NV	Suspicious US Parathyroid/FNA + Thyroid/FNA	^{99m}TC MIBI. SPECT/CT Uptake	Possible Preop. PC Suspicion	Stated Indication for Surgery	PC Location	Intraop. PC Suspicion	En Bloc Resection PC+Thyro	PC (cm) (g) IHC	NMTC (cm) Bilateral/Multifocal pTMN	Assoc. PA/PH	Outcom N/P/R	FLL-U (mo)
12.	**Lee** et al., 2016 [40]	57, F	No	Yes	Yes	∅ + No	^{99m}TC **MIBI.** Yes	Yes	PA NG	Left inferior	Yes	N/S	4.5	PTC (NS) pT?	No	-N -2xR (PC) -N	72
13.	**Song** et al., 2016 [11]	45, F	Yes	Yes	Yes	Yes + Yes	^{99m}TC **MIBI.** Yes	Yes	N/S	Left inferior	Yes	No	4.3	mPTC (0.5) pT1a(m)NxMx	NA	-N -2xR (PC) -N	6mo after III surgery
14.	**Al-Sulami**, 2015 [41]	75, F	Yes	No	Yes	∅ + No	^{99m}TC **MIBI.** Yes	Yes	N/S	Left ?	N/A	N/S	3.5 **Positive surgical margins**	3x mPTC (all < 0.5) pT1a(m)NxMx	N/A	P	24
15.	**Zakerkish** et al., 2015 [42]	21, M	Yes	No	Yes	No + ∅ (previous TT)	^{99m}TC **MIBI.** No	Yes	PC suspected –	Right ?	N/A	No **(2 years previous TT)**	N/A	mHHC (0.6) pT1a(m)NxMx	No	P -3w later death	<1
16.	**Chaychi** et al., 2010 [43]	79, F	No **hPHCa+, 6y**	No	No	Yes + Yes (FNA – PTC)	^{99m}TC **MIBI.** Yes	Yes	PHPT 2xPTC	Left superior,	N/A	Yes	5 Yes	2x PTC (2.4, 1.7) pT2(m)NxMx	No	N	6
17.	**Marcy** et al., 2009 [44]	42, F	Yes **hEBRT+**	Yes	Yes	Yes (FNA: unconclusive) + Yes (FNA – inconclusive)	^{99m}TC **MIBI.** Yes	Yes	PC suspected –	Right inferior	N/A	N/S	1.3	2x mPTC (0.8, 0.5) (m)T1aNxMx	No	N	14
18.	**Goldfarb** et al., 2009 [45]	59, M	Yes **hPHCa+, 6y**	Yes	Yes	Yes + Yes	^{99m}TC **MIBI.** Yes	Yes	PHPT –	Left ?	Yes	Yes	3.9 17 g	PTC + mPTC (3.2, 0.4) pT2(m)NxMx	PA	-h -R (PA) -N	14
19.	**Mazeh** et al., 2008 [46]	44, F	No	No	N/A	No + Yes (FNA – inconclusive)	N/A	**NO**	– NG	Left ?	N/A	Yes **(non-intended)**	1.5	PTC (NS) pT?	No	N	60
20.	**Lin** et al., 2005 [47]	38, M	Yes **hPHCa+, 6y**	Yes	Yes	No + Yes (FNA – PC)	^{99m}TC **MIBI.** ^{201}**TI Scinti** Yes	Yes	PC NG	Left inferior	Yes **(frozen section!)**	No	N/A	PTC (4) pT2(m)N + Mx	No	N	72
21.	**Kern** et al., 2004 [48]	54, F	No **Fh+**	N/A	Yes	N/A	N/A	Yes	PHPT	Right inferior	Yes	No (although PC very adherent)	2.5 7 g	2x mPT + mFTC; (0.3, 0.4, 0.3) pT1a(m)N + Mx	No	N – R + 2y later (PC distant meta) – 3y death	36
22.	**Schoretsanitis** et al., 2002 [49]	55, F	Yes **hPHCa+**	Yes	Yes	∅ + No	$^{+99m}$TC **MIBI.** Yes	Yes	PHPT NG	Left inferior	Yes	Yes	3	mPTC (0.7) pT1a(m)NxMx	PH	N	24

Table 1. *Cont.*

No.	Authors Year (Reference)	Age Sex	Suspiciou Clinical Data EBRT+ Fh+ hPHCa+	Ca > 14 mg/dL	PTH > 5x NV	Suspicious US Parathyroid/FNA + Thyroid/FNA	^{99m}TC MIBI. SPECT/CT Uptake	Possible Preop. PC Suspicion	Stated Indication for Surgery	PC Location	Intraop. PC Suspicion	En Bloc Resection PC+Thyro	PC (cm) (g) IHC	NMTC (cm) Bilateral/Multifocal pTMN	Assoc. PA/PH	Outcom N/P/R	FLL-U (mo)
23.	**Bednarek-Tupikowska** et al., 2001 [50]	42, F	Yes **hPHCa+**	Yes	Yes	Yes (FNA – no cells) + Ø (previous TT)	**^{99m}TC MIBI.** Yes **^{99m}TC MIBI** for persistent disease No	Yes	PHPT –	Left ?	Yes	**No (6 years previous TT)**	5 Yes	FC (NS) pT?	No	P	N/A
24.	**Savil** et al., 2001 [51]	47, F	No	No	N/A	Ø + Yes (FNA – PTC)	N/A	**NO**	– PTC	Left ?	No	No	3	PTC (3) pT2(m)NxMx	PH	N	12
25.	**Kurita** et al., 1979 [52]	68, M	No	No	Yes	N/A	N/A	Yes	PHPT –	Left superior	Yes	Yes	4.2 21 g	PTC (NS) pT?	No	N	24

Abbreviations: hEBRT–history of external beam radiation therapy; Fh–family history; hPHCa–history of prolonged hypercalcemia, stated or with indirect signs; PT–parathyroid tumor; NMTC–non-medullary thyroid carcinoma; PHPT–primary hyperparathyroidism; PC–parathyroid carcinoma; ePC–ectopic PC; PA–parathyroid adenoma; PH–parathyroid hyperplasia; NG–nodular goiter; PTC–papillary thyroid carcinoma; mPTC–papillary thyroid micro-carcinoma; FTC–follicular thyroid carcinoma; HCC–Hurthle cell carcinoma; NV–normal value; N/S–not stated; N/A–not available; Ø–no lesion identified; IHC–immunohistochemistry; TT–total thyroidectomy; FLL-U (mo)– follow-up (months); Outcome N/P/R–normocalcemic/persistence/recurrence.

Different symptoms of hyperparathyroidism were present in 62.5% of the 24 cases with clinical data available. Most patients had no irradiation history (92.86%), nor any relevant family history or genetic syndrome data (94.44%). In 9 cases (36%) the genetic syndromes were ruled out based on clinical, laboratory, and imagistic screening. Only one case had genetic testing for relevant family syndrome genes.

We defined the patients as having "prolonged hypercalcemia" when a long-lasting primary hyperparathyroidism (PHPT) history was stated, or when signs of possible long-lasting hypercalcemia (severe bone disease, progressive long-standing renal impairment, peptic ulcer disease, neurocognitive deficits) were described.

We defined as "data suspicious for PC" any signs and symptoms of severe hypercalcemia (nausea, vomiting, dehydration, polyuria, polydipsia, confusion, lethargy, constipation, peptic ulcer, etc.) and/or severe bone disease (severe osteoporosis, osteitis fibrosa cystica, pathologic fracture) and/or progressive renal impairment (chronic kidney disease, nephrocalcinosis) and/or neurocognitive deficits and/or palpable hard neck mass. Clinical data suspicious for PC was found in 16 patients (64%), while10 patients (36%) had no such clinical suspicion. Of those with no suspicion, seven patients were asymptomatic and three had only a neck mass palpable.

Suspicious elevated serum calcium levels ($\geq$14 mg/dL) were found in nine patients (39.13%), with 14 patients (60.87%) having elevated calcium levels but <14 mg/dL, and one patient (4.35%) presenting normal values. Most of the patients (82.61%) presented highly elevated PTH levels ($\geq$5–10-fold normal value), suspicious for PC, while four cases (17.39%) had elevated but non-suspicious PTH. One patient had no available data regarding calcium, and two regarding PTH. Non-functioning PCs with normal or slightly elevated calcium and PTH were mentioned in 4 cases.

In terms of bone disease in PC patients, parameters such as alkaline phosphatase, PO4, and renal function parameters are missing from many reports, so we decided not to include data referring to serum level creatinine or GFR.

Ultrasound (US) of the neck was not available in two cases (8%), and two patients had undergone previous thyroidectomies. Parathyroid or thyroid lesions were not identified by US in five case (21.47%) and one (4.76%) case, respectively. Parathyroid "suspicious lesions" for carcinoma were heterogeneous cystic structures with irregular or ill-defined borders and intra-nodular calcifications or signs of local invasion. Parathyroid lesions suspicious for PC were described in 50% (9 patients) of the US- identified parathyroids, the other half being diagnosed as possible Pas (18 patients in total). Associated or isolated thyroid lesions were found in 20 patients, with suspicious features being present in 15 (75%) of them, all without extrathyroidal or lymph node extension.

By US, concomitant parathyroid and thyroid lesions were identified in 15 (60%) patients, of whom six (40%) had suspicious features both for PC and thyroid carcinoma.

Fine needle aspiration (FNA) was performed on parathyroid glands for diagnostic reasons on a total of four lesions (16%). Thyroid-nodule FNA was not performed or not stated in 10 cases (40%). Of the other 15 cases which featured US suspicious thyroid nodules, FNAs were performed in only 10. Papillary thyroid carcinoma (PTC) was diagnosed by means of FNA in only three of these cases (30%), one of which was a double carcinoma. The other thyroid-nodule FNAs (seven cases, 70%) were either benign or non-conclusive.

In 21 cases (84%), ^{99m}Tc scintigraphy with MIBI (Tc-99m methoxy–isobutyl–isonitrile) (planar or SPECT/CT) was carried out to localize the affected parathyroid gland, and 17 glands (80.95%) presented radiotracer uptake. In three of these cases (17.65%) the gland had not been identified by previous neck US. In four cases (19.05%) the examination was unable to localize the affected parathyroid gland. In one case, neither ^{99m}Tc scintigraphy nor neck US were able to identify the pathological parathyroid.

"Preoperatory PC suspicion" refers to the cases in which the authors took into consideration and mentioned the PC diagnosis. A preoperative PC suspicion was mentioned in 24% (six out of 25) of the cases.

We defined "possible preoperatory PC suspicion" as at least two cumulative suspicions for PC based on available clinical-, laboratory-, US-, or FNA-based data. On these premises, in 56% of the cases (14 out of 25) the preoperatory suspicion for PC could have been justified.

"Intraoperatory PC suspicion" was based on the authors' descriptions of suspicious features for PC. Data was available in 72% of the cases and it revealed the highest suspicion (83.3%) on PC. En bloc resection was performed and stated only in 52% of the cases (13 out of 25). In the remaining 12 cases (48%), en bloc resection was either not performed (including two cases with previous total thyroidectomies), or not mentioned.

There was a preponderance of left side parathyroid lesions (12 cases, 60%), of which more than two thirds were inferior glands. Of note is one case with an ectopic left mediastinal parathyroid gland that was diagnosed and treated in the recent past by concomitant PC-NMTC operation (synchronous double PC).

Mean maximal diameter of recorded PC was 3.15 cm (range: 1.2–5 cm). Definitive PC diagnosis was achieved through histopathological examination of the resected specimens, and immune staining was additionally performed in seven cases (28%). PC-positive surgical margins were present in two of the histopathological examinations. Micropapillary (mPTC) and papillary thyroid carcinoma (PTC) were present in 23 of the 25 patients, while follicular thyroid carcinoma (FTC) and Hurthle cell carcinoma (HCC) were identified in three patients and one, respectively. Multiple DTCs were found in 52% of patients (13 out of 25 cases). Of these, the majority were mPTC (five out of 11 cases), followed by mPTC associated with PTC (three cases), and mPTC associated with FTC and with HCC (each with one case).

Isolated single-type DTCs were identified in 48% of patients (12 out of 25 cases) with PTC and mPTC having equal shares (50% respectively).

Associated parathyroid disease was documented in five of the 25 cases, as two parathyroid adenomas (PA), two cases of parathyroid hyperplasia (PH), and one ectopic mediastinal PC, see Table 2.

Table 2. Parathyroid diseases associated to PC.

Parathyroid Disease	PA	PH	Ectopic PC
No./total (%)	2/5 (40)	2/5 (40)	1/5 (20)
Calcium levels mg/dL	13.8; 14.7	11.3; 12.1	16.2
PTH levels pg/mL	441; 318	197; 249	4211

Abbreviations: Parathyroid adenomas (PA) and parathyroid hyperplasia (PH). Ectopic parathyroid carcinoma (PC).

Postoperative outcomes showed normal serum calcium level achieved in 72% of the patients (18 out of 25 cases). Persistent and recurrent disease were documented in a total of seven patients. Persistent disease was present in three of the patients (12%) and recurrence in four (16%).

Follow-up data was missing in about 25% of cases. The follow-up period was longer than two years for 11 of the total 25 patients (44%). One of the three patients with less than three months of follow-up died three weeks after surgery, due to uncontrolled hypercalcemic crisis.

There were no reported cases of postoperative hypocalcemia.

4. Discussion

The association of PC and DTC is an exceptionally rare clinical presentation, and no guidelines for diagnostic and treatment strategy are currently available.

Our review identified a total of 29 cases of associations between PC and DTC [9–12,21,33–52], but four of these we did not include in our analysis because they were not detailed as case presentations [53–55], or we could not obtain a full text article [56]. Compared with the last published review by de Falco et al. [10] from 2021, we were able to identify and analyze 14 more cases.

In our review, we identified that the mean age of patients was 54.88, but when we compared it by gender, we found that female mean age was higher than male mean age

(56.37 vs. 51). This was probably because in most cases PC was the condition that determined the patients to seek medical attendance, and not the associated DTC. The predominance of female compared to male patients (approximately 3:1), more in accordance with DTC trends than with those of PC, is an intriguing feature of the pathological association.

4.1. Etiopathology of PC and DTC

The etiopathology of neither PC nor DTC is well-known. Most PCs are sporadic, but they can also occur in the settings of genetic syndromes, such as hyperparathyroidism-jaw tumor syndrome (HPT-JT), multiple endocrine neoplasia type 1 (MEN 1) and type 2A (MEN 2A), or familial isolated hyperparathyroidism [57–59]. Genetic syndromes could be suspected based on family history and on clinical, laboratory, or imagistic findings of associated diseases. Definitive exclusion of syndromes should be based on genetic testing of relevant genes such as CDC 73, CDKN1B, MEN 1, RET, etc. [34]. In 72% of the analyzed cases, family history was mentioned, but only the case presented by Kern et al. [48] had a possible relevant history, albeit without having been further investigated. Screening by non-genetic means was carried out in 36% of the cases, and only the case from Edafe et al. [34] was genetically tested by gene sequencing and dosage analysis. None of the cases were stated to be syndromic.

Previous studies found an increased risk of benign parathyroid disease and concurrent thyroid disease in patients who had been exposed to external radiation (especially of the head and neck) [60,61]. Sporadic PC and DTC have both been linked with exposure to external radiation [17,18], but only one case in our review was found to have had external radiotherapy, 19 years before the diagnosis of associated PC and DTC.

Reports have also suggested that long-standing secondary HPT or end-stage renal disease could be associated with increased risk of PC [17,24]. In our review, we did not identify any patients with secondary HPT or end-stage renal disease. Instead, in 55% of the cases we identified clues of possible long-standing hypercalcemia in the form of relevant renal (e.g., nephrolithiasis, nephrocalcinosis, reduced renal parenchymal index) and/or skeletal involvement (e.g., osteoporosis, osteitis fibrosa cystica, subperiosteal resorption, "salt and pepper" skull, or pathologic fractures). Also, three of these patients had over six-year long histories of hypercalcemia.

4.2. Clinical Manifestations

It is important to underline that these patients were not really asymptomatic: bone markers, the level of calcium, and osteoporosis were important signs that in many cases had been neglected for years. Meanwhile, patients with severe vitamin-D deficiency have very high levels of PTH in the presence of serum calcium levels in the upper range, and these levels were not been properly evaluated in order to rule out PHPT.

DTC is largely asymptomatic [25]. PC's clinical manifestations are superimposable on those of other causes of PHPT and typically display an indolent course [62]. Still, PC should be suspected in all PHPT cases with fast onset or marked hypercalcemic symptoms or signs (such as anxiety, depression, weakness, weight loss, bone and renal disease, abdominal pain, nausea, pancreatitis, or peptic ulcer), hypercalcemic crisis, and/or a palpable neck mass [63–65]. Also, entirely asymptomatic PC [66] and non-functioning PC have been reported. Non-functioning PC usually presents at a more advanced stage with symptoms of compression or invasion in adjacent structures, such as neck mass, and/or dysphagia, hoarseness, or dyspnea [67].

In our review, seven patients (28%) were identified as asymptomatic, but three of them had a palpable neck mass on physical examination, which proved to be the PC. Also, when looking for combined clinical data (including patient family and pathological history) we identified suspicious features for PC in 64% of the patients. There were no clinical data suspicious for DTC.

4.3. Biological Features

There are no specific tumor markers for PC, nor for DTC, but in the case of PC malignancy it could be suspected in patients with severe hypercalcemia (>14 mg/dL or 3.5 mmol/L) and/or with marked PTH elevations (>5–10 times the upper normal limit or absolute levels >500 mg/dL) [2,65]. Non-secreting PCs with normal PTH and calcium levels have been reported in the literature [67–70]. In the cases reviewed in the current work, more than 50% had unsuspicious levels of calcium, but more than 80% of them had PTH-suggestive levels. As expected, there were no cases with suspicious elevated calcium and non-suspicious elevated PTH, leading us to believe that PTH is the most valuable laboratory finding for raising PC suspicion.

Alhough rare, non-functioning forms of PC may have normal [71] or minimally increased calcium and PTH levels [24,58,63]. Less than 10% of PC cases are non-functioning forms [24,42,72], and they are more likely to present as late-stage diseases, either due to the lack of symptoms and subtle evolution of the disease, or due to more aggressive tumor behavior [9,14]. Among the cases of associated parathyroid and thyroid carcinomas, four non-functioning PC were reported [12,21,46,52]. While the first three were either asymptomatic or without specific symptoms, the one described by D'Cruz et al. [12] had severe hypercalcemia symptoms, thus leading us to believe that only 12% (three of the 25 cases analyzed) were non-functioning. Consequently, despite the rarity of the disease, patients with slightly increased or even normal serum calcium and/or PTH should not be neglected from a suspected diagnosis of PC, especially if they present suspicious clinical features. The three non-functioning cases were not late-stage diseases, but that presented by Dikmen et al. [21] was unique: a double PC with one of the tumors localized ectopically in the thorax.

4.4. Imaging

Neck US, ^{99m}TC-sestamibi, and high-resolution CT or MRI are valuable adjuvants for preoperative localization of pathological parathyroids, and in the case of PC and DTC for disease staging. Furthermore, careful thyroid imaging should represent a step of great importance before any surgery performed for PHPT cases [45], seeing as most synchronous parathyroid and thyroid carcinomas are diagnosed postoperatively [10].

Retrospective reviews of preoperative US indicate that PCs present features such as larger mass size than PA, heterogeneous structure, evidence of degeneration (cystic inclusions and/or calcifications), lobulated or irregular borders, and signs of local invasion [36,51,65,73,74]. Also, studies have shown that high-volume endocrine surgeons who perform systematic preoperative neck US are proficient in identifying enlarged parathyroid glands [75,76], and have the greatest potential to identify suspicious features of both PC and DTC.

Neck US is a highly sensitive method for the detection of thyroid nodules and for the evaluation of morphological features of the nodule [77–79], which can suggest cancer suspicion and consequently the indication of FNA [80,81] for possible cancer confirmation.

In the cases reviewed, neck US was able to identify enlarged parathyroid glands and thyroid nodules in 78% and 95% of cases, respectively. PCs were quite large, with a mean maximal diameter (during the operation) of 3.15 cm recorded, which may explain the high percentage of US-identified lesions. Parathyroids suspicious for PC were described in only half of US-identified glands, and suspicious thyroid nodules were present in three quarters of patient. Concomitant parathyroid and thyroid lesions were seen in 60% of the patients with concomitant identifiable lesions (15 patients), of whom 40% had suspicious features both for PC and thyroid carcinoma. Therefore, neck US has a good potential not only to localize the affected parathyroids and to identify concomitant thyroid nodules, but also to raise the suspicion of both PC and thyroid carcinoma, thus making it a very useful tool for determining preoperatory concomitant cancer suspicion.

When ultrasound localization fails to identify an abnormal parathyroid gland, ^{99m}Tc-MIBI planar imaging or SPECT/CT may be performed to further aid preoperative localization [26], but they cannot differentiate PC from PA [82]. On the other hand, there have been false-negative results from ^{99m}Tc subtraction imaging, attributed to small lesion size, abnormal ^{99m}Tc uptake by the thyroid gland, and abnormal or absent ^{99m}Tc uptake by a parathyroid adenoma [39,83]. Also, thyroid lesions can be sestamibi-avid, therefore ultrasound examination should always be used to increase detection of synchronous thyroid carcinomas [58,84].

In 84% of the analyzed cases, ^{99m}Tc-MIBI scintigraphy was performed to localize the source of hyperparathyroidism, and 80.95% of glands were found with increased radiotracer uptake. In 17.65% of the cases, the scan was the only imagistic method able to localize the affected parathyroid gland. In 19.05%, the examinations were unable to localize the affected parathyroid gland. Of note is the case reported by Aljabri et al. [38], in which neither the scan nor neck US were able to identify the pathological parathyroid, but clinical and laboratory findings were suggestive of PC. Also to be mentioned is the complete lack in all studies of hybrid imaging evaluation (positron emission tomography with computed tomography—PET/CT or positron emission tomography with magnetic resonance imaging—PET/MRI) which might increase sensitivity, and also the very limited use of SPECT/CT (in just one case), a fact that would explain the relatively high number of negative scans.

4.5. FNA

Fine-needle aspiration of suspected PC should be avoided [26], not only because cytology cannot reliably differentiate between PA and PC, but more than that due to the documented risk of tumor seeding due to violation of the parathyroid capsule [23,27].

FNA of the parathyroid glands was undertaken in four cases. Three of them were thought to be parathyroid glands and one was a thyroid nodule, which on cytology showed features (marked nuclear pleomorphism with prominent nucleoli) suspicious of PC [46]. Of note, one FNA of the parathyroid gland by washout PTH was performed only to confirm the US-derived suspicion [34].

Thyroid nodule FNAs were not performed in 40% of the cases, for various reasons, some of which were obvious, including previous thyroidectomies (two cases) or no thyroid nodules identified on US (one case). On the other hand, only 10 of the 15 cases which featured US-suspicious thyroid nodules received FNA, and 70% of the cytology aspects were either benign or non-conclusive. A possible explanation is that the sensitivity of thyroid FNA can be affected by many factors including operational technique and the experience of the ultrasound specialist, the nodule size (lesser accuracy in small nodules), the sample retrieved (representative and adequate in cellularity), and the experience of the pathologist [85–87]. In the cases reviewed, the size of the nodules (about half of which were microcarcinomas) could have been the main source of failed FNA confirmation of cancer. Regardless of FNA results, and bearing in mind the possible concomitant association of PC and DTC, if US shows clear suspicious features of thyroid nodules on the contralateral side of the possible PC, total thyroidectomy should be performed.

In the papers reviewed, a preoperative PC suspicion was described by the authors in only 24% of cases. When we looked for possible preoperatory PC suspicion based on available clinical—, laboratory-, US- and FNA-based data, the suspicion could have been justified in 56% of the cases.

Specific guidelines for synchronous parathyroid and thyroid carcinoma treatment are lacking. The only curative treatment for PC is en bloc resection of the affected parathyroid gland with hemithyroidectomy of the ipsilateral lobe and surrounding adherent tissue or enlarged lymph nodes identified during surgery or preoperatively diagnosed [19,20,88]. Preserving the integrity of the capsule during tumoral mass removal and complete resection of affected tissues are crucial, given that recurrent PC cases have been described due to secondary local seeding produced after rupture of the tumoral capsule and neoplastic microinfiltrations in the adjacent structures [64,89]. Thus, to properly treat PC, it is helpful for the surgeon, if possible, to form a high level of cancer suspicion before and during the operation [26]. Features such as white-greyish color, firm consistency, cystic components, and foremost the adherence to surrounding structures (especially the thyroid gland) [29,30] can aid the surgeon in raising a strong PC suspicion and performing the correct operation. Intraoperatory description of lesions revealed the highest suspicion (83.3%) of PC, although, initial en bloc resection was reported to be performed in as little as 12% of case series [26].

4.6. Treatment Considerations

In cases of synchronous DTC and PC there are no treatment guidelines; an appropriate strategy would be radical surgery for PC and total thyroidectomy with resection of the enlarged lymph nodes identified during surgery or preoperatively diagnosed.

En bloc resection though was performed and stated only in 52% of the cases, one of them without any suspicion of a parathyroid pathology before or during the operation. In the remaining 12 cases (48%), en bloc resection was either not performed or not mentioned, making the 52% questionable in upper and lower directions. Two of the cases without en bloc resection (Zakerkish et al. and Bednarek-Tupikowska et al.) [42,50] had previous total thyroidectomies two and six years previously, respectively. Interestingly, the postoperative outcome was in line with persistent disease in both.

The operative details of the reviewed cases also revealed a slight preponderance (60%) of left-side PCs, more than two thirds of them affecting the inferior glands.

Even though PC can be pre– and intraoperatively suspected, definitive diagnosis can only be established after complete histopathological examination [64,89–91]. This is based on the Shantz and Castleman criteria, used since 1973, including a trabecular pattern of the parenchymal cells, capsular and vessel invasion, high mitotic rates, nuclear atypia, and a thick capsule [82,92]. Nevertheless, histopathological diagnosis of PC can sometimes be difficult, and IHC analysis that includes parafibromin, APC, galectin-3, Cyclin D, Ki67, and other markers can aid in PC diagnosis [93–97]. In 28% of cases reviewed, IHC staining was additionally performed, but the reasons for performing it were not stated by the authors. Of note is the case reported by Dikmen et al. [21] with an ectopic left mediastinal parathyroid gland diagnosed and treated as concomitant PC-DTC (synchronous double PC).

Regarding the associated DTCs, histopathological examination revealed mPTC and PTC in 84% of the cases, FTC in 12% of patients, and Hurthle cell carcinoma (HCC) in one, which is in line with the common occurrence of the disease. PTC was additionally present in associations with other histological forms in 8% of the cases. Double or multiple DTCs were found in about half of patients, the majority being mPTC in their own association or with PTCs. Of note is that one PTC excision was two years before the operation for PC and concomitant mHHC, and also the case of a double FTC, the first excised five years before the removal of the PC together with the second FTC. Associated parathyroid disease was identified in 20% of the cases with equal PAs and PH, as well as one ectopic mediastinal PC.

In addition to tumoral removal of the affected gland and other involved structures, surgical treatment aims to obtain postoperative biochemical remission, with calcium and PTH normalization [38]. No consensus on follow-up for patients with PC has been established, but periodic, life-long monitoring of calcium and PTH is recommended [62,98]. Five-year survival rates vary between 85.5% [32,48] and 90.9% [53], while 10-year survival rates are reported between 49.1% [32,48] and 77% [23]. The most important prognostic factor is the successful resection of the parathyroid tumor at the time of the initial operation [58]. Postoperative normocalcemia was achieved in 72% of the patients with PC and DTC in association. Follow-up data was missing in about 25% of cases, and only three patients had a follow-up period of more than five years. The follow-up for DTC was quite uniform, complete remission being defined by clinically negative status, negative ultrasound, and negative whole body I-131 scan with nondetectable levels of thyroglobulin and anti-thyroglobulin. Considering the mentioned synchronous cancers, efficient radical treatment for DTC and facile and specific follow-up would permit the clear depiction of the diseases in cases of persistence or recurrence.

Persistent disease was present in three patients (12%), and recurrence in four (16%). All three cases with persistent hypercalcemia and three of the four cases with recurrent disease were not subjected to en bloc resection. The one case of recurrent disease with en bloc resection [45] had recurrent hypercalcemia due to a PA missed during the operation. One of the three patients with persistent disease died three weeks after surgery due to an uncontrolled hypercalcemic crisis. This was the case reported by Zakerkish et al. [42], which had some intriguing features, being the youngest age (21 years old) of all patients, with no sestamibi uptake on scintigraphy, and two years previously total thyroidectomy for HCC (the only such thyroid carcinoma variant present). Although two of the cases [39,41] had positive surgical margins on histopathological examination, only one [41] had persistent disease two years after surgery.

5. Conclusions

Associated PC and DTC is an exceptionally rare condition, and the preoperative diagnostic and proper treatment of patients is a substantial challenge. In most cases, pre- and intraoperative suspicious features are there to be identified. Neck US has the greatest potential to raise both DTC and PC suspicion, especially if it is performed preoperatively in a highly specialized endocrine tumor center by a multidisciplinary team. Preoperative suspicion, combined with the expert's intraoperative identification of suspicious features, can lead to correct surgical treatment of the condition. To date, comprehensive follow-up data and centralized documentation of these rare cases have been lacking, and are of great importance to enable future studies and improvements in diagnosis and treatment.

Author Contributions: Conceptualization, R.S. and D.P.; methodology, R.S. and M.P.; software, M.P.; validation, D.P. and V.M.; formal analysis, A.P.; investigation, D.P.; writing—original draft preparation, R.S. and M.P.; writing—review and editing, A.P. and D.P.; visualization, D.P.; supervision, V.M. All authors have read and agreed to the published version of the manuscript.

Funding: This research did not receive any funding.

Data Availability Statement: The authors confirm that the data supporting the findings of this study are available within the article.

Conflicts of Interest: The authors declare no conflict of interest.

References

1. Fraser, W.D. Hyperparathyroidism. *Lancet* **2009**, *374*, 145–158. [CrossRef]
2. Cetani, F.; Pardi, E.; Marcocci, C. Update on parathyroid carcinoma. *J. Endocrinol. Investig.* **2016**, *39*, 595–606. [CrossRef] [PubMed]
3. Fernandes, J.M.P.; Paiva, C.; Correia, R.; Polónia, J.; Moreira da Costa, A. Parathyroid carcinoma: From a case report to a review of the literature. *Int. J. Surg. Case Rep.* **2018**, *42*, 214–217. [CrossRef] [PubMed]
4. National Comprehensive Cancer Network. NCCN clinical practice guidelines in oncology, thyroid carcinoma. Nccn.org Web site. Available online: https://www.nccn.org/professionals/physician_gls/pdf/thyroid.pdf (accessed on 29 April 2022).
5. Lorenz, K.; Schneider, R.; Elwerr, M. Thyroid Carcinoma: Do We Need to Treat Men and Women Differently? *Visc. Med.* **2020**, *36*, 10–14. [CrossRef]
6. Li, M.; Dal Maso, L.; Vaccarella, S. Global trends in thyroid cancer incidence and the impact of overdiagnosis. *Lancet Diabetes Endocrinol.* **2020**, *8*, 468–470. [CrossRef]
7. Kilfoy, B.A.; Devesa, S.S.; Ward, M.H.; Zhang, Y.; Rosenberg, P.S.; Holford, T.R.; Anderson, W.F. Gender is an age-specific effect modifier for papillary cancers of the thyroid gland. *Cancer Epidemiol. Biomark. Prev.* **2009**, *18*, 1092–1100. [CrossRef]
8. Siegel, R.L.; Miller, K.D.; Jemal, A. Cancer Statistics, 2017. *CA Cancer J. Clin.* **2017**, *67*, 7–30. [CrossRef]
9. Lam-Chung, C.E.; Rodriguez-Orihuela, D.L.; Anda Gonzalez, J.; Gamboa-Dominguez, A. An Unusual Simultaneous Existence of Parathyroid Carcinoma and Papillary Thyroid Carcinoma: Case Report and Review of Literature. *Case Rep. Endocrinol.* **2020**, *2020*, 2128093. [CrossRef]
10. De Falco, N.; Santangelo, G.; Chirico, F.; Cangiano, A.; Sommella, M.G.; Cosenza, A.; Ronchi, A.; Accardo, M.; Pellino, G.; Parmeggiani, D.; et al. Synchronous intrathyroidal parathyroid carcinoma and thyroid carcinoma: Case report and review of the literature. *BMC Endocr. Disord.* **2021**, *21*, 60. [CrossRef]
11. Song, C.; Wang, J.; Cai, X.; Gao, L. Synchronous parathyroid carcinoma and papillary thyroid carcinoma: A case study and review of literature. *Int. J. Clin. Exp. Pathol.* **2016**, *9*, 302–309.

12. D'cruz, R.T.; Seet, J.E.; Parameswaran, R. Synchronous symptomatic parathyroid carcinoma and parathyroid adenoma with incidental follicular thyroid carcinoma. *Ann. R. Coll. Surg. Engl.* **2020**, *102*, e192–e195. [CrossRef] [PubMed]
13. Preda, C.; Branisteanu, D.; Armasu, I.; Danila, R.; Velicescu, C.; Ciobanu, D.; Covic, A.; Grigorovici, A. Coexistent papillary thyroid carcinoma diagnosed in surgically treated patients for primary versus secondary hyperparathyroidism: Same incidence, different characteristics. *BMC Surg.* **2019**, *19*, 94. [CrossRef]
14. Javadi, H.; Jallalat, S.; Farrokhi, S.; Semnani, S.; Mogharrabi, M.; Riazi, A.; Nabipour, I.; Moshtaghi, D.; Assadi, M. Concurrent papillary thyroid cancer and parathyroid adenoma as a rare condition: A case report. *Nucl. Med. Rev. Cent. East. Eur.* **2012**, *15*, 153–155. [PubMed]
15. Colin, I.M.; Denef, J.F.; Lengele, B.; Many, M.C.; Gerard, A.C. Recent insights into the cell biology of thyroid angiofollicular units. *Endocr. Rev.* **2013**, *34*, 209–238. [CrossRef] [PubMed]
16. Cheung, L.; Howlett, D.; El Teraifi, H.; Kirkland, P. Association of synchronous medullary and papillary thyroid carcinomas with primary hyperparathyroidism: First case report and literature review. *J. Laryngol. Otol.* **2014**, *128*, 565–568. [CrossRef] [PubMed]
17. Koea, J.B.; Shaw, J.H. Parathyroid cancer: Biology and management. *Surg. Oncol.* **1999**, *8*, 155–165. [CrossRef]
18. Schneider, A.B.; Recant, W.; Pinsky, S.M.; Ryo, U.Y.; Bekerman, C.; Shore-Freedman, E. Radiation-induced thyroid carcinoma. Clinical course and results of therapy in 296 patients. *Ann. Intern. Med.* **1986**, *105*, 405–412. [CrossRef]
19. Schulte, K.M.; Talat, N.; Galata, G.; Gilbert, J.; Miell, J.; Hofbauer, L.C.; Barthel, A.; Diaz-Cano, S.; Bornstein, S.R. Oncologic resection achieving r0 margins improves disease-free survival in parathyroid cancer. *Ann. Surg. Oncol.* **2014**, *21*, 1891–1897. [CrossRef]
20. Digonnet, A.; Carlier, A.; Willemse, E.; Quiriny, M.; Dekeyser, C.; de Saint Aubain, N.; Lemort, M.; Andry, G. Parathyroid carcinoma: A review with three illustrative cases. *J. Cancer* **2011**, *2*, 532–537. [CrossRef]
21. Dikmen, K.; Bostanci, H.; Gobut, H.; Yildiz, A.; Ertunc, O.; Celik, A.; Akin, M.; Taneri, F. Nonfunctional double parathyroid carcinoma with incidental thyroid micropapillary carcinoma: A rare case. *Pan Afr. Med. J.* **2017**, *27*, 241. [CrossRef]
22. Paschke, R.; Lincke, T.; Müller, S.P.; Kreissl, M.C.; Dralle, H.; Fassnacht, M. The Treatment of Well-Differentiated Thyroid Carcinoma. *Dtsch. Ärzteblatt Int.* **2015**, *112*, 452–458. [CrossRef] [PubMed]
23. Wei, C.H.; Harari, A. Parathyroid carcinoma: Update and guidelines for management. *Curr. Treat. Options Oncol.* **2012**, *13*, 11–23. [CrossRef]
24. Shane, E.; Bilezikian, J.P. Parathyroid carcinoma: A review of 62 patients. *Endocr. Rev.* **1982**, *3*, 218–226. [CrossRef] [PubMed]
25. Tuttle, R.M.; Ball, D.W.; Byrd, D.; Dilawari, R.A.; Doherty, G.M.; Duh, Q.Y.; Ehya, H.; Farrar, W.B.; Haddad, R.I.; Kandeel, F.; et al. Thyroid carcinoma. *J. Natl. Compr. Cancer Netw.* **2010**, *8*, 1228–1274. [CrossRef]
26. Long, K.L.; Sippel, R.S. Current and future treatments for parathyroid carcinoma. *Int. J. Endocr. Oncol.* **2018**, *5*, IJE06. [CrossRef]
27. Agarwal, G.; Dhingra, S.; Mishra, S.K.; Krishnani, N. Implantation of parathyroid carcinoma along fine needle aspiration track. *Langenbecks Arch. Surg.* **2006**, *391*, 623–626. [CrossRef]
28. Spinelli, C.; Bonadio, A.G.; Berti, P.; Materazzi, G.; Miccoli, P. Cutaneous spreading of parathyroid carcinoma after fine needle aspiration cytology. *J. Endocrinol. Investig.* **2000**, *23*, 255–257. [CrossRef] [PubMed]
29. Hoelting, T.; Weber, T.; Werner, J.; Herfarth, C. Surgical treatment of parathyroid carcinoma (Review). *Oncol. Rep.* **2001**, *8*, 931–934. [CrossRef] [PubMed]
30. Ippolito, G.; Palazzo, F.F.; Sebag, F.; De Micco, C.; Henry, J.F. Intraoperative diagnosis and treatment of parathyroid cancer and atypical parathyroid adenoma. *Br. J. Surg.* **2007**, *94*, 566–570. [CrossRef]
31. Lee, P.K.; Jarosek, S.L.; Virnig, B.A.; Evasovich, M.; Tuttle, T.M. Trends in the incidence and treatment of parathyroid cancer in the United States. *Cancer* **2007**, *109*, 1736–1741. [CrossRef]
32. Hundahl, S.A.; Fleming, I.D.; Fremgen, A.M.; Menck, H.R.; The American College of Surgeons Commission on Cancer and the American Cancer Society. Two hundred eighty-six cases of parathyroid carcinoma treated in the U.S. between 1985–1995: A National Cancer Data Base Report. *Cancer* **1999**, *86*, 538–544. [CrossRef]
33. Kalthoum, M.; Nacef, I.B.; Mekni, S.; Rojbi, I.; Mechirgui, N.; Khiari, K. Synchronous parathyroid carcinoma and papillary thyroid carcinoma: A case report. *Endocr. Abstr.* **2020**, *70*, EP95. [CrossRef]
34. Edafe, O.; Debono, M.; Tahir, F.; Balasubramanian, S.P. Simultaneous presentation of parathyroid carcinoma, papillary thyroid cancer and ACTH-independent hypercortisolism due to benign cortical adenoma. *BMJ Case Rep.* **2019**, *12*, e230438. [CrossRef] [PubMed]
35. Kuzu, F.; Cinkaya, A.; Ekici, M.; Kodaz, H.; Deger, A. Synchronous parathyroid carcinoma and multifocal papillary thyroid carcinoma: A case report. *Eurasian J. Med. Oncol.* **2017**, *1*, 49–52.
36. Baek, C.-O.; Kim, K.H.; Song, S.K. Synchronous parathyroid carcinoma and papillary thyroid carcinoma in a patient with long-standing schizophrenia. *Korean J. Intern. Med. (Korean Assoc. Intern. Med.)* **2017**, *32*, 1104–1107. [CrossRef]
37. Demir, D. A very rare case, coexistence of invasive thyroid papillary carcinoma and parathyroid carcinoma in hypercalcemic 29 year old woman how is treated for urolithiasis. *Diabetes Metab. Disord.* **2017**, *4*, 1–3. [CrossRef]
38. Aljabri, K.S.; Bokhari, S.A.; Alshareef, M.A.; Khan, P.M.; Abdulhafez, D.A.; Aljabri, B.K. Coexistence of parathyroid cancer and papillary thyroid cancer: A case report with a review of the literature. *EC Endocrinol. Metab. Res.* **2017**, *81*, 98–102.
39. Neslihan, S.A.; Sibel, G.; Ebru, T.; Hüseyin, Ç.; Atakan, S.; Nermin, T.; Armağan, T. A Rare Co-Occurrence of Parathyroid and Papillary Thyroid Carcinoma. *ARC J. Clin. Case Rep.* **2016**, *2*, 1–4.
40. Lee, H.W.; Kim, H.J.; Moon, J.S. A case of recurred parathyroid carcinoma with multiple lymph node metastasis: Consurrent with papillary thyroid cancer. In Proceedings of the 18th European Congress of Endocrinology 2016, Munich, Germany, 28–31 May 2016.

41. Al-Sulami, S.S. Parathyroid carcinoma: Atypical presentation and coexistence with papillary thyroid cancer. *Saudi J. Med. Med. Sci.* **2015**, *3*, 245–246. [CrossRef]
42. Zakerkish, M.; Rajaei, E.; Dargahi, M.; Bahadoram, M. A Rare Constellation of Hürthle Cell Thyroid Carcinoma and Parathyroid Carcinoma. *J. Clin. Diagn. Res.* **2015**, *9*, OD08–OD10. [CrossRef]
43. Chaychi, L.; Belbruno, K.; Golding, A.; Memoli, V. Unusual manifestation of parathyroid carcinoma in the setting of papillary thyroid cancer. *Endocr. Pract.* **2010**, *16*, 664–668. [CrossRef] [PubMed]
44. Marcy, P.Y.; Thariat, J.; Sudaka, A.; Poissonnet, G. Synchronous parathyroid and papillary thyroid carcinomas. *Thyroid* **2009**, *19*, 1131–1133. [CrossRef] [PubMed]
45. Goldfarb, M.; O'Neal, P.; Shih, J.L.; Hartzband, P.; Connolly, J.; Hasselgren, P.O. Synchronous parathyroid carcinoma, parathyroid adenoma, and papillary thyroid carcinoma in a patient with severe and long-standing hyperparathyroidism. *Endocr. Pract.* **2009**, *15*, 463–468. [CrossRef]
46. Mazeh, H.; Kouniavsky, G.; Schneider, D.F.; Makris, K.I.; Sippel, R.S.; Dackiw, A.P.; Chen, H.; Zeiger, M.A. Intrathyroidal parathyroid glands: Small, but mighty (a Napoleon phenomenon). *Surgery* **2012**, *152*, 1193–1200. [CrossRef] [PubMed]
47. Lin, S.D.; Tu, S.T.; Hsu, S.R.; Chang, J.H.; Yang, K.T.; Yang, L.H. Synchronous parathyroid and papillary thyroid carcinoma. *J. Chin. Med. Assoc.* **2005**, *68*, 87–91. [CrossRef]
48. Kern, M.; Lee, G.; Robbins, P.; Bynevelt, M.; Watson, P. Intracranial metastatic parathyroid carcinoma. Case report and review of the literature. *J. Neurosurg.* **2004**, *101*, 1065–1069. [CrossRef]
49. Schoretsanitis, G.; Melissas, J.; Kafousi, M.; Karkavitsas, N.; Tsiftsis, D.D. Synchronous parathyroid and papillary thyroid carcinoma: A case report. *Am. J. Otolaryngol.* **2002**, *23*, 382–385. [CrossRef]
50. Bednarek-Tupikowska, G.; Tołłoczko, T.; Tupikowski, W.; Bogdańska, M.; Karwacki, J.; Medraś, M.; Milewicz, A. Coexistence of parathyroid carcinoma and non-medullary carcinoma of the thyroid. *Med. Sci. Monit.* **2001**, *7*, 448–456.
51. Savli, H.; Sevinc, A.; Sari, R.; Ozen, S.; Buyukberber, S.; Ertas, E. Occult parathyroid carcinoma in a patient with papillary thyroid carcinoma and Hashimoto's thyroiditis. *J. Endocrinol. Investig.* **2001**, *24*, 42–44. [CrossRef]
52. Kurita, S.; Mihashi, S.; Hirano, M.; Nakashima, T.; Tanimura, A. [Hyperfunctioning parathyroid carcinoma combined with papillary carcinoma of the thyroid gland–report of a case (author's transl)]. *Nihon Gan Chiryo Gakkai Shi* **1979**, *14*, 1127–1135.
53. Basceken, S.I.; Genc, V.; Ersoz, S.; Sevim, Y.; Celik, S.U.; Bayram, I.K. Is local resection sufficient for parathyroid carcinoma? *Clinics* **2015**, *70*, 247–249. [CrossRef]
54. Tseleni-Balafouta, S.; Gakiopoulou, H.; Kavantzas, N.; Agrogiannis, G.; Givalos, N.; Patsouris, E. Parathyroid proliferations: A source of diagnostic pitfalls in FNA of thyroid. *Cancer* **2007**, *111*, 130–136. [CrossRef] [PubMed]
55. Christmas, T.J.; Chapple, C.R.; Noble, J.G.; Milroy, E.J.; Cowie, A.G. Hyperparathyroidism after neck irradiation. *Br. J. Surg.* **1988**, *75*, 873–874. [CrossRef]
56. Amoodi, H.A.; Makki, F.M.; Mark Taylor, S.; Bullock, M.J.; Hart, R.D.; Trites, J.R. Synchronous Thyroid/parathyroid Carcinomas. *J. Otolaryngol. Head Neck Surg.* **2010**, *39*, E42–E47. [PubMed]
57. Torresan, F.; Iacobone, M. Clinical Features, Treatment, and Surveillance of Hyperparathyroidism-Jaw Tumor Syndrome: An Up-to-Date and Review of the Literature. *Int. J. Endocrinol.* **2019**, *2019*, 1761030. [CrossRef]
58. Kebebew, E. Parathyroid carcinoma. *Curr. Treat. Options Oncol.* **2001**, *2*, 347–354. [CrossRef]
59. Sharretts, J.M.; Simonds, W.F. Clinical and molecular genetics of parathyroid neoplasms. *Best Pract. Res. Clin. Endocrinol. Metab.* **2010**, *24*, 491–502. [CrossRef]
60. McMullen, T.; Bodie, G.; Gill, A.; Ihre-Lundgren, C.; Shun, A.; Bergin, M.; Stevens, G.; Delbridge, L. Hyperparathyroidism after irradiation for childhood malignancy. *Int. J. Radiat. Oncol. Biol. Phys.* **2009**, *73*, 1164–1168. [CrossRef]
61. Rasmuson, T.; Damber, L.; Johansson, L.; Johansson, R.; Larsson, L.G. Increased incidence of parathyroid adenomas following X-ray treatment of benign diseases in the cervical spine in adult patients. *Clin. Endocrinol.* **2002**, *57*, 731–734. [CrossRef]
62. Ferraro, V.; Sgaramella, L.I.; Di Meo, G.; Prete, F.P.; Logoluso, F.; Minerva, F.; Noviello, M.; Renzulli, G.; Gurrado, A.; Testini, M. Current concepts in parathyroid carcinoma: A single centre experience. *BMC Endocr. Disord.* **2019**, *19* (Suppl. S1), 46. [CrossRef]
63. Dudney, W.C.; Bodenner, D.; Stack, B.C., Jr. Parathyroid carcinoma. *Otolaryngol. Clin. North Am.* **2010**, *43*, 441–453. [CrossRef] [PubMed]
64. Givi, B.; Shah, J.P. Parathyroid carcinoma. *Clin. Oncol. (R. Coll. Radiol.)* **2010**, *22*, 498–507. [CrossRef] [PubMed]
65. Machado, N.N.; Wilhelm, S.M. Parathyroid Cancer: A Review. *Cancers* **2019**, *11*, 1676. [CrossRef]
66. Busaidy, N.L.; Jimenez, C.; Habra, M.A.; Schultz, P.N.; El-Naggar, A.K.; Clayman, G.L.; Asper, J.A.; Diaz, E.M., Jr.; Evans, D.B.; Gagel, R.F.; et al. Parathyroid carcinoma: A 22-year experience. *Head Neck* **2004**, *26*, 716–726. [CrossRef] [PubMed]
67. Piciu, D.; Irimie, A.; Kontogeorgos, G.; Piciu, A.; Buiga, R. Highly aggressive pathology of non-functional parathyroid carcinoma. *Orphanet J. Rare Dis.* **2013**, *8*, 115. [CrossRef]
68. Wilkins, B.J.; Lewis, J.S., Jr. Non-functional parathyroid carcinoma: A review of the literature and report of a case requiring extensive surgery. *Head Neck Pathol.* **2009**, *3*, 140–149. [CrossRef]
69. Poppe, K.; Pipeleers-Marichal, M.; Flamen, P.; Bossuyt, A.; Lamote, J.; Vanhaelst, L.; Velkeniers, B. Non-secreting atypical parathyroid adenoma. *J. Endocrinol. Investig.* **2001**, *24*, 107–110. [CrossRef]
70. Sen, M.; Nagaoka, R.; Kazusaka, H.; Matsui, M.; Saitou, M.; Sugitani, I.; Sakatani, T.; Kameyama, K. Non-functioning oxyphilic parathyroid carcinoma: A case report. *Surg. Case Rep.* **2021**, *7*, 119. [CrossRef]

71. Wang, L.; Han, D.; Chen, W.; Zhang, S.; Wang, Z.; Li, K.; Gao, Y.; Zou, S.; Yang, A. Non-functional parathyroid carcinoma: A case report and review of the literature. *Cancer Biol. Ther.* **2015**, *16*, 1569–1576. [CrossRef]
72. Gao, W.C.; Ruan, C.P.; Zhang, J.C.; Liu, H.M.; Xu, X.Y.; Sun, Y.P.; Wang, Q. Nonfunctional parathyroid carcinoma. *J. Cancer Res. Clin. Oncol.* **2010**, *136*, 969–974. [CrossRef]
73. Harari, A.; Clark, O.H.; Duh, Q.-Y.; Kebebew, E.; Gosnell, J.E. Parathyroid carcinoma. In *Textbook of Endocrine Surgery*, 3rd ed.; Shen, W.T., Ed.; Jaypee Brothers Medical Publishers: New Delhi, India, 2016; pp. 927–935.
74. Nam, M.; Jeong, H.S.; Shin, J.H. Differentiation of parathyroid carcinoma and adenoma by preoperative ultrasonography. *Acta Radiol.* **2017**, *58*, 670–675. [CrossRef] [PubMed]
75. Jabiev, A.A.; Lew, J.I.; Solorzano, C.C. Surgeon-performed ultrasound: A single institution experience in parathyroid localization. *Surgery* **2009**, *146*, 569–577. [CrossRef] [PubMed]
76. Arora, S.; Balash, P.R.; Yoo, J.; Smith, G.S.; Prinz, R.A. Benefits of surgeon-performed ultrasound for primary hyperparathyroidism. *Langenbecks Arch. Surg.* **2009**, *394*, 861–867. [CrossRef] [PubMed]
77. Brito, J.P.; Gionfriddo, M.R.; Al Nofal, A.; Boehmer, K.R.; Leppin, A.L.; Reading, C.; Callstrom, M.; Elraiyah, T.A.; Prokop, L.J.; Stan, M.N.; et al. The accuracy of thyroid nodule ultrasound to predict thyroid cancer: Systematic review and meta-analysis. *J. Clin. Endocrinol. Metab.* **2014**, *99*, 1253–1263. [CrossRef]
78. Smith-Bindman, R.; Lebda, P.; Feldstein, V.A.; Sellami, D.; Goldstein, R.B.; Brasic, N.; Jin, C.; Kornak, J. Risk of thyroid cancer based on thyroid ultrasound imaging characteristics: Results of a population-based study. *JAMA Intern. Med.* **2013**, *173*, 1788–1796. [CrossRef]
79. Lee, J.Y.; Baek, J.H.; Ha, E.J.; Sung, J.Y.; Shin, J.H.; Kim, J.H.; Lee, M.K.; Jung, S.L.; Lee, Y.H.; Ahn, H.S.; et al. 2020 Imaging Guidelines for Thyroid Nodules and Differentiated Thyroid Cancer: Korean Society of Thyroid Radiology. *Korean J. Radiol.* **2021**, *22*, 840–860. [CrossRef]
80. Russ, G.; Bonnema, S.J.; Erdogan, M.F.; Durante, C.; Ngu, R.; Leenhardt, L. European Thyroid Association Guidelines for Ultrasound Malignancy Risk Stratification of Thyroid Nodules in Adults: The EU-TIRADS. *Eur. Thyroid J.* **2017**, *6*, 225–237. [CrossRef]
81. Tessler, F.N.; Middleton, W.D.; Grant, E.G.; Hoang, J.K.; Berland, L.L.; Teefey, S.A.; Cronan, J.J.; Beland, M.D.; Desser, T.S.; Frates, M.C.; et al. ACR Thyroid Imaging, Reporting and Data System (TI-RADS): White Paper of the ACR TI-RADS Committee. *J. Am. Coll. Radiol.* **2017**, *14*, 587–595. [CrossRef]
82. Lavely, W.C.; Goetze, S.; Friedman, K.P.; Leal, J.P.; Zhang, Z.; Garret-Mayer, E.; Dackiw, A.P.; Tufano, R.P.; Zeiger, M.A.; Ziessman, H.A. Comparison of SPECT/CT, SPECT, and planar imaging with single- and dual-phase (99m)Tc-sestamibi parathyroid scintigraphy. *J. Nucl. Med.* **2007**, *48*, 1084–1089. [CrossRef]
83. Kumar, R.; Gupta, R.; Malhotra, A. Brown tumor of the sternum: A potential source of false-positive Tl-201 and Tc-99m subtraction imaging in the mediastinum. *Clin. Nucl. Med.* **2000**, *25*, 44–47. [CrossRef]
84. Guo, R.; Wang, J.; Zhang, M.; Zhang, M.; Meng, H.; Zhang, Y.; Li, B. Value of 99mTc-MIBI SPECT/CT parathyroid imaging and ultrasonography for concomitant thyroid carcinoma. *Nucl. Med. Commun.* **2017**, *38*, 676–682. [CrossRef]
85. Shrestha, M.; Crothers, B.A.; Burch, H.B. The impact of thyroid nodule size on the risk of malignancy and accuracy of fine-needle aspiration: A 10-year study from a single institution. *Thyroid* **2012**, *22*, 1251–1256. [CrossRef] [PubMed]
86. Pitman, M.B.; Abele, J.; Ali, S.Z.; Duick, D.; Elsheikh, T.M.; Jeffrey, R.B.; Powers, C.N.; Randolph, G.; Renshaw, A.; Scoutt, L. Techniques for thyroid FNA: A synopsis of the National Cancer Institute Thyroid Fine-Needle Aspiration State of the Science Conference. *Diagn. Cytopathol.* **2008**, *36*, 407–424. [CrossRef] [PubMed]
87. Malheiros, D.C.; Canberk, S.; Poller, D.N.; Schmitt, F. Thyroid FNAC: Causes of false-positive results. *Cytopathology* **2018**, *29*, 407–417. [CrossRef] [PubMed]
88. Asare, E.A.; Sturgeon, C.; Winchester, D.J.; Liu, L.; Palis, B.; Perrier, N.D.; Evans, D.B.; Winchester, D.P.; Wang, T.S. Parathyroid Carcinoma: An Update on Treatment Outcomes and Prognostic Factors from the National Cancer Data Base (NCDB). *Ann. Surg. Oncol.* **2015**, *22*, 3990–3995. [CrossRef]
89. Sandelin, K.; Tullgren, O.; Farnebo, L.O. Clinical course of metastatic parathyroid cancer. *World J. Surg.* **1994**, *18*, 594–598, discussion 599. [CrossRef]
90. Fernandez-Ranvier, G.G.; Khanafshar, E.; Jensen, K.; Zarnegar, R.; Lee, J.; Kebebew, E.; Duh, Q.Y.; Clark, O.H. Parathyroid carcinoma, atypical parathyroid adenoma, or parathyromatosis? *Cancer* **2007**, *110*, 255–264. [CrossRef]
91. Al-Kurd, A.; Mekel, M.; Mazeh, H. Parathyroid carcinoma. *Surg. Oncol.* **2014**, *23*, 107–114. [CrossRef]
92. Schantz, A.; Castleman, B. Parathyroid carcinoma. A study of 70 cases. *Cancer* **1973**, *31*, 600–605. [CrossRef]
93. Shattuck, T.M.; Välimäki, S.; Obara, T.; Gaz, R.D.; Clark, O.H.; Shoback, D.; Wierman, M.E.; Tojo, K.; Robbins, C.M.; Carpten, J.D.; et al. Somatic and germ-line mutations of the HRPT2 gene in sporadic parathyroid carcinoma. *N. Engl. J. Med.* **2003**, *349*, 1722–1729. [CrossRef]
94. Hosny Mohammed, K.; Siddiqui, M.T.; Willis, B.C.; Zaharieva Tsvetkova, D.; Mohamed, A.; Patel, S.; Sharma, J.; Weber, C.; Cohen, C. Parafibromin, APC and MIB-1 are useful markers for distinguishing parathyroid carcinomas from adenomas. *Appl. Immunohistochem. Mol. Morphol.* **2017**, *25*, 731–735. [CrossRef]
95. Hu, Y.; Liao, Q.; Cao, S.; Gao, X.; Zhao, Y. Diagnostic performance of parafibromin immunohistochemical staining for sporadic parathyroid carcinoma: A meta-analysis. *Endocrine* **2016**, *54*, 612–619. [CrossRef] [PubMed]

96. Bergero, N.; De Pompa, R.; Sacerdote, C.; Gasparri, G.; Volante, M.; Bussolati, G.; Papotti, M. Galectin-3 expression in parathyroid carcinoma: Immunohistochemical study of 26 cases. *Hum. Pathol.* **2005**, *36*, 908–914. [CrossRef] [PubMed]
97. Vasef, M.A.; Brynes, R.K.; Sturm, M.; Bromley, C.; Robinson, R.A. Expression of cyclin D1 in parathyroid carcinomas, adenomas, and hyperplasias: A paraffin immunohistochemical study. *Mod. Pathol.* **1999**, *12*, 412–416. [PubMed]
98. Salcuni, A.S.; Cetani, F.; Guarnieri, V.; Nicastro, V.; Romagnoli, E.; de Martino, D.; Scillitani, A.; Cole, D.E.C. Parathyroid carcinoma. *Clin. Endocrinol. Metab.* **2018**, *32*, 877–889. [CrossRef] [PubMed]

medicina

MDPI

Article

Absolute and Functional Iron Deficiency in Colon Cancer: A Cohort Study

Vlad Făgărășan [1], David Andraș [1,*], Giorgiana Amarinei [1], Radu Ioan Seicean [1], Vasile Virgil Bințințan [1], George Calin Dindelegan [1] and Calin Ioan Căinap [2]

1 1st Surgical Clinic, Department of General Surgery, Iuliu Hațieganu University of Medicine and Pharmacy, 400012 Cluj Napoca, Romania

2 Department of Medical Oncology, Iuliu Hațieganu University of Medicine and Pharmacy, 400012 Cluj Napoca, Romania

* Correspondence: andrasdavid88@gmail.com

Abstract: *Background and Objectives*: Iron is an essential micronutrient for many biological functions and has been found to be intimately linked to cancer biology. Although the effects of increased dietary iron consumption in the development of CRC have been previously investigated in several cohort studies, the available evidence on the involvement of iron deficiency in this process is relatively scarce. Previously published papers did not analyze specific outcomes, such as the presence of biologically aggressive histopathological characteristics, that are associated with the subtypes of iron deficiency. The purpose of this study was to investigate the connection between the development of colorectal cancer and the presence of functional iron deficiency (FID), which is defined as insufficient biological availability of iron in the presence of adequate storage reserves, or absolute iron deficiency (AID), which is defined as severely depleted iron storage levels. *Materials and Methods*: Our paper represents a single center registry-based cohort study. Iron levels were routinely evaluated upon diagnosis of CRC and the collected data were coupled with patient- and tumor-specific data (2018–2022). Spearman's correlation coefficient and the chi-squared test were used to analyze the association. *Results*: Out of 129 patients, 75 (58.13%) were anemic. AID was identified in 26.35% of cases and FID was encountered in 51.16% of cases. A statistically significant association between FID and lymphatic invasion was encountered. An analysis of the correlation demonstrated a significant association between anemia and right-sided tumor location. *Conclusions*: Functional iron deficiency seems to be independently associated with lymphatic invasion. Although a statistically significant correlation with the T or N stage was not demonstrated, the analysis suggested a potential positive relationship between the presence of FID and more aggressive tumor characteristics.

Keywords: absolute iron deficiency; functional iron deficiency; colorectal cancer

Citation: Făgărășan, V.; Andraș, D.; Amarinei, G.; Seicean, R.I.; Bințințan, V.V.; Dindelegan, G.C.; Căinap, C.I. Absolute and Functional Iron Deficiency in Colon Cancer: A Cohort Study. *Medicina* **2022**, *58*, 1202. https://doi.org/10.3390/medicina58091202

Academic Editor: Roberto Cirocchi

Received: 20 June 2022
Accepted: 30 August 2022
Published: 1 September 2022

1. Introduction

Colorectal cancer (CRC) is the third most frequently diagnosed and the second most deadly malignant tumor, and the incidence of this disease is expected to increase in the near future [1]. Although CRC is encountered more frequently in highly developed countries, there is a rising incidence in developing nations due to the adoption of "western" dietary habits [2]. In addition, an increased incidence in early-onset CRC has been observed in recent years, possibly due to the widespread introduction of screening programs [3]. Despite recent advances in surgical techniques and the application of increasingly efficient neoadjuvant treatment regimens, CRC continues to be associated with significant mortality rates, especially in advanced stages [4]. The pathogenesis of intestinal malignancies is multifactorial, with both environmental and genetic factors playing a significant role in the development of this disease [5]. Among the multitude of factors implicated in colorectal carcinogenicity, emerging evidence has demonstrated that reduced dietary iron availability and low systemic iron levels may influence colorectal tumorigenesis [6,7]. Although the effects of increased dietary iron consumption in the development of CRC have been previously investigated in several cohort studies, the available evidence on the involvement of iron deficiency in this process is relatively scarce [8–11]. Iron is an essential micronutrient with a significant influence on various biological functions, several of which are intimately connected to the development of malignant tumors [10,12]. Iron participates in a wide variety of metabolic processes such as erythropoiesis, oxygen transportation, deoxyribonucleic acid (DNA) synthesis, adenosine triphosphate generation, immunological functions, and REDOX cycling. The catalytic form of iron mediates the production of reactive oxygen species via the Fenton reaction and generates oxygen free radicals, resulting in oxidative DNA damage and promoting malignant transformation [12]. Furthermore, iron is necessary for appropriate immunological functions, potentially altering the tumor microenvironment and influencing immune cell-mediated cancer surveillance, both of which may promote tumor development [6]. Therefore, due to the complex interactions of this essential element, a delicate balance of dietary iron must be maintained in order to avoid the negative consequences associated with excessive or deficient intake on the proliferation and development of malignant tumors.

An unexplained diagnosis of iron deficiency anemia (IDA) may be the first indicator for an underlying malignancy in over 8% of cases, especially in the elderly population [13]. Colorectal cancer is frequently associated with the development of IDA, especially right-sided tumors, which have been associated with an unfavorable prognosis. [14] Possible causes of IDA in colorectal cancer include: continuous occult bleeding from the tumor bed, reduced luminal absorption of iron and altered iron metabolism due to the presence of chronic inflammation. Chronic intestinal bleeding results in a severe reduction in or absence of iron reserves, which is defined as absolute iron deficiency (AID) [6]. Chronic inflammation in the presence of malignant tumors leads to a reduction in biologically available iron through an increased production of hepcidin, which causes reduced uptake of iron from the intestine as well as impaired iron release from macrophages, thus resulting in functional iron deficiency (FID) [15].

The purpose of this study was to investigate the correlation between the progression of colorectal cancer and the presence of iron deficiency, either functional or absolute, for a cohort of patients who underwent surgery in a single institution by determining the characteristics of tumors as evidenced by postoperative histopathological analysis.

2. Materials and Methods

2.1. Patient Selection and Data Collection

Clinical data from 351 patients diagnosed with colon cancer who underwent open or laparoscopic colon resections between January 2018 and April 2022 at the 1st Surgical Clinic, Emergency County Clinical Hospital of Cluj-Napoca were extracted from the hospital records and introduced into an electronic database. All patients with resectable tumors located in the cecum, ascending colon, hepatic flexure, transverse colon, splenic flexure, descending colon, or sigmoid colon were included in the analysis. Patients were excluded from the study if they had received any blood product transfusions or iron administration up to six months prior to diagnosis. In addition, patients that underwent surgery in an emergency setting (tumor perforation or bowel obstruction) were also excluded from the analysis.

Data retrieved from the hospital database included: demographical information; location of the tumor; Union for International Cancer Control, Tumor Node Metastasis (UICC TNM) staging classification according to the postoperative histopathological analysis report; tumor grade; and number of positive lymph nodes. Tumor grade was defined as: low grade or well differentiated (G1), intermediate grade or moderately differentiated (G2), or high grade or poorly differentiated (G3). Patients with tumors located in the cecum, ascending colon, or hepatic flexure of the transverse colon were classified in the right colon category, while patients with tumors located in the splenic flexure, descending colon, and sigmoid colon were classified in the left colon category. Tumors located in the recto-sigmoid junction were considered as superior rectal cancer and were excluded from the analysis. Patients with appendiceal tumors were also excluded from the study.

Hemoglobin, ferritin, transferrin, and serum iron levels were recorded preoperatively 1 to 10 days prior to the surgical intervention. Transferrin saturation (TSAT) was calculated according to the following formula [16]:

$$\text{TSAT} = (\text{Serum iron concentration}/\text{Serum transferrin concentration}) \times 70.9$$

Anemia was defined according to the World Health Organization criteria as follows: Hb < 13 g/dL (8.1 mmol/L) for adult males and Hb < 12 g/dL (7.4 mmol/L) for non-pregnant adult females [17]. Iron deficiency was classified as AID if plasma ferritin levels were <30 ng/mL and FID if plasma ferritin levels were $\geq$30 ng/mL and TSAT was <20% [18]. The primary outcome was functional iron deficiency. Secondary outcomes were absolute iron deficiency and anemia.

2.2. Statistical Analysis

Statistical analysis was performed using GraphPad Prism version 9.3.0 for Windows, GraphPad Software, San Diego, CA, USA. Descriptive statistics were calculated as means and percentages for continuous variables. Categorical variables were expressed as contingency tables and an analysis was performed using chi-squared and Fisher's exact tests to compare differences between observed and expected frequencies. Spearman's correlation coefficient was used to measure the degree of correlation between independent variables. The statistical significance level was set to 0.05 in this study.

3. Results

Data were imported from the hospital database on 351 patients who underwent open or laparoscopic colon resections between January 2018 and April 2022. A total of 129 (36.75%) patients met the inclusion criteria and had complete data sets, and thus were included for analysis.

The cohort included 75 (58%) male patients and 54 (42%) female patients. The mean age of all included patients was 69 years old, ranging from 41 to 89 years old. The mean age of patients according to the type of iron deficiency is presented in Table 1.

Table 1. Mean age of patients according to type of iron deficiency.

Type of Deficiency	Mean ± SD (Years)
FID	68.20 ± 9.09
AID	70.15 ± 11.72
Anemia	69.00 ± 10.19
No anemia	70.22 ± 9.34
Overall	69.51 ± 9.82

The mean concentration of hemoglobin was 11.8 mg/dL (CI 95% 11.4–12.2); 58.13% (75) of patients were anemic, while 41.87% (54) of patients had normal levels of hemoglobin. The mean serum iron concentration was 47.4 μg/mL (CI 95% 41.5–53.3). The mean serum ferritin value was 90.1 ng/mL (CI 95% 74.1–106). Absolute iron deficiency was identified in 26.35% (34) of cases and corrected prior to the surgical intervention. Serum transferrin levels had a mean of 255 mg/dl (CI 95% 245–265). The transferrin saturation coefficient was calculated for each patient and had a mean value of 13.5% (CI 95% 11.8–15.2), ranging from 2.82% to 51.8%. Functional iron deficiency was encountered in 51.16% (66) of cases. The tumor-specific pathological characteristics of the included cases are presented in Table 2.

Table 2. Tumor-specific pathological characteristics of the cohort.

Variable	Number of Cases (Percentage)
Location of tumor	
Right colon	45 (34.88%)
Cecum	11 (8.52%)
Ascending	34 (26.35%)
Transverse colon	23 (17.82%)
Left colon	61 (47.28%)
Descending	23 (17.82%)
Sigmoid	38 (29.45%)
Differentiation	
G1	46 (35.6%)
G2	69 (53.5%)
G3	14 (10.9%)
T stage	
T1	3 (2.32%)
T2	24 (18.61%)
T3	79 (61.25%)
T4	23 (17.82%)
N stage	
N0	79 (61.24%)
N1	29 (22.48%)
N2	21 (16.28%)
M stage	
M0	118 (91.48%)
M1	11 (8.52%)
Lymphatic invasion	
L0	90 (69.77%)
L1	39 (30.23%)
Venous invasion	
V0	103 (79.85%)
V1	26 (20.15%)
Perineural invasion	
Pn0	102 (79.07%)
Pn1	27 (20.93%)

Although the correlation analysis demonstrated a weak positive relationship between the presence of FID and more aggressive tumor characteristics (advanced T stage—T4, N stage—N2), the results were not statistically significant. The correlation analysis did not demonstrate a significant association between FID or AID and tumor grade. The results of the correlation matrix (Spearman's correlation coefficient) are presented in Figures 1–3.

A contingency table analysis using chi-squared and Fisher's exact tests showed a statistically significant association between FID and lymphatic invasion (OR 2.364; CI 95% 1.019–5.172; *p*-value 0.0451). None of the other investigated parameters demonstrated a statistically significant association with FID.

The correlation analysis showed a significant association between anemia and right-sided tumor location (Spearman's r coefficient 0.37; $p = 0.00002$; CI 95% 0.2026–0.5117). In addition, anemia was associated with absolute iron deficiency (Spearman r coefficient 0.328; $p = 0.0001$; CI 95% 0.1600–0.4784). The correlation matrix for this analysis is presented in Figure 4.

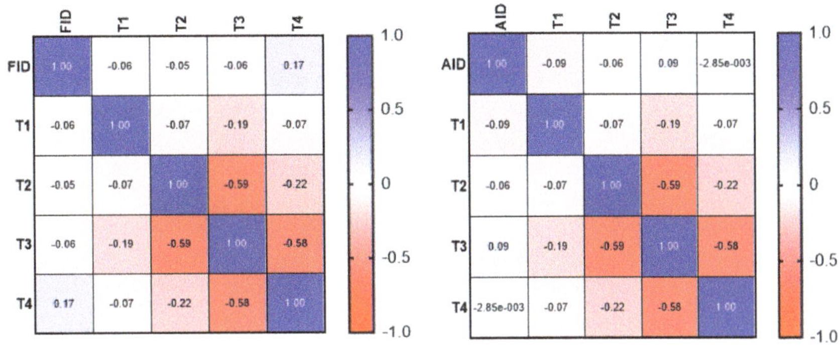

(a) (b)

Figure 1. Comparison between FID and AID: (**a**) correlation matrix for FID and T stage; (**b**) correlation matrix for AID and T stage.

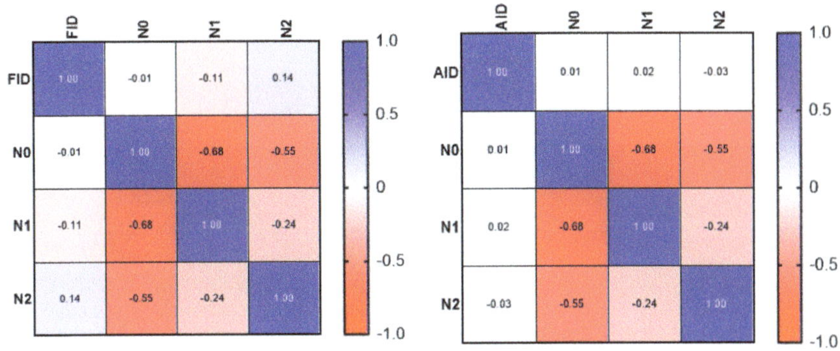

(a) (b)

Figure 2. Comparison between FID and AID: (**a**) correlation matrix for FID and N stage; (**b**) correlation matrix for AID and N stage.

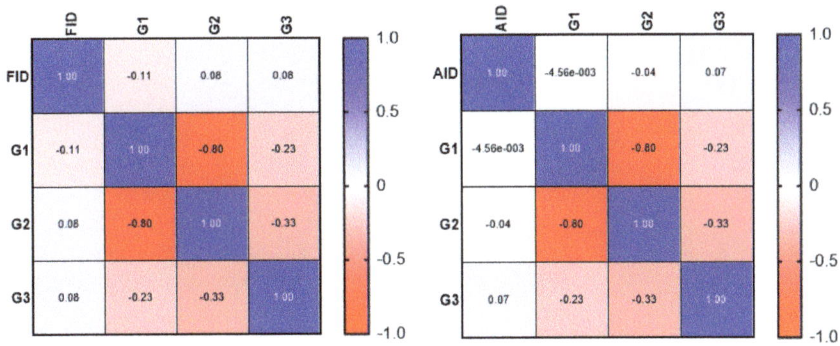

(a) (b)

Figure 3. Comparison between FID and AID: (**a**) correlation matrix for FID and tumor grade; (**b**) correlation matrix for AID and tumor grade.

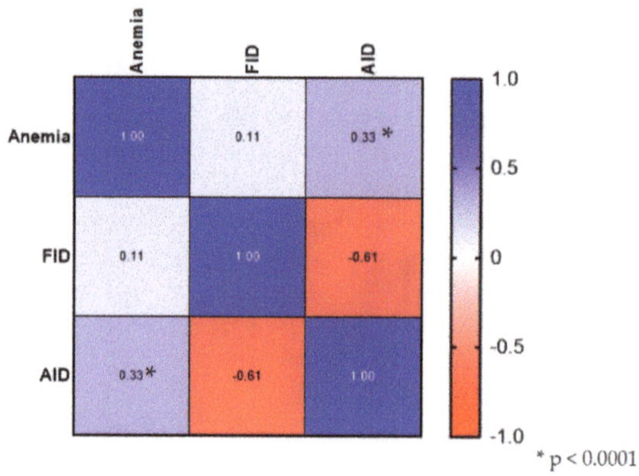

Figure 4. Analysis of correlation between anemia, FID, and AID.

4. Discussion

The present study included 66 (51.16%) patients with functional iron deficiency and 34 (26.35%) patients with absolute iron deficiency. While FID is determined by insufficient iron availability for incorporation into erythroid precursors despite the availability of sufficient reserves, AID is determined by severely reduced iron deposits in the bone marrow, liver, and spleen [19]. The principal cause of FID in cancer is the release of pro-inflammatory cytokines such as IL-6, IL-1, TNF-α, and IFN-γ, which amplify hepcidin synthesis and thereby reduce the quantity of iron released into the circulation [7,20,21]. The increased incidence of FID in our cohort could suggest an important inflammatory component that may influence, or be caused by, carcinogenesis and tumor proliferation.

The practical importance of differentiating between AID and FID concerns the way in which iron supplementation is administered in each case [22]. The treatment of AID requires immediate iron administration regardless of the presence of anemia. Conversely, iron supplementation is recommended in the treatment of FID only in symptomatic patients with anemia and should be withheld in patients with elevated ferritin levels [23]. The choice of oral versus intravenous iron therapy is also significant, since oral absorption of iron is reduced in FID through inflammation-related, IL-6-increased hepcidin production in the duodenum, and therefore the effectiveness of this administration method is reduced [24].

Although anemia is a well-known complication of malignant tumors and has been extensively studied, the prevalence of iron deficiency in malignant tumors has been relatively overlooked in the literature [9,10,23,24]. In our study, anemia was identified in 63.2% of patients with FID and 84% of patients with AID. In line with previous studies, a multivariate analysis demonstrated a significant association between tumor location (cecum, ascending colon, and hepatic flexure) and anemia [25–27]. Although the sample size for our cohort was relatively small, a statistically significant association was established between anemia and AID. This could be due to occult blood loss, which is more frequent in right-sided colon cancer. Interestingly, FID appeared to have a lesser influence on anemia despite the fact that the number of cases with FID in our cohort was almost double compared to the number of cases with AID.

Previous studies have demonstrated that solid tumors were more frequently associated with both functional and absolute iron deficiency in comparison with hematological malignancies [28,29]. An observational study by Ludwig et al. conducted on 1528 patients with various types of malignant tumors reported an increased incidence of iron deficiency in pancreatic (63.2%), colorectal (51.9%), and pulmonary (50.7%) cancers. Furthermore, the prevalence of ID was significantly associated with cancer stage ($p = 0.001$) and disease status ($p = 0.001$) [23]. Several studies demonstrated that ID with or without anemia was associated with a poorer prognosis, lower disease-free survival, and a reduced response to oncological treatment [7,18,30]. However, to the best of our knowledge, the majority of previously published papers did not analyze specific outcomes associated with AID and FID.

Although our results did not show a statistically significant correlation with either the T stage or N stage, most likely due to the relatively small sample size of the cohort, this finding might be explained by an increased inflammatory response in the presence of a significantly increased tumor burden. However, in our cohort, functional iron deficiency was independently associated with the presence of lymphatic invasion. This association suggests that patients with functional iron deficiency and colon cancer may develop alterations in the immunosurveillance mechanisms and immune cell differentiation processes, which ultimately lead to unfavorable responses to treatment and a poorer prognosis for these patients.

The main strength of the present study was the fact that complete sets of data were available for all cases included in the analysis. The primary limitation of the study was the sample size. Therefore, in assessing the correlations between absolute/functional iron deficiency and tumor-specific pathological factors, the small sample size did not allow us to draw firm conclusions on associations. Despite this fact, our findings suggested that lymphatic invasion seems to be independently associated with functional iron deficiency. Further studies based on a larger number of cases are required in order to further clarify these issues.

5. Conclusions

Iron deficiency was highly prevalent in our cohort of patients diagnosed with CRC. Anemia was more frequently encountered in patients with right-sided tumor locations. Functional iron deficiency was significantly associated with higher odds for lymphatic invasion. A large proportion of patients with a normal hemoglobin level were also found to be iron deficient. Further studies on functional and absolute iron deficiency are required in order to more accurately assess possible correlations with colorectal cancer.

Author Contributions: Conceptualization, V.F. and D.A.; methodology, V.F.; software, V.F.; validation, R.I.S., V.V.B., G.C.D., and C.I.C.; formal analysis, V.F.; investigation, V.F. and D.A.; resources, V.F.; data curation, V.F., D.A., and G.A.; writing—original draft preparation, V.F.; writing—review and editing, V.F. and G.A.; visualization, C.I.C.; supervision, R.I.S., V.V.B., and G.C.D.; project administration, C.I.C. All authors have read and agreed to the published version of the manuscript.

Funding: This research received no external funding.

Institutional Review Board Statement: The study was approved by the Ethics Committee of the "Iuliu Hatieganu" University of Medicine and Pharmacy, Cluj Napoca, Romania (approval code: DEP68 on 11 March 2022).

Informed Consent Statement: Informed consent was obtained from all subjects involved in the study.

Data Availability Statement: Data supporting the reported results can be obtained via email from the corresponding author.

Acknowledgments: This article was a part of an "Iuliu Hatieganu" University of Medicine and Pharmacy, Cluj Napoca, research grant (4626/01.10.2021).

Conflicts of Interest: The authors declare no conflict of interest.

References

1. Sung, H.; Ferlay, J.; Siegel, R.L.; Laversanne, M.; Soerjomataram, I.; Jemal, A.; Bray, F. Global Cancer Statistics 2020: GLOBOCAN Estimates of Incidence and Mortality Worldwide for 36 Cancers in 185 Countries. *CA Cancer J. Clin.* **2021**, *71*, 209–249. [CrossRef]
2. Xi, Y.; Xu, P. Global colorectal cancer burden in 2020 and projections to 2040. *Transl. Oncol.* **2021**, *14*, 101174. [CrossRef]
3. Levin, T.R.; Corley, D.A.; Jensen, C.D.; Schottinger, J.E.; Quinn, V.P.; Zauber, A.G.; Lee, J.K.; Zhao, W.K.; Udaltsova, N.; Ghai, N.R.; et al. Effects of Organized Colorectal Cancer Screening on Cancer Incidence and Mortality in a Large Community-Based Population. *Gastroenterology* **2018**, *155*, 1383–1391.e5. [CrossRef]
4. Dekker, E.; Tanis, P.J.; Vleugels, J.L.A.; Kasi, P.M.; Wallace, M.B. Colorectal cancer. *Lancet* **2019**, *394*, 1467–1480. [CrossRef]

5. Schmitt, M.; Greten, F.R. The inflammatory pathogenesis of colorectal cancer. *Nat. Rev. Immunol.* **2021**, *21*, 653–667. [CrossRef] [PubMed]
6. Phipps, O.; Brookes, M.J.; Al-Hassi, H.O. Iron deficiency, immunology, and colorectal cancer. *Nutr. Rev.* **2021**, *79*, 88–97. [CrossRef] [PubMed]
7. Aksan, A.; Farrag, K.; Aksan, S.; Schroeder, O.; Stein, J. Flipside of the Coin: Iron Deficiency and Colorectal Cancer. *Front. Immunol.* **2021**, *12*, 635899. [CrossRef] [PubMed]
8. Nelson, R.L. Dietary iron and colorectal cancer risk. *Free Radic. Biol. Med.* **1992**, *12*, 161–168. [CrossRef]
9. Ashmore, J.H.; Rogers, C.J.; Kelleher, S.L.; Lesko, S.M.; Hartman, T.J. Dietary Iron and Colorectal Cancer Risk: A Review of Human Population Studies. *Crit. Rev. Food Sci. Nutr.* **2016**, *56*, 1012–1020. [CrossRef] [PubMed]
10. Brown, R.A.M.; Richardson, K.L.; Kabir, T.D.; Trinder, D.; Ganss, R.; Leedman, P.J. Altered Iron Metabolism and Impact in Cancer Biology, Metastasis, and Immunology. *Front. Oncol.* **2020**, *10*, 476. [CrossRef] [PubMed]
11. Hara, A.; Sasazuki, S.; Inoue, M.; Iwasaki, M.; Shimazu, T.; Sawada, N.; Yamaji, T.; Takachi, R.; Tsugane, S.; Japan Public Health Center–Based Prospective Study Group. Zinc and heme iron intakes and risk of colorectal cancer: A population-based prospective cohort study in Japan. *Am. J. Clin. Nutr.* **2012**, *96*, 864–873. [CrossRef] [PubMed]
12. Toyokuni, S. Iron-induced carcinogenesis: The role of redox regulation. *Free Radic. Biol. Med.* **1996**, *20*, 553–566. [CrossRef]
13. Stone, H.; Almilaji, O.; John, C.; Smith, C.; Surgenor, S.L.; Ayres, L.; Williams, E.J.; Snook, J. The dedicated iron deficiency anaemia clinic: A 15-year experience. *Front. Gastroenterol.* **2022**, *13*, 20–24. [CrossRef] [PubMed]
14. Almilaji, O.; Parry, S.D.; Docherty, S.; Snook, J. Evidence for improved prognosis of colorectal cancer diagnosed following the detection of iron deficiency anaemia. *Sci. Rep.* **2021**, *11*, 13055. [CrossRef] [PubMed]
15. D'Angelo, G. Role of hepcidin in the pathophysiology and diagnosis of anemia. *Blood Res.* **2013**, *48*, 10–15. [CrossRef] [PubMed]
16. Eleftheriadis, T.; Liakopoulos, V.; Antoniadi, G.; Stefanidis, I. Which is the best way for estimating transferrin saturation? *Ren. Fail.* **2010**, *32*, 1022–1023. [CrossRef] [PubMed]
17. Cappellini, M.D.; Motta, I. Anemia in Clinical Practice-Definition and Classification: Does Hemoglobin Change With Aging? *Semin. Hematol.* **2015**, *52*, 261–269. [CrossRef] [PubMed]
18. Ploug, M.; Kroijer, R.; Qvist, N.; Lindahl, C.H.; Knudsen, T. Iron deficiency in colorectal cancer patients: A cohort study on prevalence and associations. *Color. Dis.* **2020**, *23*, 853–859. [CrossRef] [PubMed]
19. Gafter-Gvili, A.; Schechter, A.; Rozen-Zvi, B. Iron Deficiency Anemia in Chronic Kidney Disease. *Acta Haematol.* **2019**, *142*, 44–50. [CrossRef] [PubMed]
20. Grotto, H.Z. Anaemia of cancer: An overview of mechanisms involved in its pathogenesis. *Med. Oncol.* **2008**, *25*, 12–21. [CrossRef] [PubMed]
21. Busti, F.; Marchi, G.; Ugolini, S.; Castagna, A.; Girelli, D. Anemia and iron deficiency in cancer patients: Role of iron replacement therapy. *Pharmaceuticals* **2018**, *11*, 94. [CrossRef] [PubMed]
22. Ludwig, H.; Evstatiev, R.; Kornek, G.; Aapro, M.; Bauernhofer, T.; Buxhofer-Ausch, V.; Fridrik, M.; Geissler, D.; Geissler, K.; Gisslinger, H.; et al. Iron metabolism and iron supplementation in cancer patients. *Wien. Klin. Wochenschr.* **2015**, *127*, 907–919. [CrossRef] [PubMed]
23. Wilson, M.J.; Dekker, J.W.T.; Harlaar, J.J.; Jeekel, J.; Schipperus, M.; Zwaginga, J.J. The role of preoperative iron deficiency in colorectal cancer patients: Prevalence and treatment. *Int. J. Color. Dis.* **2017**, *32*, 1617–1624. [CrossRef]
24. Beale, A.L.; Penney, M.D.; Allison, M.C. The prevalence of iron deficiency among patients presenting with colorectal cancer. *Colorectal Dis.* **2005**, *7*, 398–402. [CrossRef]
25. Sadahiro, S.; Suzuki, T.; Tokunaga, N.; Mukai, M.; Tajima, T.; Makuuchi, H.; Saito, T. Anemia in patients with colorectal cancer. *Int. J. Color. Dis.* **1998**, *33*, 488–494. [CrossRef]
26. Väyrynen, J.P.; Tuomisto, A.; Väyrynen, S.A.; Klintrup, K.; Karhu, T.; Mäkelä, J.; Herzig, K.-H.; Karttunen, T.J.; Mäkinen, M.J. Preoperative anemia in colorectal cancer: Relationships with tumor characteristics, systemic inflammation, and survival. *Sci. Rep.* **2018**, *8*, 1126. [CrossRef] [PubMed]
27. Almilaji, O.; Parry, S.; Docherty, S.; Snook, J. Colorectal Cancer and the Blood Loss Paradox. *Front. Gastroenterol.* **2022**, *13*, 381–385. [CrossRef]
28. Ludwig, H.; Müldür, E.; Endler, G.; Hübl, W. Prevalence of iron deficiency across different tumors and its association with poor performance status, disease status and anemia. *Ann. Oncol.* **2013**, *24*, 1886–1892. [CrossRef] [PubMed]
29. Hashemi, S.M.; Mashhadi, M.A.; Mohammadi, M.; Ebrahimi, M.; Allahyari, A. Absolute and Functional Iron Deficiency Anemia among Different Tumors in Cancer Patients in South Part of Iran, 2014. *Int. J. Hematol. Stem Cell Res.* **2017**, *11*, 192–198.
30. Gvirtzman, R.; Livovsky, D.M.; Tahover, E.; Goldin, E.; Koslowsky, B. Anemia can predict the prognosis of colorectal cancer in the pre-operative stage: A retrospective analysis. *World J. Surg. Oncol.* **2021**, *19*, 341. [CrossRef] [PubMed]

medicina

MDPI

Review

What Is Different in Acute Hematologic Malignancy-Associated ARDS? An Overview of the Literature

Mihail Cotorogea-Simion [1,†], Bogdan Pavel [2,†], Sebastian Isac [1,2,*], Teodora Telecan [3,4], Irina-Mihaela Matache [2], Anca Bobirca [5], Florin-Teodor Bobirca [6], Razvan Rababoc [7] and Gabriela Droc [1]

1 Department of Anesthesiology and Intensive Care I, Fundeni Clinical Institute, 022328 Bucharest, Romania
2 Department of Physiology, Faculty of Medicine, Carol Davila University of Medicine and Pharmacy, 020021 Bucharest, Romania
3 Department of Urology, Iuliu Hatieganu University of Medicine and Pharmacy, 400012 Cluj-Napoca, Romania
4 Department of Urology, Municipal Hospital, 400139 Cluj-Napoca, Romania
5 Department of Rheumatology, Dr. I. Cantacuzino Hospital, 073206 Bucharest, Romania
6 Department of General Surgery, Dr. I. Cantacuzino Hospital, 073206 Bucharest, Romania
7 Department of Internal Medicine II, Fundeni Clinical Institute, 022328 Bucharest, Romania
* Correspondence: sebastian.isac@umfcd.ro
† These authors contributed equally to this work.

Abstract: *Background and Objectives*: Acute hematologic malignancies are a group of heterogeneous blood diseases with a high mortality rate, mostly due to acute respiratory failure (ARF). Acute respiratory distress syndrome (ARDS) is one form of ARF which represents a challenging clinical condition. The paper aims to review current knowledge regarding the variable pathogenic mechanisms, as well as therapeutic options for ARDS in acute hematologic malignancy patients. *Data collection*: We provide an overview of ARDS in patients with acute hematologic malignancy, from an etiologic perspective. We searched databases such as PubMed or Google Scholar, including articles published until June 2022, using the following keywords: ARDS in hematologic malignancy, pneumonia in hematologic malignancy, drug-induced ARDS, leukostasis, pulmonary leukemic infiltration, pulmonary lysis syndrome, engraftment syndrome, diffuse alveolar hemorrhage, TRALI in hematologic malignancy, hematopoietic stem cell transplant ARDS, radiation pneumonitis. We included relevant research articles, case reports, and reviews published in the last 18 years. *Results:* The main causes of ARDS in acute hematologic malignancy are: pneumonia-associated ARDS, leukostasis, leukemic infiltration of the lung, pulmonary lysis syndrome, drug-induced ARDS, radiotherapy-induced ARDS, diffuse alveolar hemorrhage, peri-engraftment respiratory distress syndrome, hematopoietic stem cell transplantation-related ARDS, transfusion-related acute lung injury. *Conclusions:* The short-term prognosis of ARDS in acute hematologic malignancy relies on prompt diagnosis and treatment. Due to its etiological heterogeneity, precision-based strategies should be used to improve overall survival. Future studies should focus on identifying the relevance of such etiologic-based diagnostic strategies in ARDS secondary to acute hematologic malignancy.

Keywords: acute hematologic malignancy; acute respiratory distress syndrome; pneumonia; leukostasis; diffuse alveolar hemorrhage; engraftment syndrome; radiation recall pneumonitis; transfusion-related acute lung injury

Citation: Cotorogea-Simion, M.; Pavel, B.; Isac, S.; Telecan, T.; Matache, I.-M.; Bobirca, A.; Bobirca, F.-T.; Rababoc, R.; Droc, G. What Is Different in Acute Hematologic Malignancy-Associated ARDS? An Overview of the Literature. *Medicina* **2022**, *58*, 1215. https://doi.org/10.3390/medicina58091215

Academic Editors: Nicolae Crisan and Călin Căinap

Received: 27 July 2022
Accepted: 31 August 2022
Published: 3 September 2022

1. Introduction

Hematologic malignancies are a diverse group of pathologies, which affect an estimated 548.8 per 100,000 people, comprising roughly 20% of all types of neoplasia [1,2]. They can be divided, according to the most common subtypes, into leukemias, Hodgkin's lymphomas, non-Hodgkin's lymphomas, multiple myeloma, myelodysplastic syndromes, and myeloproliferative disorders, each with its origin, pathogenic mechanisms, incidence, burden on health, and mortality. In 2018, leukemia had an incidence of 407,000 cases and claimed 309,000 lives [3]. There is a marked heterogeneity even within this subpopulation. For example, the age-standardized incidence rate (ASIR) of leukemia as a whole decreased by approximately 9% between 1990 and 2017, mostly due to reductions in acute lymphocytic leukemia and chronic myeloid leukemia, whereas the values for chronic lymphocytic leukemia and acute myeloid leukemia increased [3].

The therapeutic options for hematologic malignancies can be divided into chemotherapy, radiotherapy, immunotherapy, targeted therapy, and hematopoietic stem cell transplantation (HSCT) [4–6]. Immunotherapy uses compounds that augment the natural defense capabilities of the organism, such as increasing interferon production or enhancing antibody-dependent cellular cytotoxicity [5]. Targeted therapy refers to drugs which disrupt the metabolic pathways specific to the cancer cell and its genetic abnormalities, discriminately targeting rapidly dividing cells [4].

The most common cause for ICU admission among patients with hematologic malignancies is acute respiratory failure (ARF), which also accounts for most non-relapse deaths in such cases [2]. Furthermore, acute hematologic malignancies are particularly prone to developing respiratory complications [1]. According to the literature, the main causes of ARF could be related to the injury of the lung parenchyma, pneumonia with opportunistic agents, chemo- or radiotherapy-induced ARF, transfusion-related acute lung injury, leukemic infiltration of the lung, tumor lysis pneumopathy, pulmonary alveolar proteinosis, and diffuse alveolar hemorrhage or engraftment syndrome. Other causes include chest wall and pleura damage, thromboembolic events, or neuromuscular origin (paraneoplastic syndromes, metabolic encephalopathy, sedative-induced) [2].

Acute respiratory distress syndrome (ARDS) is a particularly relevant entity, due to its high mortality. ARDS is defined, according to the Berlin criteria, by the presence of: (i) hypoxemia ($PaO_2/FiO_2 \leq 300$ mmHg with PEEP or CPAP ≥ 5 cm H_2O), (ii) respiratory failure not completely attributable to cardiac failure or fluid overload, (iii) onset within one week of exposure to a known risk factor, and (iv) bilateral opacities on chest X-ray or CT scan which cannot be fully explained by atelectasis, nodules, or pleural effusions [7]. In patients with hematologic malignancy, ARDS can result from either direct or indirect lung injury. Both mechanisms follow, however, a similar pattern. First, damage to pulmonary epithelial and endothelial cells leads to inflammation, apoptosis, and necrosis of alveolar type I and II cells, and to increased permeability of the alveolar-capillary membrane, causing fluid exudation and hemorrhage in the alveoli [7,8]. Then, during the proliferative phase, the barrier's integrity is reestablished, and the excess fluid is reabsorbed [7,8]. Lastly, as an optional occurrence, fibrotic tissue begins to appear in the lung, with unfavorable consequences in terms of mortality or recovery (Figure 1) [7,8].

ARDS management is largely supportive [7]. Most of the ARDS therapies target optimizing mechanical ventilation and obtaining an appropriate end-expiratory pressure, to prevent alveolar atelectrauma and additional fluid buildup in the alveoli, while simultaneously keeping plateau pressure at acceptable levels, to avoid barotrauma [7]. Maintaining a minimal change in flow variability between respiratory cycles could prevent ergotrauma [9,10]. Regarding the positive end-expiratory pressure (PEEP), meta-analyses have come up with conflicting results in terms of the effectiveness of higher PEEP compared to lower values, with discernible benefits only in selected patient subgroups [11–13]. One therapeutic intervention widely accepted is ventilation using low tidal volumes (4–6 mL/kg predicted body weight), which appears to be effective regardless of the putative ARDS triggering factor, accompanied by reduced circulating interleukin-6 levels [7,11–13]. The use of prone positioning aims to improve matching between ventilation and perfusion by taking advantage of gravity and intrathoracic organ repositioning, with promising results in severe ARDS in terms of oxygenation, and reduced mortality [7,14,15]. Neuromuscular blocking agents administered for short periods also improve oxygenation, while reducing mortality, barotrauma, and overall requirement for mechanical ventilation in severe ARDS cases [16]. These drugs also reduce the level of circulating inflammatory mediators [7,17]. Other specialized interventions have presented some promising results and are subject to further research, such as inhaled prostaglandins or nitric oxide, CO_2 removal strategies such as CO_2 removal filters attached to dialysis machines, and venovenous extracorporeal membrane oxygenation [18–21].

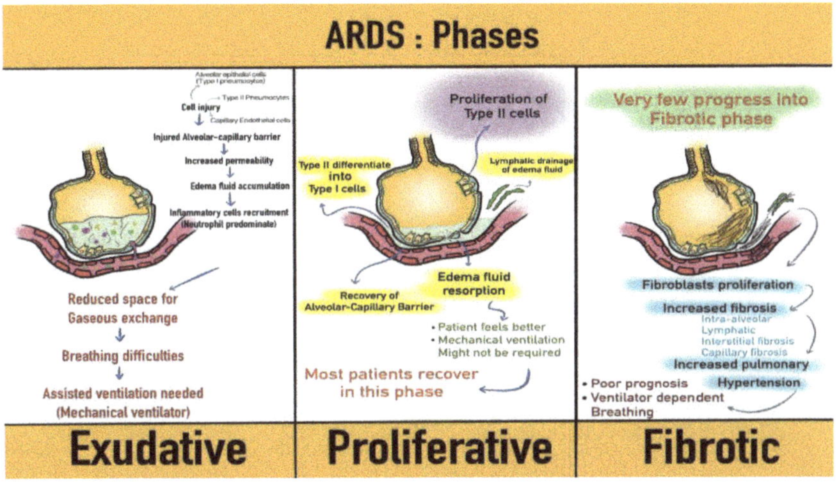

Figure 1. The progression of histopathological findings in acute respiratory distress syndrome (ARDS).

Studies attempting to prove the potential usefulness of targeted therapeutic measures in reducing mortality have largely come up with negative results [17,22–24]. With the dawn of personalized medicine, this has been attributed to the heterogeneity of ARDS, in terms of its clinical, physiologic, biochemical, and radiologic features, which also leads to variable responses to different therapeutic options, especially in patients with hematologic malignancy [25–27].

When referring to clinical classification, differences were noted when determining the cause of thoracic stiffness, with direct ARDS causing increased elastance in the lung parenchyma, whereas indirect ARDS exhibits higher elastance in the chest wall [26]. Within the broad group of indirect risk factors, higher mortality rates were noted in sepsis- vs. non-sepsis ARDS, further supporting the idea of a need to tailor the management to each individual case, instead of applying a cookie-cutter treatment to all patients [25]. A higher mortality rate was also noticed in cases of late onset ARDS (days 3–4 after the triggering injury), while early-onset patients exhibited higher plasma levels of biomarkers, suggestive alveolar–endothelial barrier disruption [26].

Biochemical inhomogeneity in ARDS patients stems from the blood and bronchoalveolar lavage fluid (BALF) profiles of various inflammatory endothelial and epithelial markers [26,28]. The main biochemical changes in ARDS are displayed in Figure 2.

The phenotypes most relevant and potentially useful for clinical practice are the so-called "hyperinflammatory" and "hypoinflammatory" phenotypes. The hyperinflammatory phenotype exhibits higher levels of circulating proinflammatory interleukins (IL-6, IL-8, IL-10), PAI-1, soluble receptors for TNF-α and advanced glycation end-products, and lower levels of plasma C protein and surfactant proteins in the BALF (Figure 2) [25,28]. This classification is relevant to clinical practice in terms of estimating patient mortality and therapeutic strategy. Thus, a study by Famous et al. found that hyperinflammatory phenotype cases benefit from a liberal fluid therapy strategy, while hypoinflammatory ARDS patients had a lower mortality when assigned to a fluid-conservative approach [29]. Additionally, treatment plans might be influenced by which phenotype the patient fits in. The HARP-2 trial observed that those with a hyperinflammatory phenotype had a better 28-day survival when receiving 80 mg Simvastatin [30]. The same could not be said about a 40 mg loading/20 mg maintenance dose of Rosuvastatin. This could come down to differences in drug bioavailability and intracellular concentrations [30]. Moreover, tailoring ARDS management according to the patients' inflammatory phenotype in hematologic malignancy is challenging due to the malignant cellular clones' abnormal response to inflammation.

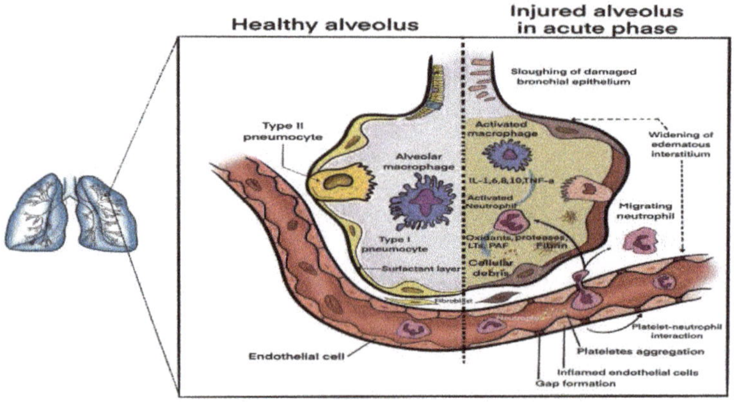

Figure 2. Biochemical changes in acute respiratory distress syndrome (ARDS)-damaged lung tissue.

From a radiological standpoint, ARDS can be divided into focal or non-focal [25]. Non-focal ARDS appears on thoracic CT scans as diffuse alveolar opacity, and is correlated with higher mortality, while focal ARDS have patches of parenchyma with loss of aeration, particularly in the lower lobes and dependent regions [25,26]. This dichotomy influences therapeutic management and expected mortality. In cases of focal ARDS, resorting to a low PEEP and normal tidal volume strategy, combined with prone positioning to optimize ventilation/perfusion ratio, might have beneficial effects [25,26]. This observation relies on the idea that using high PEEP in lungs with inhomogeneous compliance would lead to overdistension of already compliant areas, while the damaged zones would not benefit from the greater pressure [25,26]. On the other hand, non-focal ARDS benefits from recruitment maneuvers, meant to expand the collapsed alveoli, combined with higher PEEP to keep them open [25,26].

This review focuses on the main causes of ARDS in hematologic patients, considering the challenges of a precision-based approach in associated blood malignancy.

2. Data Collection

We used, for the narrative review, articles appearing in various databases such as PubMed or Google Scholar until June 2022. We used as key words ARDS in hematologic malignancy, pneumonia in hematologic malignancy, drug induced ARDS, leukostasis, pulmonary leukemic infiltration, pulmonary lysis syndrome, engraftment syndrome, diffuse alveolar hemorrhage, TRALI in hematologic malignancy, hematopoietic stem cell transplant ARDS, radiation pneumonitis, providing a total number of 11,261 articles. We included research articles, case reports and reviews referring to various respiratory complications in hematologic malignancy. We added relevant articles published in the last 19 years.

3. ARDS in Acute Hematologic Malignancy-Specific Causes

3.1. Pneumonia-Associated ARDS

Owing to the disease itself, as well as its therapeutic options, patients with acute hematologic malignancy find themselves immunosuppressed, particularly neutropenic (<500 neutrophiles/mm^3) [31]. Febrile neutropenia patients suffer most often from pulmonary complications, with a rate of 15–20% [31]. Pneumonia occurs in 13–31% of patients undergoing induction chemotherapy and up to 80% of those receiving hematopoietic stem cell transplants [31]. Other studies revealed that ARDS secondary to pneumonia is the main reason for ICU admission in cancer patients [32,33]. The study by Azoulay et al. also noted that, of all cases of pneumonia-induced ARDS, 58.16% were of bacterial origin and 32.18% were fungal, with 9.67% of infections caused by *Pneumocystis jirovecii* [32]. A review by Evans and Ost found that mortality among leukemia patients ranges between 25 and 80%, standing at 90% in the case of stem cell transplantation patients [33]. While most patients with hematologic malignancy tend to have frequent contact with hospitals, one must remember the possibility of community-acquired pneumonia in neutropenic individuals. Commonly incriminated bacterial agents are *Streptococcus pneumoniae* (especially in patients with dysfunctional humoral immunity) and *pyogenes*, *Staphylococcus aureus*, *Pseudomonas* spp., non-fermentative Gram-negative bacilli (including *Moraxella* and *Stenotrophomonas*), *meningococcus*, and atypical germs, such as *Mycoplasma*, *Chlamydophila*, and *Legionella* [2,33]. Influenza, parainfluenza, and adenoviruses represent the bulk of viral community-acquired pneumonia [2,33]. Nosocomial pneumonia (healthcare-associated, hospital-acquired, and ventilator-associated, with similar causative microbes) is most often produced by multidrug resistant bacteria, along with *Enterobacteriaceae*, *Nocardia* spp., and Mycobacteria—both *M. tuberculosis* and atypical [33]. Other etiological agents are fungi (mostly *Aspergillus* spp., with mucormycosis on the rise in recent years), *Pneumocystis*, and viruses (respiratory syncytial virus, metapneumovirus, varicella zoster and human herpes virus, cytomegalovirus, SARS-CoV-2) [32,33]. A secondary analysis of the Global Initiative for MRSA Pneumonia (GLIMP) database discovered that hematological malignancies were significantly more often associated with community-acquired pneumonia caused by fungi and non-influenza viruses [34].

Cytomegalovirus is a member of the herpesvirus family, and its prevalence in the general population exceeds 80% in Europe and North America, being close to omnipresent in Africa and Asia, owing to its multiple transmission paths (blood, saliva, breast milk, sexual contact) and the persistent infection it causes [35]. Cytomegalovirus uses macrophages and CD34+ cells, which include hematopoietic stem cells as reservoirs [35,36]. CD34+ cells include those cells used in stem cell transplant procedures [36]. Infection reactivation and progression to clinical disease depends on the proper reconstitution of the various T-cell subtypes following HSCT [36]. Unfortunately, the fact that the process is dependent on proper thymus function, which is impaired in hematologic malignancies, means that physiological proportions of T-cell subtypes (mainly the CD4:CD8 ratio) cannot be reached, leading to compromised anti-cytomegalovirus protection [36].

Aspergillus can disseminate in the lungs either through the blood vessels, causing infarction in the surrounding tissues, or by way of the airways, being frequently incriminated in patients with hematological malignancies [37]. However, the prevalence of infection does not strictly correlate with the depth of the immune dysfunction, which suggests that a genetic component might also be involved [38]. Multiple studies have investigated the effects which the single-nucleotide polymorphisms (SNP) in genes coding for components of the immune system have on the risk of developing invasive pulmonary aspergillosis. While many of the investigated polymorphisms had no statistically significant impact on invasive aspergillosis rates in the investigated stem cell transplant recipients, there were some which had deleterious consequences on the host's defensive capabilities, such as: (1) Toll-like receptor (TLR) 1, 3, 4, 5, and 6; (2) IL-4 receptor; (3) Plasminogen; (4) Vascular endothelial growth factor (VEGF); and (5) IL-8 [38–44]. TLR-4 and 5 were unexpected results, since they bind lipopolysaccharides, usually found in bacterial walls, and of which Aspergillus has none [38]. The IL-4 receptor polymorphism (rs2107356) has also been associated with multiple myeloma, gastric cancer, thymoma-associated myasthenia gravis, and graft dysfunction after kidney transplantation [45,46]. Plasminogen bound to Aspergillus active conidia, potentially facilitating its entry in the organism and enhancing tissue damage [42]. The relationship between angiogenesis and aspergillosis appeared to be bidirectional—VEGF inhibition increases the susceptibility to the disease, with the fungus producing toxins capable of inhibiting angiogenesis [43,44]. However, the products of some

SNPs had favorable interactions with Aspergillus, presenting protective effects against the infection. SNP-induced interferon-γ overproduction enhances fungicidal capabilities in macrophages [39]. IL-23 receptor hinders Aspergillus clearance by neutrophils and leads to chronic inflammation due to the IL-23/Th17 pathway [47]. IL-17 promotes fungal germination through its inhibition of indoleamine 2,3-dioxygenase, which also plays a role in subverting the function of T-helper type 1 cells [40,47]. T-helper type 1 cells are responsible for the production of interferon-γ. Thus, decreasing IL-17 production through the mutant IL-23 receptor leads to a lower likelihood of invasion by *Aspergillus*, as well as to protection against graft-versus-host disease (GVHD) in stem cell transplant recipients [40,47,48]. Lower levels of the anti-inflammatory IL-10 lead to resistance to invasive aspergillosis [49]. Finally, IL-12 has also a protective effect, by involving the T-helper type 1 response [49].

Over the past 2 years, SARS-CoV-2 has proven to be a serious challenge to healthcare systems worldwide, having infected over 270 million people and claiming 5.3 million lives worldwide by the end of 2021 [50]. Thus, it became unavoidable that some of the hematologic malignancy patients would contract it as well. A small observational study in China comparing COVID-19 rates among healthcare providers and hospitalized patients with hematological malignancy found no difference. However, ARDS rates were higher, as was mortality, among patients with hematologic neoplasia [51]. This finding is in accordance with the "cytokine storm" theory of COVID-19 pathogenesis, wherein the virus potently activates T-helper type 1 lymphocytes, leading to increased production of IL-6, one of the inflammatory markers incriminated in the onset of ARDS [52]. However, Suárez-Garcia et al. have shown that immunosuppression tends to be paradoxically associated with worse outcomes in COVID-19 pneumonia (ARDS, ICU admission, death) [53]. When compared to other viruses, such as influenza, SARS-CoV-2 appears to be more capable of inducing potentially fatal severe inflammatory responses in hematological malignancy patients [54].

Pneumonia-induced ARDS management includes, beyond standard ARDS supportive measures, antimicrobial therapy with the goal of eradicating the causative agent. The most recent ECIL guidelines recommend an escalation/de-escalation empirical approach. The choice of antimicrobials should be based on the risk of the patient having contracted a resistant germ, as well as the resistance profile of the commonly encountered microbes in the local healthcare setting [55]. For patients without prior resistant pathogen infection or risk of complicated disease evolution, empiric therapy consists of initial monotherapy (piperacillin/tazobactam 4/0.5 g IV q6h or ceftazidime 2 g IV q8h or cefepime 2 g IV q8h or q12h), for a duration of 7–8 days [55,56]. Should this prove ineffective, with deteriorating patient status or proven microbial resistance, the regimen should be changed. The recommended antibiotic regimen must provide broader coverage: carbapenems (meropenem 1 g IV q8h or imipenem 500 mg IV q6h) or antipseudomonal beta-lactams combined with aminoglycosides (e.g., ceftazidime 2 g IV q8h + amikacin 20 mg/kg IV q24h) or with colistin (e.g., ceftazidime 2 g IV q8h + colistin 9 million UI IV loading dose, then 3 million UI IV q8h) [55,56]. Vancomycin should be added if the hospital reports high rates of methicillin-resistant *S. aureus* [55,56]. De-escalation strategies follow the inverse steps and are recommended when bacteriological results are available, in case the patient had been infected or colonized with resistant germs, exhibits signs of unfavorable evolution (hypotension, shock), or if the healthcare center often deals with multidrug-resistant germs [55]. For de-escalation, the course of antibiotic treatment lasts for 14 days [55]. Should the patient suffer from *Legionella* pneumonia, treatment choice is 500 mg levofloxacin IV q12h for 21 days [2,56].

In case of viral pneumonia, etiological treatment is available for influenza viruses, adenoviruses, respiratory syncytial virus, and cytomegalovirus. Influenza virus often responds to oseltamivir 75 mg p.o. q12h or, alternatively, zanamivir [2]. For adenovirus, the treatment of choice is cidofovir in varying reported doses, mostly 5 mg/kg IV weekly for 3 weeks, or 1 mg/kg IV 3 times weekly for 3 weeks in patients with underlying renal dysfunction [57]. For respiratory syncytial virus, available treatment options include ribavirin (aerosols 2 g over 2 h q8h, or 6 g over 18 h, 7–10 days) and RSV immunoglobulins [2]. For cytomegalovirus, ganciclovir 5 mg/kg IV q12h or foscarnet 60 mg/kg IV q8h are recommended [2].

For fungal lung infections, the regimens vary based on the causative agent. Invasive pulmonary aspergillosis requires voriconazole 6 mg/kg i.v. on day 1, then 4 mg/kg q12h (p.o. if renal dysfunction) or amphotericin B (1 mg/kg qd if deoxycholate or 3 mg/kg qd if liposomal, depending on kidney function), in case of no response to voriconazole [2,58]. Mucormycosis exhibits reduced susceptibility to voriconazole [2]. Amphotericin B 5–10 mg/kg qd should be administered until resolution of either clinical or radiologic features, or of the immunosuppression [2]. Trimethoprim-sulfamethoxazole 5 mg/kg i.v. or p.o. (nearly 100% bioavailability) q8h or 3.75 mg/kg q6h for 21 days is the treatment of choice for *Pneumocystis jirovecii* [2]. Alternatively, clindamycin/primaquine or pentamidine can be used for severe cases [2,58]. Additionally, patients infected with *P. jirovecii* are prone to developing pneumothorax and pneumomediastinum, which is challenging for mechanically ventilated patients [59,60].

3.2. Leukostasis, Leukemic Infiltration of the Lung, Pulmonary Lysis Syndrome

Leukostasis is mostly a complication of myeloid leukemias (acute myelomonocytic or monocytic, and chronic during the blast crisis), especially those with leukocyte counts over 50,000/mm^3, although the degree of hyperleukocytosis does not necessarily correlate with the severity of the symptoms [61,62]. Leukostasis consists of white blood cell build-up inside small vessels, not only in the lungs, but also brain and heart, among other places, which explains the symptoms associated with this condition (acute respiratory failure, acute myocardial infarction, right ventricular overload, headaches, dizziness, tinnitus, coma, intracranial bleeding, peripheral ischemia, mesenteric infarction, priapism etc.) [61,63,64]. Leukostasis occurs not just due to increased viscosity and low flow in the pulmonary circulation, but also because of cytokines (mostly IL-1 and TNF-α) released by the pathologic cells [61,64]. The cytokine buildup leads to increased expression of adhesion molecules on endothelial cells (such as selectins and ICAM-1), leukocyte aggregation and activation, and secretion of matrix metalloproteinases, causing endothelial damage, increased vascular permeability, and subsequent extravasation of fluid, blood, and leukemic cells [61,64,65]. This migration from the intravascular to the interstitial and alveolar spaces is the basis for the radiologic opacities and the hypoxic respiratory failure that constitute hallmarks of ARDS [64]. However, the aforementioned hypoxemia has yet another causative mechanism—the occlusion of pulmonary capillary vessels, mimicking a pulmonary embolism, which explains how patients with histologically diagnosed leukostasis can be hypoxemic, yet show no abnormalities on chest X-rays [64].

Leukemic infiltration of the lung entails blasts building up in the pulmonary extravascular space, without any other discernible causes (infectious, hydrostatic, or otherwise) [66]. Leukostasis causes migration of leukemic cells into the interstitium. Thus, leukostasis and leukemic pulmonary infiltration are not two separate entities, but rather two sides of the same coin [64]. The two seem to occur more often in myeloid leukemia patients, but infiltration, unlike leukostasis, is even less correlated with hyperleukocytosis. However, it should be suspected in patients with a blast ratio exceeding 40% of total peripheral blood leukocytes [61,67]. The symptomatology is rather sparse, with the patient usually complaining about cough, fever, and dyspnea [61]. Imagistic findings include thickening of the interlobular septa or the bronchovascular bundles, as well as non-systematized "ground-glass" opacities [61,64,66].

Management of ARDS in the case of leukostasis and leukemic infiltration includes, beyond general supportive measures, therapies meant to reduce blood viscosity and deplete the number of leukocytes in the pulmonary vasculature and parenchyma [61]. Thus, patients should receive generous infusions of isotonic intravenous solutions, while avoiding blood transfusions for as long as the patient's clinical status does not call for urgent action [61]. For leukodepletion, clinicians can resort to either chemotherapy, in the shape of hydroxyurea, or leukapheresis, a process in which the patient's blood is run through a device containing a centrifuge, which mechanically separates cellular elements from plasma, before being fed back into the blood vessels [61,68]. In case of rapid degradation of clinical status, leukapheresis should be the preferred option, as it leads to a quicker drop in leukocyte levels [68]. However, in promyelocytic leukemia, this procedure should not be used, for two reasons: due to the disseminated intravascular coagulation which contraindicates apheresis, and because of the usually lower than normal white blood cell count [68,69]. Leukapheresis does not influence the long-term survival [68,70–74].

Hydroxyurea has been used as the mainstay of leukodepletion therapy for many decades, as it effectively lowers the number of leukocytes, with a low occurrence of acute tumor lysis syndrome, albeit over a longer period of time (24–48 h after initiation of therapy) [61,75]. Some studies have proven its usefulness in preventing short-term mortality [61,68,76]. Hydroxyurea acts by inhibiting ribonucleotide reductase, preventing the synthesis of deoxyribonucleotides, and halting the cell cycle in the S phase [77].

Due to the involvement of cytokines in leukostasis, corticosteroids have proven themselves useful in reducing leukemic pulmonary infiltration, mortality, and relapse incidence, as well as improving overall and disease-free survival [62,78]. Corticoids act by binding to cytoplasmic glucocorticoid receptors, which then relocate to the nucleus, exerting effects predominantly by induction of gene transcription (selective acetylation of histones) and increasing the expression of anti-inflammatory products such as IL-10 and IκB-α (inhibitor of NFκB) [79,80]. Conversely, they also bind and inhibit other proteins that act as histone acetyltransferases and activators, but for proinflammatory genes, thus switching them off [79]. Another mechanism relies on destabilization of mRNA molecules that encode for inflammatory proteins [77].

Pulmonary lysis syndrome, also known as acute lysis pneumopathy, occurs after initiation of cytostatic treatment [61,67]. However, it does not owe its effects to its direct mechanism of action, but rather to the massive destruction of tumor cells, which release their cytotoxic contents, producing diffuse alveolar damage or lung hemorrhage [61,67]. The incriminated components include reactive oxygen species, enzymes, and damage-associated molecular patterns (cellular components such as DNA, histones, heat shock proteins, uric acid) [61,81]. White blood cell count appears inconsequential, as cases have been reported in patients with fewer than 50,000 leukocytes per mm^3 [61]. The clinical presentation is typical for acute respiratory failure, while imaging typically shows bilateral "ground-glass" opacities [74]. Manifestations usually appear within 48 h of induction of therapy, but exceptions were noted by studies by both Azoulay et al. and Kunitomo et al., at 15 and 14 days, respectively [82,83]. Management of pulmonary lysis syndrome-induced ARDS includes corticosteroids and supportive therapy. There has been some debate regarding the usefulness of chemotherapy cessation [2,61,74].

3.3. Drug-Induced ARDS

Drugs administered in the treatment of hematological malignancy could lead to ARDS not only through their intrinsic action towards the lung, but also by way of their interaction with the neoplastic cells. The incidence varies from 0.1 to 15% [84].

The pathophysiology of drug-induced ARDS is complex, with incriminated mechanisms ranging from idiosyncratic reactions to anaphylaxis, capillary leak syndrome, or reactive oxygen species and inflammatory cytokine production [85]. The relevant drugs incriminated in lung damage leading to ARDS are: bleomycin, mitomycin-C, cyclophosphamide, gemcitabine, cytarabine, GM-CSF, and vinca alkaloids [86–89]. Bleomycin and mitomycin-C increase reactive oxygen species production [86,87]. Gemcitabine increases cytokine release [86,87]. Cytarabine has a direct toxic effect [88]. GM-CSF increases neutrophil adhesion to lung endothelium, due to higher expression of glycoproteins, and superoxide production [88]. Vinca alkaloids cause endothelial dysfunction by disrupting the organization of tubulin [89].

Of particular interest is all-trans retinoic acid (ATRA), which is used in acute promyelocytic leukemia, where a chromosomal translocation leads to a change in the function of the retinoic acid receptor [90]. Consequently, the gene responsible for cell maturation and differentiation no longer responds to physiological ATRA doses [90]. Thus, the myeloid cells remain trapped in their promyelocyte stage [90]. Another drug used in acute promyelocytic leukemia is arsenic trioxide, which increases degradation of the mutant receptor in the lysosome [91]. The administration of either of these drugs could lead to retinoic acid syndrome, a particular type of drug-induced ARDS. This is an entity which appears in 2 to 31% of patients treated with such drugs, mostly when treatment consists of these alone, during the induction phase, usually 10 days after the initiation of the treatment [92]. When ATRA binds to the retinoic acid receptor, immature cells are forced to differentiate. This changes the profile of secreted cytokines (IL-1β, IL-6, IL-8), increasing expression of lymphocyte function-associated antigen 1 (a molecule involved in the migration of leukocytes), intercellular adhesion molecule 1, matrix metalloproteinase 9, and cathepsin G [93,94]. These changes increase the vascular permeability and facilitate lung infiltration [93,94]. Some of the cytokines are also involved in altering hemostasis [92,93]. Thus, retinoic acid syndrome-associated ARDS manifests itself as leukemic infiltration of the lung and alveolar hemorrhage. Management of this pathology consists of intravenous dexamethasone (10 mg i.v. q12h), stopping the administration of the incriminated drug, and the addition of a different cytostatic agent in cases of leukocytosis [93].

3.4. Radiotherapy-Induced ARDS

Management options for hematologic malignancies go beyond pharmacological means. Radiation therapy is also useful, especially in lymphomas, where irradiating affected lymph nodes in selected patients leads to excellent 5-year survival and relapse rates [95]. Its use also extends to leukemia, but more as prophylactic, post-chemotherapy, or palliative therapy [95]. However, body tissues are also susceptible to radiation damage, with the lungs being the most sensitive of the thoracic organs and radiation pneumonitis occurring at doses as low as 15–16 Gy [95]. The radiation-induced death of endo- and epithelial cells leads to a vicious circle of inflammation, increased vascular permeability, and cytokine release, while infiltrating macrophages amplify tissue damage by producing reactive oxygen and nitrogen species and cytokines [96]. The cytokine milieu varies with the time elapsed since the pulmonary injury [96]. The first 2 weeks are characterized by high levels of TNF-α, IL-1 and -6, fibroblastic and platelet-derived growth factors. On a tissular level, this stage is characterized by vascular congestion and intra-alveolar edema, leukocyte infiltration, and pneumocyte apoptosis [96]. In later stages (about 6–8 weeks after the original insult), TGF-β1 expression increases, while vascular and alveolar linings begin to detach, leading to capillary lumen reduction and thrombi formation, and to alveolar collapse with associated fibrin exudation and hyaline membranes, respectively [96].

Radiation pneumonitis could occur months, even years, after radiotherapy [97,98]. In such patients, the triggering factor was proven to be a round of chemotherapy, although cases have been reported where immunotherapy was incriminated instead [96,99]. The mechanisms involved in radiation recall pneumonitis are still being investigated. However, postulated theories include: (1) constant subliminal inflammatory cytokine secretion; (2) changes in local stem cell function, either increased turnover (which increases their susceptibility to antineoplastic agents) or reduced proliferation; (3) accumulation of the anticancer drug due to local changes in angiogenesis and vascular permeability [98,99]. The severity of symptoms does not appear to be correlated with the time elapsed between radio- and chemotherapy [96]. One could mistake radiation recall for chemotherapy-induced lung damage; however, in the case of radiation recall pneumonitis, the ground-glass opacities and infiltrates conform to the shape of the previously irradiated areas [98].

Treatment of ARDS induced by radiation therapy consists mainly of intravenous corticosteroids [96]. Furthermore, some prophylactic options exist, which dampen the effects radiation has on the lung tissue: pentoxifylline, with its TNF-α and IL-1 suppressing action leading to an improvement in symptoms, and amifostine, which acts by scavenging free radicals and by inducing tissular hypoxia, with protective effects [96,100].

3.5. Hematopoietic Stem Cell Transplantation-Related ARDS

While chemo- and radiotherapy have remission and symptom control as their goals, stem cell transplantation has been used with curative effects [101]. There are multiple types of transplantation, but the two most widely used are autologous and allogeneic.

Autologous transplant involves harvesting stem cells from the patient, either directly from the bone marrow or from the blood after marrow stimulation [102]. Then the patient undergoes myeloablative therapy, which destroys the malignant cells, along with their own hematopoietic cells, and has the harvested stem cells reimplanted, in hope that they would resume their function [102]. While considered a curative therapy option, relapse rates remain high, mostly due to the stem cell harvest contamination by neoplastic cells [103].

Allogenic HSCT requires a donor, related to the patient or not, with HLA antigen matching [104]. While the complications and non-relapse mortality rate of allogenic HSCT is worse than that of the autologous one, lower relapse rates offset the difference, leading to similar long-term survival [104]. The benefit of allogenic grafts is an immune reaction mediated by minor histocompatibility antigens, which prevents the subsequent growth of leukemic cells [105]. The minor histocompatibility antigen is usually expressed on cells belonging to the immune system, including the malignant ones [105]. However, the donor cells sometimes react with epithelial cells, which also express such antigens, leading to graft vs. host disease [105]. GVHD occurs due to pre-existent damage to the host tissues, through the underlying disease or the preconditioning chemotherapy, which leads to an elevated state of inflammation in the body, culminating in ARDS [106]. Of note is the occurrence of GVHD in autologous stem cell transplant recipients, in spite of the complete cellular antigen matching [107,108]. The putative mechanism is the loss of self-reactive cell suppression, either through direct regulatory T cell expression inhibition (caused by specific agents, such as thalidomide derivatives), or through poor thymus function owing to cytotoxic therapy [107].

Other mechanisms related to hematopoietic SCT, which led to ARDS, are diffuse alveolar hemorrhage, peri-engraftment respiratory distress syndrome (PERDS), and cryptogenic organizing pneumonia [37].

Diffuse alveolar hemorrhage is an exclusion diagnosis, being defined as lung hemorrhage-induced ARDS in the absence of any infection within 1 week after hematopoietic stem cell transplantation [107]. The diffuse alveolar hemorrhage can last between 1 week and 1 month, during the engraftment period, when neutrophil production increases, causing them to flow towards the pulmonary vasculature [37]. To establish the diagnosis, bronchoalveolar lavages must be performed [109]. The bronchoalveolar lavage must appear increasingly bloody as time passes or contain macrophages which are loaded with hemosiderin in proportion higher than 20% [109]. The initial pulmonary lesion is caused by high-dose radiotherapy, releasing host antigens into the circulation, which are then recognized by donor T cells, in the case of allogeneic stem cell transplants, leading to their activation and inflammatory cytokine production [109].

Diffuse alveolar hemorrhage can also occur in autologous transplant recipients [107,109]. The T cells might be activated by compounds such as lipopolysaccharides, which end up in the bloodstream following gastrointestinal epithelium damage, as is the case in mucositis (caused by melphalan, a drug used in the treatment of multiple myeloma) or GVHD [107,109]. GVHD leads to alveolitis, manifesting as alveolar hemorrhage and increased counts of alveolar leukocytes, regardless of post-transplantation leukopenia [110]. Consequently, the endothelial swelling and medial hyperplasia leads to narrower vessel lumina, increasing the extravasation of erythrocytes into the lung parenchyma [110]. Outcomes for diffuse alveolar hemorrhage appear remarkably poor, with reported mortality rates between 64 and 100% [109,110]. Supportive therapy includes platelet transfusions, clotting factor (recombinant factor VIIa) or antifibrinolytic drug intake, and ventilatory support, while corticosteroids are largely unhelpful [109,110]. The recombinant factor VIIa, particularly when administered locally, through bronchoscopy, overcomes any preexisting tissue factor pathway inhibitors and leads to bleeding control [110].

Engraftment syndrome (ES) occurs at a reported rate of 7–90% in patients receiving HSCT, most often after autologous HSCTs [111]. ES occurs within 4 days of engraftment, which is defined as the first of 3 consecutive days when neutrophil levels maintain themselves at over 500/mm^3, and it is caused by the engrafted neutrophils' production of inflammatory cytokines and degranulation, leading to systemic endothelial damage [112,113]. A particular manifestation of ES is periengraftment respiratory distress syndrome, whose reported incidence rates vary from 2.5 to 25% [109]. The risk factors for PERDS are female gender, a quick immune function recovery, autologous HSCT, less intensive preconditioning therapy or GM-CSF instead of G-CSF, and the need of preengraftment platelet transfusions [109,114]. PERDS has a very similar clinical presentation to that of acute GVHD and, while self-limited, can be severe enough to warrant corticosteroid therapy (3 days of 1–2 mg/kg iv methylprednisolone q12h), along with supportive therapy [37,109].

3.6. TRALI in Patients with Acute Hematologic Malignancy

Since bone marrow suppression in hematological malignancies is either a consequence of the disease itself, or a desired effect of medication (as is the case in myeloablative therapy), cytopenia in at least one blood cell line occurs in almost all patients [115]. Once the levels of a particular component reach a critically low value, blood product transfusions should be performed [116,117]. The transfusion threshold for hemoglobin and platelets is 7 g hemoglobin/100 mL whole blood and 10×10^3 platelets/mm^3 whole blood, respectively (numbers apply in the absence of active bleeding) [116,117]. While their usefulness cannot be understated, transfusions are plagued by side effects, with transfusion-related acute lung injury (TRALI) being one of the most important issues when evaluating the prognosis of these patients.

TRALI is defined as pulmonary edema occurring within 6 h post-transfusion, in the absence of other ARDS-precipitating factors or evidence of circulatory overload [118]. It occurs in roughly 1 in every 1000 blood product recipients, with the incidence being 50 to 80 times higher in intensive care settings [118]. However, patients with hematologic malignancy develop this complication less frequently than other patient categories, most likely due to the associated neutropenia [118]. Mortality stands at approximately 10%, while mechanical ventilation requirement occurs in 70 to 90% of cases [119]. Risk factors associated with TRALI are the total number of administered blood products, previously pregnant donors, chronic alcohol and tobacco use, pretransfusion shock, and positive fluid balance [118,119]. The occurrence of TRALI can most often be attributed to the presence of leukocyte antibodies in the plasma contained in blood products [118]. The antibodies bind their corresponding recipient antigens and trigger an immune reaction culminating in resident neutrophil activation, capillary leak, and lung injury [118].

TRALI in neutropenic patients might occur due to antibodies binding directly to endothelial cells, which are then damaged by reactive oxygen species produced through the activation of the complement cascade or monocytes [118]. It has been proven that even platelets are capable of secreting proinflammatory mediators, while also migrating into alveolae, augmenting leukocytic infiltration [118]. Finally, erythrocyte transfusion bags might carry significant amounts of proinflammatory factors (reactive oxygen species, cytokines, etc.), which can trigger acute lung injury in the recipient [118,119].

TRALI is self-limited, with patients recovering within 3 to 4 days. In some cases, corticosteroids or diuretic therapy might prove useful [2]. In order to reduce the occurrence rate, mitigation strategies have been implemented: collecting plasma-rich products only from male donors or nulliparous females, antibody screening in female thrombocyte apheresis donors (since this product contains a significant amount of plasma), or pooling together plasma and platelets from multiple donors, to dilute or neutralize any residual antibodies [119].

4. Conclusions

ARDS in acute hematologic malignancy represents a real therapeutic challenge, mainly due to the etiological heterogeneity. Moreover, the short-term prognosis relies on prompt diagnosis and treatment. Further precision-based strategies aiming to overcome this heterogeneity should be developed, in the hope of aiding clinicians establishing a diagnosis more accurately and rapidly. Future studies should focus on identifying the relevance of such approaches in ARDS secondary to acute hematologic malignancy.

Author Contributions: Conceptualization, S.I. and M.C.-S.; methodology, M.C.-S. and B.P.; software, R.R., A.B.; validation, B.P., S.I. and G.D.; formal analysis, T.T., F.-T.B.; investigation, R.R. and T.T.; resources, M.C.-S. and B.P.; data curation, I.-M.M., A.B.; writing—original draft preparation, M.C.-S. and T.T.; writing—review and editing, B.P., S.I., G.D.; visualization, I.-M.M., F.-T.B.; supervision, S.I. and G.D.; project administration, G.D. All authors have read and agreed to the published version of the manuscript.

Funding: This research received no external funding.

Institutional Review Board Statement: Not applicable.

Informed Consent Statement: Not applicable.

Data Availability Statement: Not applicable.

Conflicts of Interest: The authors declare no conflict of interest.

References

1. Li, J.; Smith, A.; Crouch, S.; Oliver, S.; Roman, E. Estimating the Prevalence of Hematological Malignancies and Precursor Conditions Using Data from Haematological Malignancy Research Network (HMRN). *Cancer Causes Control* **2016**, *27*, 1019–1026. [CrossRef]
2. Vadde, R.; Pastores, S.M. Management of Acute Respiratory Failure in Patients with Hematological Malignancy. *J. Intensive Care Med.* **2016**, *31*, 627–641. [CrossRef] [PubMed]
3. Keykhaei, M.; Masinaei, M.; Mohammadi, E.; Azadnajafabad, S.; Rezaei, N.; Saeedi Moghaddam, S.; Rezaei, N.; Nasserinejad, M.; Abbasi-Kangevari, M.; Malekpour, M.-R.; et al. A Global, Regional, and National Survey on Burden and Quality of Care Index (QCI) of Hematologic Malignancies; Global Burden of Disease Systematic Analysis 1990–2017. *Exp. Hematol. Oncol.* **2021**, *10*, 11. [CrossRef] [PubMed]
4. Shukry, S.; Hariri, F.; Al-Nehmi, A.W. Target Therapy in Hematological Malignancies. In *Advances in Hematologic Malignancies*; Hamid, G.A., Ed.; IntechOpen: Rijeka, Croatia, 2019.
5. Wiernik, P.H. Treatment of Hematologic Neoplasms with New Immunomodulatory Drugs (IMiDs). *Curr. Treat. Options Oncol.* **2009**, *10*, 1–15. [CrossRef] [PubMed]
6. Tseng, Y.D.; Ng, A.K. Hematologic Malignancies. *Hematol. Oncol. Clin. N. Am.* **2020**, *34*, 127–142. [CrossRef]
7. Fanelli, V.; Vlachou, A.; Ghannadian, S.; Simonetti, U.; Slutsky, A.S.; Zhang, H. Acute Respiratory Distress Syndrome: New Definition, Current and Future Therapeutic Options. *J. Thorac. Dis.* **2013**, *5*, 326–334. [CrossRef]
8. Meyer, N.J.; Gattinoni, L.; Calfee, C.S. Acute Respiratory Distress Syndrome. *Lancet* **2021**, *398*, 622–637. [CrossRef]
9. Marini, J.J.; Crooke, P.S.; Tawfik, P.; Chatburn, R.L.; Dries, D.J.; Gattinoni, L. Intracycle Power and Ventilation Mode as Potential Contributors to Ventilator-Induced Lung Injury. *Intensive Care Med. Exp.* **2021**, *9*, 55. [CrossRef]
10. Gattinoni, L.; Tonetti, T.; Cressoni, M.; Cadringher, P.; Herrmann, P.; Moerer, O.; Protti, A.; Gotti, M.; Chiurazzi, C.; Carlesso, E.; et al. Ventilator-Related Causes of Lung Injury: The Mechanical Power. *Intensive Care Med.* **2016**, *42*, 1567–1575. [CrossRef]
11. Ball, L.; Serpa Neto, A.; Trifiletti, V.; Mandelli, M.; Firpo, I.; Robba, C.; Gama de Abreu, M.; Schultz, M.J.; Patroniti, N.; Rocco, P.R.M.; et al. Effects of Higher PEEP and Recruitment Manoeuvres on Mortality in Patients with ARDS: A Systematic Review, Meta-Analysis, Meta-Regression and Trial Sequential Analysis of Randomized Controlled Trials. *Intensive Care Med. Exp.* **2020**, *8*, 39. [CrossRef]
12. Guo, L.; Xie, J.; Huang, Y.; Pan, C.; Yang, Y.; Qiu, H.; Liu, L. Higher PEEP Improves Outcomes in ARDS Patients with Clinically Objective Positive Oxygenation Response to PEEP: A Systematic Review and Meta-Analysis. *BMC Anesthesiol.* **2018**, *18*, 172. [CrossRef] [PubMed]
13. Shao, S.; Kang, H.; Qian, Z.; Wang, Y.; Tong, Z. Effect of Different Levels of PEEP on Mortality in ICU Patients without Acute Respiratory Distress Syndrome: Systematic Review and Meta-Analysis with Trial Sequential Analysis. *J. Crit. Care* **2021**, *65*, 246–258. [CrossRef]

14. Mancebo, J.; Fernández, R.; Blanch, L.; Rialp, G.; Gordo, F.; Ferrer, M.; Rodríguez, F.; Garro, P.; Ricart, P.; Vallverdú, I.; et al. A Multicenter Trial of Prolonged Prone Ventilation in Severe Acute Respiratory Distress Syndrome. *Am. J. Respir. Crit. Care Med.* **2006**, *173*, 1233–1239. [CrossRef] [PubMed]
15. Sud, S.; Friedrich, J.O.; Taccone, P.; Polli, F.; Adhikari, N.K.J.; Latini, R.; Pesenti, A.; Guérin, C.; Mancebo, J.; Curley, M.A.Q.; et al. Prone Ventilation Reduces Mortality in Patients with Acute Respiratory Failure and Severe Hypoxemia: Systematic Review and Meta-Analysis. *Intensive Care Med.* **2010**, *36*, 585–599. [CrossRef] [PubMed]
16. Papazian, L.; Forel, J.-M.; Gacouin, A.; Penot-Ragon, C.; Perrin, G.; Loundou, A.; Jaber, S.; Arnal, J.-M.; Perez, D.; Seghboyan, J.-M.; et al. Neuromuscular Blockers in Early Acute Respiratory Distress Syndrome. *N. Engl. J. Med.* **2010**, *363*, 1107–1116. [CrossRef]
17. Duggal, A.; Ganapathy, A.; Ratnapalan, M.; Adhikari, N. Pharmacological Treatments for Acute Respiratory Distress Syndrome: Systematic Review. *Minerva Anestesiol.* **2015**, *81*, 567–588.
18. Parekh, M.; Abrams, D.; Brodie, D. Extracorporeal Techniques in Acute Respiratory Distress Syndrome. *Ann. Transl. Med.* **2017**, *5*, 296. [CrossRef]
19. Puri, N.; Dellinger, R.P. Inhaled Nitric Oxide and Inhaled Prostacyclin in Acute Respiratory Distress Syndrome: What Is the Evidence? *Crit. Care Clin.* **2011**, *27*, 561–587. [CrossRef]
20. Fanelli, V.; Costamagna, A.; Ranieri, V.M. Chapter 124—Extracorporeal Carbon Dioxide Removal. In *Critical Care Nephrology (Third Edition)*; Ronco, C., Bellomo, R., Kellum, J.A., Ricci, Z., Eds.; Elsevier: Philadelphia, PA, USA, 2019; pp. 755–759.e1. ISBN 978-0-323-44942-7.
21. Aokage, T.; Palmér, K.; Ichiba, S.; Takeda, S. Extracorporeal Membrane Oxygenation for Acute Respiratory Distress Syndrome. *J. Intensive Care* **2015**, *3*, 17. [CrossRef]
22. Spragg, R.G.; Lewis, J.F.; Walmrath, H.-D.; Johannigman, J.; Bellingan, G.; Laterre, P.-F.; Witte, M.C.; Richards, G.A.; Rippin, G.; Rathgeb, F.; et al. Effect of Recombinant Surfactant Protein C-Based Surfactant on the Acute Respiratory Distress Syndrome. *N. Engl. J. Med.* **2004**, *351*, 884–892. [CrossRef]
23. Spragg, R.G.; Taut, F.J.H.; Lewis, J.F.; Schenk, P.; Ruppert, C.; Dean, N.; Krell, K.; Karabinis, A.; Günther, A. Recombinant Surfactant Protein C-Based Surfactant for Patients with Severe Direct Lung Injury. *Am. J. Respir. Crit. Care Med.* **2011**, *183*, 1055–1061. [CrossRef] [PubMed]
24. Steinberg, K.P.; Hudson, L.D.; Goodman, R.B.; Hough, C.L.; Lanken, P.N.; Hyzy, R.; Thompson, B.T.; Ancukiewicz, M. Efficacy and Safety of Corticosteroids for Persistent Acute Respiratory Distress Syndrome. *N. Engl. J. Med.* **2006**, *354*, 1671–1684. [CrossRef] [PubMed]
25. Matthay, M.A.; Arabi, Y.M.; Siegel, E.R.; Ware, L.B.; Bos, L.D.J.; Sinha, P.; Beitler, J.R.; Wick, K.D.; Curley, M.A.Q.; Constantin, J.-M.; et al. Phenotypes and Personalized Medicine in the Acute Respiratory Distress Syndrome. *Intensive Care Med.* **2020**, *46*, 2136–2152. [CrossRef]
26. Sinha, P.; Calfee, C.S. Phenotypes in Acute Respiratory Distress Syndrome: Moving towards Precision Medicine. *Curr. Opin. Crit. Care* **2019**, *25*, 12–20. [CrossRef]
27. Wick, K.D.; McAuley, D.F.; Levitt, J.E.; Beitler, J.R.; Annane, D.; Riviello, E.D.; Calfee, C.S.; Matthay, M.A. Promises and Challenges of Personalized Medicine to Guide ARDS Therapy. *Crit. Care* **2021**, *25*, 404. [CrossRef]
28. Binnie, A.; Tsang, J.L.Y.; dos Santos, C.C. Biomarkers in Acute Respiratory Distress Syndrome. *Curr. Opin. Crit. Care* **2014**, *20*, 47–55. [CrossRef]
29. Famous, K.R.; Delucchi, K.; Ware, L.B.; Kangelaris, K.N.; Liu, K.D.; Thompson, B.T.; Calfee, C.S. Acute Respiratory Distress Syndrome Subphenotypes Respond Differently to Randomized Fluid Management Strategy. *Am. J. Respir. Crit. Care Med.* **2017**, *195*, 331–338. [CrossRef]
30. Heijnen, N.F.L.; Bergmans, D.C.J.J.; Schnabel, R.M.; Bos, L.D.J. Targeted Treatment of Acute Respiratory Distress Syndrome with Statins-a Commentary on Two Phenotype Stratified Re-Analysis of Randomized Controlled Trials. *J. Thorac. Dis.* **2019**, *11*, S296–S299. [CrossRef]
31. Mokart, D.; van Craenenbroeck, T.; Lambert, J.; Textoris, J.; Brun, J.-P.; Sannini, A.; Chow-Chine, L.; Hamouda, S.; Fouché, L.; Ettori, F.; et al. Prognosis of Acute Respiratory Distress Syndrome in Neutropenic Cancer Patients. *Eur. Respir. J.* **2012**, *40*, 169–176. [CrossRef]
32. Azoulay, E.; Lemiale, V.; Mokart, D.; Pène, F.; Kouatchet, A.; Perez, P.; Vincent, F.; Mayaux, J.; Benoit, D.; Bruneel, F.; et al. Acute Respiratory Distress Syndrome in Patients with Malignancies. *Intensive Care Med.* **2014**, *40*, 1106–1114. [CrossRef]
33. Evans, S.E.; Ost, D.E. Pneumonia in the Neutropenic Cancer Patient. *Curr. Opin. Pulm. Med.* **2015**, *21*, 260–271. [CrossRef] [PubMed]
34. Di Pasquale, M.F.; Sotgiu, G.; Gramegna, A.; Radovanovic, D.; Terraneo, S.; Reyes, L.F.; Rupp, J.; González del Castillo, J.; Blasi, F.; Aliberti, S.; et al. Prevalence and Etiology of Community-Acquired Pneumonia in Immunocompromised Patients. *Clin. Infect. Dis.* **2019**, *68*, 1482–1493. [CrossRef]
35. Al Mana, H.; Yassine, H.M.; Younes, N.N.; Al-Mohannadi, A.; Al-Sadeq, D.W.; Alhababi, D.; Nasser, E.A.; Nasrallah, G.K. The Current Status of Cytomegalovirus (CMV) Prevalence in the MENA Region: A Systematic Review. *Pathogens* **2019**, *8*, 213. [CrossRef]
36. Alonso-Álvarez, S.; Colado, E.; Moro-García, M.A.; Alonso-Arias, R. Cytomegalovirus in Haematological Tumours. *Front. Immunol.* **2021**, *12*, 703256. [CrossRef]

37. Wah, T.M.; Moss, H.A.; Robertson, R.J.H.; Barnard, D.L. Pulmonary Complications Following Bone Marrow Transplantation. *BJR* **2003**, *76*, 373–379. [CrossRef] [PubMed]
38. de Oliveira Cunha, D.; Leão-Cordeiro, J.A.B.; de Paula, H.d.C.; Ataides, F.S.; Saddi, V.A.; Vilanova-Costa, C.A.S.T.; Silva, A.M.T.C. Association between Polymorphisms in the Genes Encoding Toll-like Receptors and Dectin-1 and Susceptibility to Invasive Aspergillosis: A Systematic Review. *Rev. Soc. Bras. Med. Trop.* **2018**, *51*, 725–730. [CrossRef]
39. Lupiañez, C.B.; Canet, L.M.; Carvalho, A.; Alcazar-Fuoli, L.; Springer, J.; Lackner, M.; Segura-Catena, J.; Comino, A.; Olmedo, C.; Ríos, R.; et al. Polymorphisms in Host Immunity-Modulating Genes and Risk of Invasive Aspergillosis: Results from the AspBIOmics Consortium. *Infect. Immun.* **2015**, *84*, 643–657. [CrossRef]
40. Romani, L. Immunity to Fungal Infections. *Nat. Rev. Immunol.* **2011**, *11*, 275–288. [CrossRef]
41. Grube, M.; Loeffler, J.; Mezger, M.; Krüger, B.; Echtenacher, B.; Hoffmann, P.; Edinger, M.; Einsele, H.; Andreesen, R.; Holler, E. TLR5 Stop Codon Polymorphism Is Associated with Invasive Aspergillosis after Allogeneic Stem Cell Transplantation. *Med. Mycol.* **2013**, *51*, 818–825. [CrossRef]
42. Zaas, A.K.; Liao, G.; Chien, J.W.; Weinberg, C.; Shore, D.; Giles, S.S.; Marr, K.A.; Usuka, J.; Burch, L.H.; Perera, L.; et al. Plasminogen Alleles Influence Susceptibility to Invasive Aspergillosis. *PLoS Genet.* **2008**, *4*, e1000101. [CrossRef]
43. Smith, N.L.D.; Hankinson, J.; Simpson, A.; Denning, D.W.; Bowyer, P. Reduced Expression of TLR3, TLR10 and TREM1 by Human Macrophages in Chronic Cavitary Pulmonary Aspergillosis, and Novel Associations of VEGFA, DENND1B and PLAT. *Clin. Microbiol Infect.* **2014**, *20*, O960–O968. [CrossRef] [PubMed]
44. Ben-Ami, R.; Lewis, R.E.; Leventakos, K.; Kontoyiannis, D.P. Aspergillus Fumigatus Inhibits Angiogenesis through the Production of Gliotoxin and Other Secondary Metabolites. *Blood* **2009**, *114*, 5393–5399. [CrossRef]
45. Jiang, P.; Yue, Y.-X.; Hong, Y.; Xie, Y.; Gao, X.; Gu, C.-K.; Hao, H.-J.; Qin, Y.; Ding, X.-J.; Song, M.; et al. IL-4Rα Polymorphism Is Associated With Myasthenia Gravis in Chinese Han Population. *Front. Neurol.* **2018**, *9*, 529. [CrossRef] [PubMed]
46. Lee, H.J.; Kim, T.H.; Kang, S.W.; Kim, Y.H.; Kim, S.K.; Chung, J.-H.; Kim, Y.G.; Moon, J.Y.; Lee, S.H.; Ihm, C.G.; et al. Association Interleukin-4 and Interleukin-4 Receptor Gene Polymorphism and Acute Rejection and Graft Dysfunction After Kidney Transplantation. *Transplant Proc.* **2016**, *48*, 813–819. [CrossRef]
47. Cunha, C.; Rodrigues, F.; Zelante, T.; Aversa, F.; Romani, L.; Carvalho, A. Genetic Susceptibility to Aspergillosis in Allogeneic Stem-Cell Transplantation. *Med. Mycol.* **2011**, *49* (Suppl. 1), S137–S143. [CrossRef] [PubMed]
48. de Oliveira Malacco, N.L.S.; Rachid, M.A.; da Silva Gurgel, I.L.; Moura, T.R.; Sucupira, P.H.F.; de Sousa, L.P.; de Souza, D.d.G.; de Castro Russo, M.; Teixeira, M.M.; Soriani, F.M. Eosinophil-Associated Innate IL-17 Response Promotes Aspergillus Fumigatus Lung Pathology. *Front. Cell Infect. Microbiol.* **2018**, *8*, 453. [CrossRef]
49. de Boer, M.G.J.; Jolink, H.; Halkes, C.J.M.; van der Heiden, P.L.J.; Kremer, D.; Falkenburg, J.H.F.; van de Vosse, E.; van Dissel, J.T. Influence of Polymorphisms in Innate Immunity Genes on Susceptibility to Invasive Aspergillosis after Stem Cell Transplantation. *PLoS ONE* **2011**, *6*, e18403. [CrossRef]
50. Chang, D.; Chang, X.; He, Y.; Tan, K.J.K. The Determinants of COVID-19 Morbidity and Mortality across Countries. *Sci. Rep.* **2022**, *12*, 5888. [CrossRef]
51. He, W.; Chen, L.; Chen, L.; Yuan, G.; Fang, Y.; Chen, W.; Wu, D.; Liang, B.; Lu, X.; Ma, Y.; et al. COVID-19 in Persons with Haematological Cancers. *Leukemia* **2020**, *34*, 1637–1645. [CrossRef]
52. Hu, B.; Huang, S.; Yin, L. The Cytokine Storm and COVID-19. *J. Med. Virol.* **2021**, *93*, 250–256. [CrossRef]
53. Suárez-García, I.; Perales-Fraile, I.; González-García, A.; Muñoz-Blanco, A.; Manzano, L.; Fabregate, M.; Díez-Manglano, J.; Aizpuru, E.F.; Fernández, F.A.; García, A.G.; et al. In-Hospital Mortality among Immunosuppressed Patients with COVID-19: Analysis from a National Cohort in Spain. *PLoS ONE* **2021**, *16*, e0255524. [CrossRef] [PubMed]
54. Modemann, F.; Niederwieser, C.; Weisel, K.; Bokemeyer, C.; Fiedler, W.; Ghandili, S. COVID-19 and Seasonal Influenza: A Comparative Analysis in Patients with Hematological Malignancies. *Leuk. Lymphoma* **2022**, *63*, 664–671. [CrossRef] [PubMed]
55. Blennow, O.; Ljungman, P. The Challenge of Antibiotic Resistance in Haematology Patients. *Br. J. Haematol.* **2016**, *172*, 497–511. [CrossRef]
56. Aguilar-Guisado, M.; Espigado, I.; Martín-Peña, A.; Gudiol, C.; Royo-Cebrecos, C.; Falantes, J.; Vázquez-López, L.; Montero, M.I.; Rosso-Fernández, C.; de la Luz Martino, M.; et al. Optimisation of Empirical Antimicrobial Therapy in Patients with Haematological Malignancies and Febrile Neutropenia (How Long Study): An Open-Label, Randomised, Controlled Phase 4 Trial. *Lancet Haematol.* **2017**, *4*, e573–e583. [CrossRef]
57. Ganapathi, L.; Arnold, A.; Jones, S.; Patterson, A.; Graham, D.; Harper, M.; Levy, O. Use of Cidofovir in Pediatric Patients with Adenovirus Infection. *F1000 Research* **2016**, *5*, 758. [CrossRef]
58. Cordonnier, C.; Pautas, C.; Maury, S.; Vekhoff, A.; Farhat, H.; Suarez, F.; Dhédin, N.; Isnard, F.; Ades, L.; Kuhnowski, F.; et al. Empirical versus Preemptive Antifungal Therapy for High-Risk, Febrile, Neutropenic Patients: A Randomized, Controlled Trial. *Clin. Infect. Dis.* **2009**, *48*, 1042–1051. [CrossRef]
59. She, W.H.; Chok, K.S.H.; Li, I.W.S.; Ma, K.W.; Sin, S.L.; Dai, W.C.; Fung, J.Y.Y.; Lo, C.M. Pneumocystis Jirovecii-Related Spontaneous Pneumothorax, Pneumomediastinum and Subcutaneous Emphysema in a Liver Transplant Recipient: A Case Report. *BMC Infect. Dis.* **2019**, *19*, 66. [CrossRef] [PubMed]
60. Roux, A.; Gonzalez, F.; Roux, M.; Mehrad, M.; Menotti, J.; Zahar, J.-R.; Tadros, V.-X.; Azoulay, E.; Brillet, P.-Y.; Vincent, F. Update on Pulmonary Pneumocystis Jirovecii Infection in Non-HIV Patients. *Med. Mal. Infect.* **2014**, *44*, 185–198. [CrossRef]

61. Vincent, F. Leukostasis, Infiltration and Pulmonary Lysis Syndrome Are the Three Patterns of Leukemic Pulmonary Infiltrates. In *Pulmonary Involvement in Patients with Hematological Malignancies*; Azoulay, E., Ed.; Springer: Berlin/Heidelberg, Germany, 2011; pp. 509–520; ISBN 978-3-642-15742-4.
62. Azoulay, É.; Canet, E.; Raffoux, E.; Lengliné, E.; Lemiale, V.; Vincent, F.; de Labarthe, A.; Seguin, A.; Boissel, N.; Dombret, H.; et al. Dexamethasone in Patients with Acute Lung Injury from Acute Monocytic Leukaemia. *Eur. Respir. J.* **2012**, *39*, 648–653. [CrossRef]
63. Choi, M.H.; Jung, J.I.; Chung, W.D.; Kim, Y.-J.; Lee, S.-E.; Han, D.H.; Ahn, M.I.; Park, S.H. Acute Pulmonary Complications in Patients with Hematologic Malignancies. *Radiographics* **2014**, *34*, 1755–1768. [CrossRef]
64. Stefanski, M.; Jamis-Dow, C.; Bayerl, M.; Desai, R.J.; Claxton, D.F.; Van de Louw, A. Chest Radiographic and CT Findings in Hyperleukocytic Acute Myeloid Leukemia: A Retrospective Cohort Study of 73 Patients. *Medicine* **2016**, *95*, e5285. [CrossRef]
65. Bewersdorf, J.P.; Zeidan, A.M. Hyperleukocytosis and Leukostasis in Acute Myeloid Leukemia: Can a Better Understanding of the Underlying Molecular Pathophysiology Lead to Novel Treatments? *Cells* **2020**, *9*, 2310. [CrossRef] [PubMed]
66. Fayed, M.; Evans, T.; Abdulhaq, H. Leukemic Infiltration in the Settings of Acute Respiratory Failure. *Oxf. Med. Case Rep.* **2019**, *2019*, 482–485. [CrossRef] [PubMed]
67. Wu, Y.-K.; Huang, Y.-C.; Huang, S.-F.; Huang, C.-C.; Tsai, Y.-H. Acute Respiratory Distress Syndrome Caused by Leukemic Infiltration of the Lung. *J. Formos Med. Assoc.* **2008**, *107*, 419–423. [CrossRef]
68. Hölig, K.; Moog, R. Leukocyte Depletion by Therapeutic Leukocytapheresis in Patients with Leukemia. *Transfus. Med. Hemother.* **2012**, *39*, 241–245. [CrossRef]
69. McDonnell, M.H.; Smith, E.T.; Lipford, E.H.; Gerber, J.M.; Grunwald, M.R. Microgranular Acute Promyelocytic Leukemia Presenting with Leukopenia and an Unusual Immunophenotype. *Hematol./Oncol. Stem Cell Ther.* **2017**, *10*, 35–38. [CrossRef]
70. Choi, M.H.; Choe, Y.H.; Park, Y.; Nah, H.; Kim, S.; Jeong, S.H.; Kim, H.O. The Effect of Therapeutic Leukapheresis on Early Complications and Outcomes in Patients with Acute Leukemia and Hyperleukocytosis: A Propensity Score-Matched Study. *Transfusion* **2018**, *58*, 208–216. [CrossRef] [PubMed]
71. Kuo, K.H.M.; Callum, J.L.; Panzarella, T.; Jacks, L.M.; Brandwein, J.; Crump, M.; Curtis, J.E.; Gupta, V.; Lipton, J.H.; Minden, M.D.; et al. A Retrospective Observational Study of Leucoreductive Strategies to Manage Patients with Acute Myeloid Leukaemia Presenting with Hyperleucocytosis. *Br. J. Haematol.* **2015**, *168*, 384–394. [CrossRef]
72. Shallis, R.M.; Stahl, M.; Bewersdorf, J.P.; Hendrickson, J.E.; Zeidan, A.M. Leukocytapheresis for Patients with Acute Myeloid Leukemia Presenting with Hyperleukocytosis and Leukostasis: A Contemporary Appraisal of Outcomes and Benefits. *Null* **2020**, *13*, 489–499. [CrossRef]
73. Rinaldi, I.; Sari, R.M.; Tedhy, V.U.; Winston, K. Leukapheresis Does Not Improve Early Survival Outcome of Acute Myeloid Leukemia with Leukostasis Patients—A Dual-Center Retrospective Cohort Study. *J. Blood Med.* **2021**, *12*, 623–633. [CrossRef]
74. Kato, A.; Ono, Y.; Nagahata, Y.; Yamauchi, N.; Tabata, S.; Yonetani, N.; Matsushita, A.; Ishikawa, T. The Need for Continuing Chemotherapy for Leukemic Cell Lysis Pneumopathy in Patients with Acute Myelomonocytic/Monocytic Leukemia. *Intern. Med.* **2013**, *52*, 1217–1221. [CrossRef] [PubMed]
75. Korkmaz, S. The Management of Hyperleukocytosis in 2017: Do We Still Need Leukapheresis? *Transfus. Apher. Sci.* **2018**, *57*, 4–7. [CrossRef] [PubMed]
76. Mamez, A.-C.; Raffoux, E.; Chevret, S.; Lemiale, V.; Boissel, N.; Canet, E.; Schlemmer, B.; Dombret, H.; Azoulay, E.; Lengliné, E. Pre-Treatment with Oral Hydroxyurea Prior to Intensive Chemotherapy Improves Early Survival of Patients with High Hyperleukocytosis in Acute Myeloid Leukemia. *Null* **2016**, *57*, 2281–2288. [CrossRef] [PubMed]
77. Latagliata, R.; Spadea, A.; Cedrone, M.; Di Giandomenico, J.; De Muro, M.; Villivà, N.; Breccia, M.; Anaclerico, B.; Porrini, R.; Spirito, F.; et al. Symptomatic Mucocutaneous Toxicity of Hydroxyurea in Philadelphia Chromosome-Negative Myeloproliferative Neoplasms: The Mister Hyde Face of a Safe Drug. *Cancer* **2012**, *118*, 404–409. [CrossRef] [PubMed]
78. Bertoli, S.; Picard, M.; Bérard, E.; Griessinger, E.; Larrue, C.; Mouchel, P.L.; Vergez, F.; Tavitian, S.; Yon, E.; Ruiz, J.; et al. Dexamethasone in Hyperleukocytic Acute Myeloid Leukemia. *Haematol* **2018**, *103*, 988–998. [CrossRef]
79. Barnes, P.J. Corticosteroid Effects on Cell Signalling. *Eur. Respir. J.* **2006**, *27*, 413. [CrossRef]
80. Brattsand, R.; Linden, M. Cytokine Modulation by Glucocorticoids: Mechanisms and Actions in Cellular Studies. *Aliment. Pharm.* **1996**, *10* (Suppl. 2), 81–90, discussion 91–92. [CrossRef]
81. Vénéreau, E.; Ceriotti, C.; Bianchi, M.E. DAMPs from Cell Death to New Life. *Front. Immunol.* **2015**, *6*, 422. [CrossRef] [PubMed]
82. Kunitomo, Y.; Lee, S.; Avery, C.C.; Valda Toro, P.L.; Cohen, A.J.; Ehtashimi-Afshar, S.; Kahn, P.A.; Siddon, A.; Boddu, P.; Datta, R.; et al. Indolent Presentations of Leukemic Lung Disease in Acute Myeloid Leukemia. *medRxiv* **2020**. [CrossRef]
83. Azoulay, E.; Fieux, F.; Moreau, D.; Thiery, G.; Rousselot, P.; Parrot, A.; Le Gall, J.-R.; Dombret, H.; Schlemmer, B. Acute Monocytic Leukemia Presenting as Acute Respiratory Failure. *Am. J. Respir. Crit. Care Med.* **2003**, *167*, 1329–1333. [CrossRef]
84. Tvsvgk, T.; Handa, A.; Kumar, K.; Mutreja, D.; Subramanian, S. Chemotherapy-Associated Pulmonary Toxicity-Case Series from a Single Center. *South Asian J. Cancer* **2021**, *10*, 255–260. [CrossRef] [PubMed]
85. Dhokarh, R.; Li, G.; Schmickl, C.N.; Kashyap, R.; Assudani, J.; Limper, A.H.; Gajic, O. Drug-Associated Acute Lung Injury: A Population-Based Cohort Study. *Chest* **2012**, *142*, 845–850. [CrossRef] [PubMed]
86. Matsuno, O. Drug-Induced Interstitial Lung Disease: Mechanisms and Best Diagnostic Approaches. *Respir. Res.* **2012**, *13*, 39. [CrossRef]
87. Vahid, B.; Marik, P.E. Pulmonary Complications of Novel Antineoplastic Agents for Solid Tumors. *Chest* **2008**, *133*, 528–538. [CrossRef]

88. Lee-Chiong, T.J.; Matthay, R.A. Drug-Induced Pulmonary Edema and Acute Respiratory Distress Syndrome. *Clin. Chest Med.* **2004**, *25*, 95–104. [CrossRef]
89. Tanvetyanon, T.; Garrity, E.R.; Albain, K.S. Acute Lung Injury Associated with Vinorelbine. *J. Clin. Oncol.* **2006**, *24*, 1952–1953. [CrossRef]
90. Lo-Coco, F.; Cicconi, L.; Voso, M.T. Progress and Criticalities in the Management of Acute Promyelocytic Leukemia. *Oncotarget* **2017**, *8*, 99221–99222. [CrossRef]
91. Lång, E.; Grudic, A.; Pankiv, S.; Bruserud, Ø.; Simonsen, A.; Bjerkvig, R.; Bjørås, M.; Bøe, S.O. The Arsenic-Based Cure of Acute Promyelocytic Leukemia Promotes Cytoplasmic Sequestration of PML and PML/RARA through Inhibition of PML Body Recycling. *Blood* **2012**, *120*, 847–857. [CrossRef]
92. Cardinale, L.; Asteggiano, F.; Moretti, F.; Torre, F.; Ulisciani, S.; Fava, C.; Rege-Cambrin, G. Pathophysiology, Clinical Features and Radiological Findings of Differentiation Syndrome/All-Trans-Retinoic Acid Syndrome. *World J. Radiol.* **2014**, *6*, 583–588. [CrossRef]
93. Patatanian, E.; Thompson, D.F. Retinoic Acid Syndrome: A Review. *J. Clin. Pharm. Ther.* **2008**, *33*, 331–338. [CrossRef]
94. Larson, R.S.; Tallman, M.S. Retinoic Acid Syndrome: Manifestations, Pathogenesis, and Treatment. *Best Pract. Res. Clin. Haematol.* **2003**, *16*, 453–461. [CrossRef]
95. Lee, C.K. Evolving Role of Radiation Therapy for Hematologic Malignancies. *Hematol. Oncol. Clin. N. Am.* **2006**, *20*, 471–503. [CrossRef] [PubMed]
96. Arroyo-Hernández, M.; Maldonado, F.; Lozano-Ruiz, F.; Muñoz-Montaño, W.; Nuñez-Baez, M.; Arrieta, O. Radiation-Induced Lung Injury: Current Evidence. *BMC Pulm. Med.* **2021**, *21*, 9. [CrossRef] [PubMed]
97. Burris, H.A., 3rd; Hurtig, J. Radiation Recall with Anticancer Agents. *Oncologist* **2010**, *15*, 1227–1237. [CrossRef]
98. Ding, X.; Ji, W.; Li, J.; Zhang, X.; Wang, L. Radiation Recall Pneumonitis Induced by Chemotherapy after Thoracic Radiotherapy for Lung Cancer. *Radiat. Oncol.* **2011**, *6*, 24. [CrossRef]
99. Riviere, P.; Sumner, W.; Cornell, M.; Sandhu, A.; Murphy, J.D.; Hattangadi-Gluth, J.; Bruggeman, A.; Kim, S.S.; Randall, J.M.; Sharabi, A.B. Radiation Recall Pneumonitis After Treatment With Checkpoint Blockade Immunotherapy: A Case Series and Review of Literature. *Front. Oncol.* **2021**, *11*, 662954. [CrossRef] [PubMed]
100. Kouvaris, J.R.; Kouloulias, V.E.; Vlahos, L.J. Amifostine: The First Selective-Target and Broad-Spectrum Radioprotector. *Oncologist* **2007**, *12*, 738–747. [CrossRef]
101. Horowitz, M.; Schreiber, H.; Elder, A.; Heidenreich, O.; Vormoor, J.; Toffalori, C.; Vago, L.; Kröger, N. Epidemiology and Biology of Relapse after Stem Cell Transplantation. *Bone Marrow Transplant.* **2018**, *53*, 1379–1389. [CrossRef]
102. Muraro, P.A.; Martin, R.; Mancardi, G.L.; Nicholas, R.; Sormani, M.P.; Saccardi, R. Autologous Haematopoietic Stem Cell Transplantation for Treatment of Multiple Sclerosis. *Nat. Rev. Neurol.* **2017**, *13*, 391–405. [CrossRef]
103. Pettengell, R. Autologous Stem Cell Transplantation in Follicular Non-Hodgkin's Lymphoma. *Bone Marrow Transplant.* **2002**, *29* (Suppl. 1), S1–S4. [CrossRef]
104. Takami, A. Hematopoietic Stem Cell Transplantation for Acute Myeloid Leukemia. *Int. J. Hematol.* **2018**, *107*, 513–518. [CrossRef] [PubMed]
105. Bleakley, M.; Riddell, S.R. Molecules and Mechanisms of the Graft-versus-Leukaemia Effect. *Nat. Rev. Cancer* **2004**, *4*, 371–380. [CrossRef] [PubMed]
106. Choi, S.W.; Levine, J.E.; Ferrara, J.L.M. Pathogenesis and Management of Graft-versus-Host Disease. *Immunol. Allergy Clin. N. Am.* **2010**, *30*, 75–101. [CrossRef] [PubMed]
107. Lazarus, H.M.; Sommers, S.R.; Arfons, L.M.; Fu, P.; Ataergin, S.A.; Kaye, N.M.; Liu, F.; Kindwall-Keller, T.L.; Cooper, B.W.; Laughlin, M.J.; et al. Spontaneous Autologous Graft-versus-Host Disease in Plasma Cell Myeloma Autograft Recipients: Flow Cytometric Analysis of Hematopoietic Progenitor Cell Grafts. *Biol. Blood Marrow Transplant.* **2011**, *17*, 970–978. [CrossRef]
108. Hammami, M.B.; Talkin, R.; Al-Taee, A.M.; Schoen, M.W.; Goyal, S.D.; Lai, J.-P. Autologous Graft-Versus-Host Disease of the Gastrointestinal Tract in Patients With Multiple Myeloma and Hematopoietic Stem Cell Transplantation. *Gastroenterol. Res.* **2018**, *11*, 52–57. [CrossRef] [PubMed]
109. Haider, S.; Durairajan, N.; Soubani, A.O. Noninfectious Pulmonary Complications of Haematopoietic Stem Cell Transplantation. *Eur. Respir. Rev.* **2020**, *29*, 190119. [CrossRef]
110. Park, J.A. Treatment of Diffuse Alveolar Hemorrhage: Controlling Inflammation and Obtaining Rapid and Effective Hemostasis. *Int. J. Mol. Sci.* **2021**, *22*, 793. [CrossRef]
111. Spitzer, T.R. Engraftment Syndrome: Double-Edged Sword of Hematopoietic Cell Transplants. *Bone Marrow Transplant.* **2015**, *50*, 469–475. [CrossRef]
112. Uncu Ulu, B.; Yiğenoğlu, T.N.; Şahin, D.; Başcı, S.; İskender, D.; Adaş, Y.; Atasever Akkaş, E.; Hacıbekiroğlu, T.; Kızıl Çakar, M.; Dal, M.S.; et al. Does Total Body Irradiation Have a Favorable Impact on Thrombocyte Engraftment as per Neutrophil Engraftment in Allogeneic Stem Cell Transplantation? *Cureus* **2021**, *13*, e19462. [CrossRef]
113. Sheth, V.; Jain, R.; Gore, A.; Ghanekar, A.; Saikia, T. Engraftment Syndrome: Clinical Features and Predictive Factors in Autologous Stem Cell Transplant. *Indian J. Hematol. Blood Transfus.* **2018**, *34*, 448–453. [CrossRef]
114. Wieruszewski, P.M.; May, H.P.; Peters, S.G.; Gajic, O.; Hogan, W.J.; Dierkhising, R.A.; Alkhateeb, H.B.; Yadav, H. Characteristics and Outcome of Periengraftment Respiratory Distress Syndrome after Autologous Hematopoietic Cell Transplant. *Ann. ATS* **2021**, *18*, 1013–1019. [CrossRef] [PubMed]

115. Jha, A. Spectrum of Hematological Malignancies and Peripheral Cytopenias. *J. Nepal. Health Res. Counc.* **2013**, *11*, 273–278. [PubMed]
116. Franchini, M.; Marano, G.; Mengoli, C.; Pupella, S.; Vaglio, S.; Muñoz, M.; Liumbruno, G.M. Red Blood Cell Transfusion Policy: A Critical Literature Review. *Blood Transfus.* **2017**, *15*, 307–317. [CrossRef]
117. Estcourt, L.J.; Birchall, J.; Allard, S.; Bassey, S.J.; Hersey, P.; Kerr, J.P.; Mumford, A.D.; Stanworth, S.J.; Tinegate, H.; the British Committee for Standards in Haematology. Guidelines for the Use of Platelet Transfusions. *Br. J. Haematol.* **2017**, *176*, 365–394. [CrossRef] [PubMed]
118. Roubinian, N. TACO and TRALI: Biology, Risk Factors, and Prevention Strategies. *Hematol. Am. Soc. Hematol. Educ. Program.* **2018**, *2018*, 585–594. [CrossRef]
119. Vossoughi, S.; Gorlin, J.; Kessler, D.A.; Hillyer, C.D.; Van Buren, N.L.; Jimenez, A.; Shaz, B.H. Ten Years of TRALI Mitigation: Measuring Our Progress. *Transfusion* **2019**, *59*, 2567–2574. [CrossRef]

medicina

MDPI

Article

Loss of Expression of Antiangiogenic Protein FKBPL in Endometrioid Endometrial Carcinoma: Implications for Clinical Practice

Danilo D. Obradović [1,*], Nataša M. Milić [2], Nenad Miladinović [3], Lana McClements [4] and Dejan M. Oprić [1]

1 Institute of Pathology, Faculty of Medicine, University of Belgrade, 11000 Belgrade, Serbia
2 Institute for Medical Statistics and Informatics, Faculty of Medicine, University of Belgrade, 11000 Belgrade, Serbia
3 Department of Pathology, Clinical Hospital Centre Zemun, 11000 Belgrade, Serbia
4 Faculty of Science, School of Life Sciences, University of Technology Sydney, Sydney, NSW 2007, Australia
* Correspondence: danilo.obradovic@med.bg.ac.rs or drdaniloobradovic@gmail.com

Abstract: *Background and Objectives*: FK506 binding protein like (FKBPL) is a member of the immunophilin family, with anti-angiogenic effects capable of inhibiting the migration of endothelial cells and blood vessel formation. Its role as an inhibitor of tumor growth and angiogenesis has previously been shown in studies with breast and ovarian cancer. The role of FKBPL in angiogenesis, growth, and carcinogenesis of endometrioid endometrial carcinoma (EEC) is still largely unknown. The aim of this study was to examine the expression of FKBPL in EEC and benign endometrial hyperplasia (BEH) and its correlation with the expression of vascular endothelial factor-A (VEGF-A) and estrogen receptor alpha (ERα). *Materials and Methods*: Specimens from 89 patients with EEC and 40 patients with BEH, as well as histological, clinical, and demographic data, were obtained from the Clinical Hospital Centre Zemun, Belgrade, Serbia over a 10-year period (2010–2020). Immunohistochemical staining of the tissue was performed for FKBPL, VEGF-A, and ERα. Slides were analyzed blind by two pathologists, who measured the intensity of FKBPL and VEGF-A expression and used the Allred score to determine the level of ERα expression. *Results*: Immunohistochemical analysis showed moderate to high intensity of FKBPL expression in 97.5% ($n = 39$) of samples of BEH, and low or no expression in 93.3% ($n = 83$) of cases of EEC. FKBPL staining showed a high positive predictive value (98.8%) and a high negative predictive value for malignant diagnosis (86.7%). The difference in FKBPL expression between EEC and BEH was statistically significant ($p < 0.001$), showing a decrease in intensity and loss of expression in malignant tissues of the endometrium. FKBPL expression was positively correlated with ERα expression (intensity, percentage and high Allred score values) and negatively correlated with the expression of VEGF-A ($p < 0.05$ for all). *Conclusions*: FKBPL protein expression demonstrated a significant decrease in FKBPL in EEC in comparison to BEH tissue, with a high predictive value for malignancy. FKBPL might be emerging as a significant protein with antiangiogenic and antineoplastic effects, showing great promise for the diagnostic and therapeutic applications of its therapeutic derivatives in gynecological oncology.

Keywords: FKBPL; endometrioid endometrial carcinoma; angiogenesis; VEGF-A; estrogen receptor alpha

Citation: Obradović, D.D.; Milić, N.M.; Miladinović, N.; McClements, L.; Oprić, D.M. Loss of Expression of Antiangiogenic Protein FKBPL in Endometrioid Endometrial Carcinoma: Implications for Clinical Practice. *Medicina* **2022**, *58*, 1330. https://doi.org/10.3390/medicina58101330

Academic Editors: Nicolae Crisan and Călin Căinap

Received: 8 August 2022
Accepted: 19 September 2022
Published: 22 September 2022

1. Introduction

Endometrial carcinoma (EC) is the most common gynecological malignant tumor and the fourth most common malignancy in women in the United States, with a rising incidence and mortality rate [1]. The highest incidence of EC is observed in post-menopausal women, although nearly 20% of women are diagnosed before menopause, and approximately 5% of women are diagnosed before the age of 40. In addition to EC, the most frequently found hyperplastic change of the endometrium is endometrial hyperplasia without atypia/benign endometrial hyperplasia (BEH). BEH is a hormonally induced change of the endometrium with a 1–3% risk of developing EC. The main risk factors for the development of EC are high levels of circulating estrogens, exposure to exogenous estrogens, obesity, late menopause, nulliparity, a history of polycystic ovary syndrome, tamoxifen therapy, and Lynch's syndrome. Histologically, EC is classified as endometrioid, serous, clear cell, undifferentiated, dedifferentiated, mixed carcinoma, and carcinosarcoma of the uterine corpus [2–6]. Endometrioid endometrial carcinoma (EEC) accounts for 85% of cases. EEC is a highly estrogen-dependent tumor, with a plethora of evidence supporting estrogen stimulation unopposed by progesterone as one of the central mechanisms in its carcinogenesis. Estrogen receptor alpha (ERα) is the main factor through which estrogens stimulate mitogenic and proliferative activity in healthy endometrium and EEC [7]. It is shown that estradiol (E2) stimulates the proliferation and migration of endometrial cancer stem cells (CSC) through ERα. Paradoxically, the presence of ERα positivity in EECs is shown to be a positive prognostic marker, as opposed to ERα negative EECs that are more aggressive tumors with poorer prognoses, which might be due to the dedifferentiation of tumor cells [8].

The growth of EEC, before the tumor exceeds a volume of 2 mm^3, is independent of vascularization, obtaining nutrients and oxygen by diffusion, but for further growth formation of new blood vessels, it is necessary to support the metabolic requirements of neoplastic tissue [9]. Blood vessel growth occurs in the form of vasculogenesis and angiogenesis. Vasculogenesis occurs via the differentiation of angioblasts, and it is characteristic of embryonal development, while angiogenesis is growth from previously existing blood vessels. Angiogenesis is a physiologically highly regulated process, mainly stimulated by tissue hypoxia. Tumor growth is characterized by a loss of control over angiogenesis by the constant production of pro-angiogenic and decrease in anti-angiogenic factors, resulting in the rapid formation of the blood vessels, called the "angiogenic switch" [9].

The vascular endothelial growth factor (VEGF) family is a group of pro-angiogenic ligands that establish their effects through specific receptors. VEGFs have a central role in the process of physiological and pathological angiogenesis. VEGF-A is the most prominent and frequently studied angiogenic factor, shown to have increased expression in EC following the dedifferentiation of tumor cells. Increased VEGF-A expression in EC is a marker of a worse prognosis. Estrogen stimulates angiogenesis through VEGF-A expression in EC [9–11].

FK506 binding protein like—FKBPL—is a divergent member of the immunophilin family. It is a well-established anti-angiogenic protein, exhibiting its effects by targeting the cell surface receptor, CD44, on actively migrating endothelial cells, thus inhibiting migration and blood vessel formation. FKBPL's role as a negative regulator of tumor growth, metastasis, and angiogenesis is established in studies of breast cancer (BC), where FKBPL's high expression has been associated with a better prognosis of BC. FKBPL and its peptide derivative, ALM201, are shown to decrease the migration and invasion of breast cancer stem cells (CSC) and inhibit the growth of mammospheres of endocrine therapy-resistant breast CSC [12,13]. FKBPL is a part of the HSP90/ERα co-chaperone complex. A stable overexpression of FKBPL in breast CSCs is followed by an increased estrogen dependence on growth and by a higher sensitivity to tamoxifen therapy. A high expression of FKBPL is followed by a decrease in protein levels of ERα in breast CSCs [14]. Studies on ovarian cancer have shown that FKBPL therapeutic peptide derivatives stimulate the differentiation of ovarian CSCs and decrease their numbers while delaying tumor initiation and the growth of highly vascularized xenografts, in conclusion, establishing their antitumor effect through the disruption of angiogenesis [15]. On the other hand, reports on FKBPL's role in EC are lacking. In a single study examining the genetic profile of EC, it was reported that the expression of FKBPL was observed in stromal cells [16]. According to available data, the expression of FKBPL was found in the epithelium of endometrial glands showing moderate and high intensities of cytoplasmic, membranous, and nuclear expression, with typically pronounced luminal positivity; no expression was reported in stromal cells [17]. In EEC tumor cells, data showed expression varying from none to low and rarely of moderate intensity, whereas the information on FKBPL expression within stromal cells of EEC was not reported [18]. The aim of this study was to examine the expression of FKBPL and its correlation with the expression of VEGFR and ERα in BEH and EEC.

2. Materials and Methods

2.1. Collection of Samples and Data

Samples were obtained from the Clinical Hospital Centre Zemun, Belgrade, Serbia, over a 10-year period (2010–2020). The study enrolled a total of 89 patients undergoing hysterectomy who were diagnosed with EEC and 40 patients undergoing explorative curettage with a diagnosis of BEH. Inclusion criteria for patients diagnosed with EC included: performed hysterectomy, the availability of clinical staging data, and the availability of paraffin blocks containing tumor tissue covering at least 5 mm^2. Exclusion criteria included: other histological types of EC, an insufficient amount of tumor tissue for immunohistochemical staining and analysis, other malignant diseases, and previous oncologic therapy. Additional parameters analyzed for patients with EEC were: histological tumor grade, tumor stage (according to Tumor-Node-Metastases (TNM)–based staging and International Federation of Gynecology and Obstetrics (FIGO) classification), depth of myometrial invasion (DMI), and the presence of lymphovascular invasion (LVI). The ethics committee of the Clinical Hospital Centre Zemun approved this study (reference number: 12/1, 29 April 2021).

2.2. Immunohistochemical Staining

Tissue sections (4 µm thick) were deparaffinized and dehydrated. Antigen retrieval was performed using Tris-buffer pH 9.0 for FKBPL, and citrate buffer pH 6.0 for VEGF and ER, in a water bath for 30 min at 95 °C. In order to block endogenous peroxidase, slides were treated with a solution containing 3% hydrogen peroxide in phosphate-buffered saline (PBS) for 10 min, and non-specific antigen binding was blocked using a 1% bovine albumin serum (BSA) solution in PBS for 30 min. The primary antibodies used were: rabbit-polyclonal anti-FKBPL (catalog number: 10060-1-AP, Proteintech, 1:800), mouse monoclonal anti-VEGF (VG-1, Santa Cruz, Dallas, TX, USA, 1:100), and mouse monoclonal anti-estrogen receptor α (clone 6F11, Novocastra, Newcastle upon Tyne, UK, 1:100) for a 1 h incubation period at room temperature. After incubation, the slides were washed with PBS, and then streptavidin-horseradish peroxidase (HRP) was applied for 30 min. EnVision Detection System (DAKO, Jena, Germany) was applied, using 3,3′-diaminobenzidine as a substrate chromogen and counterstained with hematoxylin. Negative controls were treated by the same protocol, with a difference of using 1% BSA in PBS instead of a primary antibody. For positive external control for FKBPL, we used thyroid tissue.

2.3. Histopathological Evaluation

An analysis of immunohistochemical staining of FKBPL and VEGF-A was performed on 5 fields of magnification 100× and intensity was graded 0–3: 0—negative, 1—low, 2—moderate, 3—high intensity [11,12]. ERα expression was assessed through the Allred score, a well-established method for the quantification of ER in breast cancer, and suggested as a useful predictive factor in EC [19]. The Allred score was obtained as the sum of average intensity of stained nuclei, graded 0–3 (0—negative, 1—low, 2—moderate, 3—high intensity) and as the proportion of positive stained nuclei, graded 0–5 (0—negative, 1—less than 1%, 2—1–10%, 3—11–33%, 4—34–66% and 5—67–100% positive nuclei). The total value of the Allred score was in the range of 0–8, with an established cut-off value for positivity at a score value of 3. All slides had a blind analysis performed by two pathologists with 10 and 25 years of experience in the pathology of the female reproductive system (D.D.O. and D.M.O.).

2.4. Statistical Analysis

Numerical data are expressed as mean with standard deviation or as median with interquartile range. Categorical data are presented by absolute numbers with percentages and analyzed using a Chi-square test and Fisher exact test. For continuous variables, the Student t-test or the Mann–Whitney U test was used. The reliability of double-blinded readings was assessed by the Cronbach alpha coefficient. Measures of the diagnostic accuracy of FKBPL in discriminating between benign hyperplasia and endometrial carcinoma were determined by the receiver operating characteristic (ROC) analysis, and the cutoff level was determined. Sensitivity was defined as the % of patients with endometrial carcinoma who have no or low FKBPL expression (lower than the cut-off) (FKBPL no or low expression carcinoma patients/number of all carcinoma patients). Specificity was defined as the % of patients with hyperplasia who have FKBPL moderate or high expression levels (higher than the cut-off) (FKBPL moderate and high expression hyperplasia patients/number of hyperplasia patients). The positive predictive value was defined as the % of no or low FKBPL expression patients who have carcinoma (FKBPL no or low expression carcinoma patients/number of all FKBPL no or low expression patients). A negative predictive value was defined as the % of FKBPL moderate or high expression patients who have hyperplasia (FKBPL moderate or high expression hyperplasia patients/number of all FKBPL moderate or high expression patients). Correlations were examined by correlation coefficients according to the data scale used in the analyses (nominal by nominal, nominal by ordinal, and ordinal by scale). In all tests, $p < 0.05$ was considered statistically significant. Statistical analysis was performed using IBM SPSS statistical software (SPSS for Windows, release 25.0, SPSS, Chicago, IL, USA).

3. Results

3.1. Study Population

Clinical and histological data of cancer patients are presented in Table 1. The mean age of patients diagnosed with EC was 65.58 ± 8.59 years, and 45.23 ± 5.79 years for patients with BEH. Most patients with EEC were G1 and G2 (95.6%) and were diagnosed at an early-stage T1 (69.7%). EECs were predominantly limited to and did not extend beyond the uterus (89.9%).

Table 1. Clinical and histological data of cancer patients.

Variable		EEC (n = 89)
Grade (%)	G1	40.4 (n = 36)
	G2	55.1 (n = 49)
	G3	4.5 (n = 4)
TNM stage (%)	T1A	33.7 (n = 30)
	T1B	36 (n = 32)
	T2	20.2 (n = 18)
	T3	10.1 (n = 9)
FIGO stage (%)	IA	31.5 (n = 28)
	IB	36 (n = 32)
	II	20.2 (n = 18)
	III	11.2 (n = 10)
	IV	1.1 (n = 1)
Myometrial invasion depth (%)	Less than half	37.1 (n = 33)
	One half or more	62 (n = 56)

Table 1. *Cont.*

Variable		EEC (n = 89)
Lymphovascular invasion (%)	Yes	33.7 (n = 30)
	No	66.3 (n = 59)

Tumor-Node-Metastases (TNM)–based staging, International Federation of Gynecology and Obstetrics (FIGO) staging system.

3.2. Expression of FKBPL, VEGF-A and ERα in EEC and BEH

3.2.1. Reliability of Double-Blind Reading

Chronbach's alpha coefficient presented a high level of agreement between results obtained by a double-blind reading of immunohistochemical staining for FKBPL, ERα, and VEGF-A (Table 2).

Table 2. Reliability of double-blind reading.

Variable	Cronbach's Alpha	Internal Consistency
FKBPL	0.994	Excellent
ERα (intensity)	0.997	Excellent
ERα (percentage)	0.987	Excellent
Allred score	0.996	Excellent
VEGF-A	0.970	Excellent

FK506-binding protein-like (FKBPL), vascular endothelial growth factor-A (VEGF-A), estrogen receptor alpha (ERα).

3.2.2. Difference of FKBPL Expression in EEC and BEH

There was a significant difference in FKBPL expression between EEC and BEH ($p < 0.001$), which demonstrated a decrease in intensity and loss of expression within the tissue sections with malignant changes of the endometrium (Table 3, Figure 1).

Table 3. Expression of FKBPL, VEGF-A, and ERα in BEH and EEC.

Variable			Benign Endometrial Hyperplasia n = 40	Endometrioid Endometrial Carcinoma n = 89	Significance (p)
FKBPL	0/1		2.5% (n = 1)	93.3% (n = 83)	<0.001
	2/3		97.5% (n = 39)	6.7% (n = 6)	
ERα	Intensity	1	0% (n = 0)	46.1% (n = 41)	<0.001
		2/3	100% (n = 40)	53.9% (n = 48)	
	Percentage	<5 (≤66%)	20.0 (n = 8)	78.7% (n = 70)	<0.001
		5 (67–100%)	80.0 (n = 32)	21.3% (n = 19)	
	Allred Score	<7	10% (n = 4)	71.9% (n = 64) *	<0.001
		7/8	90% (n = 36)	28.1% (n = 25)	
VEGF-A	0		95% (n = 38)	57.3% (n = 51)	<0.001
	1/2		5% (n = 2)	42.7% (n = 38)	

Benign endometrial hyperplasia (BEH), endometrial endometrioid carcinoma (EEC). * 1 sample of endometrioid endometrial carcinoma had Allred score <3.

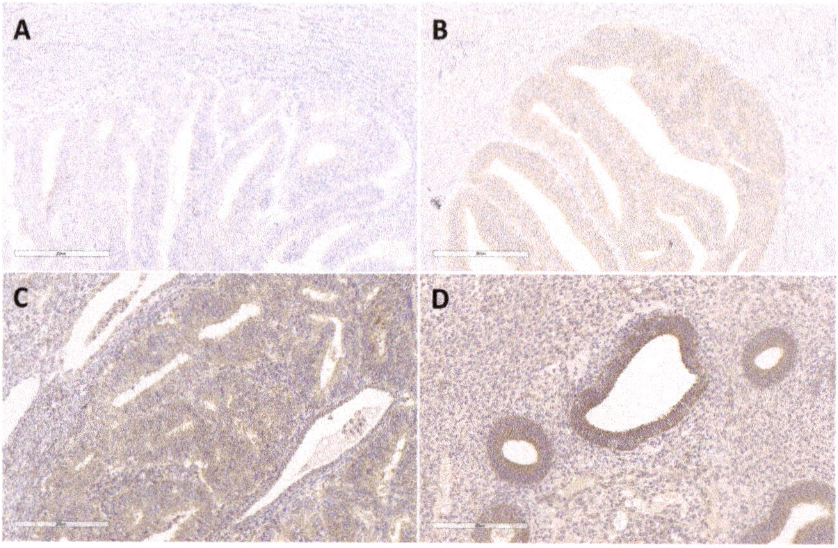

Figure 1. Immunohistochemical staining for FKBPL in endometrioid endometrial carcinoma showing no expression—0 (**A**), low expression—1 (**B**), moderate expression—2 (**C**), and immunohistochemical staining for FKBPL in benign endometrial hyperplasia showing high expression—3 (**D**). Magnification ×200.

There was no significant correlation between FKBPL expression and the histological grade or clinical stage of the tumor or depth of myometrial invasion, or lymphovascular invasion ($p > 0.05$).

3.2.3. Measures of Diagnostic Accuracy for FKBPL Expression

Immunohistochemical analysis showed a moderate to high intensity of FKBPL expression in 97.5% ($n = 39$) of the samples of BEH, and low or no expression in 93.3% ($n = 83$) of the cases of EEC. FKBPL staining showed a high positive predictive value (98.8%), and a high negative predictive value for malignant diagnosis (86.7%) (Table 4).

Table 4. Measures of diagnostic accuracy for FKBPL expression.

Variable	Sensitivity	Positive Predictive Value	Specificity	Negative Predictive Value	Accuracy
FKBPL	93.3%	98.8%	97.5%	86.7%	94.6%

3.2.4. Expression of ERα in EEC and BEH

ERα expression was found to be of moderate to high intensity in all samples of BEH, showing high values for the Allred score in 90% ($n = 36$) of the cases. ERα positivity in the EEC samples was determined to be of low intensity in 46.1% ($n = 41$), moderate intensity in 26.95% ($n = 24$), and high intensity in 26.95% ($n = 24$). In the group of patients with EEC, 78.7% ($n = 70$) showed expression of ERα in less than 66% of tumor glands, while 21.3% ($n = 19$) showed positive reaction in over 67% of tumor glands. High Allred score values were seen in 28.1% ($n = 25$), while moderate score values were observed in 68.6% ($n = 61$) of the EEC samples, whereas 2.2% ($n = 2$) showed low score values. Overall, there was a significant decrease ($p < 0.001$) in ERα expression between the EEC and BEH samples (Figure 2, Table 3).

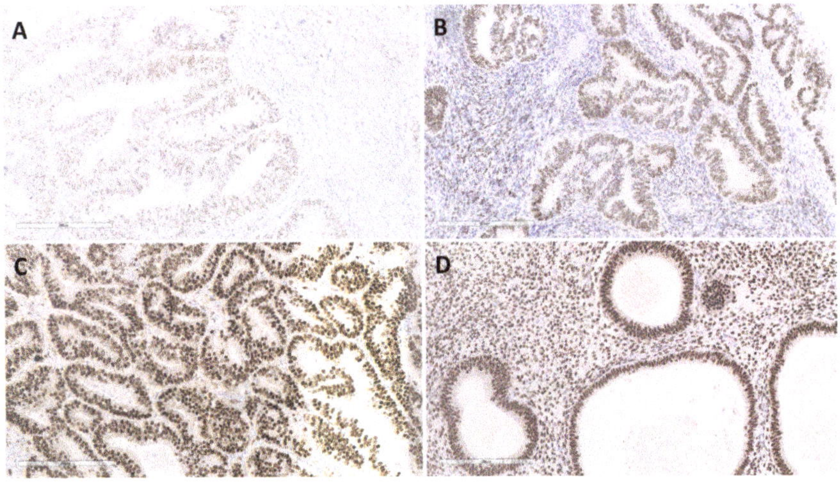

Figure 2. Immunohistochemical staining for ERα in endometrioid endometrial carcinoma showing low expression—1 (**A**), moderate expression—2 (**B**), high expression—3 (**C**), and immunohistochemical staining for ERα in benign endometrial hyperplasia showing high expression—3 (**D**). Magnification ×200.

The intensity of FKBPL expression, observed on the complete sample set from BEH and EEC patients, was in moderate positive correlation ($p < 0.05$) with the parameters of ERα expression (intensity, percentage and high Allred score values). The Allred score showed the strongest correlation.

3.2.5. Expression of VEGF-A in EEC and BEH

Immunohistochemical expression of VEGF-A was not demonstrated in 95% ($n = 38$) of BEHs, whereas a low and moderate intensity of expression was determined in 42.7% ($n = 38$) of EECs (Figure 3, Table 3).

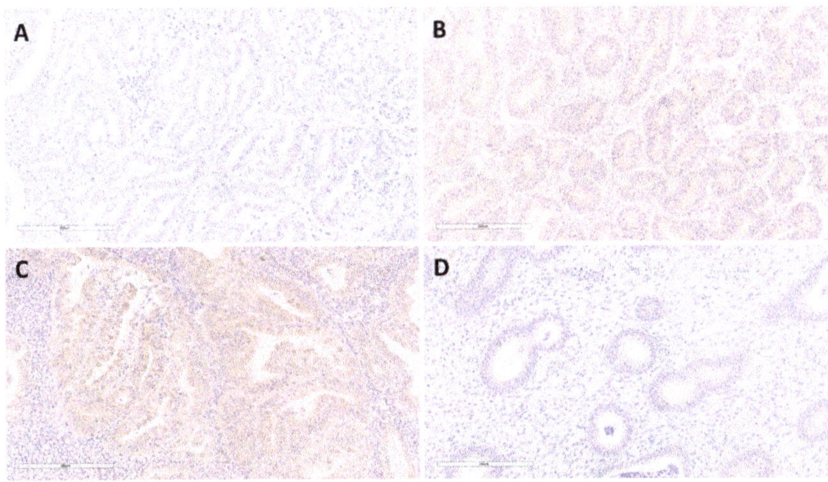

Figure 3. Immunohistochemical staining for VEGF-A in endometrioid endometrial carcinoma showing no expression—0 (**A**), low expression—1 (**B**), moderate expression—2 (**C**), and immunohistochemical staining for VEGF-A in benign endometrial hyperplasia showing no expression—0 (**D**). Magnification ×200.

The intensity of FKBPL expression, observed on the complete sample set of BEHs and EECs, was in moderate negative correlation ($p < 0.05$) with the intensity of expression of VEGF-A.

4. Discussion

4.1. Expression of FKBPL in EEC and BEH

This study demonstrates the presence of low intensity or loss of expression of FKBPL in 93.3% of ECs, and a moderate to high intensity of expression in 97.5% of BEHs, suggesting that the intensity of FKBPL expression could be considered as a potential diagnostic biomarker in routine endometrial curettage examination with high positive and negative predictive value, differentiating between benign and malignant changes of the endometrium.

These findings are in accordance with studies analyzing the expression of FKBPL in a medium of healthy cell culture and in cell culture of BC, where FKBPL was detected in the medium of healthy cell lines and was absent in the medium of BC culture unless it was experimentally overexpressed. The study concluded that FKBPL is an endogenous secreted antiangiogenic protein, whose downregulation in neoplastic cells allows uncontrolled tumor growth potential. The same study showed that the administration of purified recombinant FKBPL and its peptide derivative, AD-01, inhibited the migration of tumor cells [20]. A similar study reported a low level of secretion of FKBPL in the BC cell line when compared with human microvascular endothelial cells (HMEC-1), and a decreased secretion of FKBPL by HMEC-1 when cultivated under hypoxic conditions. Part of this study examined tumor growth in a FKBPL knockdown mouse model, confirming increased vascular sprouting and tumor growth [21]. In our study, we showed no or low expression of FKBPL in EECs, which was substantially lower than in BEHs and in agreement with previously reported data. The lack of correlation of FKBPL with tumor grade, clinical stage, DMI, and LVI, might be due to limiting factors of the study, including that most of the EECs were of low grade (grade 1 and 2) and stage I, producing a group of biologically similar tumors. These limitations correspond to usual practice circumstances, since the vast majority of EEC cases are diagnosed at an early stage due to a typical symptom of abnormal uterine bleeding that warrants further examination of patients and timely diagnosis. The clinical and biological similarity of this experimental group is also influenced by the FIGO grading system, classifying ECs as: well, moderately, and poorly differentiated (grades 1, 2, and 3), which are grouped as low grade (grade 1 and 2) and high grade (grade 3) in cases of EEC [6]. In addition, this distinction is aligned to the traditional classification of ECs according to Bokhman as type I, low-grade endometrioid, dominantly estrogen-dependent, with more favorable prognosis, and type II, endometrial carcinomas of non-endometrioid morphology, including undifferentiated and dedifferentiated ECs [3–5].

4.2. Expression of ERα and FKBPL

Our findings showed a decrease in ERα expression in EECs compared with BEHs ($p < 0.001$). In addition, there was a moderate positive correlation between the intensity of the expression of FKBPL and ERα when BEHs and EECs were combined. These findings are interesting given that the previous reports showed FKBPL downregulating ERα expression in BC, leading to decreasing phosphorylation of ERα, and increasing sensitivity to tamoxifen; in addition, FKBPL expression was upregulated by estrogen treatment [14]. Therefore, FKBPL and ERα expression are expected to be negatively correlated. A possible explanation for these differences might be due to the fact that FKBPL expression is stimulated by estrogen, whereas in our study, most of our patients were of postmenopausal age, with likely lower circulating estrogen levels. Limitations of our study include the lack of data on the potential use of estrogen replacement therapy, body mass index, and the inability to stratify patients according to the low/high ERα expression in the EECs group due to the large percentage of ERα positive tumors. Further findings might be obtained from a larger group containing more ERα negative EECs. In addition, it has been reported that the effects of ERα antagonists, including tamoxifen, a known risk factor for the development of EEC, are tissue-specific, differing between EEC and BC by inhibiting the growth of BC cells and stimulating the growth of EECs [22]. Our findings correspond to two known dilemmas surrounding EECs. The first is that estrogen stimulation of the endometrium unopposed by progesterone plays an important role in the carcinogenesis of EEC; however, EECs predominantly occur in postmenopausal women who have lower levels of circulating estrogens. This is called "the endometrial carcinoma paradox". The second dilemma to which our findings are aligned is the increased incidence of EECs in women treated with tamoxifen, which occurs in parallel to the inhibition of BC cells, despite the fact that both EEC and BC are highly estrogen-dependent tumors. Therefore, the role of ERα in EEC is still not fully understood and is somewhat contradictory. Hence, future studies should focus on elucidating this mechanism further.

4.3. Expression of VEGF-A and FKBPL

In this study, we showed an increase in VEGF-A expression in EECs compared with BEHs. In addition, when the complete sample set of BEHs and EECs was analyzed, we found a moderate negative correlation between the expression of FKBPL and VEGF-A ($p < 0.05$). Previous studies have shown that VEGF does not affect FKBPL expression, which was experimentally demonstrated ex vivo using an aortic ring assay following treatment with VEGF, where increased sprouting was indicative of increased angiogenesis from both wild-type and heterozygous FKBPL knockdown murine aortas, suggesting that FKBPL and VEGF affect angiogenesis through different pathways [21]. In light of these reports, the negative correlation that we have reported between FKBPL and VEGF-A might indicate an indirect interaction or two independent changes that complement each other as part of a pro-angiogenic switch in EEC. Also considering that previously published data relate both FKBPL and VEGF-A to estrogen and ERα [10,14], there could be a potential crosslink in the regulation of the balance between these two proteins with contrasting angiogenic effects. Several therapeutic approaches have been developed showing favorable results in blocking VEGF action, including blocking antibodies, decoy receptors, and small interfering RNA targeted at VEGF-A mRNA. However, anti-angiogenic therapy is still limited to small specific groups of patients, is associated with adverse effects, and many initially responsive patients develop resistance over time, creating a need for further understanding of the mechanisms and factors involved in the angiogenic process of EC [9,10]. An FKBPL peptide derivative, ALM-201, completed a phase I first-in-human clinical trial for ovarian cancer and other advanced solid tumors, which included two patients with EC in the safety trial group. Subsequently, ALM-201 was designated as an orphan drug for ovarian cancer by the Food and Drug Administration [23]. Further studies investigating FKBPL and its therapeutic peptide-derivative use in patients with EEC are needed in order to confirm the results obtained in our study. Considering that EEC is an evolving disease with a heterogeneous presentation and unelucidated mechanisms of carcinogenesis, there is a need for the pursuit of novel diagnostic and therapeutic approaches, which could include FKBPL-based markers and therapies that harness this emerging mechanism in EECs.

5. Conclusions

To our knowledge, we present the first comprehensive study on the immunohistochemical expression of FKBPL in EEC, showing a significant decrease in FKBPL in comparison to BEH with a high predictive value of malignancy. In addition, the study gives insight into the correlation between FKBPL expression and ERα as a known effector of endometrial proliferation, and VEGF-A as one of the central pro-angiogenic factors and target proteins for anti-angiogenic therapy in EEC patients.

Author Contributions: Conceptualization, D.D.O., N.M.M., D.M.O., L.M. and N.M.; methodology, D.D.O., N.M.M., D.M.O. and L.M.; validation, N.M.; formal analysis, D.D.O. and D.M.O.; investigation, D.D.O., D.M.O., N.M.M. and N.M.; data curation, N.M.M.; writing—original draft preparation, D.D.O., N.M.M., D.M.O., L.M. and N.M.; writing—review and editing, D.D.O., N.M.M., D.M.O., L.M. and N.M.; visualization, D.D.O. and D.M.O.; supervision, D.M.O., N.M.M. and L.M.; project administration, D.D.O. All authors have read and agreed to the published version of the manuscript.

Funding: This research received no external funding.

Institutional Review Board Statement: The study was conducted in accordance with the Declaration of Helsinki and approved by the ethics committee of Clinical Hospital Centre Zemun (reference number: 12/1, date of approval 29 April 2021).

Informed Consent Statement: Not applicable.

Data Availability Statement: Not applicable.

Acknowledgments: We thank Tracy Robinson for her contribution to our immunohistochemical staining protocol for FKBPL.

Conflicts of Interest: The authors declare no conflict of interest.

References

1. American Cancer Society. *Cancer Facts & Figures 2021*; American Cancer Society: Atlanta, GA, USA, 2021; pp. 1–72.
2. Herrington, C.S. *Female Genital Tumours*, 5th ed.; Organisation Mondiale de la Santé, Centre International de Recherche sur le Cancer, Eds.; World Health Organization Classification of Tumours; International Agency for Research on Cancer: Lyon, France, 2020.
3. Bokhman, J.V. Two Pathogenetic Types of Endometrial Carcinoma. *Gynecol. Oncol.* **1983**, *15*, 10–17. [CrossRef]
4. Sorosky, J.I. Endometrial Cancer. *Obstet. Gynecol.* **2012**, *120*, 383–397. [CrossRef] [PubMed]
5. Buhtoiarova, T.N.; Brenner, C.A.; Singh, M. Endometrial Carcinoma: Role of Current and Emerging Biomarkers in Resolving Persistent Clinical Dilemmas. *Am. J. Clin. Pathol.* **2016**, *145*, 8–21. [CrossRef] [PubMed]
6. Soslow, R.A.; Tornos, C.; Park, K.J.; Malpica, A.; Matias-Guiu, X.; Oliva, E.; Parkash, V.; Carlson, J.; McCluggage, W.G.; Gilks, C.B. Endometrial Carcinoma Diagnosis: Use of FIGO Grading and Genomic Subcategories in Clinical Practice. *Int. J. Gynecol. Pathol.* **2019**, *38*, S64–S74. [CrossRef] [PubMed]
7. Saimi, M.; Moriya, S.; Li, Z.-L.; Miyaso, H.; Nagahori, K.; Kawata, S.; Omotehara, T.; Ogawa, Y.; Hino, H.; Miyazawa, K.; et al. Cytonuclear Estrogen Receptor Alpha Regulates Proliferation and Migration of Endometrial Carcinoma Cells. *Tokai J. Exp. Clin. Med.* **2021**, *46*, 7–16. [PubMed]
8. Backes, F.J.; Walker, C.J.; Goodfellow, P.J.; Hade, E.M.; Agarwal, G.; Mutch, D.; Cohn, D.E.; Suarez, A.A. Estrogen Receptor-Alpha as a Predictive Biomarker in Endometrioid Endometrial Cancer. *Gynecol. Oncol.* **2016**, *141*, 312–317. [CrossRef] [PubMed]
9. Roškar, L.; Roškar, I.; Rižner, T.L.; Smrkolj, Š. Diagnostic and Therapeutic Values of Angiogenic Factors in Endometrial Cancer. *Biomolecules* **2021**, *12*, 7. [CrossRef] [PubMed]
10. Berger, A.A.; Dao, F.; Levine, D.A. Angiogenesis in Endometrial Carcinoma: Therapies and Biomarkers, Current Options, and Future Perspectives. *Gynecol. Oncol.* **2021**, *160*, 844–850. [CrossRef] [PubMed]
11. Dziobek, K.; Opławski, M.; Grabarek, B.O.; Zmarzły, N.; Tomala, B.; Halski, T.; Leśniak, E.; Januszyk, K.; Brus, R.; Kiełbasiński, R.; et al. Changes in the Expression Profile of VEGF-A, VEGF-B, VEGFR-1, VEGFR-2 in Different Grades of Endometrial Cancer. *Curr. Pharm. Biotechnol.* **2019**, *20*, 955–963. [CrossRef] [PubMed]
12. Nelson, L.; McKeen, H.D.; Marshall, A.; Mulrane, L.; Starczynski, J.; Storr, S.J.; Lanigan, F.; Byrne, C.; Arthur, K.; Hegarty, S.; et al. FKBPL: A Marker of Good Prognosis in Breast Cancer. *Oncotarget* **2015**, *6*, 12209–12223. [CrossRef] [PubMed]
13. McClements, L.; Annett, S.; Yakkundi, A.; O'Rourke, M.; Valentine, A.; Moustafa, N.; Alqudah, A.; Simões, B.M.; Furlong, F.; Short, A.; et al. FKBPL and Its Peptide Derivatives Inhibit Endocrine Therapy Resistant Cancer Stem Cells and Breast Cancer Metastasis by Downregulating DLL4 and Notch4. *BMC Cancer* **2019**, *19*, 351. [CrossRef] [PubMed]
14. McKeen, H.D.; Byrne, C.; Jithesh, P.V.; Donley, C.; Valentine, A.; Yakkundi, A.; O'Rourke, M.; Swanton, C.; McCarthy, H.O.; Hirst, D.G.; et al. FKBPL Regulates Estrogen Receptor Signaling and Determines Response to Endocrine Therapy. *Cancer Res.* **2010**, *70*, 1090–1100. [CrossRef]
15. Annett, S.; Moore, G.; Short, A.; Marshall, A.; McCrudden, C.; Yakkundi, A.; Das, S.; McCluggage, W.G.; Nelson, L.; Harley, I.; et al. FKBPL-Based Peptide, ALM201, Targets Angiogenesis and Cancer Stem Cells in Ovarian Cancer. *Br. J. Cancer* **2020**, *122*, 361–371. [CrossRef] [PubMed]
16. Iguchi, Y.; Ito, Y.M.; Kataoka, F.; Nomura, H.; Tanaka, H.; Chiyoda, T.; Hashimoto, S.; Nishimura, S.; Takano, M.; Yamagami, W.; et al. Simultaneous Analysis of the Gene Expression Profiles of Cancer and Stromal Cells in Endometrial Cancer: Stromal Gene Expression of Endometrial Cancer. *Genes Chromosomes Cancer* **2014**, *53*, 725–737. [CrossRef] [PubMed]
17. Tissue Expression of FKBPL-Staining in Endometrium—The Human Protein Atlas. Available online: https://www.proteinatlas.org/ENSG00000204315-FKBPL/tissue/endometrium (accessed on 4 August 2022).

18. Expression of FKBPL in Endometrial Cancer—The Human Protein Atlas. Available online: https://www.proteinatlas.org/ENSG00000204315-FKBPL/pathology/endometrial+cancer (accessed on 4 August 2022).
19. Yunokawa, M.; Yoshida, H.; Watanabe, R.; Noguchi, E.; Shimomura, A.; Shimoi, T.; Yonemori, K.; Shimizu, C.; Fujiwara, Y.; Tamura, K. Allred Score Is a Promising Predictor of Prognosis and Medroxyprogesterone Acetate Efficacy in Patients with Endometrial Cancer. *Cancer Chemother. Pharmacol.* **2017**, *80*, 127–134. [CrossRef]
20. Valentine, A.; O'Rourke, M.; Yakkundi, A.; Worthington, J.; Hookham, M.; Bicknell, R.; McCarthy, H.O.; McClelland, K.; McCallum, L.; Dyer, H.; et al. FKBPL and Peptide Derivatives: Novel Biological Agents That Inhibit Angiogenesis by a CD44-Dependent Mechanism. *Clin. Cancer Res.* **2011**, *17*, 1044–1056. [CrossRef]
21. Yakkundi, A.; Bennett, R.; Hernández-Negrete, I.; Delalande, J.-M.; Hanna, M.; Lyubomska, O.; Arthur, K.; Short, A.; McKeen, H.; Nelson, L.; et al. FKBPL Is a Critical Antiangiogenic Regulator of Developmental and Pathological Angiogenesis. *Arterioscler. Thromb. Vasc. Biol.* **2015**, *35*, 845–854. [CrossRef] [PubMed]
22. Droog, M.; Nevedomskaya, E.; Dackus, G.M.; Fles, R.; Kim, Y.; Hollema, H.; Mourits, M.J.; Nederlof, P.M.; van Boven, H.H.; Linn, S.C.; et al. Estrogen Receptor α Wields Treatment-Specific Enhancers between Morphologically Similar Endometrial Tumors. *Proc. Natl. Acad. Sci. USA* **2017**, *114*, E1316–E1325. [CrossRef] [PubMed]
23. El Helali, A.; Plummer, R.; Jayson, G.C.; Coyle, V.M.; Drew, Y.; Mescallado, N.; Harris, N.; Clamp, A.R.; McCann, J.; Swaisland, H.; et al. A First-in-Human Phase I Dose-Escalation Trial of the Novel Therapeutic Peptide, ALM201, Demonstrates a Favourable Safety Profile in Unselected Patients with Ovarian Cancer and Other Advanced Solid Tumours. *Br. J. Cancer* **2022**, *127*, 92–101. [CrossRef] [PubMed]

medicina

MDPI

Article

Hepatic Arterial Infusion Chemotherapy with Oxaliplatin Plus Raltitrexed as an Alternative Option in Advanced Hepatocellular Carcinoma Patients with Failure of, or Unsuitability for, Transarterial Chemoembolization

Yanfang Wu [1,†], Susu Zheng [1,†], Zhenzhen Zhang [1], Guobin Chen [1], Xiaochun Chen [1], Tanghui Zheng [1], Xinkun Guo [1], Hong Chen [1], Meixia Wang [1], Xiaoying Xie [1,2,*] and Boheng Zhang [1,2,3,*]

1 Department of Hepatic Oncology, Xiamen Clinical Research Center for Cancer Therapy, Zhongshan Hospital, Fudan University (Xiamen Branch), Xiamen 361015, China

2 Key Laboratory for Carcinogenesis and Cancer Invasion of Chinese Ministry of Education, The Liver Cancer Institute, Zhongshan Hospital, Shanghai Medical School, Fudan University, Shanghai 200032, China

3 Center for Evidence-Based Medicine, Shanghai Medical School, Fudan University, Shanghai 200032, China

* Correspondence: xie.xiaoying@zs-hospital.sh.cn (X.X.); zhang.boheng@zs-hospital.sh.cn (B.Z.)

† These authors contributed equally to this work.

Citation: Wu, Y.; Zheng, S.; Zhang, Z.; Chen, G.; Chen, X.; Zheng, T.; Guo, X.; Chen, H.; Wang, M.; Xie, X.; et al. Hepatic Arterial Infusion Chemotherapy with Oxaliplatin Plus Raltitrexed as an Alternative Option in Advanced Hepatocellular Carcinoma Patients with Failure of, or Unsuitability for, Transarterial Chemoembolization. *Medicina* **2022**, *58*, 1343. https://doi.org/10.3390/medicina58101343

Academic Editors: Nicolae Crisan, Călin Căinap, Roberto Montalti and Konstantinos Dimas

Received: 29 July 2022
Accepted: 20 September 2022
Published: 24 September 2022

Abstract: *Background and Objectives*: To assess the efficacy and safety of hepatic arterial infusion chemotherapy (HAIC) with oxaliplatin plus raltitrexed (HAICROX) as an alternative treatment option for advanced hepatocellular carcinoma (HCC) patients who are ineligible for, or failed, the transarterial chemoembolization (TACE) treatment. *Materials and Methods*: From July 2020 to November 2021, a total of 35 HCC patients were enrolled and received HAIC with oxaliplatin plus raltitrexed. The overall survival (OS) and time to progression (TTP) were primary and secondary endpoints, respectively. The tumor response was assessed by the modified response evaluation criteria in solid tumors (mRECIST), and the adverse events were investigated using the common terminology criteria for adverse events version 5.0 (CTCAE 5.0). *Results*: The median OS and TTP were 10 months (95% confidence interval (CI): 5.5–14.6) and 3.5 months (95% CI: 2.3–4.7), respectively. By means of multivariate analysis, anti-programmed cell death protein 1 (anti-PD-1) immunotherapy was found to be an independent prognostic factor for better survival. No patients experienced toxicity-related death. Thrombocytopenia, alanine aminotransferase (ALT), and aspartate aminotransferase (AST) elevation were the most common toxicities. No grade 3 or higher adverse events related to HAICROX were observed. *Conclusion*: HAICROX showed valuable efficacy and tolerable toxicity in advanced HCC patients who progressed on TACE or were ineligible for TACE. HAICROX is a promising treatment for advanced-stage HCC patients with TACE failure or ineligibility.

Keywords: hepatic arterial infusion chemotherapy; hepatocellular carcinoma; transarterial chemoembolization; treatment failure; unsuitability

1. Introduction

Hepatocellular carcinoma (HCC) is one of the most common malignancies and the fourth leading cause of cancer-related death worldwide [1]. HCC is also the third most common cause of cancer-related deaths in China [2]. As HCC is often asymptomatic, most patients are already in the intermediate and advanced stages when first diagnosed. The prognosis is dismal, with a median untreated survival time of 7–9 months [3]. Although tyrosine kinase inhibitors (TKIs) and immune checkpoint inhibitors are regarded as preferred treatments [4], TACE is currently recognized as the most commonly used method for the non-surgical treatment of advanced liver cancer [5–7]. However, TACE is not suitable for patients with diffuse HCC, an arterioportal/arteriovenous shunt, or main portal vein tumor thrombosis [5], who are considered as TACE-ineligible. Meanwhile, some patients are classified as TACE-refractory, defined as having refractoriness to TACE after more than two TACE procedures within 6 months [8]. However, there are currently no consistent conclusions on later-line treatments with TACE-refractory or -ineligible patients.

Hepatic arterial infusion chemotherapy (HAIC) is a method of local treatment with a high drug concentration in the liver and few systemic adverse reactions. Several studies have shown that the combination therapy of sorafenib (SORA) plus FOLFOX4-HAIC (the infusion of 5-fluorouracil, leucovorin, and oxaliplatin) showed better survival benefits compared with SORA alone [9–11]. In addition, Hsu found that HAIC with FOLFOX4 also showed acceptable outcomes in advanced HCC patients who failed, or were unsuitable for, TACE; the overall survival (OS) and progression-free survival (PFS) were 9.0 months (95% CI: 7.6–10.4) and 3.7 months (95% CI: 3.1–4.3), respectively.

However, the transarterial infusion of 5-fluorouracil needs approximately 44–48 h because 5-fluorouracil is a time-dependent chemotherapy drug. Prolonged arterial infusion is inconvenient and increases the risk of intrahepatic catheter thrombosis and displacement [12]. Conversely, raltitrexed is another thymidylate synthase inhibitor with a longer half-life than 5-Fu, reaching 198 h, and can exert a stable anti-tumor effect over a long period of time [13]. It was regarded as a better candidate for HAIC in HCC. A retrospective study showed that hepatic arterial infusion with a low dose of raltitrexed plus oxaliplatin (HAICROX) was effective in patients with advanced HCC with MPVTT; the median survival time was 8.7 months [14]. A phase II prospective study performed HAICROX in patients with intermediate and advanced-stage HCC; the ORR was 18 (51.4%) out of 35 patients, the DCR was 31 (79.5%) out of 35 patients, the median TTP was 6.7 months (95% CI: 4.6–8.8), and no treatment-related grade 3 or 4 toxicities or deaths were found [15]. However, the efficacy and safety of HAICROX treatment for TACE-refractory or -ineligible advanced HCC patients remain unknown.

Here, we carried out this study to investigate the efficacy and safety of HAICROX for advanced HCC patients with TACE failure or ineligibility.

2. Materials and Methods

2.1. Patients

The inclusion criteria were: (1) pathologically or radiologically (contrast-enhanced magnetic resonance imaging (MRI) or computed tomography (CT)) confirmed advanced HCC based on the American Association for the Study of Liver Disease criteria. (2) TACE refractoriness/failure, defined as disease progression after more than two TACE sessions (The Japan Society of Hepatology (JSH) defines refractoriness to TACE as a failure to control target lesions or the appearance of new lesions, even after two or more consecutive TACE sessions). TACE ineligibility was defined as diffuse HCC, with major portal vein cancer metastasis, a severe arterioportal/arteriovenous shunt on angiography, tumor thrombosis in the inferior vena cava or right atrium, having received TACE treatment at least once. (3) Patients had at least one measurable lesion. (4) Patients had a Child–Pugh score of ≤ 6 and an Eastern Cooperative Oncology Group (ECOG) performance status of ≤ 1.

The exclusion criteria were: uncontrollable infection; hepatic encephalopathy; gastrointestinal bleeding; refractory ascites; serum bilirubin levels > 3.0 mg/dL; ALT and AST more than 3 times the upper normal limit; albumin levels < 2.5 g/dL; platelet count < 50 × 109/L; and serum creatinine levels > 1.5 mg/dL. This study was approved by the ethical review board of our institution. Informed consent was waived for all patients. Between July 2020 and November 2021, a total of 35 patients with advanced-stage HCC, confirmed as being TACE-refractory or -ineligible and treated with HAICROX, were included. The demographic, clinical, and survival information was extracted from electronic medical records.

2.2. HAIC Procedures

For the HAIC with the oxaliplatin plus raltitrexed procedure, the Seldinger technique was performed through the femoral artery or the radial artery, and a catheter and a coaxial microcatheter were inserted into the feeding hepatic artery. The following regimens of ROX were administered: oxaliplatin (67–75 mg/m^2 continuous infusion for 4 h) and raltitrexed (2 mg/m^2 continuous infusion for 1 h). After priming chemotherapy, we removed the catheter and microcatheter, performed the treatment every 3–4 weeks, and then discontinued the treatment, as previously defined [15].

2.3. Assessment and Follow-Up

Liver function tests; routine blood tests; and determinations of cytokine, alpha-fetoprotein (AFP), and prothrombin (induced by vitamin K absence-II (PIVKA-II)) levels were performed before every treatment period, and contrast-enhanced CT or MRI of the upper abdomen was performed every 2–3 cycles of HAIC therapy to assess the treatment outcome, according to the modified response evaluation criteria in solid tumors (mRECIST). A further examination was performed if patients had suspected extrahepatic spread. As in a previous study [16], the albumin–bilirubin (ALBI) score was calculated using the following formula: ALBI score = (0.66 × log10 total bilirubin (μmol/L)) + (−0.085 × serum albumin (g/L)). The last follow-up date was 13 March 2022.

The ORR was defined as the number of patients who achieved either a partial response (PR) or a complete response (CR). Additionally, the disease control (DCR) was calculated as the sum of the CR, PR, and stable disease (SD). The primary endpoint of our study was OS, and the secondary endpoint was TTP. In this study, the OS was defined as the period from the initiation of HAIC treatment to death or the last known follow-up, and the TTP was defined as the time from the start of HAIC therapy to disease progression (a radiological or clinical evaluation by mRECIST).

The grade of toxicity was recorded and graded according to the National Cancer Institute's common terminology criteria for adverse events (CTCAE, version 5.0).

2.4. Statistical Analysis

For baseline characteristics, the categorical variables are described as frequencies and percentages, while continuous variables are presented as the mean ± standard deviation (SD). The Kaplan–Meier method was used to estimate the TTP and OS. Univariate analyses were performed using the log-rank test. For the univariate analysis results, variables with a $p < 0.05$ were entered into the multivariate analysis. The multivariate Cox model was applied to identify independent risk factors. All statistical tests were 2-sided, and $p < 0.05$ was considered statistically significant. SPSS software (SPSS version 22.0; SPSS, Chicago, IL, USA) was used to perform the statistical analyses.

3. Results

3.1. Patients Characteristics

In total, between July 2020 and November 2021, 35 HCC patients with either TACE unsuitability or refractoriness received HAICROX treatment and were enrolled in this study. The baseline demographics and characteristics of enrolled patients are shown in Table 1.

Table 1. Clinical characteristics of 35 advanced HCC patients with TACE refractoriness or ineligibility.

Characteristic	*n*= 35
Age [1]	53.0 ± 11.7
Gender	
Male	33 (94.3%)
Female	2 (5.7%)
HBV	
Negative	5 (14.3%)
Positive	30 (85.7%)
HCV	
Negative	33 (94.3%)

Table 1. *Cont.*

Characteristic	*n*= 35
Positive	2 (5.7 %)
HBV + HCV coinfection	1 (2.9%)
Child–Pugh	
A	34 (97.1%)
B	1 (2.9%)
ECOG	
0	15 (42.9%)
1	20 (57.1%)
Tumor size (mm) [1]	83.8 ± 44.8
Tumor number	
≤3	10 (28.6%)
>3	25 (71.4%)
Tumor thrombosis	
None	14 (40%)
Inferior vena cava (IVC)	7 (20%)
Main portal vein	3 (8.6%)
Branch of the portal vein	9 (25.7%)
Distant branch of portal vein	2 (5.7%)
Extrahepatic metastasis	
Yes	19 (54.3%)
No	16 (45.7%)
Number of HAIC treatments	2.3 ± 1.1
Number of previous TACE treatments	2.7 ± 2.3
Previous TKI lines	
1st	22 (62.9%)
≥2nd	13 (37.1%)
Later-line treatment	
Anti-PD-1	22 (62.9%)
TACE	10 (28.6%)
TBIL (μmol/L) [2]	12.2 (8.5–19.2)
ALT (U/L) [2]	36 (22–48)
AST (U/L) [2]	44 (35–68)
Albumin (g/L) [1]	40.3 ± 5.2
ALP (U/L) [2]	127 (92–198.5)
GGT (U/L) [2]	160 (78–207)
WBC ($\times10^9$/L) [2]	5.76 (4.01–6.74)
Neu ($\times10^9$/L) [2]	3.5 (2.1–4.6)
PLT ($\times10^9$/L) [2]	161 (101–192)
L ($\times10^9$/L) [2]	1.1 (0.8–1.6)
NLR [2]	2.8 (1.8–5.1)
PLR [2]	128.1 (91.8–151.5)
IL-6 (pg/mL) [2]	10.4 (5.9–18.6)
IL-8 (pg/mL) [2]	36.95 (17.75–63.73)
TNF (pg/mL) [2]	8 (6.7–14.7)
IL-2R(U/mL) [2]	557.5 (340–927)
ALBI (grade1/grade2/grade3)	21 (60%)/14 (40%)/0
AFP (ng/mL) [2]	457.2 (75.3–4917)
PIVKA-II (mAU/mL) [2]	4139 (703–28,798)

TBIL, total bilirubin; ALT, alanine aminotransferase; AST, aspartate aminotransferase; ALP, alkaline phosphatase; GGT, γ-glutamyl transpeptidase; WBC, white blood cell count; Neu, neutrophil cell count; PLT, platelet count; L, lymphocyte cell count; NLR, neutrophil–lymphocyte ratio; PLR, platelet–lymphocyte ratio; IL-6, interleukin-6; IL-8, interleukin-8; TNF, tumor necrosis factor; IL-2R, interleukin-2R; ALBI, albumin–bilirubin grade; AFP, alpha-fetoprotein; PIVKA-II, prothrombin induced by vitamin K absence II. [1] Data are presented as the mean ± SD. [2] Data are presented as the median (interquartile range).

All patients received TKI treatments; among them, 30 patients were refractory to TACE and 5 patients were ineligible for TACE because of major vascular tumor thrombosis. The average age of patients was 53 ± 11.7 years, and the median follow-up period was

10 months (range: 2–17.2). Most patients with an advanced stage of HCC were male (33/35, 94.3%) and had hepatitis B (30/35, 85.7%), a small number of patients had hepatitis C (2/35, 5.7%), and just one patient (2.9%) had an HBV and HCV co-infection. Most patients (34/35, 97.1%) belonged to Child–Pugh class A, and 21 (60%) and 14 (40%) patients had liver functions of ALBI grade 1 and grade 2, respectively. Twenty (57.1%) patients had ECOG 1. Twenty-five (86.7%) patients had more than three tumors. In total, 21 (60%) patients had portal vein thrombosis (PVTT), while 7 (20%) patients had tumor thrombosis in the inferior vena cava. Nineteen (54.3%) patients had extrahepatic metastasis, including four (11.4%) patients with pulmonary metastasis, four (11.4%) patients with bone metastasis, seven (20%) patients with lymph node metastasis, three (8.6%) patients with abdominal metastasis, and one (2.9%) patient with gallbladder invasion. The median number of previous TACE sessions was 2.7 ± 2.3. Meanwhile, 13 (37.1%) patients had received more than the second line of TKI treatment and 22 (62.9%) patients had received anti-PD-1 immunotherapy.

3.2. *Efficacy and Safety*

3.2.1. Tumor Response

The mean number of HAIC treatments was 2.3 (range: 1–5). At 2 months after HAIC treatment, 4 (11.4%) patients had a PR, 16 (45.7%) patients had an SD and 15 (42.9%) patients had a PD. Accordingly, the ORR was 11.4% and the DCR was 57.1%. Table 2 lists the treatment outcome details.

Table 2. Tumor response assessed according to mRECIST after two months of HAIC treatment.

Tumor Responses	*n* = 35 (%)
CR	0
PR	4 (11.4%)
SD	16 (45.7%)
PD	15 (42.9%)
ORR	4 (11.4%)
DCR	20 (57.1%)

CR, complete response; PR, partial response; SD, stable disease; PD, progressive disease; ORR, objective response rate; DCR, disease control rate.

3.2.2. Survival Outcome

In total, 21 out of 35 (60%) patients had died at the end of follow-up (March 2022). The mean follow-up time was 10 months (range: 2–17.2). The median OS and TTP were 10 months (95% CI: 5.5–14.6) and 3.5 months (95% CI: 2.3–4.7), respectively (Figure 1). The median OS for patients who achieved a response (PR) to HAIC treatment was not reached, and the median OS for patients with SD and PD (non-respondents) was 8.6 months (95% CI: 5.5–11.8). Patients who achieved clinical benefits (CR + PR + SD) showed a longer OS and TTP compared with people without clinical benefits (median OS, not reached vs. 6.8 months, $p = 0.014$; median TTP, 6.5 months vs. 2.6 months, $p = 0.023$) (Figure 2). These results show that achieving a clinical benefit could predict the patients' survival. In addition, patients with anti-PD-1 combination therapy showed better OS and TTP than without anti-PD-1 immunotherapy (median OS, 15.8 months (95% CI 7.4–24.2) vs. 6.7 months (95% CI 3.7–9.8), $p = 0.01$; median TTP, 6.5 months (95% CI 1.6–11.4) vs. 2.1 months (95% CI 1.1–3.0), $p = 0.043$) (Figure 3).

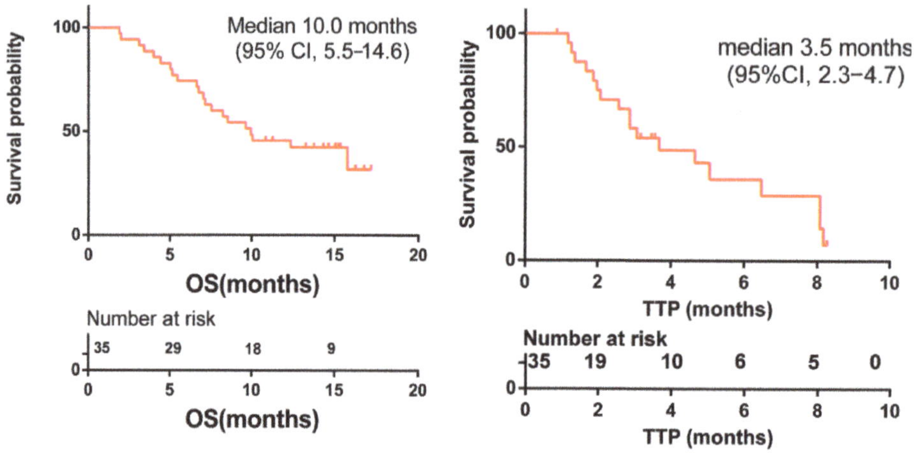

(a) (b)

Figure 1. Kaplan–Meier curves for overall survival (OS) (**a**) and time to progression (TTP) (**b**) in 35 patients undergoing HAIC with oxaliplatin plus raltitrexed (HAICROX).

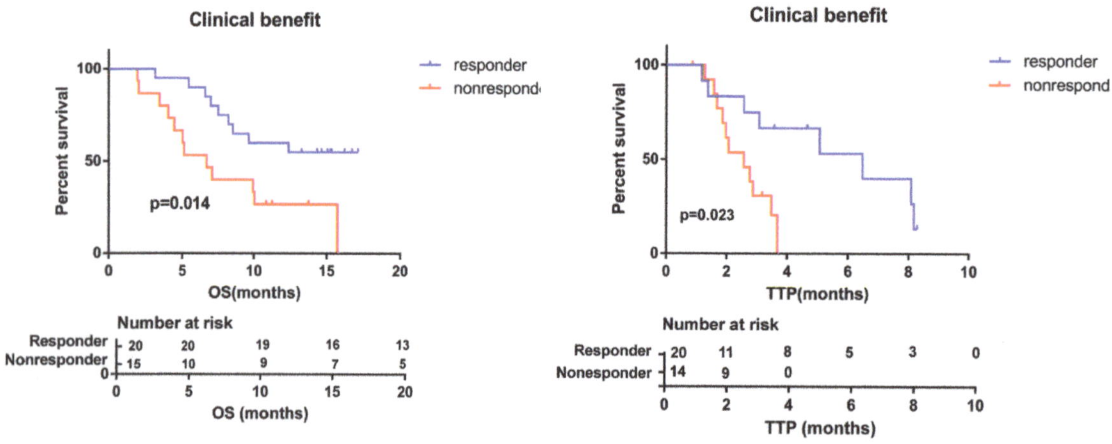

(a) (b)

Figure 2. Kaplan–Meier curves for overall survival (OS) (**a**) and TTP (**b**) for patients with clinical benefits (CR + PR + SD) and without clinical benefits (PD) in HCC patients treated with HAICROX, who experienced TACE treatment failure or unsuitability.

3.2.3. Safety and Toxicity

No treatment-related deaths occurred, and no patients experienced grade 3 or 4 toxicities. The most common toxicities associated with HAIC included AST and ALT elevation (both 34.3%), thrombocytopenia (17.1%), bilirubin elevation (8.6%), general weakness (8.6%), dyspepsia/anorexia (2.9%), nausea/vomiting (2.9%), gastrointestinal (GI) bleeding (2.9%), hyponatremia (2.9%), ascites aggravation (2.9%), and hepatic encephalopathy (2.9%). HAIC-related toxicities are detailed in Table 3.

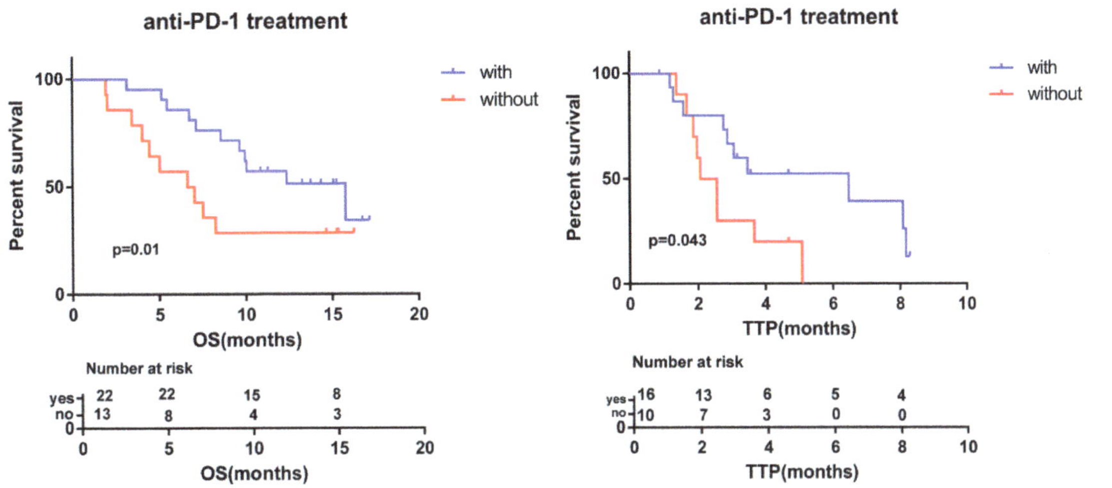

(**a**) (**b**)

Figure 3. Kaplan–Meier curves for OS (**a**) and TTP (**b**) in patients, stratified by treatment with or without anti-PD-1 immunotherapy.

Table 3. HAICROX associated with adverse events in 35 HCC patients.

Adverse Event	Grades I *n* (%)	Grades II *n* (%)	Grades III *n* (%)	Grades IV *n* (%)
Thrombocytopenia	4 (11.4%)	2 (5.7%)	0	0
Dyspepsia/anorexia	1 (2.9%)	0	0	0
Nausea/vomiting	1 (2.9%)	0	0	0
GI bleeding	0	1 (2.9%)	0	0
Fatigue	3 (2.9%)	0	0	0
General weakness	3 (8.6%)	0	0	0
AST elevation	11 (31.4%)	1 (2.9%)	0	0
ALT elevation	11 (31.4%)	1 (2.9%)	0	0
Bilirubin elevation	3 (8.6%)	0	0	0
Hyponatremia	1 (2.9%)	0	0	0
Ascites aggravation	1 (2.9%)	0	0	0
Hepatic encephalopathy	1 (2.9%)	0	0	0

GI bleeding, gastrointestinal bleeding; AST, aspartate aminotransferase; ALT, alanine aminotransferase.

3.3. Prognostic Factors

In the univariate analysis, tumor size, total bilirubin (TBIL), albumin (ALB), ALBI-grade, AST, GGT, IL-8, PIVKA-II, and treatment with anti-PD-1 immunotherapy were significantly associated with OS. Multivariate Cox analysis demonstrated that treatment with anti-PD-1 immunotherapy was an independent prognostic factor for OS (details are presented in Table 4). In the univariate analysis, AST and receiving later-line TACE or anti-PD-1 treatment were significantly associated with TTP. However, according to multivariate Cox analysis, there were no independent prognostic factors for TTP (Supplementary Table S1).

Table 4. Prognostic factors associated with overall survival in 35 patients (* $p < 0.05$ is considered statistically significant).

	Univariate Analysis			Multivariate Analysis		
	HR	95% CI	*p* Value	HR	95% CI	*p* Value
Gender (M/F)	0.545	0.21–1.414	0.212			
Age (<50 vs. ≥50)	0.61	0.257–1.452	0.081			
Etiology (HBC vs. HCV vs. unknown)	0.128	0.013–1.286	0.173			
TB (μmol/L)	1.066	1.014–1.121	0.007 *	1.076	0.978–1.185	0.134
Albumin (g/L)	0.879	0.800–0.966	0.013 *	1.013	0.822–1.248	0.905
ALBI-grade	2.71	1.105–6.647	0.029 *	2.982	0.259–33.046	0.385
ALT (U/L)	1.024	0.999–1.049	0.059			
AST (U/L)	1.018	1.009–1.026	<0.001 *	1.019	0.998–1.040	0.082
ALP (U/L)	1.008	1.003–1.014	0.093			
GGT (U/L)	1.004	1.001–1.008	0.027 *	0.997	0.988–1.007	0.554
WBC ($\times 10^9$/L)	0.996	0.795–1.248	0.971			
PLT ($\times 10^9$/L)	0.997	0.989–1.005	0.437			
Neu ($\times 10^9$/L)	1.094	0.791–1.512	0.558			
L ($\times 10^9$/L)	0.541	0.391–1.637	0.113			
NLR	1.094	0.853–1.403	0.48			
PLR	0.999	0.993–1.006	0.867			
IL-6	1.01	0.985–1.035	0.444			
IL-8	1.008	1.002–1.014	0.007 *	0.998	0.986–1.009	0.689
IL-2R	1.001	1–1.002	0.142			
TNF	1.004	0.996–1.013	0.284			
AFP (<400 mg/L, ≥400 mg/L)	2.242	0.924–5.439	0.741			
PIVKA-II (<400 mAU/mL, ≥400 mAU/mL)	2.286	0.668–7.828	0.001 *	0.554	0.095–3.236	0.512
Tumor size (mm)	1.01	1–1.02	0.04 *	1.01	0.992–1.028	0.28
Tumor number (1–3/>3)	1.739	0.634–4.766	0.282			
Tumor thrombosis (yes, no)	1.15	0.469–2.819	0.759			
Extrahepatic metastasis	1.2	0.508–2.837	0.677			
TACE times (1–2/>2)	1.115	0.461–2.696	0.809			
TKIs lines (1st/>2nd)	1.395	0.590–3.3	0.449			
ECOG (0/1)	0.632	0.268–1.491	0.295			
Child–Pugh class (A/B)	3.318	0.404–27.279	0.264			
Times of HAIC (1, 2/>2)	0.722	0.449–1.161	0.179			
Later-line treatment						
TACE (0, ≥1)	1.395	0.590–3.300	0.449			
Anti-PD-1 (yes, no)	0.329	0.135–0.802	0.014 *	0.267	0.075–0.953	0.042 *

TBIL, total bilirubin; ALT, alanine aminotransferase; AST, aspartate aminotransferase; ALP, alkaline phosphatase; GGT, γ-glutamyl transpeptidase; WBC, white blood cell count; Neu, neutrophil cell count; PLT, platelet count; L, lymphocyte cell count; NLR, neutrophil–lymphocyte ratio; PLR, platelet–lymphocyte ratio; IL-6, interleukin-6; IL-8, interleukin-8; TNF, tumor necrosis factor; IL-2R, interleukin-2R; ALBI, albumin–bilirubin grade; AFP, alpha-fetoprotein; PIVKA-II, prothrombin induced by vitamin K absence-II; TACE, transarterial chemoembolization; TKIs, tyrosine kinase inhibitors; ECOG, Eastern Cooperative Oncology Group; HAIC, hepatic arterial infusion chemotherapy; anti-PD-1, anti-programmed cell death protein 1 immunotherapy.

4. Discussion

TACE still represents a mainstay of treatment and is often the first-line therapy in patients with advanced HCC, but TACE failure or ineligibility remains a challenge for clinicians. This study provides a later-line option for advanced HCC patients with TACE refractoriness/failure or ineligibility. In this study, all patients received TKI treatment and belonged to the TACE-failure or -ineligible advanced HCC group. The median OS and TTP were 10 months (95% CI: 5.5–14.6) and 3.5 months (95% CI: 2.3–4.7), respectively, similar to previous studies [9].

Although TKI- and anti-PD-1-based treatments are regarded as basal therapies for advanced HCC according to many international HCC guidelines, TACE is the most frequently utilized treatment [7,17]. SHARP and Oriental studies showed that sorafenib could delay advanced HCC tumor progress (median TTP, 2.8–5.5 months) and prolong patient survival (median OS, 6.5–10.7 months) [18,19]. Meanwhile, lenvatinib was not inferior to sorafenib in advanced HCC (median OS, 13.6 months vs. 12.3 months) [20]. Additionally, some studies showed that the survival time of TACE treatment was comparable to sorafenib alone [21,22]. Moreover, some reviews have demonstrated that TACE plus sorafenib/lenvatinib was more effective than sorafenib/lenvatinib or TACE alone in advanced HCC patients [23–26]. In addition, patients with diffuse HCC, portal vein tumor thrombosis, or an arterioportal/arteriovenous shunt are not suitable for TACE. Furthermore, some HCC patients became TACE-refractory after repeatedly undergoing TACE treatment. More importantly, multiple TACE treatments may cause liver function abnormalities. For these patients, switching to other TKIs or PD-1 inhibitor treatments may be a good choice, such as using sorafenib [27], lenvatinib [28], and apatinib [29,30]. However, some patients may not withstand TKI treatment, and the treatment efficacy was poor (only prolonging OS by 3–6 months). Other TKIs, for example, regorafenib [31], apatinib [32], and cabozantinib [33], may also be selected for advanced HCC. However, the treatment effects were also limited, and the ORR was 2–18.8% [34]. Although several PD-1/PD-L1 inhibitors served as second-line treatments for unresectable HCC patients with a median OS of 13.9–15.6 months [35–37], they are not suitable for patients with a high tumor burden and a short life expectancy of less than 3 months.

HAIC has been used classically for unresectable colorectal liver metastases [38], and is also widely performed in HCC patients. Some studies showed that HAIC with FOLFOX4 is superior to sorafenib for advanced HCC patients [39,40]. JSH consensus statements proposed that HAIC is another choice for patients with a liver function of Child–Pugh class B or worse at the time of TACE failure/refractoriness [41]. In the present study, instead of 5-FU, raltitrexed was used in the treatment of HAIC, which shortened the infusion time, improved patient comfort, and reduced hospitalization hours. A phase II prospective study showed that raltitrexed-plus-oxaliplatin-based HAIC led to a higher ORR (51.4%, 18 of 35 patients) and was considered safe and tolerable in patients with unresectable HCC [15]. Another retrospective study highlighted the efficacy of low-dose continuous HAICROX for advanced HCC patients with MPVTT, where the median survival time was 8.7 months [14]. However, the survival benefits of HAICROX treatments as a later-line therapy in TACE-TKI-failed or -refractory HCC remain unknown. In this study, we treated TACE-failed or -ineligible advanced HCC patients with HAIC with raltitrexed plus oxaliplatin. Taking the patient's liver function and physical strength into consideration, the dosage of raltitrexed and oxaliplatin was reduced by one-third compared to the dosage in previous HAICROX studies [15]. The ORR and DCR were significantly lower in our study compared to previous results (ORR, 51.4%, 18 of 35 patients; DCR, 88.6%, 31 of 35 patients) [15], but our patients belonged to a less favorable subset of clinical baseline characteristics, such as a greater tumor burden (71.4% of patients with more than three tumors) and they were at a later BCLC stage (all of them were BCLC stage cases, 60% of patients had vascular invasion, and 54.3% of patients with extrahepatic metastasis). In fact, all of our patients were classified as TACE-failed (85.7% of patients) or TACE-unsuitable (14.3% of patients) and had received first or second lines of TKI therapy. However, our survival outcomes, including OS and TTP, were 10.0 months (95% CI: 5.5–14.6) and 3.5 months, respectively (95% CI: 2.3–4.7), comparable to those receiving HAIC with modified FOLFOX, as reported by Hsu SJ [9]. Of note, the patients in our study who received clinical benefits from HAICROX had improved OS and TTP compared to non-responders (median OS, not reached vs. 6.8 months, $p = 0.014$; median TTP, 6.5 months vs. 2.6 months, $p = 0.023$).

The results from the Check Mate 459 [42] and KEYNOTE-240 [36] trials show that anti-PD-1 immunotherapy with nivolumab or pembrolizumab as a single agent did not meet the setting primary endpoints (OS). However, in this study, anti-PD-1 treatment combination therapy was identified as an independent prognostic factor, and the patients who accepted anti-PD-1 immunotherapy and HAICROX plus TKI treatments achieved better survival outcomes; the median OS was 15.8 vs. 6.7 months ($p = 0.01$; HR = 0.329; 95% CI: 0.135–0.802) and the median TTP was 6.5 vs. 2.1 months ($p = 0.043$; HR 0.324; 95% CI:

0.113–0.926), which implied that the combination therapy may be a valuable choice and should be verified in our future studies. Consistent with another study, HAIC combined with anti-PD-1 immunotherapy (HAICAP) was superior to HAIC treatment alone for advanced hepatocellular carcinoma [43]. The potential mechanism may be that HAIC can improve anti-tumor immunity by releasing the neoplasm antigens from killed liver tumor cells [44] and increase PD-1/PD-L1 expression in the tumor microenvironment (TME). The combination of anti-PD1 therapy and locoregional therapy may also reduce the proportion of Tregs and the aggregation of myeloid-derived suppressor cells (MDSCs) [45].

Compared to previous research [14,15], HAICROX in TACE-refractory or -failed advanced HCC showed good tolerance and safety in our study. The most frequent adverse events were thrombocytopenia (17.1%), alanine aminotransferase (ALT), and aspartate aminotransferase (AST) elevation (34.29%, both), which can be managed and alleviated through dose reduction. Overall, these results show that HAICROX is a promising treatment for advanced HCC patients who failed or are unsuited to TACE combined with TKI treatments, and the subsequent HAICROX therapy prolonged the OS and TTP of advanced-stage HCC patients to some extent.

However, this study has several limitations. Firstly, this is a retrospective study, and all clinical data were obtained from a single medical center. Secondly, the categories of PD-1 inhibitors varied, which might influence the uniformity of the treatment procedure. Thirdly, our study was a single-arm study, and the sample size was small. Thus, more randomized controlled trials are required to verify the efficacy of HAICROX in this clinical setting.

In conclusion, this result highlighted that HAICROX largely prolonged the OS and TTP of advanced HCC patients who progressed on TACE-based treatment or were unsuitable/ineligible for TACE. The results provide a promising option for the later-line treatment for the majority of advanced HCC patients with TACE failure or ineligibility.

5. Conclusions

HAICROX showed valuable efficacy and tolerable toxicity, and it is an alternative option for advanced-stage HCC patients with TACE failure or ineligibility.

Supplementary Materials: The following supporting information can be downloaded at: https://www.mdpi.com/article/10.3390/medicina58101343/s1, Table S1: Univariate and Multivariate Analyses of Prognostic Factors for TTP of 35 patients.

Author Contributions: Conceptualization, Y.W. and B.Z.; methodology, Z.Z., S.Z. and B.Z.; software, Z.Z., Y.W. and M.W.; validation, T.Z. and H.C.; formal analysis, Y.W. and S.Z.; investigation, Y.W., Z.Z. and S.Z.; resources, Y.W., S.Z., Z.Z., X.C., G.C., T.Z., H.C. and X.G.; data curation, S.Z.; writing—original draft preparation, Y.W.; writing—review and editing, S.Z. and B.Z.; visualization, Y.W.; supervision, B.Z.; project administration, Y.W., X.X. and B.Z.; funding acquisition, Y.W., Z.Z. and B.Z. All authors have read and agreed to the published version of the manuscript.

Funding: This research was funded in part by the Xiamen Medical and Health Key Project (3502Z20191105 to Boheng Zhang), the Foundation of the Bethune Charitable Foundation (BCF-MSD-HCC-20201127-001 to Zhenzhen Zhang), and the Foundation of the Xiamen Branch, Zhongshan Hospital, Fudan University of China (2020ZSXMYS02 to Yanfang Wu).

Institutional Review Board Statement: The study was conducted in accordance with the Declaration of Helsinki, and approved by the Ethics Committee of Zhongshan Hospital, Fudan University (Xiamen Branch) (protocol # B2019-026, 23 March 2020).

Informed Consent Statement: Written informed consent was obtained from the patients to publish this paper.

Data Availability Statement: The data presented in this study are available on request from the corresponding author.

Conflicts of Interest: The authors declare no conflict of interest.

References

1. Bray, F.; Ferlay, J.; Soerjomataram, I.; Siegel, R.L.; Torre, L.A.; Jemal, A. Global cancer statistics 2018: GLOBOCAN estimates of incidence and mortality worldwide for 36 cancers in 185 countries. *CA Cancer J Clin.* **2018**, *68*, 394–424. [CrossRef]
2. Chen, W.; Zheng, R.; Baade, P.D.; Zhang, S.; Zeng, H.; Bray, F.; Jemal, A.; Yu, X.Q.; He, J. Cancer statistics in China, 2015. *CA Cancer J Clin.* **2016**, *66*, 115–132. [CrossRef] [PubMed]
3. Giannini, E.G.; Farinati, F.; Ciccarese, F.; Pecorelli, A.; Rapaccini, G.L.; Di Marco, M.; Benvegnù, L.; Caturelli, E.; Zoli, M.; Borzio, F.; et al. Prognosis of untreated hepatocellular carcinoma. *Hepatology* **2015**, *61*, 184–190. [CrossRef] [PubMed]
4. European Association for the Study of the Liver. EASL Clinical Practice Guidelines: Management of hepatocellular carcinoma. *J. Hepatol.* **2018**, *69*, 182–236. [CrossRef] [PubMed]
5. Park, J.W.; Chen, M.; Colombo, M.; Roberts, L.R.; Schwartz, M.; Chen, P.J.; Kudo, M.; Johnson, P.; Wagner, S.; Orsini, L.S.; et al. Global patterns of hepatocellular carcinoma management from diagnosis to death: The BRIDGE Study. *Liver Int.* **2015**, *35*, 2155–2166. [CrossRef]
6. Llovet, J.M.; Real, M.I.; Montaña, X.; Planas, R.; Coll, S.; Aponte, J.; Ayuso, C.; Sala, M.; Muchart, J.; Solà, R.; et al. Barcelona Liver Cancer Group. Arterial embolisation or chemoembolisation versus symptomatic treatment in patients with unresectable hepatocellular carcinoma: A randomised controlled trial. *Lancet* **2002**, *359*, 1734–1739. [CrossRef]
7. Lencioni, R.; de Baere, T.; Soulen, M.C.; Rilling, W.S.; Geschwind, J.F. Lipiodol transarterial chemoembolization for hepatocellular carcinoma: A systematic review of efficacy and safety data. *Hepatology* **2016**, *64*, 106–116. [CrossRef] [PubMed]
8. Jeon, M.Y.; Kim, H.S.; Lim, T.S.; Han, D.H.; Kim, B.K.; Park, J.Y.; Kim, D.Y.; Ahn, S.H.; Choi, G.H.; Choi, J.S.; et al. Refractoriness to transarterial chemoembolization in patients with recurrent hepatocellular carcinoma after curative resection. *PLoS ONE* **2019**, *14*, e0214613. [CrossRef] [PubMed]
9. Hsu, S.J.; Xu, X.; Chen, M.P.; Zhao, Z.Y.; Wang, Y.; Yin, X.; Zhang, L.; Ge, N.L.; Chen, Y.; Wang, Y.H.; et al. Hepatic Arterial Infusion Chemotherapy with Modified FOLFOX as an Alternative Treatment Option in Advanced Hepatocellular Carcinoma Patients with Failed or Unsuitability for Transarterial Chemoembolization. *Acad. Radiol.* **2021**, *28* (Suppl. S1), S157–S166. [CrossRef] [PubMed]
10. Lyu, N.; Kong, Y.; Mu, L.; Lin, Y.; Li, J.; Liu, Y.; Zhang, Z.; Zheng, L.; Deng, H.; Li, S.; et al. Hepatic arterial infusion of oxaliplatin plus fluorouracil/leucovorin vs. sorafenib for advanced hepatocellular carcinoma. *J. Hepatol.* **2018**, *69*, 60–69. [CrossRef] [PubMed]
11. Lyu, N.; Lin, Y.; Kong, Y.; Zhang, Z.; Liu, L.; Zheng, L.; Mu, L.; Wang, J.; Li, X.; Pan, T.; et al. FOXAI: A phase II trial evaluating the efficacy and safety of hepatic arterial infusion of oxaliplatin plus fluorouracil/leucovorin for advanced hepatocellular carcinoma. *Gut* **2018**, *67*, 395–396. [CrossRef] [PubMed]
12. Kudo, M.; Ueshima, K.; Yokosuka, O.; Ogasawara, S.; Obi, S.; Izumi, N.; Aikata, H.; Nagano, H.; Hatano, E.; Sasaki, Y.; et al. Sorafenib plus low-dose cisplatin and fluorouracil hepatic arterial infusion chemotherapy versus sorafenib alone in patients with advanced hepatocellular carcinoma (SILIUS): A randomised, open label, phase 3 trial. *Lancet Gastroenterol. Hepatol.* **2018**, *3*, 424–432. [CrossRef]
13. Hagner, N.; Joerger, M. Cancer chemotherapy: Targeting folic acid synthesis. *Cancer Manag. Res.* **2010**, *2*, 293–301. [CrossRef]
14. Zhu, L.Z.; Xu, S.; Qian, H.L. Transarterial embolization and low-dose continuous hepatic arterial infusion chemotherapy with oxaliplatin and raltitrexed for hepatocellular carcinoma with major portal vein tumor thrombus. *World J. Gastroenterol.* **2018**, *24*, 2501–2507. [CrossRef] [PubMed]
15. Chen, S.; Zhang, K.; Liu, W.; Yu, W. Hepatic arterial infusion of oxaliplatin plus raltitrexed in patients with intermediate and advanced stage hepatocellular carcinoma: A phase II, single-arm, prospective study. *Eur. J. Cancer* **2020**, *134*, 90–98. [CrossRef] [PubMed]
16. Johnson, P.J.; Berhane, S.; Kagebayashi, C.; Satomura, S.; Teng, M.; Reeves, H.L.; O'Beirne, J.; Fox, R.; Skowronska, A.; Palmer, D.; et al. Assessment of liver function in patients with hepatocellular carcinoma: A new evidence-based approach-the ALBI grade. *J. Clin. Oncol.* **2015**, *33*, 550–558. [CrossRef] [PubMed]
17. Llovet, J.M.; Bruix, J. Systematic review of randomized trials for unresectable hepatocellular carcinoma: Chemoembolization improves survival. *Hepatology* **2003**, *37*, 429–442. [CrossRef]
18. Llovet, J.M.; Ricci, S.; Mazzaferro, V.; Hilgard, P.; Gane, E.; Blanc, J.F.; de Oliveira, A.C.; Santoro, A.; Raoul, J.L.; Forner, A.; et al. Sorafenib in advanced hepatocellular carcinoma. *N. Engl. J. Med.* **2008**, *359*, 378–390. [CrossRef]
19. Cheng, A.L.; Kang, Y.K.; Chen, Z.; Tsao, C.J.; Qin, S.; Kim, J.S.; Luo, R.; Feng, J.; Ye, S.; Yang, T.S.; et al. Efficacy and safety of sorafenib in patients in the Asia-Pacific region with advanced hepatocellular carcinoma: A phase III randomised, double-blind, placebo-controlled trial. *Lancet Oncol.* **2009**, *10*, 25–34. [CrossRef]
20. Kudo, M.; Finn, R.S.; Qin, S.; Han, K.H.; Ikeda, K.; Piscaglia, F.; Baron, A.; Park, J.W.; Han, G.; Jassem, J.; et al. Lenvatinib versus sorafenib in first-line treatment of patients with unresectable hepatocellular carcinoma: A randomised phase 3 non-inferiority trial. *Lancet* **2018**, *391*, 1163–1173. [CrossRef]
21. Nishikawa, H.; Osaki, Y.; Iguchi, E.; Takeda, H.; Nakajima, J.; Matsuda, F.; Sakamoto, A.; Henmi, S.; Hatamaru, K.; Saito, S.; et al. Comparison of the efficacy of transcatheter arterial chemoembolization and sorafenib for advanced hepatocellular carcinoma. *Exp. Ther. Med.* **2012**, *4*, 381–386. [CrossRef] [PubMed]
22. Pinter, M.; Hucke, F.; Graziadei, I.; Vogel, W.; Maieron, A.; Königsberg, R.; Stauber, R.; Grünberger, B.; Müller, C.; Kölblinger, C.; et al. Advanced-stage hepatocellular carcinoma: Transarterial chemoembolization versus sorafenib. *Radiology* **2012**, *263*, 590–599. [CrossRef] [PubMed]

23. Wu, F.X.; Chen, J.; Bai, T.; Zhu, S.L.; Yang, T.B.; Qi, L.N.; Zou, L.; Li, Z.H.; Ye, J.Z.; Li, L.Q. The safety and efficacy of transarterial chemoembolization combined with sorafenib and sorafenib mono-therapy in patients with BCLC stage B/C hepatocellular carcinoma. *BMC Cancer* **2017**, *17*, 645. [CrossRef] [PubMed]
24. Zhang, X.; Wang, K.; Wang, M.; Yang, G.; Ye, X.; Wu, M.; Cheng, S. Transarterial chemoembolization (TACE) combined with sorafenib versus TACE for hepatocellular carcinoma with portal vein tumor thrombus: A systematic review and meta-analysis. *Oncotarget* **2017**, *8*, 29416–29427. [CrossRef]
25. Zhang, X.P.; Wang, K.; Li, N.; Zhong, C.Q.; Wei, X.B.; Cheng, Y.Q.; Gao, Y.Z.; Wang, H.; Cheng, S.Q. Survival benefit of hepatic resection versus transarterial chemoembolization for hepatocellular carcinoma with portal vein tumor thrombus: A systematic review and meta-analysis. *BMC Cancer* **2017**, *17*, 902. [CrossRef]
26. Wu, J.Y.; Yin, Z.Y.; Bai, Y.N.; Chen, Y.F.; Zhou, S.Q.; Wang, S.J.; Zhou, J.Y.; Li, Y.N.; Qiu, F.N.; Li, B.; et al. Lenvatinib Combined with Anti-PD-1 Antibodies Plus Transcatheter Arterial Chemoembolization for Unresectable Hepatocellular Carcinoma: A Multicenter Retrospective Study. *J Hepatocell Carcinoma* **2021**, *8*, 1233–1240. [CrossRef]
27. Kodama, K.; Kawaoka, T.; Aikata, H.; Uchikawa, S.; Inagaki, Y.; Hatooka, M.; Morio, K.; Nakahara, T.; Murakami, E.; Tsuge, M.; et al. Comparison of clinical outcome of hepatic arterial infusion chemotherapy and sorafenib for advanced hepatocellular carcinoma according to macrovascular invasion and transcatheter arterial chemoembolization refractory status. *J. Gastroenterol. Hepatol.* **2018**, *33*, 1780–1786. [CrossRef]
28. Shimose, S.; Kawaguchi, T.; Tanaka, M.; Iwamoto, H.; Miyazaki, K.; Moriyama, E.; Suzuki, H.; Niizeki, T.; Shirono, T.; Nakano, M.; et al. Lenvatinib prolongs the progression-free survival time of patients with intermediate-stage hepatocellular carcinoma refractory to transarterial chemoembolization: A multicenter cohort study using data mining analysis. *Oncol Lett.* **2020**, *20*, 2257–2265. [CrossRef]
29. Cao, Y.; Ouyang, T.; Xiong, F.; Kan, X.; Chen, L.; Liang, B.; Zheng, C. Efficacy of apatinib in patients with sorafenib-transarterial chemoembolization refractory hepatocellular carcinoma: A retrospective study. *Hepatol. Int.* **2021**, *15*, 1268–1277. [CrossRef]
30. Qiu, Z.; Shen, L.; Chen, S.; Qi, H.; Cao, F.; Xie, L.; Fan, W. Efficacy Of Apatinib In Transcatheter Arterial Chemoembolization (TACE) Refractory Intermediate And Advanced-Stage Hepatocellular carcinoma: A Propensity Score Matching Analysis. *Cancer Manag. Res.* **2019**, *11*, 9321–9330. [CrossRef]
31. Bruix, J.; Qin, S.; Merle, P.; Granito, A.; Huang, Y.H.; Bodoky, G.; Pracht, M.; Yokosuka, O.; Rosmorduc, O.; Breder, V.; et al. Regorafenib for patients with hepatocellular carcinoma who progressed on sorafenib treatment (RESORCE): A randomised, double-blind, placebo-controlled, phase 3 trial. *Lancet* **2017**, *389*, 56–66. [CrossRef]
32. Qin, S.; Li, Q.; Gu, S.; Chen, X.; Lin, L.; Wang, Z.; Xu, A.; Chen, X.; Zhou, C.; Ren, Z.; et al. Apatinib as second-line or later therapy in patients with advanced hepatocellular carcinoma (AHELP): A multicentre, double-blind, randomised, placebo-controlled, phase 3 trial. *Lancet Gastroenterol. Hepatol.* **2021**, *6*, 559–568. [CrossRef]
33. Abou-Alfa, G.K.; Meyer, T.; Cheng, A.L.; El-Khoueiry, A.B.; Rimassa, L.; Ryoo, B.Y.; Cicin, I.; Merle, P.; Chen, Y.; Park, J.W.; et al. Cabozantinib in Patients with Advanced and Progressing Hepatocellular Carcinoma. *N. Engl. J. Med.* **2018**, *379*, 54–63. [CrossRef] [PubMed]
34. Da Fonseca, L.G.; Reig, M.; Bruix, J. Tyrosine Kinase Inhibitors and Hepatocellular Carcinoma. *Clin. Liver Dis.* **2020**, *24*, 719–737. [CrossRef] [PubMed]
35. Yau, T.; Hsu, C.; Kim, T.Y.; Choo, S.P.; Kang, Y.K.; Hou, M.M.; Numata, K.; Yeo, W.; Chopra, A.; Ikeda, M.; et al. Nivolumab in advanced hepatocellular carcinoma: Sorafenib-experienced Asian cohort analysis. *J. Hepatol.* **2019**, *71*, 543–552. [CrossRef] [PubMed]
36. Finn, R.S.; Ryoo, B.Y.; Merle, P.; Kudo, M.; Bouattour, M.; Lim, H.Y.; Breder, V.; Edeline, J.; Chao, Y.; Ogasawara, S.; et al. Pembrolizumab As Second-Line Therapy in Patients With Advanced Hepatocellular Carcinoma in KEYNOTE-240: A Randomized, Double-Blind, Phase III Trial. *J. Clin. Oncol.* **2020**, *38*, 193–202. [CrossRef]
37. Qin, S.; Ren, Z.; Meng, Z.; Chen, Z.; Chai, X.; Xiong, J.; Bai, Y.; Yang, L.; Zhu, H.; Fang, W.; et al. Camrelizumab in patients with previously treated advanced hepatocellular carcinoma: A multicentre, open-label, parallel-group, randomised, phase 2 trial. *Lancet Oncol.* **2020**, *21*, 571–580. [CrossRef]
38. Kemeny, N.E.; Niedzwiecki, D.; Hollis, D.R.; Lenz, H.J.; Warren, R.S.; Naughton, M.J.; Weeks, J.C.; Sigurdson, E.R.; Herndon, J.E., 2nd.; Zhang, C.; et al. Hepatic arterial infusion versus systemic therapy for hepatic metastases from colorectal cancer: A randomized trial of efficacy, quality of life, and molecular markers (CALGB 9481). *J. Clin. Oncol.* **2006**, *24*, 1395–1403. [CrossRef]
39. Long, G.B.; Xiao, C.W.; Zhao, X.Y.; Zhang, J.; Li, X. Effects of hepatic arterial infusion chemotherapy in the treatment of hepatocellular carcinoma: A meta-analysis. *Medicine* **2020**, *99*, e20745. [CrossRef]
40. Liu, M.; Shi, J.; Mou, T.; Wang, Y.; Wu, Z.; Shen, A. Systematic review of hepatic arterial infusion chemotherapy versus sorafenib in patients with hepatocellular carcinoma with portal vein tumor thrombosis. *J. Gastroenterol. Hepatol.* **2020**, *35*, 1277–1287. [CrossRef]
41. Kudo, M.; Kawamura, Y.; Hasegawa, K.; Tateishi, R.; Kariyama, K.; Shiina, S.; Toyoda, H.; Imai, Y.; Hiraoka, A.; Ikeda, M.; et al. Management of Hepatocellular Carcinoma in Japan: JSH Consensus Statements and Recommendations 2021 Update. *Liver Cancer* **2021**, *10*, 181–223. [CrossRef] [PubMed]

42. Yau, T.; Park, J.W.; Finn, R.S.; Cheng, A.L.; Mathurin, P.; Edeline, J.; Kudo, M.; Harding, J.J.; Merle, P.; Rosmorduc, O.; et al. Nivolumab versus sorafenib in advanced hepatocellular carcinoma (CheckMate 459): A randomised, multicentre, open-label, phase 3 trial. *Lancet Oncol.* **2022**, *23*, 77–90. [CrossRef]
43. Mei, J.; Li, S.H.; Li, Q.J.; Sun, X.Q.; Lu, L.H.; Lin, W.P.; Zheng, L.; Chen, M.S.; Shi, M.; Wei, W.; et al. Anti-PD-1 Immunotherapy Improves the Efficacy of Hepatic Artery Infusion Chemotherapy in Advanced Hepatocellular Carcinoma. *J. Hepatocell. Carcinoma* **2021**, *8*, 167–176. [CrossRef] [PubMed]
44. Galluzzi, L.; Buqué, A.; Kepp, O.; Zitvogel, L.; Kroemer, G. Immunogenic cell death in cancer and infectious disease. *Nat. Rev. Immunol.* **2017**, *17*, 97–111. [CrossRef]
45. Noman, M.Z.; Desantis, G.; Janji, B.; Hasmim, M.; Karray, S.; Dessen, P.; Bronte, V.; Chouaib, S. PD-L1 is a novel direct target of HIF-1α, and its blockade under hypoxia enhanced MDSC-mediated T cell activation. *J. Exp. Med.* **2014**, *211*, 781–790. [CrossRef]

medicina

MDPI

Article

PD-L1 Gene Polymorphisms rs822336 G>C and rs822337 T>A: Promising Prognostic Markers in Triple Negative Breast Cancer Patients

Andreas-Evangelos Makrantonakis [1], Eleni Zografos [2,*], Maria Gazouli [2], Konstantinos Dimitrakakis [3], Konstantinos G. Toutouzas [4], Constantinos G. Zografos [4], Despoina Kalapanida [5], Andriani Tsiakou [6], George Samelis [7] and Flora Zagouri [5,*]

1 Second Department of Medical Oncology, Theagenion Cancer Hospital, 546 44 Thessaloniki, Greece
2 Department of Basic Medical Sciences, Laboratory of Biology, School of Medicine, National and Kapodistrian University of Athens, 115 27 Athens, Greece
3 Department of Obstetrics and Gynaecology, Alexandra Hospital, Medical School, National and Kapodistrian University of Athens, 115 28 Athens, Greece
4 1st Propaedeutic Surgical Department, Hippokrateio Hospital, National and Kapodistrian University of Athens, 115 27 Athens, Greece
5 Department of Clinical Therapeutics, Alexandra Hospital, Medical School, National and Kapodistrian University of Athens, 115 28 Athens, Greece
6 First Department of Dermatology, Syggros Hospital, School of Medicine, National and Kapodistrian University of Athens, 161 21 Athens, Greece
7 Department of Oncology, Hippocrateion Hospital, National and Kapodistrian University of Athens, 115 27 Athens, Greece
* Correspondence: el_zogra@hotmail.com (E.Z.); florazagouri@yahoo.co.uk (F.Z.)

Citation: Makrantonakis, A.-E.; Zografos, E.; Gazouli, M.; Dimitrakakis, K.; Toutouzas, K.G.; Zografos, C.G.; Kalapanida, D.; Tsiakou, A.; Samelis, G.; Zagouri, F. PD-L1 Gene Polymorphisms rs822336 G>C and rs822337 T>A: Promising Prognostic Markers in Triple Negative Breast Cancer Patients. *Medicina* **2022**, *58*, 1399. https://doi.org/10.3390/medicina58101399

Academic Editors: Nicolae Crisan and Călin Căinap

Received: 21 July 2022
Accepted: 28 September 2022
Published: 6 October 2022

Abstract: *Background and Objectives*: Triple-negative breast cancer (TNBC) is a highly heterogeneous subtype that is associated with unresponsiveness to therapy and hence with high mortality rates. In this study we aimed to investigate the prognostic role of the rs822336 G>C and rs822337 T>A polymorphisms of the PD-L1 (Programmed Death-Ligand 1) in TNBC patients. *Materials and methods:* Formalin-fixed paraffin-embedded tissues from 114 TNBC patients and blood samples from 124 healthy donors were genotyped, and subsequently extensive statistical analysis was performed in order to investigate the clinical value of these polymorphism in TNBC. *Results:* Regarding rs822336 G>C, we found that the CG genotype was the most common among women that harbored Stage IV breast tumors (81.8%; $p = 0.022$), recurred (38.9%; $p = 0.02$) and died (66.7%; $p = 0.04$). Similarly, the rs822337 T>A genotype AA is associated with worse prognosis, since it was the most common genotype among stage IV tumors (72.7%; $p = 0.04$) and in TNBC patients that relapsed (75%; $p = 0.021$) and died (81.5%; $p = 0.004$). Our statistical analysis revealed that the rs822336 G>C genotype CG and the rs822337 T>A allele AA are strongly associated with inferior DFS and OS intervals. Moreover, it was revealed that women harboring mutated genotypes of both SNPs had shorter disease-free (Kaplan–Meier; $p = 0.037$, Cox analysis; $p = 0.04$) and overall (Kaplan–Meier; $p = 0.025$, Cox analysis; $p = 0.03$) survival compared to patients having normal genotype of at least one SNP. Multivariate analysis also showed that the presence of mutated genotypes of both SNPs is a strong and independent marker for predicting shorter DFS ($p = 0.02$) and OS ($p = 0.008$). *Conclusion:* Our study revealed that PD-L1 rs822336 G>C and rs822337 T>A polymorphisms were differentially expressed in our cohort of TNBC patients, and that this distribution was associated with markers of unfavorable prognosis and worse survival.

Keywords: breast cancer; breast cancer prognosis; immune checkpoints; molecular tumor markers; PD-L1; polymorphisms

1. Introduction

The strong interaction between the immune system and malignant transformation is a recently recognized aspect of cancer pathogenesis that opened new horizons in cancer patients' management, since new opportunities have arisen regarding the recognition of novel immuno-oncological biomarkers and therapeutic targets. Already, many immunological concepts have been successfully translated into novel clinical treatment strategies that have radically changed the therapeutic landscape of many cancer types, including breast cancer. For example, the recent FDA approvals of the monoclonal antibody pembrolizumab and the antibody-drug conjugate acituzumab govitecan for the treatment of triple negative breast cancer (TNBC), were considered important milestones in the management of these patients [1,2].

TNBC is the most aggressive breast cancer subtype and often is associated with unresponsiveness to therapy and hence with high rates of mortality [3]. It is now accepted that the heterogeneity of TNBC renders the current therapeutic approaches inadequate as they are unable to target the different molecular pathways that are implicated in TNBC pathogenesis. Mounting evidence suggests that different degrees of immunogenic activity contribute to the phenotypic and clinical heterogeneity of TNBC [4–6]. This subtype exhibits the highest tumor immunogenicity of all breast cancer subtypes [7–11] and nowadays ongoing research efforts are focused on the thorough characterization of TNBC immune-associated features so that immune signatures can be incorporated into the established molecular subtyping, refining thus the routine prognostic and therapeutic-decision approaches. Hereto, several independent groups have studied the immune-landscape of TNBC along with the expression of immune-related genes and the results support the notion that TNBC can be subclassified in distinct subtypes according to their immunogenomic profile. This separation of strongly immunogenic tumors from the weakly ones hold promises for a more personalized prognosis and treatment in terms of immunotherapy [12–17].

Programmed death ligand-1 (PD-L1) is one of the molecules that is steadily included in the majority of these studies. The PD-L1 is an immune checkpoint molecule that is at the forefront of breast cancer research since it seems that not only it contributes to breast neoplastic transformation, but is also a clinically useful biomarker. More specifically PD-L1 expression influences TNBC prognosis [18–21], a potential that was clearly shown by the accelerated FDA approval of the immune checkpoint inhibitor atezolizumab, despite its 2021 withdrawal due to the failure of a subsequent clinical trial [22]. PD-L1 is encoded by the *CD274* gene that is located in chromosome 9p24.1 [23]. The advent of high-throughput sequencing technologies facilitated the recognition of several single nucleotide polymorphisms in *PD-L1* and paved the way for the identification of potential, novel cancer biomarkers. A recent study suggested that the SNPs rs4143815 and rs2282055 may serve as useful biomarkers for the efficacy of nivolumab in NSCLC patients [24]. Moreover it has been found that certain SNPs of *PD-L1* are associated with NSCLC outcome [24], whereas other data suggest that some SNPs of *PD-L1* have clinical value in gastric adenocarcinoma [25] and NSCLC [26].

By the virtue of (1) the significant role of PD-L1 in prognosis and in guiding optimal treatment decisions of TNBC patients and (2) the fruitful results regarding the potential value of SNPs of *PD-L1* as prognostic and predictive markers, we aimed to examine the presence of the rs822336 G>C and rs822337 T>A in 114 FFPE (formalin-fixed paraffin-embedded) tissues obtained from TNBC patients and to correlate the SNP status with established clinicopathological parameters and the survival of the patients.

2. Materials and Methods

2.1. Clinical Characteristics of the Patients

The study cohort consisted of 114 FFPEs obtained from women with TNBC patients as well as of 124 blood samples from healthy donors. Detailed medical history, demographic data, clinicopathologic characteristics and follow-up survival information were collected for each patient, for statistical analysis (Table 1).

Table 1. Clinicopathological characteristics of triple-negative breast cancer patients (n = 114).

Variant	Number of Patients	Variant	Number of Patients
Age		Ki-67 Index	
≥57	57 (50.0%)	Positive	58 (50.9%)
<57	56 (49.1%)	Negative	15 (13.1%)
Unknown	1(0.9%)	Unknown	41 (36.0%)
TNM Stage		Grade	
I	20 (17.5%)	I	2 (1.7%)
II	41 (35.9%)	II	10 (8.8%)
III	24 (21.1%)	III	95 (83.4%)
IV	22 (19.4%)	Unknown	7 (6.1%)
Unknown	7 (6.1%)		
Tumor Type		Lymph nodes	
Ductal	87 (76.3%)	N0	66 (57.9%)
Lobular	10 (8.8%)	N1	13 (11.4%)
Medular	8 (7%)	N2	9 (7.9%)
Other	9 (7.9%)	N3	19 (16.7%)
Unknown	0 (0%)	Unknown	7 (6.1%)
Disease Progression		Death	
Yes	24 (21.1%)	Yes	21 (18.4%)
No	90 (78.9%)	No	93 (81.6%)
Unknown	0 (0%)	Unknown	0 (0%)

The current study was designed according to the most recent guidelines for reporting tumor biomarkers and was approved by the ethical committee of the "Hippokration", University Hospital of Athens. Informed consent was obtained from all study participants. Moreover, research procedures of the study comply with the ethical standards of the World Medical Association Declaration of Helsinki.

2.2. Genotyping

DNA from paraffin-embedded breast tissues of patients and DNA from the blood of healthy controls was extracted from samples using the commercial Nucleopsin Tissue kit (Macherey-Nagel, Germany) according to the manufacturer's instructions. Genotyping was performed using allele-specific PCR (polymerase chain reaction). We used the primers (5′-ACTCTCAGTCATGCAGAAAAC-3′ and 5′-ACTCTCAGTCATGCAGAAAAG-3′) with the last nucleotide complementary to the allelic variant substitution base on the point mutation in question of the gene, and a common primer (5′-AAGATGGAGTCAAACAGGG-3′). The amplified PCR products of 239 bp were then digested using restriction enzymes and analyzed by 2% agarose gel electrophoresis in the presence of FastGene 100 bp DNA ladder (NIPPON Genetics Europe, 52349 Düren, Germany) using ethidium bromide staining and ultraviolet visualization.

2.3. Statistical Analysis

Genotype frequencies were analyzed with the x^2 test with Yate's correction using S-Plus (v.6.2 Insightful, Seattle, WA, USA) software. Odds ratios (ORs) and 95% confidence intervals (95% CIs) were calculated using GraphPad (version 300, GraphPad Software, San Diego, CA, USA). All p values were two-sided and p values < 0.05 were considered significant. The survival curves were constructed using the Kaplan-Meier method and comparison of two survival curves was performed using the log-rank test. The influence of each variable on survival was analyzed by the multivariate analysis of Cox proportional hazard model. The comparisons were performed using GraphPad version 3.00 (GraphPad Software Inc., San Diego, CA, USA).

3. Results

3.1. Differential Distribution of rs822336 G>C and rs822337 T>A between TNBC Patients and Healthy Controls

Initially, we compared the distribution of rs822336 G>C and rs822337 T>A genotypes in 114 TNBC patients and 124 healthy controls. Regarding rs822336 G>C, it was found that the GG genotype was the most common both in patients (48.2%) and in healthy controls (55.6%). In the patients' cohort, the second most abundant genotype was the CG (44.7%), whereas the CC genotype was the less common (7%). The same pattern was observed in the group of healthy donors and the corresponding percentages for the CG and CC genotype were 34.7% and 9.7%, respectively (Figure 1A). Likewise, the distribution of rs822337 T>A genotypes among patients and healthy controls was similar. Briefly, the most abundant genotype in both groups was the AA (patients: 53.5%, controls: 46%), following the AT (patients: 36.8%, controls: 41.1%) and the TT genotypes (patients: 9.6%, controls: 12.9%) (Figure 1B). However, the comparison of the differential presentation of the rs822336 G>C and rs822337 T>A genotypes among patients and controls, using the chi-square test, revealed no statistically significant difference (rs822336 G>C, $p = 0.27$; rs822337 T>A, $p = 0.47$).

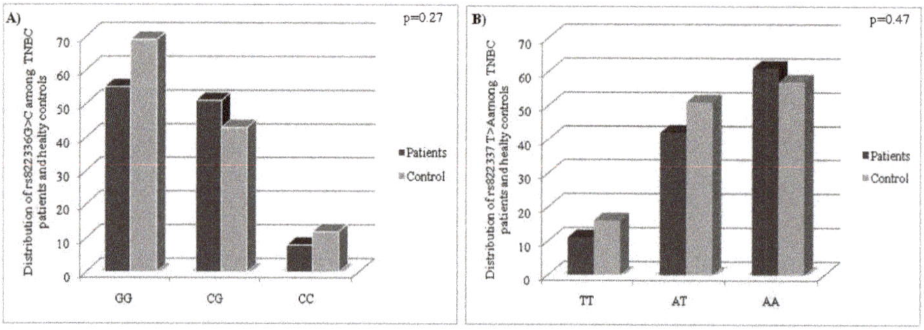

Figure 1. Distribution of the GG/CG/CC (rs822336 G>C) (**A**) and AA/AT/TT (rs822337 T>A) (**B**) alleles among TNBC patients and healthy controls.

3.2. Association of rs822336 G>C and rs822337 T>A with Patients' Clinical Variables

According to our statistical analysis, both of the studied PD-L1 SNPs demonstrated significant association with TNM stage (rs822336 G>C, $p = 0.022$; rs822337 T>A, $p = 0.04$), DFS (rs822336 G>C, $p = 0.02$; rs822337 T>A, $p = 0.021$) and OS (rs822336 G>C, $p = 0.04$; rs822337 T>A, $p = 0.004$) status. Specifically, regarding the rs822336 G>C, we found that the majority of patients with CG genotype harbored Stage IV breast tumors (81.8%) (Figure 2A). Moreover, this genotype was the most common among the women that recurred (38.9%) and died (66.7%) (Figure 2B and 2C, respectively). Similarly, we found that the rs822337 T>A genotype AA was associated with worse prognosis since it was the most common genotype among stage IV tumors (72.7%) (Figure 3A) as well as in TNBC patients that relapsed (75%) and died (81.5%) (Figure 3B and 3C, respectively).

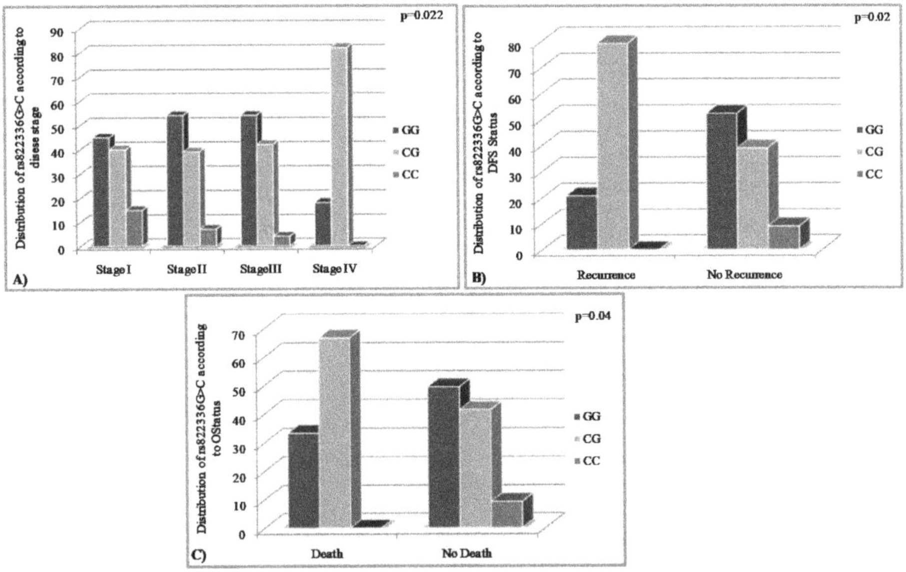

Figure 2. Distribution of the rs822336 G>C according to TNM Stage (**A**), DFS (**B**) and OS Status (**C**).

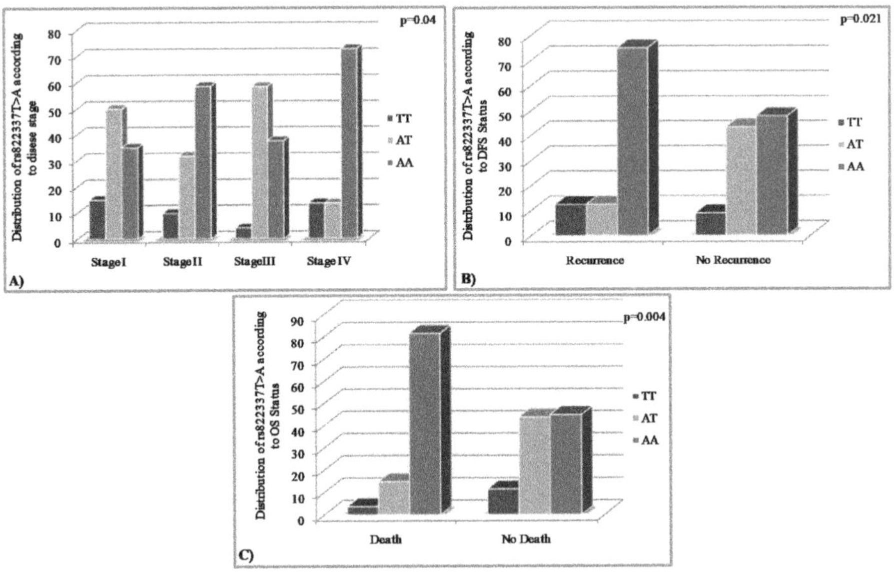

Figure 3. Distribution of the rs822337 T>A according to TNM Stage (**A**), DFS (**B**) and OS Status (**C**).

3.3. Survival Analysis and Prognostic Significance of rs822336 G>C and rs822337 T>A in TNBC Patients

Survival analysis was performed by Kaplan–Meier analysis and Cox proportional hazards regression models. Disease-free and overall survival information was available for 114 and 108 TNBC patients, respectively, and among them 24 women relapsed (21.1%) and 27 patients died (23.7%). Kaplan–Meier analysis corroborated the abovementioned chi-square test results, since it disclosed that the rs822336 G>C genotype CG was strongly associated with inferior DFS ($p = 0.002$) and OS ($p = 0.009$) intervals compared to the others genotypes (Figure 4).

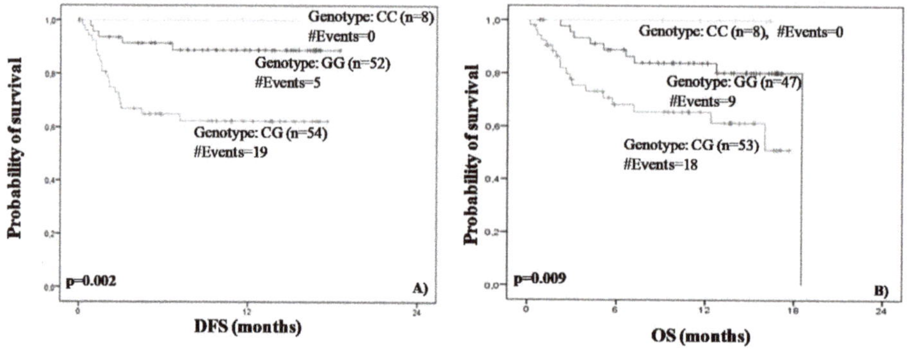

Figure 4. Kaplan–Meier disease-free (**A**) and overall (**B**) survival curves. *p* values were calculated by log-rank test.

Univariate and multivariate Cox regression analysis confirmed that the rs822336 G>C genotype CG is a marker of unfavorable prognosis in TNBC, since it was found that it is associated with increased risk of relapse (HR = 4.06, 9% CI = 1.51–4.88, $p = 0.005$) and death (HR = 2.74, 95% CI = 1.18–6.32, $p = 0.018$). Indeed, women harboring the CG allele were 4.06 and 2.74 time more likely to relapse and die, respectively (Table 2).

Table 2. Cox univariate regression analysis of rs822336 G>C for the prediction of disease-free (DFS) and overall survival (OS).

	Disease-Free Survival (DFS) ($n = 114$)		
Variable	**HR** [a]	**95% CI** [b]	***p* Value**
rs822336			
GG	1.00		
CG	4.06	1.51–4.88	0.005
CC	0.00	0.00	
	Overall Survival (OS) ($n = 108$)		
Variable	**HR** [a]	**95% CI** [b]	***p* Value**
rs822336			
GG	1.00		
CG	2.74	1.18–6.32	0.018
CC	0.00	0.00	

[a] Hazard Ratio, [b] Confidence interval of the estimated HR. Bold value indicates statistical significance

By performing the same statistical analysis, we found that rs822337 T>A allele AA was associated with worse survival in terms of DFS and OS. Briefly, Kaplan–Meier analysis demonstrated that women with AA genotype are characterized by shorter DFS ($p = 0.024$) and OS ($p = 0.004$) (Figure 5). In concordance with the abovementioned results, Cox regression analysis showed that the rs822337 T>A genotype AA was correlated with high risk of recurrence (HR = 1.02, 95% CI = 0.30–2.45, $p = 0.04$) and death (HR = 4.04, 95% CI = 0.54–3.24, $p = 0.01$) (Table 3).

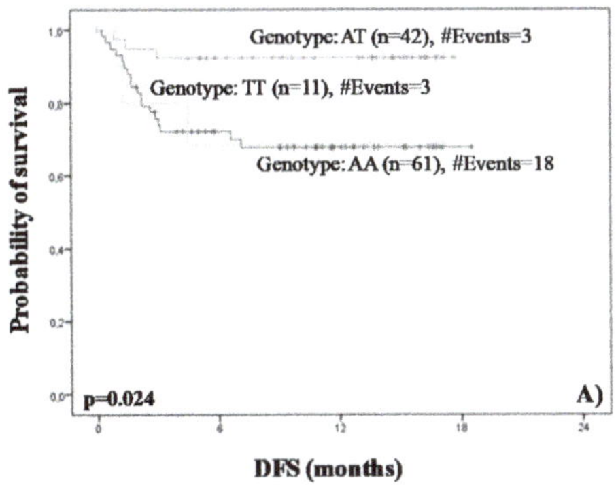

Figure 5. Kaplan–Meier disease-free (**A**) and overall (**B**) survival curves. *p* values were calculated by log-rank test.

Table 3. Cox univariate regression analysis of rs822337 T>A for the prediction of disease-free (DFS) and Overall survival (OS).

Disease-Free Survival (DFS) (n = 107)			
Variable	**HR** [a]	**95% CI** [b]	***p* Value**
rs822337			
TT	1.00		
AT	0.22	0.04–1.08	0.04
AA	1.02	0.30–2.45	
Overall Survival (OS) (n = 108)			
Variable	**HR** [a]	**95% CI** [b]	***p* Value**
rs822337			
TT	1.00		
AT	0.87	0.10–2.75	0.01
AA	4.04	0.54–3.24	

[a] Hazard Ratio, [b] Confidence interval of the estimated HR. Bold value indicates statistical significance

Based on these results we deemed it interesting to assess the prognostic value of the studied PD-L1 SNPs, after patient categorization based on the combination of the rs822336 G>C and rs822337 T>A genotypes (Table 4). The Kaplan-Meier analysis showed that the combination of the CG and AA genotypes are significantly associated with inferior DFS (p = 0.002) and OS (p = 0.002) intervals (Figure 6). The same conclusion was drawn from the Cox regression analysis, according to which TNBC patients harboring both the CG and AA genotypes demonstrated worse prognosis in terms of DFS (HR = 5.89, 95% CI = 1.35–8.65, p = 0.018) and OS (HR = 2.78, 95% CI = 1.02–7.65, p = 0.04). Indeed, these women were characterized by almost 6- and 3-times higher risk of relapse and death, respectively.

We then investigated the prognostic performance of the rs822336 G>C and rs822337 T>A after the following patients' categorization: Group I/II/III/VI (Group A) and Group IV/V/VII/VIII (Group B). According to Kaplan–Meier curves, women belonging to group B and harboring mutated genotypes of both SNPs (i.e., CG+AA/CG+AT/CC+AA/CC+AT) had shorter disease-free (p = 0.037) (Figure 6C) and overall (p = 0.025) (Figure 6D) survival compared to patients having a normal genotype of at least one SNP.

Table 4. Patient stratification based on the combination of the rs822336 G>C and rs822337 T>A genotypes.

Group	rs822336 Genotype	rs822337 Genotype	Total	Recurrence	Death
I	GG	AA	21	2 (9.5%)	6 (28.6%)
II	GG	AT	24	2 (8.3%)	2 (8.3%)
III	GG	TT	7	1 (14.3%)	1 (14.3%)
IV	CG	AA	36	16 (44.4%)	16 (44.4%)
V	CG	AT	14	1 (7.1%)	2 (14.2%)
VI	CG	TT	4	2 (50%)	0
VII	CC	AA	4	0	0
VIII	CC	AT	4	0	0
IX	CC	TT	-	-	-

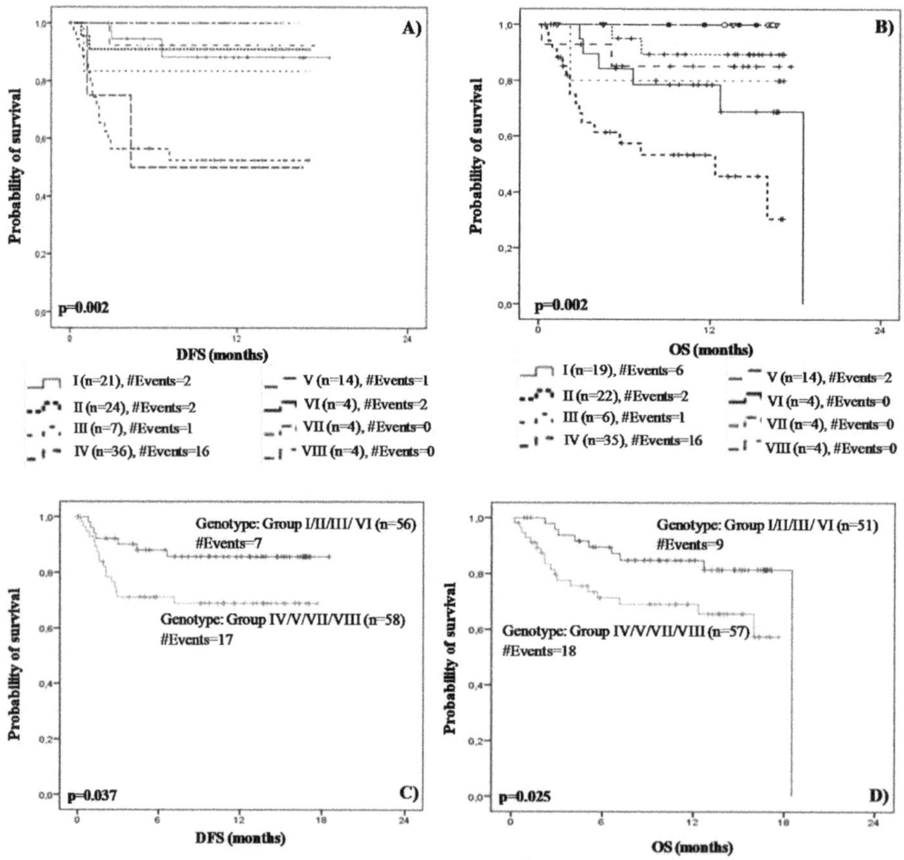

Figure 6. Kaplan–Meier disease-free (**A**,**C**) and overall (**B**,**D**) survival curves, after patients' stratification. p values were calculated by log-rank test.

Cox univariate analysis confirmed these results since patients belonging to Group B exhibited inferior DFS compared to those belonging to Group A (HR = 2.45, 95%CI = 1.03–5.98, p = 0.04) (Table 5). Taking a step further, we evaluated the independence of the studied SNPs in predicting unfavorable outcomes in TNBC patients by developing a Cox multivariate proportional-hazard regression model adjusted for the combination of the rs822336 G>C and rs822337 T>A genotypes, tumor grade, patients' age, lymph node status and histological type. According to this model, the presence of mutated genotypes of both SNPs was a strong and independent marker of worse prognosis in terms of DFS (HR = 2.89, 95% CI = 1.13–7.87, p = 0.02) (Table 5). A similar analysis for the association of the presence of mutated genotypes of both SNPs with patients' overall survival demonstrated that patients belonging to Group B exhibited shorter OS compared to those belonging to Group A (HR = 2.52, 95% CI = 1.09–5.80, p = 0.03) (Table 6). Moreover, the presence of mutated genotypes of both SNPs was a strong and independent marker for predicting shorter OS (HR = 3.44, 95% CI = 1.37–8.61, p = 0.008) (Table 6).

Table 5. Cox univariate and multivariate regression analysis for the prediction of disease-free survival (DFS) after stratification of study cohort according to combination of the rs822336 G>C and rs822337 T>A genotypes.

	Univariate Analysis (n = 107)		
Variable	**HR [a]**	**95% CI [b]**	***p* Value**
Group			
A	1.00		0.04
B	2.45	1.03–5.98	
	Multivariate Analysis (n = 98)		
Variable	**HR [a]**	**95% CI [b]**	***p* Value**
Group			
A	1.00		0.02
B	2.89	1.13–7.87	
Grade	1.37	0.37–5.11	0.63
Age	0.63	0.26–1.50	0.29
Lymph node status	1.35	0.96–1.90	0.07
Histology	1.15	0.75–1.76	0.52

[a] Hazard Ratio, [b] Confidence interval of the estimated HR. Bold value indicates statistical significance.

Table 6. Cox univariate and multivariate regression analysis for the prediction of overall survival (OS) after stratification of study cohort according to combination of the rs822336 G>C and rs822337 T>A genotypes.

	Univariate Analysis (n = 108)		
Variable	**HR [a]**	**95% CI [b]**	***p* Value**
Group			
A	1.00		0.03
B	2.52	1.09–5.80	
	Multivariate Analysis (n = 98)		
Variable	**HR [a]**	**95% CI [b]**	***p* Value**
Group			
A	1.00		0.008
B	3.44	1.37–8.61	
Grade	0.94	0.28–3.10	0.92
Age	1.88	0.79–4.45	0.15
Lymph node status	1.63	1.19–2.22	0.002
Histology	0.86	0.55–1.34	0.51

[a] Hazard Ratio, [b] Confidence interval of the estimated HR. Bold value indicates statistical significance.

4. Discussion

The advent of high-throughput genomic technologies inaugurated a new era of molecular genetics, where previously unnoticed genome elements, such as SNPs, represent a new route to assess the etiology of cancer. The possible causative role of SNPs in cancer fueled much interest in recognizing novel cancer biomarkers and therapeutic targets in this class of genetic polymorphisms. Specifically, for TNBC, several studies suggest that SNPs can be used as therapeutic targets and/or prognostic/predictive markers [10,27,28].

As the PD-1/PD-L1 axis has crucial mechanisms as well as a clinical role in TNBC, *PD-1* and *PD-L1* genes merit further investigation for their potential exploitation in the clinical setting of oncology as markers of predicting the 1. risk of TNBC development, 2. course of disease and 3. response to therapy [29].

Considering that mounting evidence indicates that PD-L1 SNPs can be exploited as prognostic markers, we examined the presence of the *PD-L1* SNPs rs822336 G>C and rs822337 T>A in 114 FFPE (formalin-fixed paraffin-embedded) tissues obtained from TNBC patients, and we correlated the SNP status with the clinicopathological and survival data of the patients. The rs822336G>C and rs822337T>A polymorphisms are situated on the promoter region close to the transcription start site. The rs822336C-rs822337A haplotypes are associated with a significant reduction of promoter activity, and it has been found that the rs822336C and rs822337A haplotypes are related with reduced PD-L1 expression at protein level [30]. Both SNPs are located at sites that play a pivotal role in the activation of the promoter, and the region of rs822337T>A represents a significant interaction site between the promoters of *PD-L1* and NF-κB [31], whereas it seems that the A allele abolishes the bonding site of the transcription factors SPIB and FOXO3 [32].

These SNPs have been studied for their clinical potential in NSCLC and it has been reported that the co-existence of the rs822336C and rs822337A haplotypes is associated with worse survival, as well as with significantly lower promoter activity and thus with down-regulation of *PD-L1* [30]. Specifically, the rs822337T>A polymorphism has been studied as a predictive marker in NSCLC patients after first line paclitaxel-cisplatin chemotherapy, but no statistical significant association was found [33]. A recent study confirmed the abovementioned results regarding the rs822336C polymorphism, since it was found not only that patients with GC or CC alleles demonstrated inferior OS intervals but also that the rs822336C haplotype seems to lead to decreased *PD-L1* expression and/or to the malfunction of the protein [34]. The clinical significance of rs822336G>C has also been investigated in patients with gastric cancer, but the results are inconsistent with those reported for NSCLC as it was found that the rs822336 CC genotype is an independent prognostic marker of favorable prognosis in gastric cancer [35]. This discrepancy is attributed to the different biology of these cancer types.

The first step of our study was the analysis of the differential distribution of the rs822336 G>C and rs822337 T>A polymorphisms among TNBC patients and healthy controls. The statistical analysis demonstrated that both the patients and the healthy controls exhibited the same expressional pattern of the rs822336 G>C and rs822337 T>A polymorphisms. In more detail, we found that not only in patients but also in healthy controls the most abundant allele of the rs822336 G>C was the GG followed by the CG and the CC ($p = 0.27$). Similarly, regarding the rs822337 T>A polymorphism, no significant difference was observed regarding the differential distribution of this polymorphism in the two study cohorts ($p = 0.47$), since in both groups the most abundant genotype was the AA (patients: 53.5%, controls: 46%), followed by the AT (patients: 36.8%, controls: 41.1%) and then the TT genotype (patients: 9.6%, controls: 12.9%).

Next, we aimed to study the association of the rs822336 G>C and rs822337 T>A polymorphisms with important clinicopathological data. By performing the chi square test, we found regarding the rs822336 G>C polymorphism that the majority of patients with the CG genotype harbored Stage IV breast tumors (81.8%), as well as that this genotype was the most common among the women that recurred (38.9%) and died (66.7%). A similar association with markers of worse prognosis was observed for the rs822337 T>A polymorphism as well. More specifically, it was found that AA was the most common genotype among stage IV tumors (72.7%) as well as in TNBC patients that relapsed (75%) and died (81.5%). Overall, these results underline that the rs822336C and rs822337A haplotypes, which are associated with decreased expression levels of PD-L1, are significantly correlated with markers of unfavorable prognosis. This observation can be interpreted based on the known data regarding the mechanistic and clinical role of these SNPs. In more detail, as mentioned above, Lee et al. stated that rs822336C and rs822337A haplotypes are related with reduced PD-L1 expression [30], whereas at the same time it has been found that breast cancer patients with decreased PD-L1 expression are characterized by worse survival rates [36] as well as that in TNBC the PD-L1 downregulation is associated with unfavorable prognosis [37]. Hence, by combining these data we can conclude that the rs822336C and rs822337A haplotypes can act as indicators of unfavorable prognosis since they can lead to reduced expression of the PD-L1.

The final step of our analysis was the performance of a thorough survival analysis in order to examine the association of the SNPs' distribution with disease-free (DFS) and the overall survival (OS). Both Kaplan–Meier and Cox regression analysis demonstrated that the rs822336 G>C genotype CG is strongly associated with inferior DFS (Kaplan–Meier analysis $p = 0.002$, Cox analysis, $p = 0.005$) and OS (Kaplan–Meier analysis; $p = 0.009$, Cox analysis; $p = 0.018$) intervals compared to the others genotypes (Figure 4). Moreover, a univariate Cox analysis revealed that women harboring the CG allele were 4.06 and 2.74 time more likely to relapse and die, respectively. The same statistical analysis revealed that rs822337 T>A allele AA is associated with worse survival in terms of DFS and OS. According to Kaplan–Meier analysis, patients with the AA genotype are characterized by shorter DFS ($p = 0.024$) and OS ($p = 0.004$). Cox regression analysis confirmed these results; it was found that rs822337 T>A genotype AA is marginally correlated with high risk of recurrence ($p = 0.04$) but most importantly with a four-fold higher risk of death ($p = 0.01$).

Taking a step further, we examined the prognostic significance of the rs822336 G>C and rs822337 T>A polymorphisms after patients' categorization according to Table 6. Kaplan–Meier and Cox regression analyses showed that the combination of the CG and AA genotypes is significantly associated with inferior DFS (Kaplan–Meier; $p = 0.002$, Cox analysis; $p = 0.018$) and OS (Kaplan–Meier; $p = 0.002$, Cox analysis; $p = 0.04$) intervals. Moreover, Kaplan–Meier and univariate Cox analysis revealed that women harboring mutated genotypes of both SNPs (i.e., CG+AA/CG+AT/CC+AA/CC+AT) had shorter disease-free (Kaplan–Meier; $p = 0.037$, Cox analysis; $p = 0.04$) and overall (Kaplan–Meier; $p = 0.025$, Cox analysis; $p = 0.03$) survival, compared to patients having normal genotypes of at least one SNP. Multivariate analysis adjusted for significant clinipathological features (i.e., tumor grade, patients age, lymph node status and histological type) also showed that the presence of mutated genotypes of both SNPs is a strong and independent marker for predicting shorter DFS (HR = 2.89, 95% CI = 1.13–7.87, $p = 0.02$) and OS (HR = 3.44, 95% CI = 1.37–8.61, $p = 0.008$). Overall, our statistical analysis reinforced the abovementioned significant association of the rs822336C and rs822337A haplotypes with unfavorable prognosis, since it was found that these haplotypes (individually or in combination) are markers of shorter DFS and OS. Our results are in line with a previous study, according to which NSCLC patients harboring the rs822336CC and rs822337AA gonotypes demonstrated a marginally significant decreased OS, whereas the same group concluded that the rs822336C and rs822337A haplotypes were associated with inferior overall survival intervals [30].

A noteworthy biological implication of our approach is the interplay between polymorphisms and the glycosylation patterns of PD-L1. Studies have shown that TNBC cells have higher levels of glycosylated PD-L1 [38]. This feature affects the quality of immunohistochemical (IHC) staining procedures, generating more false negative results, and potentially excluding from therapy tumors that would likely be responsive to anti-PD-L1 therapies [39,40]. In fact, the underestimation of PD-L1 expression in breast tumor tissues has been suggested as a potential underlying cause for the reduced benefit of atezolizumab plus nab-paclitaxel in the post-market phase-III trial that led to its withdrawal from the indication in TNBC in 2021 [38]. Specifically, in the IMpassion130 study, patients were stratified using standard IHC methods, whereas a preceding de-glycosylation process of breast tissue samples has been proposed as a more accurate approach to assess PD-L1 expression and therefore atezolizumab effectiveness [24]. Interestingly, N-glycosylation occurs after the transfer of a glycan to an asparagine (Asn) amino acid side-chain acceptor [41]. Since Asn is coded by synonymous codons AAC and AAT, the studied rs822337 AT genotype may affect glycosylation (and concurrently response) to anti-PD-L1 therapy. Since AT was one of the most abundant PD-L1 genotypes in both groups included in our study, these results are worthy of attention and can serve as a basis for further investigation on the relationship between the level of PD-L1 glycosylation and response to anti-PD-L1 therapy.

Our study represents a first attempt to investigate the clinical impact in TNBC of two known *PD-L1* polymorphisms which have attracted much interest recently, due to their implication in expressional regulation of PD-L1, a molecule that according to mounting evidence is in the core not only of certain cancer-related events but also of clinical management of several cancer types, including TNBC. According to our data the rs822336C and rs822337A haplotypes are associated with markers of unfavorable prognosis in TNBC patients as well as with inferior DFS and OS intervals. Moreover, the presence of mutated genotypes of both SNPs is a strong and independent marker for predicting shorter DFS and OS.

5. Conclusions

In conclusion, PD-L1 rs822336 G>C and rs822337 T>A polymorphisms are differentially expressed in TNBCs, and this distribution is associated with markers of unfavorable prognosis and with worse patient survival. Considering how polymorphisms exert not only transcriptional but also a wide range of post-translational effects, including affecting the level of *PD-L1* glycosylation, our findings may have biological and clinical implications in the elucidation of TNBC's aggressive phenotype. Further well-designed investigations in a larger cohort of patients should be performed to elucidate the underlying mechanisms by which these SNPs affect the prognosis of TNBC patients, and to verify our results.

Author Contributions: Conceptualization, F.Z. and A.-E.M.; methodology, M.G.; validation, C.G.Z., D.K. and A.T.; investigation, A.-E.M.; writing—original draft preparation, A.-E.M. and E.Z.; writing—review and editing, E.Z.; visualization, F.Z.; M.G. supervision, M.G., K.D., K.G.T., G.S.; project administration, F.Z.; funding acquisition, A.-E.M. All authors have read and agreed to the published version of the manuscript.

Funding: This work was supported by the Hellenic and international Society of Molecular targeted and Personalized Treatments (H.I.S.M.T.P.T). The funder of the H.I.S.M.T.P.T was involved in the approval of this article.

Institutional Review Board Statement: The Declaration of Helsinki was approved by the Institutional Review Board of Hippokration General Hospital, Athens, Greece (protocol code 37/15-5-17 and date of 15 Feb 2022 approval).

Informed Consent Statement: Informed consent was obtained from all subjects involved in the study.

Conflicts of Interest: The authors declare no conflict of interest.

Abbreviations

PD-L1	Programmed death-Ligand 1
TNBC	Triple negative breast cancer
FFPEs	Formalin-fixed paraffin-embedded
DFS	Disease-free survival
OS	Overall survival
PD-1	Programmed death 1 receptor
FDA	Food and Drug Administration
SNPs	Single-nucleotide polymorphism
IHC	Immunohistochemistry
PCR	Polymerase chain reaction
NSCLC	Non-small cell lung cancer

References

1. Bardia, A.; Mayer, I.A.; Vahdat, L.T.; Tolaney, S.M.; Isakoff, S.J.; Diamond, J.R.; O'Shaughnessy, J.; Moroose, R.L.; Santin, A.D.; Abramson, V.G.; et al. Sacituzumab Govitecan-hziy in Refractory Metastatic Triple-Negative Breast Cancer. *N. Engl. J. Med.* **2019**, *380*, 741–751. [CrossRef] [PubMed]
2. Schmid, P.; Cortes, J.; Pusztai, L.; McArthur, H.; Kümmel, S.; Bergh, J.; Denkert, C.; Park, Y.H.; Hui, R.; Harbeck, N.; et al. Pembrolizumab for Early Triple-Negative Breast Cancer. *N. Engl. J. Med.* **2020**, *382*, 810–821. [CrossRef] [PubMed]
3. Ovcaricek, T.; Frkovic, S.G.; Matos, E.; Mozina, B.; Borstnar, S. Triple negative breast cancer—Prognostic factors and survival. *Radiol. Oncol.* **2011**, *45*, 46–52. [CrossRef]
4. Loi, S.; Drubay, D.; Adams, S.; Pruneri, G.; Francis, P.; Lacroix-Triki, M.; Joensuu, H.; Dieci, M.V.; Badve, S.; Demaria, S.; et al. Tumor-Infiltrating Lymphocytes and Prognosis: A Pooled Individual Patient Analysis of Early-Stage Triple-Negative Breast Cancers. *J. Clin. Oncol.* **2019**, *37*, 559–569. [CrossRef] [PubMed]
5. Denkert, C. The immunogenicity of breast cancer–molecular subtypes matter. *Ann. Oncol. Off. J. Eur. Soc. Med. Oncol.* **2014**, *25*, 1453–1455. [CrossRef]
6. Denkert, C.; von Minckwitz, G.; Darb-Esfahani, S.; Lederer, B.; Heppner, B.; Weber, K.E.; Budczies, J.; Huober, J.; Klauschen, F.; Furlanetto, J.; et al. Tumour-infiltrating lymphocytes and prognosis in different subtypes of breast cancer: A pooled analysis of 3771 patients treated with neoadjuvant therapy. *Lancet Oncol.* **2018**, *19*, 40–50. [CrossRef]
7. Solinas, C.; Carbognin, L.; De Silva, P.; Criscitiello, C.; Lambertini, M. Tumor-infiltrating lymphocytes in breast cancer according to tumor subtype: Current state of the art. *Breast* **2017**, *35*, 142–150. [CrossRef]
8. Liu, Z.; Li, M.; Jiang, Z.; Wang, X. A Comprehensive Immunologic Portrait of Triple-Negative Breast Cancer. *Transl. Oncol.* **2018**, *11*, 311–329. [CrossRef]
9. Nagarajan, D.; McArdle, S. Immune Landscape of Breast Cancers. *Biomedicines* **2018**, *6*, 20. [CrossRef]
10. Bareche, Y.; Venet, D.; Ignatiadis, M.; Aftimos, P.; Piccart, M.; Rothe, F.; Sotiriou, C. Unravelling triple-negative breast cancer molecular heterogeneity using an integrative multiomic analysis. *Ann. Oncol.* **2018**, *29*, 895–902. [CrossRef]
11. Safonov, A.; Jiang, T.; Bianchini, G.; Győrffy, G.; Karn, T.; Hatzis, C.; Pusztai, L. Immune Gene Expression Is Associated with Genomic Aberrations in Breast Cancer. *Cancer Res.* **2017**, *77*, 3317–3324. [CrossRef] [PubMed]
12. He, Y.; Jiang, Z.; Chen, C.; Wang, X. Classification of triple-negative breast cancers based on Immunogenomic profiling. *J. Exp. Clin. Cancer Res.* **2018**, *37*, 327. [CrossRef] [PubMed]
13. Romero-Cordoba, S.; Meneghini, E.; Sant, M.; Iorio, M.; Sfondrini, L.; Paolini, B.; Agresti, R.; Tagliabue, E.; Bianchi, F. Decoding Immune Heterogeneity of Triple Negative Breast Cancer and Its Association with Systemic Inflammation. *Cancers* **2019**, *11*, 911. [CrossRef] [PubMed]
14. Mei, J.; Zhao, J.; Fu, Y. Molecular classification of breast cancer using the mRNA expression profiles of immune-related genes. *Sci. Rep.* **2020**, *10*, 4800. [CrossRef]
15. Lehmann, B.; Bauer, J.; Chen, X.; Sanders, M.; Chakravarthy, A.; Shyr, Y.; Pietenpol, J.A. Identification of human triple-negative breast cancer subtypes and preclinical models for selection of targeted therapies. *J. Clin. Investig.* **2011**, *121*, 2750–2767. [CrossRef]
16. Bonsang-Kitzis, H.; Sadacca, B.; Hamy-Petit, A.; Moarii, M.; Pinheiro, A.; Laurent, C.; Reyal, F. Biological network-driven gene selection identifies a stromal immune module as a key determinant of triple-negative breast carcinoma prognosis. *Oncoimmunology* **2015**, *5*, e1061176. [CrossRef]
17. Burstein, M.; Tsimelzon, A.; Poage, G.; Covington, K.; Contreras, A.; Fuqua, S.A.W.; Savage, M.I.; Osborne, C.K.; Hilsenbeck, S.G.; Chang, J.C.; et al. Comprehensive genomic analysis identifies novel subtypes and targets of triple-negative breast cancer. *Clin. Cancer Res.* **2015**, *21*, 1688–1698. [CrossRef]
18. Beckers, R.; Selinger, C.; Vilain, R.; Madore, J.; Wilmott, J.; Harvey, K.; Holliday, A.; Cooper, C.L.; Robbins, E.; Gillett, D.; et al. Programmed death ligand 1 expression in triple-negative breast cancer is associated with tumour-infiltrating lymphocytes and improved outcome. *Histopathology* **2016**, *69*, 25–34. [CrossRef]
19. Mittendorf, E.A.; Philips, A.V.; Meric-Bernstam, F.; Qiao, N.; Wu, Y.; Harrington, S.; Su, X.; Wang, Y.; Gonzalez-Angulo, A.M.; Chawla, A.; et al. PD-L1 Expression in Triple-Negative Breast Cancer. *Cancer Immunol. Res.* **2014**, *2*, 361–370. [CrossRef]
20. Cimino-Mathews, A.; Thompson, E.; Taube, J.M.; Ye, X.; Lu, Y.; Meeker, A.; Xu, H.; Sharma, R.; Lecksell, K.; Cornish, T.C.; et al. PD-L1 (B7-H1) expression and the immune tumor microenvironment in primary and metastatic breast carcinomas. *Hum. Pathol.* **2016**, *47*, 52–63. [CrossRef]
21. Li, X.; Li, M.; Lian, Z.; Zhu, H.; Kong, L.; Wang, P.; Yu, J. Prognostic Role of Programmed Death Ligand-1 Expression in Breast Cancer: A Systematic Review and Meta-Analysis. *Target Oncol.* **2016**, *11*, 753–761. [CrossRef] [PubMed]
22. Schmid, P.; Adams, S.; Rugo, H.S.; Schneeweiss, A.; Barrios, C.H.; Iwata, H.; Diéras, V.; Hegg, R.; Im, S.-A.; Wright, G.S.; et al. Atezolizumab and Nab-Paclitaxel in Advanced Triple-Negative Breast Cancer. *N. Engl. J. Med.* **2018**, *379*, 2108–2121. [CrossRef] [PubMed]
23. Dong, H.; Zhu, G.; Tamada, K.; Chen, L. B7-H1, a third member of the B7 family, co-stimulates T-cell proliferation and interleukin-10 secretion. *Nat. Med.* **1999**, *5*, 1365–1369. [CrossRef] [PubMed]
24. Nomizo, T.; Ozasa, H.; Tsuji, T.; Funazo, T.; Yasuda, Y.; Yoshida, H.; Yagi, Y.; Sakamori, Y.; Nagai, H.; Hirai, T.; et al. Clinical Impact of Single Nucleotide Polymorphism in PD-L1 on Response to Nivolumab for Advanced Non-Small-Cell Lung Cancer Patients. *Sci. Rep.* **2017**, *7*, 45124. [CrossRef]

25. Wang, W.; Li, F.; Mao, Y.; Zhou, H.; Sun, J.; Li, R.; Liu, C.; Chen, W.; Hua, D.; Zhang, X. A miR-570 binding site polymorphism in the B7-H1 gene is associated with the risk of gastric adenocarcinoma. *Hum. Genet.* **2013**, *132*, 641–648. [CrossRef]
26. Ma, Y.; Liu, X.; Zhu, J.; Li, W.; Guo, L.; Han, X.; Song, B.; Cheng, S.; Jie, L. Polymorphisms of co-inhibitory molecules (CTLA-4/PD-1/PD-L1) and the risk of non-small cell lung cancer in a Chinese population. *Int. J. Clin. Exp. Med.* **2015**, *8*, 16585.
27. Häberle, L.; Hein, A.; Rübner, M.; Schneider, M.; Ekici, A.B.; Gass, P.; Hartmann, A.; Schulz-Wendtland, R.; Beckmann, M.W.; Lo, W.Y.; et al. Predicting Triple-Negative Breast Cancer Subtype Using Multiple Single Nucleotide Polymorphisms for Breast Cancer Risk and Several Variable Selection Methods. *Geburtshilfe Frauenheilkd* **2017**, *77*, 667–678. [CrossRef]
28. Stevens, K.N.; Vachon, C.M.; Couch, F.J. Genetic Susceptibility to Triple-Negative Breast Cancer. *Cancer Res.* **2013**, *73*, 2025–2030. [CrossRef]
29. Ren, H.-T.; Li, Y.-M.; Wang, X.-J.; Kang, H.-F.; Jin, T.-B.; Ma, X.-B.; Liu, X.H.; Wang, M.; Liu, K.; Xu, P.; et al. PD-1 rs2227982 Polymorphism Is Associated With the Decreased Risk of Breast Cancer in Northwest Chinese Women: A Hospital-Based Observational Study. *Medicine* **2016**, *95*, e3760. [CrossRef]
30. Lee, S.Y.; Jung, D.K.; Choi, J.E.; Jin, C.C.; Hong, M.J.; Do, S.K.; Kang, H.G.; Lee, W.K.; Seok, Y.; Lee, E.B.; et al. Functional polymorphisms in PD-L1 gene are associated with the prognosis of patients with early stage non-small cell lung cancer. *Gene* **2017**, *599*, 28–35. [CrossRef]
31. Huang, G.; Wen, Q.; Zhao, Y.; Gao, Q.; Bai, Y. NF-κB Plays a Key Role in Inducing CD274 Expression in Human Monocytes after Lipopolysaccharide Treatment. *PLoS ONE* **2013**, *8*, e61602. [CrossRef] [PubMed]
32. Makuku, R.; Khalili, N.; Razi, S.; Keshavarz-Fathi, M.; Rezaei, N. Current and Future Perspectives of PD-1/PDL-1 Blockade in Cancer Immunotherapy. *J. Immunol. Res.* **2021**, *2021*, 6661406. [CrossRef] [PubMed]
33. Lee, S.Y.; Jung, D.K.; Choi, J.E.; Jin, C.C.; Hong, M.J.; Do, S.K.; Kang, H.G.; Lee, W.K.; Seok, Y.; Lee, E.B.; et al. PD-L1 polymorphism can predict clinical outcomes of non-small cell lung cancer patients treated with first-line paclitaxel-cisplatin chemotherapy. *Sci. Rep.* **2016**, *6*, 25952. [CrossRef] [PubMed]
34. Zhao, X.; Peng, Y.; Li, X.; Zhang, S.; Peng, Y.; Shan, J.; Li, M.; Dai, N.; Feng, Y.; Xu, C.; et al. The association of PD-L1 gene polymorphisms with non-small-cell lung cancer susceptibility and clinical outcomes in a Chinese population. *Int. J. Clin. Exp. Pathol.* **2020**, *13*, 2130. [PubMed]
35. Wu, Y.; Zhao, T.; Jia, Z.; Cao, D.; Cao, X.; Pan, Y.; Zhao, D.; Zhang, B.; Jiang, J. Polymorphism of the programmed death-ligand 1 gene is associated with its protein expression and prognosis in gastric cancer. *J. Gastroenterol. Hepatol.* **2019**, *34*, 1201–1207. [CrossRef] [PubMed]
36. Baptista, M.; Sarian, L.; Derchain, S.; Pinto, G.; Vassallo, J. Prognostic significance of PD-L1 and PD-L2 in breast cancer. *Hum. Pathol.* **2016**, *47*, 78–84. [CrossRef] [PubMed]
37. Planes-Laine, G.; Rochigneux, P.; Bertucci, F.; Chrétien, A.-S.; Viens, P.; Sabatier, R.; Gonçalves, A. PD-1/PD-L1 Targeting in Breast Cancer: The First Clinical Evidences are Emerging—A Literature Review. *Cancers* **2019**, *11*, 1033. [CrossRef]
38. Ou-Yang, F.; Li, C.-L.; Chen, C.-C.; Shen, Y.-C.; Moi, S.-H.; Luo, C.-W.; Xia, W.Y.; Wang, Y.N.; Lee, H.H.; Wang, L.H.; et al. De-glycosylated membrane PD-L1 in tumor tissues as a biomarker for responsiveness to atezolizumab (Tecentriq) in advanced breast cancer patients. *Am. J. Cancer Res.* **2022**, *12*, 123.
39. Lee, H.H.; Wang, Y.N.; Xia, W.; Chen, C.H.; Rau, K.M.; Ye, L.; Wei, Y.; Chou, C.K.; Wang, S.C.; Yan, M.; et al. Removal of N-linked glycosylation enhances PD-L1 detection and predicts anti-PD-1/PD-L1 therapeutic efficacy. *Cancer Cell* **2019**, *36*, 168. [CrossRef]
40. Mei, J.; Xu, J.; Yang, X.; Gu, D.; Zhou, W.; Wang, H.; Liu, C. A comparability study of natural and deglycosylated PD-L1 levels in lung cancer: Evidence from immunohistochemical analysis. *Mol. Cancer* **2021**, *20*, 11. [CrossRef]
41. Helenius, A.; Aebi, M. Intracellular functions of N-linked glycans. *Science* **2001**, *291*, 2364–2369. [CrossRef] [PubMed]

medicina

Article

The Role of miR-375-3p, miR-210-3p and Let-7e-5p in the Pathological Response of Breast Cancer Patients to Neoadjuvant Therapy

Lorena Alexandra Lisencu [1], Andrei Roman [2], Simona Visan [3], Eduard-Alexandru Bonci [1,4], Andrei Pașca [1,4], Emilia Grigorescu [5], Elena Mustea [5], Andrei Cismaru [6], Alexandru Irimie [1,4], Cosmin Lisencu [1,4], Loredana Balacescu [3,7,*], Ovidiu Balacescu [3,7] and Oana Tudoran [3,7,*]

1 Department of Oncological Surgery and Gynecological Oncology, "Iuliu Hațieganu" University of Medicine and Pharmacy, 400012 Cluj-Napoca, Romania
2 Department of Radiology, The Oncology Institute "Prof. Dr. Ion Chiricuță", 400015 Cluj-Napoca, Romania
3 Department of Genetics, Genomics and Experimental Pathology, The Oncology Institute "Prof. Dr. Ion Chiricuță", 400015 Cluj-Napoca, Romania
4 Department of Surgical Oncology, The Oncology Institute "Prof. Dr. Ion Chiricuță", 400015 Cluj-Napoca, Romania
5 Department of Pathological Anatomy, County Emergency Hospital, 400347 Cluj-Napoca, Romania
6 Research Center for Functional Genomics, Biomedicine and Translational Medicine, University of Medicine and Pharmacy "Iuliu Hatieganu", 400037 Cluj-Napoca, Romania
7 11th Department of Medical Oncology, University of Medicine and Pharmacy "Iuliu Hatieganu", 400012 Cluj-Napoca, Romania
* Correspondence: lbalacescu@iocn.ro (L.B.); oana.tudoran@iocn.ro (O.T.); Tel.: +40-264590638 (O.T.)

Citation: Lisencu, L.A.; Roman, A.; Visan, S.; Bonci, E.-A.; Pașca, A.; Grigorescu, E.; Mustea, E.; Cismaru, A.; Irimie, A.; Lisencu, C.; et al. The Role of miR-375-3p, miR-210-3p and Let-7e-5p in the Pathological Response of Breast Cancer Patients to Neoadjuvant Therapy. *Medicina* **2022**, *58*, 1494. https://doi.org/10.3390/medicina58101494

Academic Editor: Maria Rosaria De Miglio

Received: 9 July 2022
Accepted: 13 October 2022
Published: 20 October 2022

Abstract: *Background and Objectives*: Prediction of response to therapy remains a continuing challenge in treating breast cancer, especially for identifying molecular tissue markers that best characterize resistant tumours. Microribonucleic acids (miRNA), known as master modulators of tumour phenotype, could be helpful candidates for predicting drug resistance. We aimed to assess the association of miR-375-3p, miR-210-3p and let-7e-5p in breast cancer tissues with pathological response to neoadjuvant therapy (NAT) and clinicopathological data. *Material and methods*: Sixty female patients diagnosed with invasive breast cancer at The Oncology Institute "Ion Chiricuță", Cluj-Napoca, Romania (IOCN) were included in this study. Before patients received any treatment, fresh breast tissue biopsies were collected through core biopsy under echographic guidance and processed for total RNA extraction and miRNA quantification. The Cancer Genome Atlas Breast Invasive Carcinoma (TCGA-BRCA) database was used as an independent external validation cohort. *Results*: miR-375-3p expression was associated with more differentiated tumours, hormone receptor presence and lymphatic invasion. According to the Miller–Payne system, a higher miR-375-3p expression was calculated for patients that presented with intermediate versus (vs.) no pathological response. Higher miR-210-3p expression was associated with an improved response to NAT in both Miller–Payne and RCB evaluation systems. Several druggable mRNA targets were correlated with miR-375-3p and miR-210-3p expression, with upstream analysis using the IPA knowledge base revealing a list of possible chemical and biological targeting drugs. Regarding let-7e-5p, no significant association was noticed with any of the analysed clinicopathological data. *Conclusions*: Our results suggest that tumours with higher levels of miR-375-3p are more sensitive to neoadjuvant therapy compared to resistant tumours and that higher miR-210-3p expression in responsive tumours could indicate an excellent pathological response.

Keywords: breast cancer; neoadjuvant therapy; pathological complete response; miR-375-3p; let-7e-5p; MiR-210-3p

1. Introduction

Breast cancer (BC) is the most common cause of cancer among women worldwide, accounting for 15–20% of all cancer deaths in women [1]. Due to advances in diagnosis and treatment modalities, the prognosis of this disease has improved in recent years, with a 5-year survival rate of almost 90%. However, around 30% of breast cancer patients fail to respond to conventional treatments, leading to tumour progression [1].

Breast cancer is a highly heterogeneous disease; therefore, its treatment depends on multiple clinical and pathological factors such as tumour grade and hormone receptor (HR) status. Tumour characteristics and the extent of the disease direct the choice and timing of systemic treatments (chemotherapy, endocrine therapy or HER2-directed therapy). In the case of high-risk primary tumours or locally advanced breast cancer, neoadjuvant (preoperative) therapy is a frequently practical therapeutic approach as it offers the advantage of reducing the extent of surgery [2,3]. Furthermore, a tumour's response to NAT can be used to guide adjuvant treatment selection and offer prognostic information regarding patient outcome [4].

The response to neoadjuvant therapy (NAT) is usually assessed clinically and pathologically. Pathological evaluation is the gold standard, as clinical evaluation can often misevaluate the response to NAT. While several systems have been proposed to evaluate the pathological response to NAT, the most used system for prognosis prediction are the Miller–Payne (MP) and the residual cancer burden (RCB) systems [5,6]. The MP system evaluates the changes in tumour cellularity between biopsy and surgery tissue. It has five grades as follows: 1 (no change), 2 (minor reduction in tumour cells, but ≤30%), 3 (reduction in tumour cells by 30–90%), 4 (reduction in tumour cells with >90%) and 5 (no detectable tumour cells). Grades 1–4 correspond to partial pathological response (pPR) while grade 5 means pathological complete response (pCR) [7,8]. The RCB system measures the primary tumour's bidimensional size and cellularity and assesses lymph nodes' involvement. The RCB index is classified as 0 (pCR), 1 (minimal residual disease), 2 (moderate residual disease) and 3 (extensive residual disease) [8,9]. The goal of NAT is pCR, as it plays an important prognostic role in BC patients. PCR is associated with improved overall survival and disease-free survival in comparison with those that do not achieve pCR and who have an unfavourable prognosis [3]. Due to its role in the prognosis of BC patients, predicting pCR is important in order to identify those patients that would benefit from NAT.

While routine HR, HER2 receptors, grading and Ki-67 assessment remain essential for treatment guidance, non-coding ribonucleic acids (RNA) have increased in popularity as their dysregulation has been associated with breast cancer pathogenesis. Microribonucleic acids (miRNA) are small, non-coding RNAs that regulate gene expression at the post-transcriptional level by binding to target messenger RNAs (mRNAs) and triggering their degradation. One of the first papers about the role of miRNA in cancer pathology demonstrated that miRNA is a better classifier than mRNA profiling when investigating poorly differentiated tumours, opening the way for using miRNA expression as a reliable marker for cancer diagnosis, prognosis and treatment response [10]. Since then, overwhelming data have indicated that miRNAs are involved in the regulation of processes such as proliferation, apoptosis and migration of cancer cells [3], having the potential of being oncogenic (oncomirs), tumour suppressors or both [1,11]. As key modulators of oncogenesis, miRNAs have been reported to have clinical utility in the diagnostic, prognostic and therapeutic approach of breast cancer patients [1,11], making them highly attractive as biomarkers for personalized medicine [1].

Several miRNAs have been reported to have predictive power in pathological response following NAT in breast cancer [12–14], with most of these studies being focused on neoadjuvant chemotherapy (NACT). However, recent treatment guidelines [15] encourage the administration of endocrine as well as targeted therapies concurrent with, or instead of, NACT to increase tumours' sensitivity to treatment. Thus, there is an increasing need to further explore the role of these miRNAs as biomarkers of NAT response. Based on the existing literature, we have identified conflicting data regarding the prognostic role of several miRNAs. Of interest, miR-375-3p, miR-210-3p and let-7e have shown discrepancies regarding the clinical significance as prognostic biomarkers [16–20], being reported to have both increased and decreased expression associations with BC patients' response to NAT. In this study, we aimed to assess the prognostic value of these highly controversial miRNAs, miR-375-3p, miR-210-3p and let-7e-5p in breast cancer tissues by investigating their expression association with patients' pathological response to neoadjuvant therapy and clinicopathological features.

2. Materials and Methods

2.1. Breast Cancer Patients and Samples Collection

Sixty female patients diagnosed with invasive breast cancer at The Oncology Institute "Ion Chiricuță", Cluj-Napoca, Romania (IOCN) were included in this study. The study was approved by the IOCN ethical committee (Approval No. 59/29.11.2016) and by the University of Medicine and Pharmacy Iuliu Hatieganu, Cluj-Napoca, Romania (Approval No. 290/09.09.2020). All patients were informed and gave their written consent for participation in the study following the Declaration of Helsinki. Before patients received any treatment, fresh breast tissue biopsies were collected through core biopsy under echographic guidance. The first core biopsy was sent for pathologic analysis, while a second biopsy was collected in RNAlater (Invitrogen, Thermo Fisher Scientific, Waltham, MA, USA) and stored in liquid nitrogen for transcriptomic studies.

2.2. RNA Extraction

Frozen biopsies were homogenized in TriReagent Solution (Ambion, Thermo Fisher Scientific, Waltham, MA, USA) using a Miccra D-1 (Miccra GmbH, Mullheim, Germany) polytron and processed for total RNA extraction using the classic phenol–chloroform method. The RNAs were quantified using NanoDrop ND-1000 (Thermo Scientific, Wilmington, DE, USA) and 2100 Bioanalyzer (Agilent Technologies, Santa Clara, CA, USA).

2.3. miRNA Expression Evaluation

Fifty nanograms (ng) of total RNAs were pre-amplified using universal RT miRNA primers to generate cDNAs following the TaqMan Advanced miRNA cDNA Synthesis Kit protocol (Thermo Fisher Scientific, Waltham, MA, USA). Next, 1:10 v/v diluted cDNAs and specific miRNA advanced assays were amplified with TaqMan Fast Advanced Master Mix (2X) (Thermo Fisher Scientific, Waltham, MA, USA) using the Light Cycler 480 device (Roche, Basel, Switzerland) with the following PCR settings: 55 °C for 2 min to remove RNA contaminants; 95 °C for 20 s for Taq polymerase amplification; and 40 cycles of 95 °C for 3 s followed by 60 °C for 30 s for PCR amplification. The ΔΔCt method was used for miRNA relative quantification by reporting the Ct values of the miRNAs of interest to miR-16-5p Ct values.

2.4. TCGA Data Analysis

The Cancer Genome Atlas Breast Invasive Carcinoma (TCGA-BRCA) expression data (miRNA and mRNA) and their clinical information were obtained from National Cancer Institute Genomic Data Commons (NCI GDC) data portal (https://portal.gdc.cancer.gov/, accessed on 19 April 2019) and cBioPortal for Cancer Genomics (https://www.cbioportal.org/, accessed on 19 April 2019). The miRNA-seq data, expressed as reads per million and fragments per kilobase millions mRNA-seq data, were filtered and log2(x + 1)-transformed. After processing, a miRNA dataset containing 916 tumoral samples and 93 standard samples and an mRNA dataset of 983 tumoral samples were retained for subsequent analysis. Pearson correlation was used to test potential miRNAs–mRNA associations and intersected with validated miRNA–target interactions retrieved from the miRTarBase. The Ingenuity Pathway Analysis (IPA, Qiagen, Redwood City, CA, USA) upstream analysis module was used to interrogate for possible targeting drugs.

2.5. Statistical Analysis

The correlation between clinicopathological characteristics and tissue miRNAs expression was evaluated with the Mann–Whitney U test for two categorical variables or the Kruskal–Wallis test. It was followed by Dunn's multiple comparison post hoc test in the case of three or more categorical variables based on the data distribution. A p-value less than 0.05 was considered statistically significant. Fold regulation (FR) was calculated as the ratio between mean value of the interest group and the reference group.

3. Results

3.1. Patient and Tumour Characteristics

The clinicopathological features of the 60 included patients are summarized in Table 1. The median age of the patients was 60 (29–77), with most of the patients (73.33%) being over 50 years old at the time of diagnosis. Over 85% of the patients had moderately to poorly differentiated carcinomas, and 61.67% presented Ki-67 higher than 20. Most of the patients had luminal tumours (76.67%), with luminal B being the predominant subtype (46.67%). Over 85% of the patients were already in advanced clinical stages at diagnosis (>II). Of the 60 investigated patients, 49 received NAT: 32 received chemotherapy alone, 7 received hormonal therapy, 4 also received Her2 targeted therapy, while 5 received combinations of regimens. TNM staging was retained for prognostic information (primary and post-NAT surgery), while the Miller–Payne and RCB systems were used to evaluate the pathological response of the patients to NAT. According to the Miller–Payne evaluation, 14 patients did not respond to NAT, 4 presented a minor response, 13 had an intermediate response, and 13 had almost complete pathological response. According to the RCB classification system, 13 patients reached a high pathological response, 16 were therapy-resistant, and 15 had a partial response.

Table 1. Clinicopathological data of the patients included in the study.

Variable	Patients Characteristics
n = 60	
Age	
≤50	16 (26.67%)
>50	44 (73.33%)
Grading	
G1	7 (11.67%)
G2	32 (53.33%)
G3	20 (33.33%)
NA	1 (1.67%)
Ki-67	
≤20	22 (36.67%)
>20	37 (61.67%)
NA	1(1.67%)
Molecular Subtype	
Luminal A	18 (30%)
Luminal B	28 (46.67%)
HER2+	3 (5%)
TNBC [1]	8 (13.33%)
NA [2]	3(5%)
Tumour size (c [3])	
cT1	6 (10%)
cT2	29 (8.33%)
cT3	8 (13.33%)
cT4	11 (18.33%)
NA	6 (10%)

Table 1. *Cont.*

Variable	Patients Characteristics
Lymph nodes (c)	
cN0	15 (25%)
cN1	15 (25%)
cN2	21 (35%)
cN3	3 (5%)
NA	6 (10%)
Metastasis (c)	
cM0	49 (81.67%)
cM1	1 (1.67%)
NA	10 (16.67%)
Clinical stage	
Stage I	5 (8.33%)
Stage II	19 (31.67%)
Stage III	27 (45%)
Stage IV	1 (1.67%)
NA	8 (13.33%)
Tumour size (p [4])	
pT0	7 (11.67%)
pT1	23 (38.33%)
pT2	18 (30%)
pT3	1 (1.67)
NA	11 (18.33%)
Lymph nodes (p)	
pN0	26 (43.33%)
pN1	13 (21.67%)
pN2	8 (13.33%)
pN3	4 (6.67%)
NA	9 (15%)
Lymphatic invasion (p)	
L0	30 (50%)
L1	21 (35%)
NA	9 (15%)
n = 49	
Neoadjuvant therapy	
Only CT [5]	32 (65.31%)
Only ET [6]	7 (14.29%)
CT + HT	3 (6.12%)
Combinatory CT/ET/RTE [7]	2 (4.08%)
Her2+ TT [8]	4 (8.16%)
NA	1 (2.04%)
Miller–Payne system	
Grade 1	14 (28.57%)
Grade 2	4 (8.16%)
Grade 3	13 (26.53%)
Grade 4	5 (10.2%)
Grade 5	8 (16.33%)
NA	5 (10.2%)
RCB [9]	
RCB 0	8 (16.33%)
RCB-I	5 (10.2%)
RCB-II	15 (30.61%)
RCB-III	16 (32.65%)
NA	5 (10.2%)

[1] TNBC—triple negative breast cancer, [2] NA—nonassessable, [3] c—clinic, [4] p—pathologic, [5] CT—chemotherapy, [6] HT—hormonal therapy, [7] RTE—external radiotherapy, [8] TT—targeted therapy, [9] RCB—residual cancer burden.

3.2. *Investigation of miRNA Expression in Breast Cancer Specimens*

The association between tissue miRNA expression and the clinicopathological data of the patients included is presented in Table 2. No correlations were observed between the investigated miRNA expression and the age of the patients at a 50-years-of-age cut-off value. Additionally, no significant clinicopathological associations were found for let-7e-5p expression. Lower miR-375-3p expressions were associated with higher tumour grading and KI67 proliferation index. ER- and PR-positive tumours had a higher miR-375-3p expression, with significantly decreased expression in TNBC compared to luminal subtypes. Except for miR-375-3p expression with lymphatic positivity, no other significant correlations were demonstrated between tissue miRNAs expression and TNM staging, neither for clinical nor pathological evaluations. According to the Miller–Payne system, patients with a high pathological response to NAT had lower miR-375-3p and higher miR-210-3p expressions compared to intermediate- and low-responding patients. Higher miR-210-3p expression was observed for high responders versus partial responders in the RCB evaluation system.

Table 2. Tissue miRNA expression in relation to the clinicopathological data.

Clinicopathological Features	miR-375-3p	miR-210-3p	let-7e-5p
	p-Value	*p*-Value	*p*-Value
Age ≤50 vs. >50	0.116	0.709	0.682
Grading_Biopsy G1 vs. [1] G2 vs. G3	**0.007 **** **FR [2] G3 vs. G1 * = −2.42** **FR G3 vs. G2 ** = −3.08**	0.783	0.118
ER [3] [21] Negative (<1) vs. Positive (≥1)	**0.0008 ***** **FR ER+ vs. ER- = 6.13**	0.970	0.209
PgR [4] [21] Negative (<1) vs. Positive (≥1)	**0.009 **** **FR PgR+ vs. PgR- = 2.28**	0.384	0.305
Ki-67 [22] Negative (≤20) vs. Positive (>20)	**0.042 *** **FR Ki-67+ vs. Ki-67- = −1.49**	0.187	0.052
Molecular Subtype LuminalA (LumA) vs. LuminalB (LumB) vs. TNBC [5]	**0.004 **** **FR TNBC vs. LumA *** = −9.80** **FR TNBC vs. LumB ** = −7.28**	0.332	0.296
cT [6] T1 vs. T2 vs. T3 vs. T4	0.341	0.278	0.885
cN [7] Negative (N0) vs. Positive (N1 + N2 + N3)	0.799	0.879	0.350
Clinical Stage Early (Stage I + II) vs. Advanced (Stage III + Stage IV)	0.343	0.834	0.463
pT [8] T0 vs. T1 vs. T2	0.261	0.555	0.749
pN [9] Negative (N0) vs. Positive (N1 + N2 + N3)	0.469	0.815	0.929
Lymphatic Invasion L0 vs. L1	**0.043 *** **FR L1 vs. L0 = 1.94**	0.064	0.853
Miller–Payne Low (1+2) vs. Intermediate (3) vs. High (4 + 5)	**0.019 *** **FR High vs. Intermediate * = −3.03** **FR Intermediate vs. Low * = 3.14**	**0.016 *** **FR High vs. Low * = 2.54** **FR Intermediate vs. Low * = 2.06**	0.409
RCB [10] 0/I vs. II vs. III	0.416	**0.023 *** **FR 0/I vs. III ** = 2.87**	0.475

[1] vs.—versus, [2] FR—fold regulation, [3] ER—oestrogen receptor, [4] PgR—progesterone receptor, [5] TNBC—triple negative, [6] cT—clinic tumour, [7] cN—clinic lymph node, [8] pT—pathologic tumour, [9] pN—pathologic lymph node, [10] RCB—residual cancer burden. * The table shows the *p*-values for all the compared groups, and where these differences were statistically significant, the expression level of that miRNA (FR) in the interest group versus the reference group was calculated. According to data distributions, differences in expression in the case of two groups were evaluated with the Mann–Whitney test and for three groups with the Kruskal–Wallis test, followed by Dunn's multiple comparison post hoc test. In the case of three or more groups, the asterisk next to the FR value is associated with the *p*-value obtained by the Dunn test (* $p < 0.05$, ** $p < 0.01$, *** $p < 0.001$).

Analysis of the miRNAs expression in the TCGA database was performed as an independent external validation cohort. Increased miRNA expression was observed between tumour and standard samples (Supplementary Figure S1A), with luminal tumours having significantly higher miR-375-3p and lower miR-210-3p expression (Figure 1, Supplementary Table S1 and Figure S1B–D) compared to the basal-like subtype. No significant miRNA expression differences (FR cut-off > 1.5) were observed for patients with positive lymph nodes (Supplementary Figure S2A) or metastatic diseases (Supplementary Figure S2B). Patients with advanced diseases had higher miR-210-3p expression (Figure 2, Table S1), while a slight decrease in miR-210-3p expression was associated with a better survival (Figure 3, Table S1).

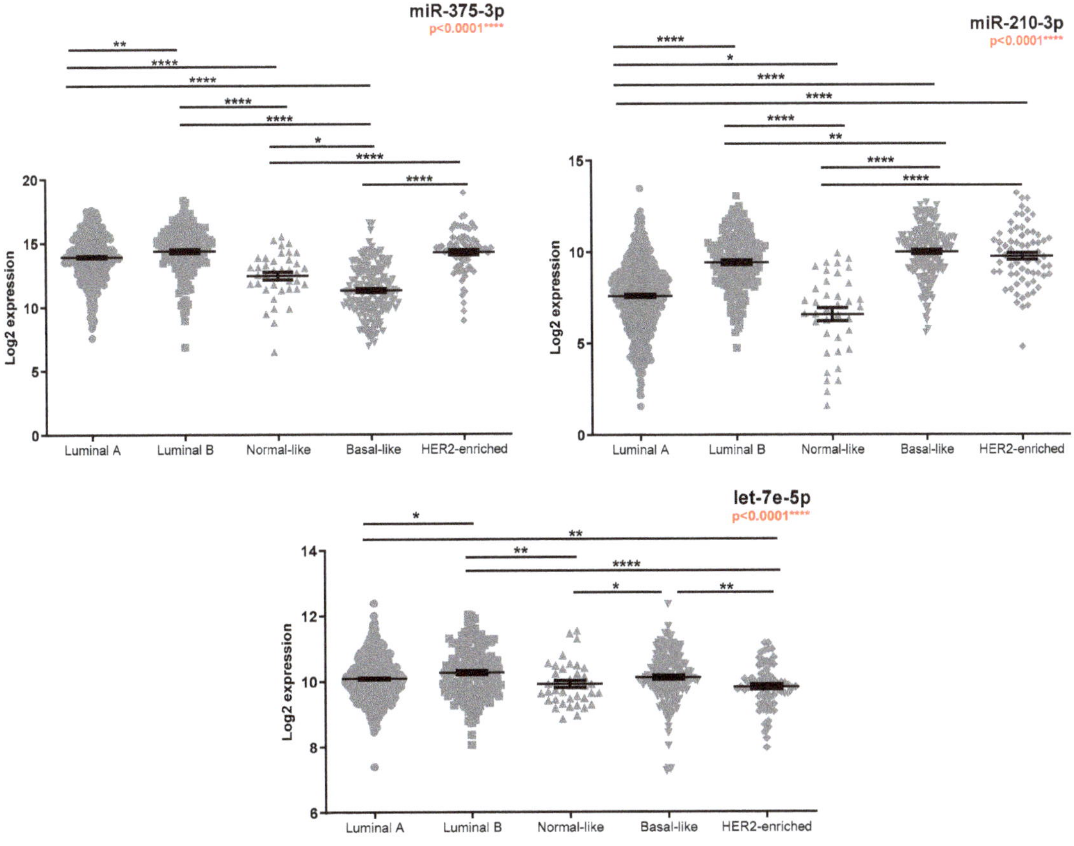

Figure 1. TCGA miR-375-3p, miR-210-3p, and let-7e-5p expression according to PAM50 classification (n = 470 luminal A, n = 162 luminal B, n = 38 normal-like, n = 159 basal-like, and n = 71 HER2-enriched). Kruskal–Wallis test was used for between-groups comparison and Dunn test for multiple comparisons (* $p < 0.05$, ** $p < 0.01$, **** $p < 0.0001$). FR values between groups of interest along with individual p-values obtained by Dunn test are reported in Table S1.

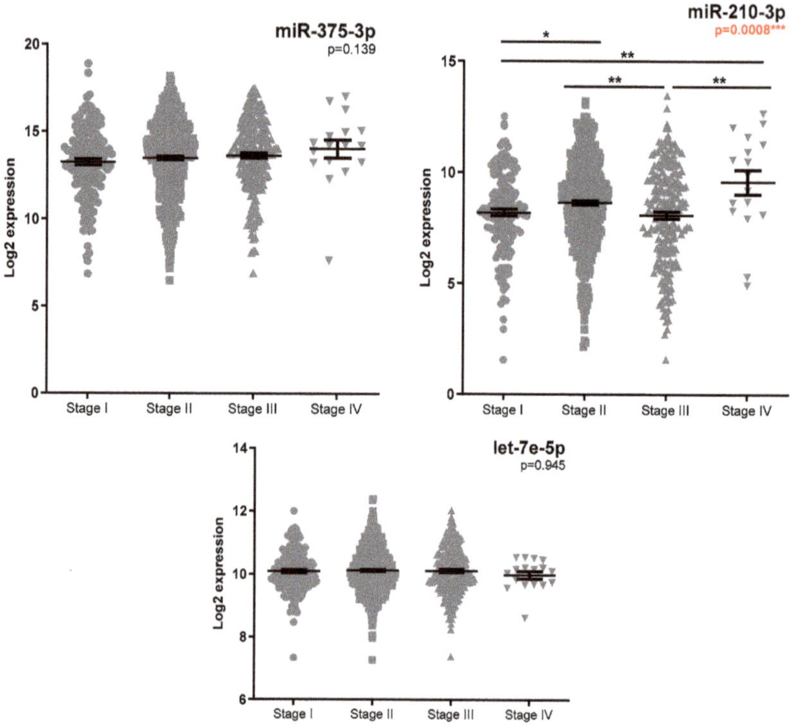

Figure 2. TCGA miR-375-3p, miR-210-3p, and let-7e-5p expression according to pathologic tumour stage (n = 158 stage I, n = 510 stage II, n = 210 stage III, and n = 17 stage IV). Kruskal–Wallis test was used for between-groups comparison and Dunn test for multiple comparisons (* $p < 0.05$, ** $p < 0.01$, *** $p < 0.001$). FR values between groups of interest along with individual p-values obtained by Dunn test are reported in Table S1.

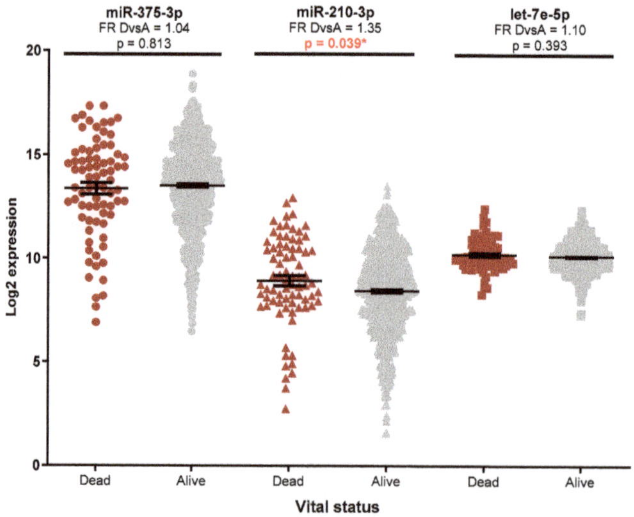

Figure 3. TCGA miR-375-3p, miR-210-3p, and let-7e-5p expression according to vital status. Groups: D—dead, with tumour or with a new tumour event (n = 77); A—alive, tumour-free, without a new event tumour (n = 724). Mann–Whitney test was used for group comparison (* $p < 0.05$).

In order to explore the possible mechanisms mediated by the investigated miRNAs, a correlation analysis between miRNA and mRNA expression in the TCGA cohort was undertaken. The significantly correlated genes were intersected with the validated mRNA targets downloaded from the miRTarBase. Only the validated target genes that were inversely correlated with miRNA expression and with a correlation coefficient under −0.3 were considered of interest (Table S2). A total of 17 possible mRNA targets for miR-375-3p and 9 for miR-210-3p were identified (Figure 4A). No significantly correlated genes were observed for let-7e-5p. The number of identified genes was too small to run a GSEA analysis; therefore, no specific molecular mechanisms could be attributed to either miRNA. Upstream analysis using the IPA knowledge base identified a list of 127 possible chemical and biological targeting drugs (Figure 4B).

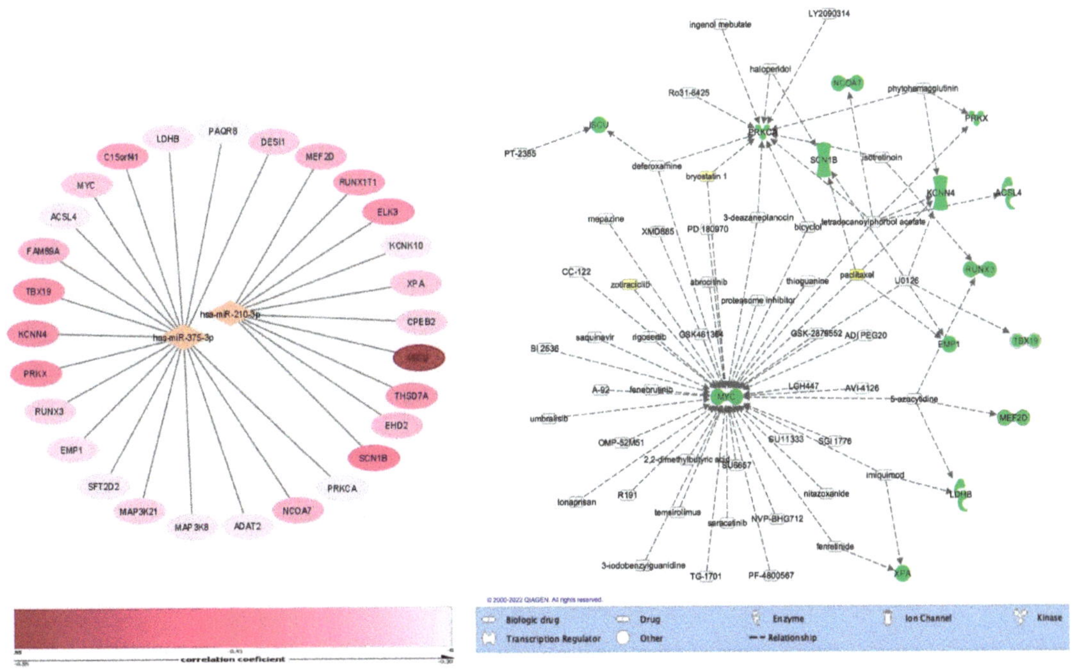

(A) (B)

Figure 4. (**A**) miR-375-3p- and miR-210-3p-correlated validated mRNA targets in the TCGA cohort. Node fill colour was continuously mapped according to the correlation coefficient, (**B**) Drugs that target mRNA downstream genes.

4. Discussion

Different miRNA signatures have been associated with response to chemo-endocrine or radiotherapy in breast cancer [11], emerging as valuable biomarkers for a personalized therapeutic approach of the disease [1,23]. This study reports the expression profiles of three miRNAs involved in the treatment response of BC patients. The analysis explored the association of miRNA expression with the pathological response to NAT and their correlation with the clinicopathological data of the patients.

The patients' pathological response to NAT was assessed using both MP and RCB systems. However, both of them have limitations. The MP system ignores the involvement of the axillary lymph nodes; therefore, the prognosis can be overrated in lymph-node-positive patients [8]. The RCB classification showed a better performance than the MP system, especially for the TNBC subtype [8], but it is limited to anatomical factors without considering the biological ones [8]. To improve its value, assessment of Ki-67 expression after treatment in combination with the RCB index might improve the prediction of survival outcomes [24]. Of the three investigated miRNAs, miR-375-3p and miR-210-3p were significantly associated with MP response, while miR-210-3p expression was also significantly associated with RCB.

MiR-375-3p is dysregulated in various types of cancer. It is involved in epithelial-to-mesenchymal transition (EMT), and is associated with increased invasiveness potential while also being correlated with refractory response to chemotherapy [25]. MiR-375 is a known tumour suppressor. In hepatocellular carcinoma (HCC), it inhibits the autophagy and tumour growth. Moreover, miR-375 promotes the release of mitochondrial apoptotic proteins, reducing the viability of HCC cells in hypoxic conditions. In HCC, miR-375 was downregulated. Autophagy is an adaptive mechanism of the tumour cells that helps them to survive in the tumour microenvironment conditions by reducing apoptosis and enhancing the elimination of the injured mitochondria [26]. Although miR-375 could be related to treatment response by mitochondria reprogramming, to date, there are no data presenting evidence about the role of mir-375 in inducing NAT response through mitochondria reprogramming in breast cancer. Despite the well-documented role as a tumour suppressor, in breast cancer, miR-375-3p is upregulated [16,27,28] and is highly expressed in hormone-receptor-positive breast tumours [29] and lymph-node-positive patients [29]. The present results are in line with these findings. The upregulated expression of miR-375-3p in breast cancer suggests a potential oncogenic activity [16]. Predicted miR-375-3p targets were downloaded from the miRTarBase and intersected with inversely correlated miRNA-mRNA genes from TCGA. Using GSEA, we interrogated Reactome, Kyoto Encyclopedia of Genes and Genomes (KEEG), and Gene Ontology (GO) databases to explore gene functionality. However, the gene list was too small to generate any significant signalling pathways associated with the identified genes. Thus, the regulatory role of miR-375-3p in breast cancer remains unclear.

Multiple miR-375-3p-mediated therapy resistance mechanisms have been described. Generally, miR-375-3p expression is downregulated in drug-resistant BC cells, while its overexpression has been shown to increase cells' sensitivity to chemo-endocrine or targeted therapy. Mir-375-3p-mediated targeting of YBX1 [30] and JAK2 genes [31] has led to increased sensitivity to Adriamycin and paclitaxel first-line treatments. In fulvestrant-resistant BC cells, the overexpression of miR-375-3p inhibited cell growth and autophagy by silencing autophagy-related proteins [30]. Furthermore, by targeting HOXB3 (17) or MDTH gene expression [32], miR-375-3p has decreased EMT, stem features and resistance to tamoxifen in ER-positive BC cells. Moreover, epigenetic silencing of miR-375-3p induced trastuzumab resistance in HER2-positive BC by targeting IGF1R [33]. All this evidence suggests that miR-375-3p might serve as a potential therapeutic approach for the treatment of resistant breast cancer and as a prognostic marker of therapy.

According to the MP evaluation system, in the present cohort, higher miR-375-3p expression was calculated for patients with intermediate response to NAT compared to nonresponders or good responders. Consistent with previous reports, lower miR-375-3p expression levels were observed in the resistant-to-NAT tumour group, while higher expression levels were associated with an improved response. Notably, low levels of miR-375 were also observed in the high-responsive group of patients.

The existing literature regarding the prognostic potential of miR-375-3p is controversial. Furthermore, most reports are based on circulating miR-375-3p levels. According to a three-year follow-up study, patients with relatively higher tumour miR-375-3p expression had a worse survival rate and less survival time, namely, a worse prognosis [34]. On the other hand, a lower expression of circulating miR-375-3p was associated with incomplete response to NAT, while increased expression was noticed in patients achieving pCR after NAT [35]. Similarly, Wu et al. [36] reported that miR-375-3p prevalence in circulation was associated with better clinical outcomes, complete response to NAC, and an absence of relapse. Simultaneously, lower levels of miR-375-3p were noted in therapy-resistant HER2-positive patients. In nonresponder luminal B HER2- patients, NAT can induce the upregulation of circulating miR-375-3p, and this change might be associated with a good response to NAT [35].

Moreover, miR-375-3p association with NAT response seems to be subtype-specific. A lower expression of miR-375-3p was correlated with an increased risk of disease relapse in luminal B patients, while in luminal A, patients with lower miR-375-3p expression were found to be more sensitive to NAC [37]. However, when comparing circulating miRNA levels with tumour levels, it should be considered that the cellular source of circulating miR-375-3p remains unknown; their prevalence may not reflect expression in the primary tumour but rather a combination with other cell types, such as immune cells.

High expression of miR-210-3p in human cancers has become a predictive marker of tumour hypoxia, increasing experimental evidence supporting its clinical relevance. In BC, miR-210-3p is overexpressed in tumour tissues, specifically in triple-negative and HER2+ tumours compared [38], while miR-210-3p upregulation was associated with drug-resistant breast cancer cells [18,39].

The literature data regarding the association of miR-210-3p expression and patient response to neoadjuvant treatment are conflicting, highlighting that both increased and decreased miR-210-3p expression have shown discrepancies regarding the clinical significance as prognostic biomarkers. In HER2-positive BC patients, higher circulating miR-210-3p levels were noticed before surgical excision in patients with residual disease and lymph node metastasis [18]. Muller et al. reported increased miR-210-3p circulating levels following NAT. However, no association between miR-210-3p levels and pCR was observed [40]. Conversely, higher circulating miR-210-3p levels have been associated with residual disease following trastuzumab-combined NAC in BC patients [18,41]. Similar results were also described in ER-positive BC patients treated with tamoxifen; miR-210-3p expression was linked with poor clinical outcomes and an increased risk of relapse [42]. A meta-analysis revealed that miR-210-3p overexpression correlated with poorer survival in TNBC patients and associated circulating miR-210 expression with resistance to doxorubicin, cyclophosphamide, cisplatin and paclitaxel [43]. When quantified in tissue samples, no clinical association with NAT response was observed in miR-210-3p expression levels in BC formalin-fixed paraffin-embedded tissues [40]. However, miR-210-3p expression was suggested to represent a marker for predicting metastasis development and a worse prognosis in patients treated with taxanes in adjuvant settings [44]. Bioinformatic analysis identified microtubule regulation, drug efflux pathways and NRF2-mediated oxidative stress response as the most significant associated pathways between miR-210-3p signalling and docetaxel resistance [44].

Analysis of the miR-210-3p expression in the present cohort was performed to further elucidate the controversy surrounding this miR's ability to predict cPCR. Patients with almost complete or partial pathological responses had significantly higher miR-210-3p expression levels than nonresponders. MiR-210-3p's role as an oncogene is well-characterized; however, it was also suggested to act as a tumour suppressor [19]. For example, in oesophageal squamous cell carcinoma, miR-210-3p has been shown to inhibit cancer cell survival and proliferation by inducing cell death and cell cycle arrest in G(1)/G(0) and G(2)/M through FGFRL1 downregulation [19]. More recently, Bar I et al. [45] showed that in TNBC patients, miR-210-3p is expressed by both tumour cells and the tumour microenvironment (TME) that is more likely to be regulated by a mechanism independent of HIF-1 alpha. MiR-210-3p has multiple functions, including the regulation of the immune response and increasing data support the concept that immunologically "hot" tumours are more responsive to chemotherapy [46].

Let-7e-5p is one of the first discovered miRNAs [47]. This is a tumour suppressor gene that targets essential pathways involved in tumorigeneses such as Janus protein tyrosine kinase (JAK), c-Myc and signal traducer and activator of transcription 3 (STAT3). The literature reports about let-7e-5p expression during NAT are scarce. In patients with a lower Ki-67 and pCR, a decrease in let-7e-5p expression was noticed after NAT [48]. Lv. J. et al. showed that let-7e-5p expression is down-regulated in chemoresistant tumours, while decreased expression was associated with a worse prognosis [20]. The present analysis did not find any significant association between let-7e-5p expression and pathological response to NAT or with the clinicopathological features of the patients, neither in the given cohort nor in the TCGA database.

Our study is limited by the patients' heterogeneity and the relatively small sample size cohort and, thus, the present analysis has low statistical power. However, to the best of our knowledge, this is the first study that illustrates a putative relationship between miR-375-3p and miR-210-3p and breast tumours' pathological response to neoadjuvant therapy. Increasing the number of patients would allow for more homogeneous groups and subsequent differential analysis based on the administered type of therapy. Of great interest, prognostic markers for patients' response to combined regimens are largely unexplored.

5. Conclusions

Based on documented mechanistic actions of the two miRNAs, our results suggest that tumours with higher levels of miR-375-3p are more sensitive to neoadjuvant therapy compared to resistant tumours, and that higher miR-210-3p expression in responsive tumours could indicate immunologically "hot" tumours. These findings suggest the potential role of these two miRNAs in stratifying BC patients that will respond to NAT. However, as data regarding these two miRNAs are controversial, further studies are needed to elucidate their complex role in mediating BC patients' response to neoadjuvant therapy.

Supplementary Materials: The following supporting information can be downloaded at: https://www.mdpi.com/article/10.3390/medicina58101494/s1, Figure S1: TCGA miR-375-3p, miR-210-3p, and let-7e-5p expression according to PAM50 classification; Figure S2: TCGA miR-375-3p, miR-210-3p, and let-7e-5p expression according to pathologic tumour stage; Table S1: Clinicopathological data of the patients included in the study; Table S2: Tissue miRNA expression in relation to the clinicopathological data.

Author Contributions: Conceptualization, L.A.L., O.B. and O.T.; methodology, S.V., A.R. and L.B.; software, O.T., O.B. and L.B.; validation, S.V. and L.B.; formal analysis, A.R. and L.B.; investigation A.R., L.A.L. and C.L.; resources, L.A.L., O.T., O.B., C.L. and A.R.; data curation, S.V.; writing—original draft preparation, L.A.L.; writing—review and editing, O.T., E.-A.B., A.P., E.G., E.M., O.B. and L.B.; visualization, A.P., A.C., A.I., C.L., E.G. and E.M., supervision, E.-A.B., A.P., A.I., A.C. and C.L.; project administration, E.-A.B., A.P., A.C. and A.I.; funding acquisition, O.T., O.B., L.A.L. and C.L. All authors have read and agreed to the published version of the manuscript.

Funding: This work was supported by the Competitiveness Operational Programme 2014–2020, Contract no. 41/02.09.2016, MySMIS 103557, Project title: Genomic and microfluidic approaches towards blocking breast cancer cell invasion and metastasis—BREASTMINCROGENOMICS and Knowledge Transfer of Biogenomics in Oncology and Related Domains in clinical applications grant—BIOGENONCO, MySMIS code: 105774, financing contract no: 10/01.09.2016.

Institutional Review Board Statement: The study was conducted in accordance with the Declaration of Helsinki, and approved by the Institutional Ethics Committee of The Oncology Institute "Ion Chiricuță", Cluj-Napoca, Romania (Approval No. 59/29.11.2016) and the University of Medicine and Pharmacy Iuliu Hatieganu, Cluj-Napoca, Romania (Approval No. 290/09.09.2020).

Informed Consent Statement: Informed consent was obtained from all subjects involved in the study.

Data Availability Statement: Not applicable.

Conflicts of Interest: The authors declare no conflict of interest.

References

1. Davey, M.G.; Lowery, A.J.; Miller, N.; Kerin, M.J. MicroRNA expression profiles and breast cancer Chemotherapy. *Int. J. Mol. Sci.* **2021**, *22*, 10812. Available online: https://pubmed.ncbi.nlm.nih.gov/34639152/ (accessed on 5 May 2022).
2. McDonald, E.S.; Clark, A.S.; Tchou, J.; Zhang, P.; Freedman, G.M. Clinical Diagnosis and Management of Breast Cancer. *J. Nucl. Med.* **2016**, *57* (Suppl. 1), 9S–16S. Available online: https://jnm.snmjournals.org/content/57/Supplement_1/9S (accessed on 21 April 2022). [CrossRef]
3. Thompson, A.M.; Moulder-Thompson, S.L. Neoadjuvant treatment of breast cancer. *Ann. Oncol.* **2012**, *23* (Suppl. 10). [CrossRef] [PubMed]
4. Sikov, W.; Boughey, J.; Al-Hilli, Z. UpToDate: General Principles of Neoadjuvant Management of Breast Cancer. 2021, pp. 1–21. Available online: https://www.uptodate.com/contents/general-principles-of-neoadjuvant-management-of-breast-cancer. (accessed on 21 August 2022).
5. Sahoo, S.; Lester, S.C. Pathology of Breast Carcinomas After Neoadjuvant Chemotherapy: An Overview With Recommendations on Specimen Processing and Reporting. *Arch. Pathol. Lab. Med.* **2009**, *133*, 633–642. Available online: https://meridian.allenpress.com/aplm/article/133/4/633/460740/Pathology-of-Breast-Carcinomas-After-Neoadjuvant (accessed on 5 May 2022). [CrossRef] [PubMed]
6. Park, C.K.; Jung, W.H.; Koo, J.S. Pathologic Evaluation of Breast Cancer after Neoadjuvant Therapy. *J. Pathol. Transl. Med.* **2016**, *50*, 173. Available online: https://www.ncbi.nlm.nih.gov/pmc/articles/PMC4876080/ (accessed on 5 May 2022). [CrossRef]
7. Shintia, C.; Endang, H.; Diani, K. Assessment of pathological response to neoadjuvant chemotherapy in locally advanced breast cancer using the Miller-Payne system and TUNEL. *Malays. J. Pathol.* **2016**, *38*, 25–32. Available online: https://pubmed.ncbi.nlm.nih.gov/27126661/ (accessed on 14 April 2022). [PubMed]
8. Wang, W.; Liu, Y.; Zhang, H.; Zhang, S.; Duan, X.; Ye, J.; Xu, L.; Zhao, J.; Cheng, Y.; Liu, Q. Prognostic value of residual cancer burden and Miller-Payne system after neoadjuvant chemotherapy for breast cancer. *Gland. Surg.* **2021**, *10*, 3211. [CrossRef] [PubMed]
9. Symmans, W.F.; Peintinger, F.; Hatzis, C.; Rajan, R.; Kuerer, H.; Valero, V.; Assad, L.; Poniecka, A.; Hennessy, B.; Green, M.; et al. Measurement of residual breast cancer burden to predict survival after neoadjuvant chemotherapy. *J. Clin. Oncol.* **2007**, *25*, 4414–4422. [CrossRef] [PubMed]
10. Lu, J.; Getz, G.; Miska, E.A.; Alvarez-Saavedra, E.; Lamb, J.; Peck, D.; Sweet-Cordero, A.; Ebert, B.L.; Mak, R.H.; Ferrando, A.A.; et al. MicroRNA expression profiles classify human cancers. *Nature* **2005**, *435*, 834–838. Available online: https://pubmed.ncbi.nlm.nih.gov/15944708/ (accessed on 11 May 2022). [CrossRef]
11. Petrović, N.; Nakashidze, I.; Nedeljković, M. Breast Cancer Response to Therapy: Can microRNAs Lead the Way? *J. Mammary Gland Biol. Neoplasia* **2021**, *26*, 157–178. Available online: https://pubmed.ncbi.nlm.nih.gov/33479880/ (accessed on 10 February 2022). [CrossRef]
12. Zhang, Z.; Zhang, H.; Yu, J.; Xu, L.; Pang, X.; Xiang, Q.; Liu, Q.; Cui, Y. miRNAs as therapeutic predictors and prognostic biomarkers of neoadjuvant chemotherapy in breast cancer: A systematic review and meta-analysis. *Breast Cancer Res. Treat.* **2022**, *194*, 483–505. Available online: https://pubmed.ncbi.nlm.nih.gov/35727379/ (accessed on 24 September 2022). [CrossRef]
13. To, N.H.; Nguyen, H.Q.; Thiolat, A.; Liu, B.; Cohen, J.; Radosevic-Robin, N.; Belkacemi, Y. Radiation therapy for triple-negative breast cancer: Emerging role of microRNAs as biomarkers and radiosensitivity modifiers. A systematic review. *Breast Cancer Res. Treat.* **2022**, *193*, 265–279. Available online: https://pubmed.ncbi.nlm.nih.gov/35397079/ (accessed on 24 September 2022). [CrossRef] [PubMed]
14. Davey, M.G.; Davey, M.S.; Richard, V.; Wyns, W.; Soliman, O.; Miller, N.; Lowery, A.; Kerin, M. Overview of MicroRNA Expression in Predicting Response to Neoadjuvant Therapies in Human Epidermal Growth Receptor-2 Enriched Breast Cancer—A Systematic Review. *Breast Cancer* **2022**, *16*. Available online: https://pubmed.ncbi.nlm.nih.gov/35340888/ (accessed on 24 September 2022). [CrossRef] [PubMed]
15. Jackisch, C.; Harbeck, N.; Huober, J.; Von Minckwitz, G.; Gerber, B.; Kreipe, H.H.; Liedtke, C.; Marschner, N.; Möbus, V.; Scheithauer, H.; et al. 14th St. Gallen International Breast Cancer Conference 2015: Evidence, Controversies, Consensus—Primary Therapy of Early Breast Cancer: Opinions Expressed by German Experts. *Breast Care* **2015**, *10*, 211. [CrossRef] [PubMed]
16. Tang, W.; Li, G.S.; Li JDi Pan, W.Y.; Shi, Q.; Xiong, D.D.; Mo, C.H.; Zeng, J.J.; Chen, G.; Feng, Z.B. The role of upregulated miR-375 expression in breast cancer: An in vitro and in silico study. *Pathol. Res. Pract.* **2020**, *216*, 152754. Available online: https://pubmed.ncbi.nlm.nih.gov/31787478/ (accessed on 9 June 2022). [CrossRef]
17. Fu, H.; Fu, L.; Xie, C.; Zuo, W.S.; Liu, Y.S.; Zheng, M.Z.; Yu, J.M. CancermiR-375 inhibits cancer stem cell phenotype and tamoxifen resistance by degrading HOXB3 in human ER-positive breast cancer. *Oncol. Rep.* **2017**, *37*, 1093–1099. Available online: http://www.spandidos-publications.com/10.3892/or.2017.5360/abstract (accessed on 11 May 2022). [CrossRef] [PubMed]
18. Jung, E.J.; Santarpia, L.; Kim, J.; Esteva, F.J.; Moretti, E.; Buzdar, A.U.; Di Leo, A.; Le, X.F.; Bast, R.C.; Park, S.T.; et al. Plasma microRNA 210 levels correlate with sensitivity to trastuzumab and tumor presence in breast cancer patients. *Cancer* **2012**, *118*, 2603–2614. [CrossRef] [PubMed]
19. Tsuchiya, S.; Fujiwara, T.; Sato, F.; Shimada, Y.; Tanaka, E.; Sakai, Y.; Shimziu, K.; Tsujimoto, G. MicroRNA-210 Regulates Cancer Cell Proliferation through Targeting Fibroblast Growth Factor Receptor-like 1 (FGFRL1). *J. Biol. Chem.* **2011**, *286*, 420. [CrossRef] [PubMed]

20. Lv, J.; Xia, K.; Xu, P.; Sun, E.; Ma, J.; Gao, S.; Zhou, Q.; Zhang, M.; Wang, F.; Chean, F.; et al. miRNA expression patterns in chemoresistant breast cancer tissues. *Biomed. Pharmacother.* **2014**, *68*, 935–942. [CrossRef] [PubMed]
21. Hammond, M.E.H.; Hayes, D.F.; Wolff, A.C.; Mangu, P.B.; Temin, S. American Society of Clinical Oncology/College of American Pathologists Guideline Recommendations for Immunohistochemical Testing of Estrogen and Progesterone Receptors in Breast Cancer. *J. Oncol. Pract.* **2010**, *6*, 195. [CrossRef]
22. Goldhirsch, A.; Winer, E.P.; Coates, A.S.; Gelber, R.D.; Piccart-Gebhart, M.; Thürlimann, B.; Senn, H.J.; Albain, K.S.; André, F.; Bergh, J.; et al. Personalizing the treatment of women with early breast cancer: Highlights of the St Gallen International Expert Consensus on the Primary Therapy of Early Breast Cancer 2013. *Ann. Oncol. Off. J. Eur. Soc. Med. Oncol.* **2013**, *24*, 2206–2223. Available online: https://pubmed.ncbi.nlm.nih.gov/23917950/ (accessed on 9 August 2022). [CrossRef] [PubMed]
23. McGuire, A.; Casey, M.C.; Waldron, R.M.; Heneghan, H.; Kalinina, O.; Holian, E.; McDermott, A.; Lowery, A.J.; Newell, J.; Dwyer, R.M.; et al. Prospective Assessment of Systemic MicroRNAs as Markers of Response to Neoadjuvant Chemotherapy in Breast Cancer. *Cancers* **2020**, *12*, 1820. Available online: https://www.mdpi.com/2072-6694/12/7/1820/htm (accessed on 13 February 2022). [CrossRef]
24. Sheri, A.; Smith, I.E.; Johnston, S.R.; A'hern, R.; Nerurkar, A.; Jones, R.L.; Hills, M.; Detre, S.; Pinder, S.E.; Symmans, W.F.; et al. Residual proliferative cancer burden to predict long-term outcome following neoadjuvant chemotherapy. *Ann. Oncol. Off. J. Eur. Soc. Med. Oncol.* **2015**, *26*, 75–80. Available online: https://pubmed.ncbi.nlm.nih.gov/25361988/ (accessed on 16 April 2022). [CrossRef]
25. Wei, J.; Lu, Y.; Wang, R.; Xu, X.; Liu, Q.; He, S.; Pan, H.; Liu, X.; Yuan, B.; Ding, Y. MicroRNA-375: Potential cancer suppressor and therapeutic drug. *Biosci. Rep.* **2021**. [CrossRef]
26. Chang, Y.; Yan, W.; He, X.; Zhang, L.; Li, C.; Huang, H.; Nace, G.; Geller, D.A.; Lin, J.; Tsung, A. miR-375 inhibits autophagy and reduces viability of hepatocellular carcinoma cells under hypoxic conditions. *Gastroenterology* **2012**, *143*, 177-87.e8. Available online: https://pubmed.ncbi.nlm.nih.gov/22504094/ (accessed on 29 September 2022). [CrossRef]
27. Rocha Simonini, P.D.S.; Breiling, A.; Gupta, N.; Malekpour, M.; Youns, M.; Omranipour, R.; Malekpour, F.; Volinia, S.; Croce, C.M.; Naimabadi, H. Epigenetically deregulated microRNA-375 is involved in a positive feedback loop with estrogen receptor α in breast cancer cells. *Cancer Res.* **2010**, *70*, 9175–9184. Available online: https://aacrjournals.org/cancerres/article/70/22/9175/559828/Epigenetically-Deregulated-microRNA-375-Is (accessed on 9 June 2022). [CrossRef]
28. Liu, J.; Wang, P.; Zhang, P.; Zhang, X.; Du, H.; Liu, Q.; Huang, B.; Qian, C.; Zhang, S.; Zhu, W.; et al. An integrative bioinformatics analysis identified miR-375 as a candidate key regulator of malignant breast cancer. *J. Appl. Genet.* **2019**, *60*, 335–346. Available online: https://pubmed.ncbi.nlm.nih.gov/31372832/ (accessed on 9 June 2022). [CrossRef] [PubMed]
29. Jonsdottir, K.; Janssen, S.R.; Da Rosa, F.C.; Gudlaugsson, E.; Skaland, I.; Baak, J.P.A.; Janssen, E.A.M. Validation of Expression Patterns for Nine miRNAs in 204 Lymph-Node Negative Breast Cancers. *PLoS ONE* **2012**, *7*. Available online: https://www.ncbi.nlm.nih.gov/pmc/articles/PMC3492447/ (accessed on 9 June 2022). [CrossRef] [PubMed]
30. Liu, S.L.; Sui, Y.F.; Lin, M.Z. MiR-375 is epigenetically downregulated due to promoter methylation and modulates multi-drug resistance in breast cancer cells via targeting YBX1. *Eur. Rev. Med. Pharmacol. Sci.* **2016**, *20*, 3223–3229. Available online: https://www.europeanreview.org/article/11237 (accessed on 12 May 2022). [PubMed]
31. Zhao, Q.; Liu, Y.; Wang, T.; Yang, Y.; Ni, H.; Liu, H.; Guo, Q.; Xi, T.; Zheng, L. MiR-375 inhibits the stemness of breast cancer cells by blocking the JAK2/STAT3 signaling. *Eur. J. Pharmacol.* **2020**, *884*, 173359. [CrossRef] [PubMed]
32. Ward, A.; Balwierz, A.; Zhang, J.D.; Küblbeck, M.; Pawitan, Y.; Hielscher, T.; Wiemann, S.; Sahin, Ö. Re-expression of microRNA-375 reverses both tamoxifen resistance and accompanying EMT-like properties in breast cancer. *Oncogene* **2012**, *32*, 1173–1182. Available online: https://www.nature.com/articles/onc2012128 (accessed on 11 May 2022). [CrossRef] [PubMed]
33. Ye, X.M.; Zhu, H.Y.; Bai, W.D.; Wang, T.; Wang, L.; Chen, Y.; Yang, A.G.; Jia, L.T. Epigenetic silencing of miR-375 induces trastuzumab resistance in HER2-positive breast cancer by targeting IGF1R. *BMC Cancer* **2014**, *14*, 134. Available online: https://bmccancer.biomedcentral.com/articles/10.1186/1471-2407-14-134 (accessed on 11 May 2022). [CrossRef] [PubMed]
34. Guan, X.; Shi, A.; Zou, Y.; Sun, M.; Zhan, Y.; Dong, Y.; Fan, Z. EZH2-Mediated microRNA-375 Upregulation Promotes Progression of Breast Cancer via the Inhibition of FOXO1 and the p53 Signaling Pathway. *Front. Genet.* **2021**, *12*, 633756. Available online: https://pubmed.ncbi.nlm.nih.gov/33854524/ (accessed on 10 June 2022). [CrossRef] [PubMed]
35. Zhang, Z.; Zhang, H.; Li, C.; Xiang, Q.; Xu, L.; Liu, Q.; Pang, X.; Zhang, W.; Zhang, H.; Zhang, S.; et al. Circulating microRNAs as indicators in the prediction of neoadjuvant chemotherapy response in luminal B breast cancer. *Thorac. Cancer* **2021**, *12*, 3396–3406. Available online: https://pubmed.ncbi.nlm.nih.gov/34751517/ (accessed on 5 May 2022). [CrossRef] [PubMed]
36. Wu, X.; Somlo, G.; Yu, Y.; Palomares, M.R.; Li, A.X.; Zhou, W.; Chow, A.; Yen, Y.; Rossi, J.J.; Gao, H.; et al. De novo sequencing of circulating miRNAs identifies novel markers predicting clinical outcome of locally advanced breast cancer. *J. Transl. Med.* **2012**, *10*, 42. Available online: https://tmu.pure.elsevier.com/en/publications/de-novo-sequencing-of-circulating-mirnas-identifies-novel-markers-2 (accessed on 5 May 2022). [CrossRef] [PubMed]
37. Chekhun, V.F.; Borikun, T.V.; Bazas, V.M.; Andriiv, A.V.; Klyusov, O.M.; Yalovenko, T.M.; Lukianova, N.Y. Association of Circulating MIR-21,-205, and-182 with Response of Luminal Breast Cancers to Neoadjuvant Fac and Ac Treatment. *Exp. Oncol.* **2020**, *42*, 162–166. Available online: https://doi.org/10.32471/exp-oncology.2312-8852.vol-42-no-3.14805 (accessed on 5 May 2022). [PubMed]

38. Evangelista, A.F.; Oliveira, R.J.; Viviane, V.A.; Rene, R.A.; Reis, R.M.; Marcia, M.M. Integrated analysis of mRNA and miRNA profiles revealed the role of miR-193 and miR-210 as potential regulatory biomarkers in different molecular subtypes of breast cancer. *BMC Cancer* **2021**, *21*, 76. Available online: https://bmccancer.biomedcentral.com/articles/10.1186/s12885-020-07731-2 (accessed on 16 April 2022). [CrossRef]
39. Chen, X.; Lu, P.; Wang D dan Yang S jin Wu, Y.; Shen, H.Y.; Zhong, S.L.; Zhao, J.H.; Tang, J.H. The role of miRNAs in drug resistance and prognosis of breast cancer formalin-fixed paraffin-embedded tissues. *Gene* **2016**, *595*, 221–226. [CrossRef] [PubMed]
40. Müller, V.; Gade, S.; Steinbach, B.; Loibl, S.; Von Minckwitz, G.; Untch, M.; Schwedler, K.; Lübbe, K.; Schem, C.; Fasching, P.A.; et al. Changes in serum levels of miR-21, miR-210, and miR-373 in HER2-positive breast cancer patients undergoing neoadjuvant therapy: A translational research project within the Geparquinto trial. *Breast Cancer Res. Treat.* **2014**, *147*, 61–68. Available online: https://link.springer.com/article/10.1007/s10549-014-3079-3 (accessed on 6 May 2022). [CrossRef] [PubMed]
41. Davey, M.G.; Davies, M.; Lowery, A.J.; Miller, N.; Kerin, M.J. The Role of MicroRNA as Clinical Biomarkers for Breast Cancer Surgery and Treatment. *Int. J. Mol. Sci.* **2021**, *22*, 8290. Available online: https://www.ncbi.nlm.nih.gov/pmc/articles/PMC8346977/ (accessed on 10 September 2022). [CrossRef] [PubMed]
42. Egeland, N.G.; Lunde, S.; Jonsdottir, K.; Lende, T.H.; Cronin-Fenton, D.; Gilje, B.; Janssen, E.A.M.; Søiland, H. The Role of MicroRNAs as Predictors of Response to Tamoxifen Treatment in Breast Cancer Patients. *Int. J. Mol. Sci.* **2015**, *16*, 24243. Available online: https://www.ncbi.nlm.nih.gov/pmc/articles/PMC4632748/ (accessed on 10 February 2022). [CrossRef]
43. Qattan, A.; Al-Tweigeri, T.; Alkhayal, W.; Suleman, K.; Tulbah, A.; Amer, S. Clinical identification of dysregulated circulating microRNAs and their implication in drug response in triple negative breast cancer (TNBC) by target gene network and meta-analysis. *Genes* **2021**, *12*, 549. Available online: https://www.ncbi.nlm.nih.gov/pmc/articles/PMC8068962/ (accessed on 11 May 2022). [CrossRef] [PubMed]
44. Pasculli, B.; Barbano, R.; Rendina, M.; Fontana, A.; Copetti, M.; Mazza, T.; Valori, V.M.; Morritti, M.; Mariello, E.; Graziano, P.; et al. Hsa-miR-210-3p expression in breast cancer and its putative association with worse outcome in patients treated with Docetaxel. *Sci. Rep.* **2019**, *9*, 14913. [CrossRef] [PubMed]
45. Bar, I.; Merhi, A.; Abdel-Sater, F.; Ben Addi, A.; Sollennita, S.; Canon, J.L.; Delrée, P. The MicroRNA miR-210 Is Expressed by Cancer Cells but Also by the Tumor Microenvironment in Triple-Negative Breast Cancer. *J. Histochem. Cytochem.* **2017**, *65*, 335. Available online: https://www.ncbi.nlm.nih.gov/pmc/articles/PMC5625853/ (accessed on 9 June 2022). [CrossRef] [PubMed]
46. Binnewies, M.; Roberts, E.W.; Kersten, K.; Chan, V.; Fearon, D.F.; Merad, M.; Coussens, L.M.; Gabrilovich, D.L.; Ostrand-Rosenberg, S.; Hedrick, C.C.; et al. Understanding the tumor immune microenvironment (TIME) for effective therapy. *Nat. Med.* **2018**, *24*, 541–550. [CrossRef]
47. Liu, K.; Zhang, C.; Li, T.; Ding, Y.; Tu, T.; Zhou, F.; Qi, W.; Chen, H.; Sun, X. Let-7a inhibits growth and migration of breast cancer cells by targeting HMGA1. *Int. J. Oncol.* **2015**, *46*, 2526–2534. Available online: http://www.spandidos-publications.com/10.3892/ijo.2015.2949/abstract (accessed on 16 April 2022). [CrossRef]
48. Gezer, U.; Keskin, S.; Iğci, A.; Tükenmez, M.; Tiryakioğlu, D.; Çetinkaya, M.; Dişci, R.; Dalay, N.; Eralp, Y. Abundant circulating microRNAs in breast cancer patients fluctuate considerably during neoadjuvant chemotherapy. *Oncol. Lett.* **2014**, *8*, 845–848. Available online: http://www.spandidos-publications.com/10.3892/ol.2014.2188/abstract (accessed on 6 May 2022). [CrossRef]

medicina

MDPI

Article

Development and Preliminary Characterization of Polyester-Urethane Microparticles Used in Curcumin Drug Delivery System for Oropharyngeal Cancer

Alexandru Chioreanu [1,†], Ion Cristian Mot [1], Delia Ioana Horhat [1,*], Nicolae Constantin Balica [1], Cristian Andrei Sarau [1], Raluca Morar [1], Eugenia Maria Domuta [2,†], Catalin Dumitru [3], Rodica Anamaria Negrean [4], Bogdan Andrei Bumbu [5], Madhavi Ravulapalli [6], Satish Alambaram [6], Raja Akshay [7] and Marius Pricop [8]

1 Department of Ear-Nose-Throat, "Victor Babes" University of Medicine and Pharmacy Timisoara, Eftimie Murgu Square 2, 300041 Timisoara, Romania
2 Surgery Department, Faculty of Medicine and Pharmacy, University of Oradea, Piata 1 Decembrie 10, 410073 Oradea, Romania
3 Department of Obstetrics and Gynecology, "Victor Babes" University of Medicine and Pharmacy Timisoara, 300041 Timisoara, Romania
4 Faculty of Medicine and Pharmacy, University of Oradea, 410073 Oradea, Romania
5 Department of Dental Medicine, Faculty of Medicine and Pharmacy, University of Oradea, 410073 Oradea, Romania
6 School of General Medicine, Bhaskar Medical College, Amdapur Road 156-162, Hyderabad 500075, India
7 Malla Reddy Institute of Medical Sciences, Suraram Main Road 138, Hyderabad 500055, India
8 Discipline of Oral and Maxillo-Facial Surgery, Faculty of Dental Medicine, "Victor Babes" University of Medicine and Pharmacy Timisoara, Eftimie Murgu Square 2, 300041 Timisoara, Romania
* Correspondence: horhat.ioana@umft.ro
† These authors contributed equally to this work.

Citation: Chioreanu, A.; Mot, I.C.; Horhat, D.I.; Balica, N.C.; Sarau, C.A.; Morar, R.; Domuta, E.M.; Dumitru, C.; Negrean, R.A.; Bumbu, B.A.; et al. Development and Preliminary Characterization of Polyester-Urethane Microparticles Used in Curcumin Drug Delivery System for Oropharyngeal Cancer. *Medicina* **2022**, *58*, 1689. https://doi.org/10.3390/medicina58111689

Academic Editors: Nicolae Crisan and Călin Căinap

Received: 4 November 2022
Accepted: 18 November 2022
Published: 21 November 2022

Abstract: *Background and Objectives*: Curcumin (Cc) as an active substance is known for its anti-inflammatory, anticoagulant, antioxidant, and anti-carcinogenic effects, together with its role in cholesterol regulation, and its use in different gastrointestinal derangements. On the other hand, curcumin can be used for its properties as an inactive substance, with Cc particles being more often tested in pharmaceutical formulations for drug delivery, with promising safety records and kinetics. The aim of this research was to obtain and characterize polyurethane microparticles that can be used as a carrier with a controlled Cc release. *Materials and Methods*: The in vitro samples were characterized by the Zetasizer procedure, and UV–Vis spectroscopy, while the in-vivo measurements on human subjects were performed by non-invasive skin assays (trans-epidermal water loss, erythema, and skin hydration). A total of 16 patients with oropharyngeal cancer stages II and III in equal proportions were recruited for participation. *Results*: The experimental values of sample characteristics using the Zetasizer identified a mean structural size of 215 nm in the polyester-urethane preparate (PU), compared to 271 nm in the curcumin-based PU. Although the size was statistically significantly different, the IPDI and Zeta potential did not differ significantly (22.91 mV vs. 23.74 mV). The average age during the study period was 57.6 years for patients in the PU group, respectively, and 55.1 years in those who received the curcumin preparations. The majority of oropharyngeal cancers were of HPV-related etiology. There were no significant side effects; 75.0% of patients in the PU group reporting no side effects, compared to 87.5% in the Cc group. The 48 h TEWL measurement at the end of the experiment found a statistically significant difference between the PU and the Cc group (2.2 $g/h/m^2$ vs. 2.6 $g/h/m^2$). The erythema assessment showed a starting measurement point for both research groups with a 5.1-unit difference. After 48 h, the difference between PU and PU_Cc was just 1.7 units (p-value = 0.576). The overall difference compared to the reference group with sodium lauryl sulfate (SLS) was statistically significant at a 95% significance level. *Conclusions*: Our findings indicate the obtaining of almost homogeneous particles with a medium tendency to form agglomerations, with a good capacity of encapsulation (around 60%), a medium release rate, and a non-irritative potential. Therefore, this polyester-urethane with Cc microparticles can be tested in other clinical evaluations.

Keywords: drug delivery systems; polyurethane; cancer management; curcumin; release kinetics

1. Introduction

When current pharmaceutical formulations do not offer the essential attributes for a particular medical purpose, drug delivery systems, which are also known as drug carriers, are used in the medical industry. It is possible for these organic or inorganic nano- and micro-structures to encapsulate the active drugs in order to modify their physicochemical properties and to increase drug absorption; by appropriately selecting the delivery system, one can increase the therapeutic effect of the encapsulated drug or can ensure its prolonged or targeted release, thereby avoiding the risks of underdosing or overdosing [1]. It has been shown in the past that altering the polyurethane chains by utilizing diethanolamine as a chain extender results in an improved cytotoxic impact on cancer cells [2]. These findings were compared to those obtained using other diethanolamine polyurethane micelles as a reference. The scientists came to the conclusion that because of this, the structure of the polyurethane carrier has a considerable impact on the loading, pH-triggered release, and intracellular delivery of the medication.

Although the process of drug release using this polyester-urethane delivery system was demonstrated using paclitaxel as the target drug, other non-reactive substances are hypothesized to be used with this method for various cancer histology and different anatomic regions. Around 500,000 people are diagnosed with oral and oropharyngeal cancers each year [3]. This makes the O-OPCs group the sixth most prevalent kind of cancer overall. There has been a concerning rise in the prevalence of O-OPCs in younger patients, notably in Africa (17.2%) and the Middle East (14.5%) [4]. A significant amount of diversity was found in the epidemiology of O-OPCs that was seen all over the globe depending on the geographical distribution, gender, and age of the people. It is possible that the incidence of oropharyngeal cancer is rising among younger age groups in the Western Hemisphere. This may be due to an increased relationship with human papillomavirus 16, which may also be responsible for the rise [5].

The general risk factors for oropharyngeal cancers are represented by gender, being twice as common in men, probably due to a higher smoking and alcohol consumption [6], while smoking independently is present in approximately 80% of those with oropharyngeal cancer, and alcohol consumption in 70% of cases [7]. Other important risk factors are patients' age, with the average onset around 62 years, two-thirds of oropharyngeal cancer patients being older than 55, exposure to UV radiation, poor diet behavior low in fruits and vegetables, and genetic risk factors. Recently, scientific attention has been distributed towards the HPV infection of the oropharyngeal region due to the significant increase in recent years in developed countries [8,9].

Although surgery is used less commonly in the treatment of oropharyngeal cancer, it is an essential component of the management of oral cavity and pharyngeal cancer in terms of the removal of the main tumor and neck lymph nodes. Surgery may be performed alone or in conjunction with radiation, chemotherapy, and immunotherapy/biotherapy to treat early-stage illness. Besides the conventional treatments, other attempts have been made, such as the use of curcumin (turmeric), with a demonstrated potential benefit in various cancers [10]. Although turmeric is often referred to as a single plant, the term defines a genus that includes nearly 100 species of perennials in the Zingiberaceae family [11]. Basically, curcumin, a yellow polyphenol, is the active turmeric ingredient, a plant that began to be used in India 4000 years ago, extracted from the processing of the turmeric rhizome. It reaches turmeric powder in a proportion of 2.5 to 4%. Due to the beneficial properties of curcumin, it can also be found in the composition of food supplements, in a much higher concentration than in powder [12]. Curcumin is a powerful anti-inflammatory agent that acts at the molecular level. Several studies showed that it could block the action of the NF-kB protein complex. Although it plays an important role in the immune system, the irregular action of the NF-kB molecule can lead to viral infections, septic shock, and inflammatory and autoimmune diseases [13].

Although there are studies affirming that curcumin is not a harmful substance to humans, some believe that improper administration can cause side effects such as nausea and diarrhea. On the other hand, curcumin is not recommended for pregnant or breastfeeding women or for patients suffering from kidney or gallbladder stones [14]. It has also been observed that turmeric may have anticoagulant properties, which is why its association with anticoagulant medication may increase the risk of bleeding [15]. Turmeric may also interact with nonsteroidal anti-inflammatory drugs (ibuprofen, naproxen). Besides the many advantages, curcumin has many disadvantages for medical uses, including poor water solubility, limited oral bioavailability, quick liver metabolism, and rapid systemic elimination [16]. Various curcumin encapsulations are used nowadays for effective delivery and a decrease in the needed dosage to achieve the desired effect. Different curcumin encapsulations showed more antioxidant activity than curcumin in its bulk form. Therefore, the main purposes of this research were to obtain a polymer-drug carrier that can deliver curcumin with a specific release rate and determine specific safety features for chemotherapy use in a sample of patients with oropharyngeal cancer.

2. Materials and Methods

2.1. Design and Ethical Considerations

The current study was designed as an experimental pilot study on human subjects, aiming to determine the characteristics of a drug delivery method using polyester-urethane microparticles for an encapsulated active agent of Curcumin (Cc). The in vitro and in vivo experiments were carried out at the "Victor Babes" University of Medicine and Pharmacy from Timisoara, Romania, following the entire jurisdiction and the principles of the Helsinki Declaration. The experiments were supported by the grant *5EXP/1244/30.01.2020* from "Victor Babes" University of Medicine and Pharmacy, Timisoara, Romania, and approved by the Ethics Committee of the "Victor Babes" University of Medicine and Pharmacy. Sixteen patients suffering from oropharyngeal cancer were recruited to participate in the current study, ingesting the Cc preparations and being assessed by a few non-invasive determinations of skin parameters such as trans-epidermal water loss, erythema, and the stratum corneum hydration have revealed the safe character of these samples to be used on humans.

2.2. The Reagents

The raw materials used in the synthesis of the polyester-urethane microparticles (PeU_MP) are 1,4-butanediol (BD) from Carl Roth GmbH (Karlsruhe, Germany), the solvent (acetone) from Chimopar S.A. (Bucuresti, Romania), while the others: polyethylene-glycol (PEG, M $\approx$ 200), the surfactant (Tween®20), isophorone-diisocyanate (IPDI), and hexamethylene-diisocyanate (HMDI) were obtained from Merck (Darmstadt, Germany); polycaprolactone diol (PC), average Mn ~ 530) was achieved from Sigma-Aldrich (St. Louis, MO, USA). All reagents were used as received, without any previous purification. Different salts of Na and K (Cl-, HPO_{42}-, H_2PO_4- and HCO_3-) and HCl 1N were obtained from Chimopar S.A. (Bucuresti, Romania), and they were used to obtain a simulated body fluid, which was used as a proper degradation medium for the drug delivery system and its release profile. The drug release medium used was phosphate buffer pH 7.4, 280 mosm/L, 0.13 M.

2.3. Chemical Synthesis

A procedure based on different steps was used to synthesize the drug delivery system [17–20], as follows: (1) An organic component based on a mixture of IPDI and HMDI in acetone (Table 1) was prepared under magnetic stirring (450 rpm) at 25 °C for 10 min; (2) A hydroxylic component based on a mixture of BD, PC, PEG, and Tween®20 in deionized water was homogenized (450 rpm) at 25 °C for 10 min; (3) The components were then mixed together (450 rpm) at 30 °C for 3 h due to the absence of any catalyst; (4) The obtained products were repeatedly washed and centrifuged, and they were dried as thin layers in Petri dishes at 65 °C for almost 48 h. The experiment was repeated two times to synthesize a sample with empty particles (PU) and another sample of PeU_MP with Cc (PU_Cc), with the chemical structure presented in Figure 1.

Table 1. The ratios between the raw materials.

Sample	Hydroxylic Comp. (mL/50 mL)				Curcumin (mg/50 mL)	Organic Comp. (mL/50 mL)	
	BD	PC	PEG	Tween		IPDI	HMDI
PU	0.40	0.60	2.15	1.50	0.00	2.15	2.85
PU_Cc	0.40	0.60	2.15	1.50	5.50	2.15	2.85

PU—Empty Particles; PU_Cc—Curcumin-filled Particles; BD—1,4-butanediol; PC—polycaprolactone diol; PEG—polyethylene-glycol; IPDI—isophorone-diisocyanate; HMDI—hexamethylene-diisocyanate.

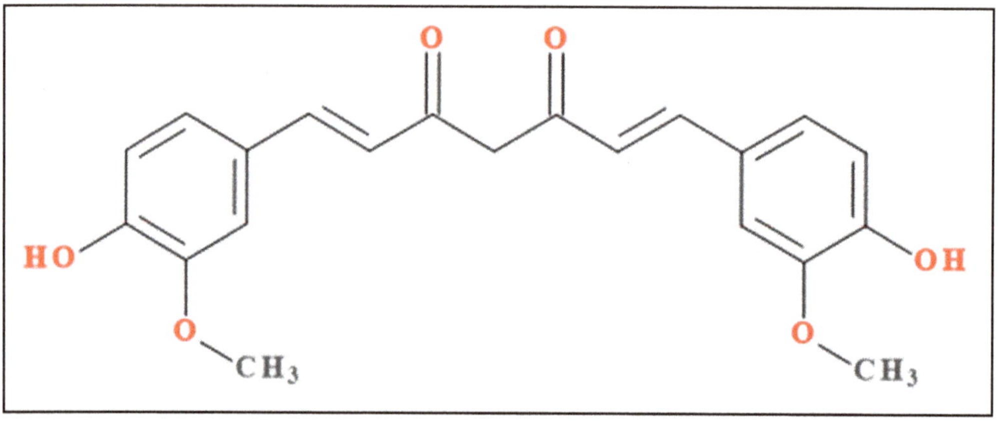

Figure 1. Chemical structure of curcumin (original picture).

2.4. Sample Calibration

The percentage of the encapsulated Cc was calculated by reporting the amount of free Cc to its total quantity that was added in that synthesis. An UVi Line 9400 Spectrophotometer (SI Analytics, Mainz, Germany), a calibration curve (Figure 2), and the Beer–Lambert law were involved in establishing the encapsulation percentage. The release rate of the synthesized carrier was evaluated by maintaining the sample PU_Cc inside a degradation medium for ten days; 1 mL medium was replaced every second day with fresh medium, and its absorption was read in triplicate; average values +/− errors were used to present the release profile, as seen in Figure 2. The size and agglomeration tendency of Peu_MP were estimated using a Vasco Particle Size Analyzer and a Wallis Zeta potential Analyzer (Cordouan Technology, Pessac, France); the following parameters were set: assessment temperature (21 ± 1 °C), the interval of time (20 ± 3 µs), number of channels (around 480), power of the laser (85–90%), acquisition mode (continuous), analysis mode (Pade–Laplace), Wallis resolution (medium), and Smoluchowski model as Henry function. The drug entrapment efficiency (DEE) was calculated with the following formula:

$$\text{DEE: (Total Drug conc.} - \text{Supernatant Drug conc.)/(Total Drug conc.)} \times 100\% \quad (1)$$

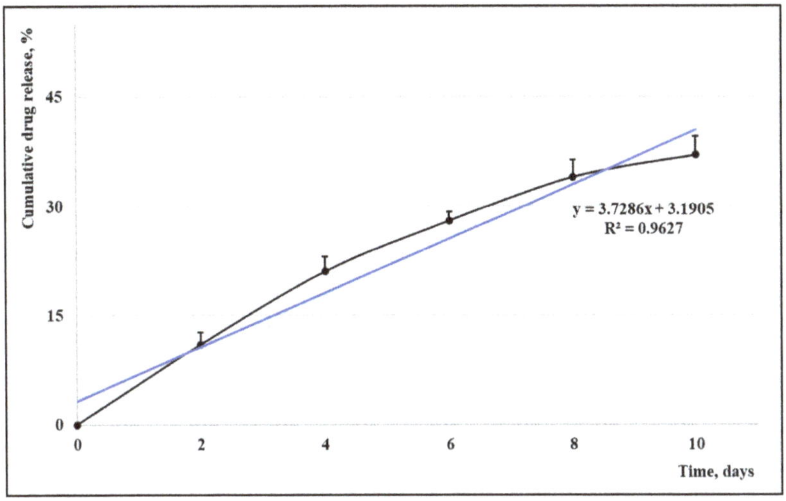

Figure 2. The release profile of Cc from sample PU_Cc.

2.5. Sample Analysis and Statistical Analysis

A fast investigation (6 measurements/48 h) were conducted with an MPA System (Courage and Khazaka, Köln, Germany) equipped with a Tewameter®TM300 probe for the evaluation of the trans-epidermal water loss, a Mexameter®MX18 probe to assess the erythema level, and a Corneometer®CM825 probe for the hydration of stratum corneum. All measurements were performed on the following samples: pure Cc, empty PeU_MP (PU), PU_Cc, and sodium lauryl sulfate (SLS) used as reference. Simple ANOVA was involved in determining whether there are any statistically significant differences between different values of more than two groups vs. SLS; * for $p \leq 0.05$, ** for $p \leq 0.01$, and *** for $p \leq 0.001$). Proportional data calculated using Fisher's exact test; Data presented using mean and standard error of the mean were compared with the Student's *t*-test.

3. Results

3.1. Characterization of the Encapsulated Active Agent

A calibration curve between 0 and 3.5 µg/mL Cc (y = 0.1506x + 0.0058; R^2 = 0.9864) was used to calculate the encapsulation efficiency and the Cc release, as described in Figure 3. A good capacity to encapsulate Cc (61.24%) was found in the case of sample PU_Cc. The drug release trend, expressed as a function of the cumulative drug concentration vs. time, is probably the main parameter that describes a drug delivery system. The drug carriers can be classified into different groups based on their release rate: fast release (a few hours/maximum one day), medium release rate (a few days/maximum one week), and respectively late release (more than ten days). The obtained PeU_MP has a medium release rate.

3.2. Characterization of the Zetasizer

Table 2 presents the particle characterization, which was obtained during the Zetasizer measurements. There were obtained microparticles with a medium homogeneity (the polydispersity index, PDI is around 0.5) and a medium tendency to form agglomerations, according to Gallardo et al. [21]. The experimental values of sample characteristics using the Zetasizer identified a mean structural size of 215 nm in the polyester-urethane preparate (PU), compared to 271 nm in the curcumin-based PU (*p*-value < 0.001). Although the size was statistically significantly different, the IPDI and Zeta potential did not differ significantly (22.91 mV vs. 23.74 mV).

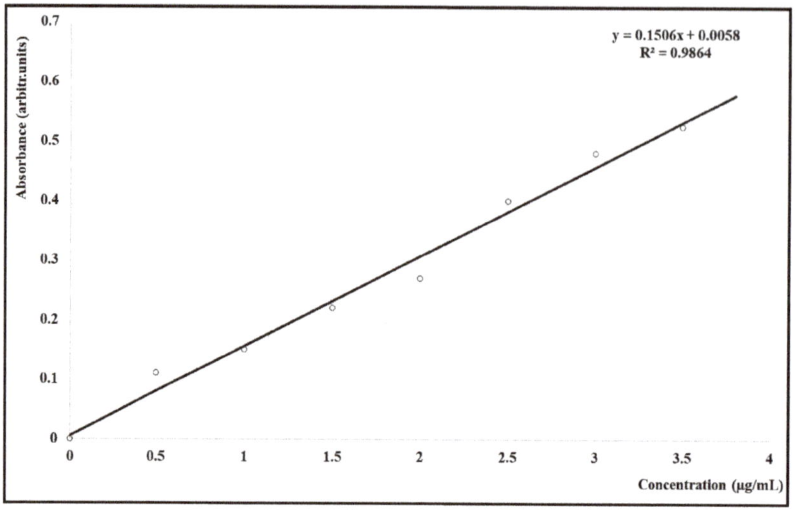

Figure 3. The calibration curve for curcumin (Cc).

Table 2. Experimental values of the Zetasizer characterization.

Sample	Size of Structures (nm)		Zeta Potential (mV)
	Mean ± SD	IPDI	
PU	215 ± 11	0.4	+22.91
PU_Cc	271 ± 19	0.6	+23.74

PU—Empty Particles; PU_Cc—Curcumin-filled Particles; SD—Standard Deviation; IPDI—isophorone-diisocyanate.

3.3. *In Vivo Characterization of Samples*

The irritation assessment is often found in many studies developed on new pharmaceutical formulations. These assays are proper for cosmetics, but many other products can be verified in the same way. Among the study participants (n = 16), there group differences between genders did not differ significantly (five men in the PU group vs. four men in the PU_Cc group, p-value = 0.614). The average age during the study period was 57.6 years for patients in the PU group, respectively 55.1 years in those who received the curcumin preparations. The majority of oropharyngeal cancers were of HPV-related etiology (75.0% in the PU group vs. 87.5% in the target group, p-value = 0.521). Lastly, half of the patients had stage II oropharyngeal cancer on the TNM staging, while the other half were stage III. There were no significant side effects, with 75.0% of patients in the PU group reporting no side effects, compared to 87.5% in the Cc group (p-value = 0.583), as described in Table 3.

Figure 4 and Table 4 present the evolutions of the three skin parameters that present important changes in irritations. As described in Figure 4a, the TEWL measurement at the beginning of the experiment indicated a non-significant difference in the average trans-epidermal water loss (1.9 g/h/m^2 in the PU group vs. 2.1 g/h/m^2 in the curcumin group, p-value = 0.386). However, the 48 h measurement at the end of the experiment found a statistically significant difference between the PU and the Cc group (2.2 g/h/m^2 vs. 2.6 g/h/m^2, p-value = 0.013). The overall difference compared to the reference group with sodium lauryl sulfate (SLS) was statistically significant at a 99% significance level.

Table 3. Background characteristics and medication dynamics among study participants.

Variables	PU (n = 8)	PU_Cc (n = 8)	p-Value
Gender			0.614
Men	5 (62.5%)	4 (50.0%)	
Women	3 (37.5%)	4 (50.0%)	
Age (mean ± SD)	57.6 ± 12.2	55.1 ± 10.9	0.672
Etiology			0.521
HPV-related	6 (75.0%)	7 (87.5%)	
Non-HPV	2 (25.0%)	1 (12.5%)	
TNM Staging			1
Stage II	4 (50.0%)	4 (50.0%)	
Stage III	4 (50.0%)	4 (50.0%)	
Self-reported side effects			0.583
0	6 (75.0%)	7 (87.5%)	
1	1 (12.5%)	1 (12.5%)	
≥2	1 (12.5%)	0 (0.0%)	

Proportional data calculated using Fisher's exact test; Data presented using mean and standard deviation were compared with the Student's t-test; SD—Standard deviation; HPV—Human Papilloma Virus; TNM—Tumor Nodes Metastasis cancer staging system.

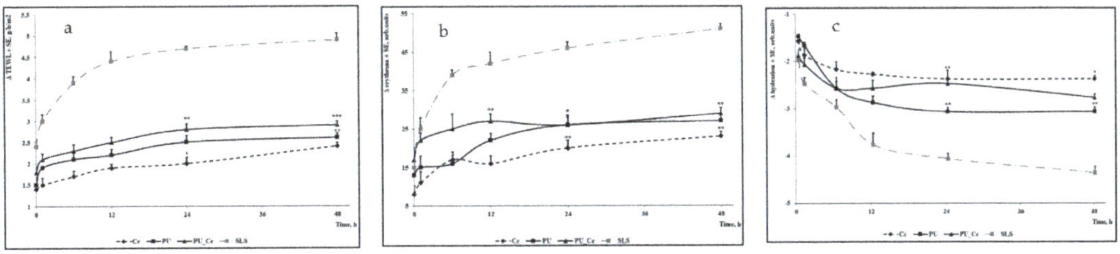

Figure 4. The evolution of the main skin parameters: (**a**) TEWL, (**b**) erythema, (**c**) skin hydration.

Table 4. Medication dynamics measured among study groups.

Variables (Mean ± SE)	PU (n = 8)	PU_Cc (n = 8)	p-Value
TEWL			
Time 0	1.9 ± 0.1	2.1 ± 0.2	0.386
Time 12	2.0 ± 0.2	2.4 ± 0.2	0.179
Time 24	2.2 ± 0.1	2.5 ± 0.2	0.201
Time 48	2.2 ± 0.1	2.6 ± 0.1	0.013
Erythema			
Time 0	15.1 ± 3.2	20.2 ± 2.4	0.223
Time 12	21.0 ± 1.9	24.3 ± 3.1	0.379
Time 24	20.3 ± 2.7	20.5 ± 2.9	0.960
Time 48	23.4 ± 2.0	25.1 ± 2.2	0.576
Skin hydration			
Time 0	−1.9 ± 0.5	−2.2 ± 0.7	0.732
Time 12	−3.0 ± 0.3	−2.6 ± 0.5	0.503
Time 24	−3.1 ± 0.2	−2.2 ± 0.4	0.063
Time 48	−3.0 ± 0.1	−2.8 ± 0.1	0.179

Data presented using mean and standard error of the mean were compared with the Student's t-test; SE—Standard error; PU—Polyurethane; PU_Cc-Polyurethane-Curcumin; TEWL—Trans-epidermal Water Loss.

The erythema evaluation presented in Figure 4b indicated a starting measurement point for both the study groups with a 5.1-unit difference, although without statistical significance (PU = 15.1 units vs. PU_Cc = 20.2 units, p-value = 0.223). After 48 h, the difference was reduced to only 1.7 units (PU = 23.4 units vs. PU_Cc = 25.1 units, p-value = 0.576). Similar to the TEWL measurement, the comparison with the reference variable (SLS) was statistically significant at the 95% significance level. Lastly, the skin hydration test showed no statistically significant differences between the polyurethane and polyurethane–curcumin group (Figure 4c). The T0 measurement had a starting point at −1.9 units in the PU group and −2.2 in the PU_Cc group (p-value = 0.732). The T48 measurement was also non-significant (PU = −3.0 vs. PU_Cc = −2.8, p-value = 0.179).

4. Discussion

4.1. Supporting Literature

The current study describes the main parameters trends of the skin being very similar for the tested samples (pure curcumin, empty polyester-urethane microparticles, PeU_MP with curcumin and sodium lauryl sulfate) but with different amplitudes that favor the curcumin-based drug delivery method. Other researchers have already used sodium lauryl sulfate (SLS) as a reference for measurements of TEWL, erythema, and skin hydration in a 25 h experiment on the skin of healthy human volunteers [22]. It is important to highlight the differences between the values recorded in the case of SLS and the values of samples with Cc for every parameter; the non-irritative potential of these samples is proven by the values of the three parameters, which are only half of the level of the same parameters in the case of SLS.

One of the reasons for developing polyester-urethane with curcumin microparticle drug delivery systems is that other particles, such as Bisphenol-A (BPA), are known to exhibit long-term harmful effects on human health [23]. Multiple investigations have shown that BPA's hypertensive impact is due to its detrimental influence on endothelial function. In our investigation, we discovered that nanomicelle curcumin produces opposing effects on blood pressure. However, we did not assess its processes. Modulating AT1 receptor (AT1R) expression in arteries and AT1R-mediated vasoconstriction likely reduces the development of hypertension. AT-1 receptor stimulates ROS generation and atherosclerosis development. After activation of angiotensin II (Ang II) by BPA, CaMKIIa was activated, which led to oxidative stress, impaired vascular relaxation, and elevated blood pressure through uncoupling of eNOS [24]. Additionally, BPA was able to activate CaMKIIa through Ang II. Consequently, phospholipase A2 is activated, and arachidonic acid levels rise, causing vasoconstriction and hypertension. Free radicals may promote nitration and inactivation of important enzymes, such as cGMP-dependent protein kinase, the enzyme responsible for NO-mediated vasorelaxation. Endothelial cell activation of endothelin-1 (ET-1) by Ang II, a vasoconstrictor peptide, is another route for BPA-induced hypertension [25]. ET-1 causes vasoconstriction by increasing intracellular calcium61 and decreasing endothelial vasodilation through eNOS deactivation. Therefore, nanomicelle curcumin may have lowered blood pressure through the aforementioned mechanisms.

In one investigation, BPA (50 mg/kg) was shown to increase the phosphorylation of p38 and JNK while lowering the phosphorylation of AKT and ERK1/2. However, nanomicelle curcumin (50 mg/kg) in cardiac tissue enhanced the protein levels of p-AKT and p-ERK while decreasing the protein levels of p-P38 and p-JNK [26]. In contrast to these findings, research indicated that curcumin induces apoptosis in H9c2 cardiac myoblast cells via boosting ROS production and JNK activation [27]. In addition, other research shows that BPA promoted the growth of MCF-7 human breast cancer cells through estrogen receptors and dysregulated p-AKT expression. Curcumin, in contrast, reduced the proliferation-promoting effects of BPA on MCF-7 cells [28]. In other research, 100 nM BPA treatment dramatically increased oxidative stress and activation of JNK, p38, and ERK in HepG2 cells. While curcumin reduced the effects of BPA, it had a greater impact on ERK [29].

In other experimental studies, Cc was used with polycarbonate products, where it was demonstrated how the distinctive attributes of Cc would have a favorable impact on the dynamics of the polycarbonate delivery products [30]. Previous research has shown, without a doubt, that curcumin's antioxidant and biocidal qualities may be put to use in the area of polymers in order to improve the features of such materials. It has been demonstrated that curcumin is a more effective melt and thermo-oxidative stabilizer of polyethylene than the synthetic phenolic antioxidant and that curcumin can be advantageously used in the synthesis of rigid polyurethane foams with enhanced mechanical, antibacterial, and anti-aging properties [31]. In particular, it has been shown that curcumin is a more efficient melt and thermo-oxidative stabilizer of polyethylene than the synthetic phenolic antioxidant, while it is anticipated that the incorporation of Cc into the production of polycarbonate may bring about benefits, particularly in terms of the antioxidant characteristics that they possess, even if more steps and more in-depth characterizations are certainly required.

Studies showed that it is possible to use Cc's or the derivatives as compounds ready to use in drug delivery systems, increasing the potential of replacing molecules of fossil origin that are concerning for their toxic and eco-toxic effects exhibited. Curcumin products had been successfully implemented recently in various treatment methods for cancer, proving their feasibility for mass production [32]. Similar to our study, it was previously attempted to observe the potential adjuvant effect of curcumin nanoparticles in oral cancer treatment [33]. The researchers added 200 nm nanocurcumin particles in addition to cetuximab for malignant cells expressing KB 3-1 and compared the compound with a reference treatment of cetuximab only. The results were promising, since relative to the cetuximab-alone treatment group, the combined therapy group saw a considerable increase in cell death, while nanocurcumin sensitization significantly increased cell death compared to the control group. Even though the industrial scalability of the proposed synthesis is complicated at the present time by the higher cost of Cc, which is currently being marketed as a food-grade or research-grade material, we are convinced that under the pressure of international legislation and the increased environmental sensitivity of consumers, natural alternatives such as Cc may become accessible also to the polymer industry, which would make a positive development for the environment. Additionally, the EU has imposed severe restrictions on the use of molecules that have been shown to be carcinogenic and endocrine disruptors, such as bisphenol-A, while simultaneously encouraging the search for safer and more environmentally friendly alternatives [34].

4.2. Study Limitations and Future Perspectives

The current study provides important data about the kinetics, stability, and safety of curcumin-based polyurethane products for chemotherapy use. However, the feasibility of the large-scale use of these products in pharmacology and oncology is yet to be determined. Similarly, the human safety record is promising, but as a pilot study, the number of subjects involved was only sixteen participants; therefore, further trials should be implemented to draw clear conclusions. From a future perspective, greater production and use of non-edible turmeric products might, in the not-too-distant future, result in a dramatic reduction in the cost of industrially used curcumin. In addition, meeting the ecological requirements may be accomplished by combining the production of bio-based polymers with the recycling of commonly generated wastes made of plastic.

5. Conclusions

In conclusion, a different preparation was developed in order to mitigate the drawbacks of polyurethane-based medication delivery methods in oncology while simultaneously increasing their therapeutic potential against HCC while maintaining a good safety profile. This paper describes the obtaining and the preliminary characterization of a synthetic drug delivery system used as a carrier for curcumin. Polyester-urethane microparticles with sizes between 215 and 271 nm have been developed, and their release rate was investigated in a simulated body fluid used as a degradation media. Additionally, the safety concerns were tested on patients suffering from oropharyngeal cancer, with very promising results, as the non-invasive determinations of skin parameters such as trans-epidermal water loss, erythema, and stratum corneum hydration have revealed the safe character of these samples to be used on humans. Despite the fact that further study is undoubtedly required, it is possible that this research will prove to be of significant relevance to the field of contemporary polyurethane-based methods for drug delivery.

Author Contributions: A.C. and E.M.D. contributed equally to this study. Conceptualization: A.C. and E.M.D.; methodology: A.C. and E.M.D. software: I.C.M.; validation: I.C.M.; formal analysis: D.I.H. and N.C.B.; investigation: D.I.H. and N.C.B.; resources: C.A.S.; data curation: C.A.S.; writing—original draft preparation: A.C. and E.M.D.; writing—review and editing: B.A.B., S.A., R.A. and M.R.; visualization: C.D. and R.M.; supervision: M.P. and R.A.N.; project administration: M.P. All authors have read and agreed to the published version of the manuscript.

Funding: This study was supported by the grant *5EXP/1244/30.01.2020* from the "Victor Babes" University of Medicine and Pharmacy, Timisoara, Romania.

Institutional Review Board Statement: The study was conducted according to the guidelines of the Declaration of Helsinki and approved by the Ethics Committee of the "Victor Babes" University of Medicine and Pharmacy in Timisoara, approved on 30 January 2020, with approval number 5EXP/1244.

Informed Consent Statement: Written informed consent has been obtained from the patients to publish this paper.

Data Availability Statement: Data available on request.

Conflicts of Interest: The authors declare no conflict of interest.

References

1. Borcan, F.; Mioc, M.; Ghiulai, R.; Pinzaru, I.; Dehelean, C.A.; Soica, C.M. A Study on the Behavior of a Polyurethane Drug Carrier in Various pH Media. *J. Med. Life* **2020**, *13*, 195–199. [CrossRef] [PubMed]
2. Yao, Y.; Xu, D.; Liu, C.; Guan, Y.; Zhang, J.; Su, Y.; Zhao, L.; Meng, F.; Luo, J. Biodegradable pH-sensitive polyurethane micelles with different polyethylene glycol (PEG) locations for anti-cancer drug carrier applications. *RSC Adv.* **2016**, *6*, 97684–97693. [CrossRef]
3. Warnakulasuriya, S. Global epidemiology of oral and oropharyngeal cancer. *Oral Oncol.* **2009**, *45*, 309–316. [CrossRef] [PubMed]
4. Alqahtani, W.S.; Almufareh, N.A.; Al-Johani, H.A.; Alotaibi, R.K.; Juliana, C.I.; Aljarba, N.H.; Alqahtani, A.S.; Almarshedy, B.; Elasbali, A.M.; Ahmed, H.G.; et al. Oral and Oropharyngeal Cancers and Possible Risk Factors Across Gulf Cooperation Council Countries: A Systematic Review. *World J. Oncol.* **2020**, *11*, 173–181. [CrossRef] [PubMed]
5. Cohan, D.M.; Popat, S.; Kaplan, S.E.; Rigual, N.; Loree, T.; Hicks, W.L. Oropharyngeal cancer: Current understanding and management. *Curr. Opin. Otolaryngol. Head Neck Surg.* **2009**, *17*, 88–94. [CrossRef]
6. Lynch, K.L.; Twesten, J.E.; Stern, A.L.; Augustson, E.M. Level of Alcohol Consumption and Successful Smoking Cessation. *Nicotine Tob. Res.* **2018**, *21*, 1058–1064. [CrossRef]
7. Huang, C.-C.; Hsiao, J.-R.; Lee, W.-T.; Lee, Y.-C.; Ou, C.-Y.; Chang, C.-C.; Lu, Y.-C.; Huang, J.-S.; Wong, T.-Y.; Chen, K.-C.; et al. Investigating the Association between Alcohol and Risk of Head and Neck Cancer in Taiwan. *Sci. Rep.* **2017**, *7*, 9701. [CrossRef]
8. Elrefaey, S.; Massaro, M.A.; Chiocca, S.; Chiesa, F.; Ansarin, M. HPV in oropharyngeal cancer: The basics to know in clinical prac-tice. *Acta Otorhinolaryngol. Ital.* **2014**, *34*, 299–309.
9. Lechner, M.; Liu, J.; Masterson, L.; Fenton, T.R. HPV-associated oropharyngeal cancer: Epidemiology, molecular biology and clinical management. *Nat. Rev. Clin. Oncol.* **2022**, *19*, 306–327. [CrossRef]
10. Zoi, V.; Galani, V.; Lianos, G.D.; Voulgaris, S.; Kyritsis, A.P.; Alexiou, G.A. The Role of Curcumin in Cancer Treatment. *Biomedicines* **2021**, *9*, 1086. [CrossRef]
11. Rahaman, M.; Rakib, A.; Mitra, S.; Tareq, A.; Emran, T.; Shahid-Ud-Daula, A.; Amin, M.; Simal-Gandara, J. The Genus *Curcuma* and Inflammation: Overview of the Pharmacological Perspectives. *Plants* **2020**, *10*, 63. [CrossRef] [PubMed]
12. Sharifi-Rad, J.; El Rayess, Y.; Rizk, A.A.; Sadaka, C.; Zgheib, R.; Zam, W.; Sestito, S.; Rapposelli, S.; Neffe-Skocińska, K.; Zielińska, D.; et al. Turmeric and Its Major Compound Curcumin on Health: Bioactive Effects and Safety Profiles for Food, Pharmaceutical, Biotechnological and Medicinal Applications. *Front. Pharmacol.* **2020**, *11*, 01021. [CrossRef] [PubMed]
13. Maiti, P.; Dunbar, G.L. Use of Curcumin, a Natural Polyphenol for Targeting Molecular Pathways in Treating Age-Related Neurodegenerative Diseases. *Int. J. Mol. Sci.* **2018**, *19*, 1637. [CrossRef]
14. Hewlings, S.J.; Kalman, D.S. Curcumin: A Review of Its Effects on Human Health. *Foods* **2017**, *6*, 92. [CrossRef] [PubMed]
15. Keihanian, F.; Saeidinia, A.; Bagheri, R.K.; Johnston, T.P.; Sahebkar, A. Curcumin, hemostasis, thrombosis, and coagulation. *J. Cell Physiol.* **2017**, *233*, 4497–4511. [CrossRef]
16. Valokola, M.G.; Karimi, G.; Razavi, B.M.; Kianfar, M.; Jafarian, A.H.; Jaafari, M.R.; Imenshahidi, M. The protective activity of nanomicelle curcumin in bisphenol A-induced cardiotoxicity following subacute exposure in rats. *Environ. Toxicol.* **2018**, *34*, 319–329. [CrossRef]
17. Heghes, A.; Soica, C.M.; Ardelean, S.; Ambrus, R.; Muntean, D.; Galuscan, A.; Dragos, D.; Ionescu, D.; Borcan, F. Influence of emulsifi-ers on the characteristics of polyurethane structures used as drug carrier. *Chem. Cent. J.* **2013**, *7*, 66. [CrossRef]
18. Danciu, C.; Borcan, F.; Șoica, C.; Zupko, I.; Csanyi, E.; Ambrus, R.; Muntean, D.; Sass, C.; Antal, D.; Toma, C.; et al. Polyurethane Microstructures-a Good or Bad in vitro Partner for the Isoflavone Genistein? *Nat. Prod. Commun.* **2015**, *10*, 951–954. [CrossRef]

19. Borcan, L.-C.; Dudas, Z.; Len, A.; Fuzi, J.; Borcan, F.; Tomescu, M.C. Synthesis and characterization of a polyurethane carrier used for a prolonged transmembrane transfer of a chili pepper extract. *Int. J. Nanomed.* **2018**, *13*, 7155–7166. [CrossRef]
20. Borcan, F.; Chirita-Emandi, A.; Andreescu, N.I.; Borcan, L.-C.; Albulescu, R.C.; Puiu, M.; Tomescu, M.C. Synthesis and preliminary characterization of polyurethane nanoparticles with ginger extract as a possible cardiovascular protector. *Int. J. Nanomed.* **2019**, *14*, 3691–3703. [CrossRef]
21. Gallardo, V.; Morales, M.E.; Ruiz, M.A.; Delgado, A.V. An experimental investigation of the stability of ethylcellulose latex: Cor-relation between zeta potential and sedimentation. *Eur. J. Pharm. Sci.* **2005**, *26*, 170–175. [CrossRef] [PubMed]
22. Gurita, V.G.; Pavel, I.Z.; Borcan, F.; Moaca, A.; Danciu, C.; Diaconeasa, Z. Toxicological Evaluation of Some Essential Oils Obtained from Se-lected Romania Lamiaceae Species in Complex with Hydroxypropyl—Gamma-cyclodextrin. *Rev. Chim. Buchar.* **2019**, *70*, 3703–3707. [CrossRef]
23. Cimmino, I.; Fiory, F.; Perruolo, G.; Miele, C.; Beguinot, F.; Formisano, P.; Oriente, F. Potential Mechanisms of Bisphenol A (BPA) Contributing to Human Disease. *Int. J. Mol. Sci.* **2020**, *21*, 5761. [CrossRef] [PubMed]
24. Saura, M.; Marquez, S.; Reventun, P.; Olea-Herrero, N.; Arenas, M.I.; Moreno-Gómez-Toledano, R.; Gómez-Parrizas, M.; Muñóz-Moreno, C.; González-Santander, M.; Zaragoza, C.; et al. Oral administration of bisphenol A induces high blood pressure through angiotensin II/CaMKII-dependent uncoupling of eNOS. *FASEB J.* **2014**, *28*, 4719–4728. [CrossRef] [PubMed]
25. Yao, Y.; Wang, W.E.; Li, M.; Ren, H.; Chen, C.; Wang, J.; Yang, J.; Zeng, C. Curcumin Exerts its Anti-hypertensive Effect by Down-regulating the AT1 Receptor in Vascular Smooth Muscle Cells. *Sci. Rep.* **2016**, *6*, 25579. [CrossRef]
26. Kia, S.J.; Basirat, M.; Saedi, H.S.; Arab, S.A. Effects of nanomicelle curcumin capsules on prevention and treatment of oral mucosits in patients under chemotherapy with or without head and neck radiotherapy: A randomized clinical trial. *BMC Complement. Med. Ther.* **2021**, *21*, 232. [CrossRef]
27. Zikaki, K.; Aggeli, I.-K.; Gaitanaki, C.; Beis, I. Curcumin induces the apoptotic intrinsic pathway via upregulation of reactive oxygen species and JNKs in H9c2 cardiac myoblasts. *Apoptosis* **2014**, *19*, 958–974. [CrossRef]
28. Li, X.; Xie, W.; Xie, C.; Huang, C.; Zhu, J.; Liang, Z.; Deng, F.; Zhu, M.; Zhu, W.; Wu, R.; et al. Curcumin modulates miR-19/PTEN/AKT/p53 axis to suppress bisphenol A-induced MCF-7 breast cancer cell proliferation. *Phytother. Res.* **2014**, *28*, 1553–1560. [CrossRef]
29. Geng, S.; Wang, S.; Zhu, W.; Xie, C.; Li, X.; Wu, J.; Zhu, J.; Jiang, Y.; Yang, X.; Li, Y.; et al. Curcumin attenuates BPA-induced insulin resistance in HepG2 cells through suppression of JNK/p38 pathways. *Toxicol. Lett.* **2017**, *272*, 75–83. [CrossRef]
30. De Leo, V.; Casiello, M.; Deluca, G.; Cotugno, P.; Catucci, L.; Nacci, A.; Fusco, C.; D'Accolti, L. Concerning Synthesis of New Biobased Polycarbonates with Curcumin in Replacement of Bisphenol A and Recycled Diphenyl Carbonate as Example of Circular Economy. *Polymers* **2021**, *13*, 361. [CrossRef]
31. Tátraaljai, D.; Kirschweng, B.; Kovács, J.; Földes, E.; Pukánszky, B. Processing stabilisation of PE with a natural antioxidant, curcumin. *Eur. Polym. J.* **2013**, *49*, 1196–1203. [CrossRef]
32. Zheng, Y.; Jia, R.; Li, J.; Tian, X.; Qian, Y. Curcumin- and resveratrol-co-loaded nanoparticles in synergistic treatment of hepatocellular carcinoma. *J. Nanobiotechnol.* **2022**, *20*, 339. [CrossRef] [PubMed]
33. Mukherjee, D.; Dash, P.; Ramadass, B.; Mangaraj, M. Nanocurcumin in Oral Squamous Cancer Cells and Its Efficacy as a Chemo-Adjuvant. *Cureus* **2022**, *14*, e24678. [CrossRef]
34. Vicente-Martínez, Y.; Caravaca, M.; Soto-Meca, A. Determination of Very Low Concentration of Bisphenol A in Toys and Baby Pacifiers Using Dispersive Liquid–Liquid Microextraction by In Situ Ionic Liquid Formation and High-Performance Liquid Chromatography. *Pharmaceuticals* **2020**, *13*, 301. [CrossRef] [PubMed]

medicina

MDPI

Case Report

Insights on Lipomatosis after Platinum-Based Chemotherapy Use in Pediatric Oncology: A Case Report

Estera Boeriu [1,2,†], Alexandra Georgiana Boc [1], Alexandra Borda [2], Rodica Anamaria Negrean [3], Bogdan Feciche [4,†], Amalia Iulia Boeriu [5], Florin George Horhat [6,*], Ion Cristian Mot [7], Ioana Delia Horhat [7], Madhavi Ravulapalli [8], Omar Sabuni [9], Abduljabar Adi [10], Adnan Anjary [11] and Smaranda Teodora Arghirescu [1,2]

1 Department of Pediatrics, "Victor Babes" University of Medicine and Pharmacy, Eftimie Murgu Square 2, 300041 Timisoara, Romania
2 Department of Oncology and Haematology, "Louis Turcanu" Emergency Clinical Hospital for Children, Iosif Nemoianu Street 2, 300011 Timisoara, Romania
3 Faculty of Medicine and Pharmacy, University of Oradea, 410073 Oradea, Romania
4 Department of Urology, Satu-Mare County Emergency Hospital, Strada Ravensburg 2, 440192 Satu-Mare, Romania
5 Klinikum Landshut, Teaching Hospital of the LMU Munich, Clinic for Anaesthesiology, Intensive Care Medicine and Pain Therapy, Robert-Koch-Strasse, 184034 Landshut, Germany
6 Multidisciplinary Research Center on Antimicrobial Resistance (MULTI-REZ), Microbiology Department, "Victor Babes" University of Medicine and Pharmacy, 300041 Timisoara, Romania
7 Department of Ear-Nose-Throat, "Victor Babes" University of Medicine and Pharmacy Timisoara, Eftimie Murgu Square 2, 300041 Timisoara, Romania
8 Bhaskar Medical College, Amdapur Road 156-162, Hyderabad 500075, India
9 Faculty of General Medicine, Altinbas University, Dilmenler Cd., 34217 Istanbul, Turkey
10 Faculty of General Medicine, Baskent University, Fatih Sultan, 06790 Ankara, Turkey
11 Faculty of General Medicine, Yeditepe University, Kayısdagı Cd., 34755 Istanbul, Turkey
* Correspondence: horhat.florin@umft.ro
† These authors contributed equally to this work.

Citation: Boeriu, E.; Boc, A.G.; Borda, A.; Negrean, R.A.; Feciche, B.; Boeriu, A.I.; Horhat, F.G.; Mot, I.C.; Horhat, I.D.; Ravulapalli, M.; et al. Insights on Lipomatosis after Platinum-Based Chemotherapy Use in Pediatric Oncology: A Case Report. *Medicina* **2022**, *58*, 1715. https://doi.org/10.3390/medicina58121715

Academic Editor: Konstantinos Dimas

Received: 17 October 2022
Accepted: 22 November 2022
Published: 23 November 2022

Abstract: Agents of platinum-based chemotherapy, such as cisplatin or carboplatin, are used in the treatment of a wide range of malignancies that affect children, such as brain tumors, osteosarcoma, neuroblastoma, hepatoblastoma, and germ cell tumors (GCTs). The Cyclophosphamide Equivalent Dose (CED) calculator for reproductive risk does not take platinum-based chemotherapy into account, despite the fact that it accounts for the majority of chemotherapy medications that are typically administered for pediatric GCTs. As a result, exposure to platinum-based drugs throughout infancy can have predictable long-term effects such as infertility, as well as other rare encounters such as lipoma formation and lipomatosis. Lipomas are the most prevalent benign soft tissue tumor subtype. They may be either solitary entities or engaged in multiple lipomatosis, which may have a familial origin or be an acquired disorder. Chemotherapy is a possible cause of lipomatosis. Chemotherapy based on cisplatin has been linked to a variety of long-term consequences, including kidney damage, neurotoxicity, and pulmonary toxicity, and may even create secondary cancers. However, lipoma development is known to occur in fewer than 1 in 100 individuals, and only a few examples of multiple cutaneous lipomatosis triggered by this therapy have been documented. Here we present a very rare case of lipomatosis in a pediatric patient with GCT under cisplatin therapy, which might be the third report of this kind affecting children.

Keywords: pediatric oncology; lipomatosis; cisplatin; platinum-based chemotherapy; germ cell tumors

1. Introduction

Lipomas are benign, most often encapsulated, noninvasive soft tissue tumors composed of mature adipocytes. They represent about half of all the adipocytic tumors in the pediatric population and, depending on their size and location, and they vary from being asymptomatic to causing various non-specific symptoms. Cytogenetic aberrations can often be described in these tumors, most of them affecting the chromosome 12q. The term "lipomatosis", on the other hand, describes tumors comprised of matured adipose tissue with an infiltrative growth pattern, ill-defined margins, and a tendency toward symmetry [1–3].

While the cause of these tumors remains mostly unknown, there are some etiologically and clinically described subtypes of lipomatosis, such as Madelung's disease associated with alcohol consumption and metabolic problems; lipomatosis dolorosa, potentially associated with improper lymph drainage; or Dercum's disease and Roch–Leri mesosomatous lipomatosis, which are considered very rare lipomatoses of unknown etiology [4,5]. There have been cases reported where it seems that the medical intervention, whether through the administered medication or surgery, might have either caused or exacerbated the growth of lipomatous tumors [6–8]. Among the incriminated drugs lies cisplatin, a platinum-based compound with a cytotoxic effect used in the treatment of various cancers [9–12]. Due to its side effects and drug resistance, cisplatin is currently used in combination with other antineoplastic agents. Through oxidative stress, high doses of this drug can induce nephrotoxicity and hepatotoxicity, demonstrated by elevated creatinine levels, hepatic enzyme levels, and bilirubin levels. Necrosis and inflammatory cell infiltration of the portal area are some of the histopathological changes observed in patients treated with cisplatin [13]. Other platinum-based agents have been demonstrated to induce hepatosteatosis, a condition that has been linked to the growth of lipomas [14,15].

In this case report, we present a very rare case of a 16-year-old girl diagnosed with a giant ovarian GCT of mixed histology with extension into the abdominal cavity and asymptomatic amputation of the contra-lateral ovary, who developed lipomatosis after chemotherapy with cisplatin. A thorough literature search was also performed to find similar cases, since this rare occurrence of peritoneal and liver lipomas as a possible and specific side effect of platinum-based chemotherapy is scarcely described in pediatric patients.

2. Case Presentation

A 16-year-old Caucasian girl was admitted to our hospital, presenting with abdominal pain and abdominal distension, fever, and a dry cough. Upon clinical examination, the patient was found moderately pale and febrile, with a body mass index (BMI) of 28.6 kg/m^2, being in the 95th percentile for girls her age, corresponding to class I obesity. The abdomen was distended and moderately painful upon palpation, the puddle sign was present, and percussion notes varied between tympanic and dull, raising suspicion of ascites. Lab values were significantly modified, suggesting a severe inflammatory process, as described in Table 1.

It was observed that CRP levels were 3215 mg/L, ferritin 1118 ng/mL, fibrinogen 629 mg/dL, and ESR was 92 mm/h, all of which ranged much higher than the normal threshold. Besides the severe inflammatory status of the patient, the girl was also suffering from anemia, with a red blood cell count (RBC) measured at 4.0 million, and a hemoglobin level of 10.5 g/dL. Apart from class I obesity, the adolescent patient had an elevated serum glucose (109 mg/dL), correlating with a metabolic syndrome. The serum lipid levels were also above the normal range for adolescents, although without a significant increase (total cholesterol = 183 mg/dL, triglycerides = 155 mg/dL). The patient did not have any documented endocrinological problems.

A chest X-ray found bilateral interstitial reticular coarseness and opacification of the right costophrenic angle, suggestive of pleural effusion. The patient was further investigated in our ultrasound department; the abdominal sonogram raised suspicion of an abdominal–pelvic tumor mass with mild ascites in the abdominal cavity, particularly pronounced in the perihepatic and peri-splenic spaces.

Surgery was considered and computer tomography (CT) was performed on the same day to better assess the extension of the tumor mass. The CT examination of the abdominal and pelvic cavities found a gigantic heterogeneous mass of 18/13/9 cm (Figure 1). The tumor did not invade the abdominal organs or retroperitoneal vascular structures, but manifested a compressive effect on the intestinal loops and urinary bladder. While the right ovary could not be visualized, a calcified mass of 26/14/12 mm was identified within the right anterolateral side of the recto-uterine pouch.

Table 1. Modified laboratory values significant for diagnosis.

Test	Value	Range	Significance
RBC (female)	$4.0 \times 10^9/mm^3$	$4.2–5.4 \times 10^9/mm^3$	↓
Hemoglobin	10.5 g/dL	11.8–15.7 g/dL	↓
Leukocytes	17.200/mm3	$4.800–10.800/mm^3$	↑
Neutrophils	13.810/mm3	$1870–8100/mm^3$	↑
Serum glucose	109 mg/dL	70–100 mg/dL	↑
Total cholesterol	183 mg/dL	<170 mg/dL	↑
Triglycerides	155 mg/dL	<150 mg/dL	↑
Serum albumin	4.1 mg/dL	3.4–5.4 g/dL	Within normal range
CRP	3215 mg/L	0–5 mg/L	↑
Ferritin	1118 ng/mL	15–150 ng/mL	↑
Fibrinogen	629 mg/dL	200–400 mg/dL	↑
Procalcitonin	0.244 ng/mL	0–0.5 ng/mL	Within normal range
ESR	92 mm/h	2–13 mm/h	↑
AFP	–	–	–
β-HCG	–	–	–
CA-125	–	–	–

Tumor markers were not determined prior to surgery; CRP—C-reactive protein; ESR-Erythrocyte sedimentation rate; AFP—Alpha-fetoprotein; β-HCG—beta human chorionic gonadotropin; CA-125—Cancer antigen 125. ↓: below normal range; ↑: above normal range.

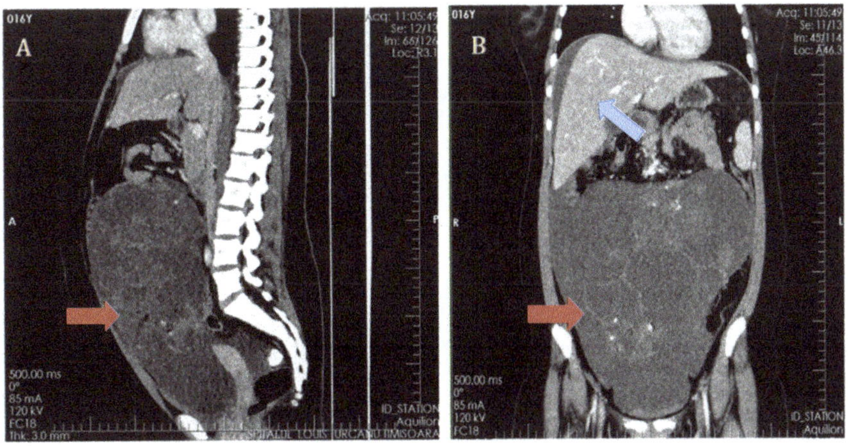

Figure 1. CT scan revealed a massive heterogeneous tumor mass of 18/13/9 (red arrow) cm with a volume of 1101 mm^3, of pelvic origin with extension in the abdominal cavity manifesting compression on the local organs. Perihepatic ascites can be easily identified (blue arrow).

The major findings could be summarized as massive mixed solid-cystic tumors occupying the abdominal and pelvic cavities of suspected ovarian origin, abdominopelvic ascites, minimal right pleural effusion, and para-aortic lymphadenopathy. The patient underwent surgery on the third day of hospitalization. A midline laparotomy found peritoneal fluid that was drained and sampled for cytology and a massive septated cystic tumor occupying the abdominal and pelvic cavity with left ovarian origin. A surgical biopsy was performed along with ablation of the ovarian tumor and epiploectomy. Upon further inspection of the pelvic cavity, a calcified amputated right ovary of approximately 3cm in diameter was found, most likely a sequela of compression due to the massive volume of 1101.43mm^3 from the tumor originating in the contralateral ovary. The patient was transferred to our intensive care unit for post-surgery monitoring awaiting the pathology and cytology results.

The cytologic examination of the peritoneal fluid identified hemorrhagic fluid with high heterogeneous cellularity of 75% neutrophils, 15% monocytes, 10% lymphocytes, reactive mesothelial cells, and several old mesothelial cells with "signet ring cell" phenotype, raising suspicion of peritoneal dissemination. Pathology findings revealed on macroscopic examination the tumor appeared as a round, encapsulated multi-cystic mass of 18.5/13.8/9.5 cm, with grey to light pink solid/gelatinous components. Immunohistochemistry (IHC) analysis found a glial fibrillary acidic protein (GFAP), alpha-fetoprotein, CD10, macro creatine kinase (MCK), and vimentin, all intensely positive, beta-HCG-negative, and KI-67-positive in 10% of solid tumor areas [16–18].

The pathology findings identified the tumor as a mixed germ cell tumor with three components, as observed in Figure 2: mature teratoma (25%), immature teratoma, high-grade G3 (60%), and yolk sack tumor (15%). The tumor was graded at stage IV (pT1cNxM1) due to evidence of metastasis, according to the TNM staging system [19], where "p" and "c" stand for staging based on the histopathological assessment of a surgical specimen and evidence acquired by clinical and paraclinical procedures, respectively.

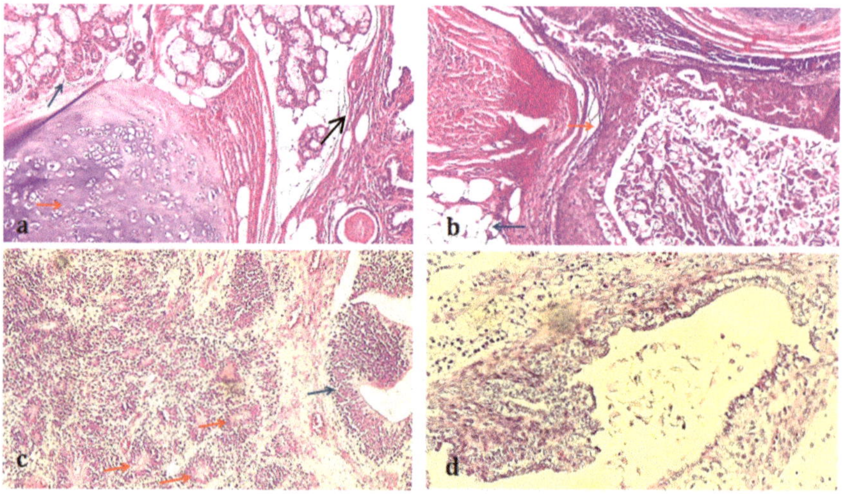

Figure 2. Histological examination revealed a tumor of germ cell origin composed of areas of mature teratoma, immature teratoma, and small areas of yolk sac tumor. (**a**). Teratoma with non-specific glandular tissue (blue arrow) and cartilaginous tissue (red arrow). (**b**). Teratoma shows areas of the dispersed squamous-stratified epithelium (red arrow), adipocytes (blue arrow), and immature smooth muscle cells (black arrow). (**c**). The major component of the tumor consisted of immature teratoma histology, with immature neuroectodermal tissue forming rosettes (red arrows) and tubules (blue arrows). (**d**). Detail of yolk sac component with glandular growth pattern.

A second pathology examination was requested, and the result indicated a high-grade immature teratoma. The final diagnosis was set after careful consideration of clinical, paraclinical, histopathological, and IHC data. The tumor was labeled stage IIIC according to the FIGO staging system that identified peritoneal metastasis [20], and a high-grade/G3 immature teratoma of the ovary [1].

Therapy and follow-up: the patient underwent chemotherapy treatment (MAKEI 2005 protocol) with four PEI cycles (cisplatin, etoposide, and ifosfamide). The chemotherapy was well tolerated with no significant events, and complete tumor remission was accomplished after the completion of the four PEI cycles. Shortly after completing the last cycle of chemotherapy, a routine MRI showed elements previously absent four months prior: a small oval peripheral intrahepatic mass of ~0.8 cm under the Glisson capsule and several smaller cysts in the perihepatic space. A PET scan was further performed and found no signs of malignancy but identified the liver mass as a possible lipoma. The tumor markers AFP and beta-HCG were within normal values; thus, a "watch and wait" approach was decided upon.

The routine check-ups involving physical examination, lab values, tumor markers, and imaging studies showed no signs of relapse or significant changes in the patient's health, apart from mild hepatic steatosis and significant weight gain (IMC = 35.4 kg/m^2), advancing the previously diagnosed obesity to class II. No endocrine causes for the patient's weight gain were found. The iatrogenic menopause following bilateral adnexectomy was well controlled by hormone substitution therapy, therefore the weight gain was attributed to genetics, an admittedly unbalanced diet, and a sedentary lifestyle.

In the 18th month since the end of chemotherapy, a routine MRI found that the perihepatic masses had increased in number and dimension. The investigation found intrahepatic, perihepatic, and subdiaphragmatic fat-containing lesions of a maximum of 2/1.5 cm (Figure 3C,D), a smaller mass of approximately 1.1/0.6 cm in the hepatorenal recess, and another of 0.4 cm in the gastrosplenic recess. Tumor markers AFP, b-HCG, and CA-125 were normal; therefore, the rare growing teratoma syndrome (GTS) was feared as a severe outcome [21,22]. Five tumor masses were removed by exploratory laparotomy two weeks later, and the pathology exam found all the lesions to be lipomas.

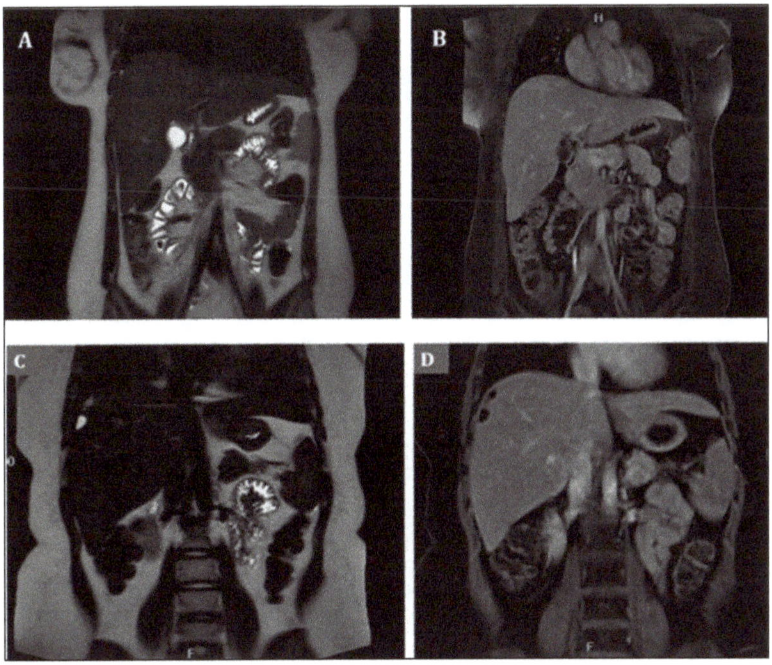

Figure 3. Coronal MRI of the abdominal cavities. (**A**). MRI T2-weighted performed after the surgical procedure, immediately before the start of chemotherapy. (**B**). Contrast-enhanced T1-weighted sequence in the same examination. (**C**). MRI T2-weighted of the abdominal cavity 18 months after completion of chemotherapy shows hyperintense perihepatic and subdiaphragmatic masses. (**D**). T1-weighted fat-suppressed (Dixon) sequence revealed the masses to display low intensity, identifying the small tumors as fat structures rather than liquid containing possible lipomas, an assumption later confirmed after surgical excision.

3. Discussion

3.1. Literature Search for Similar Findings

We report a case of a female pediatric patient that presented with systemic lipomatosis after cisplatin therapy for GCT and no familial history of lipomatosis. Lipomatosis is not an uncommon side effect of chemotherapy, and it is generally linked to the use of corticosteroids [23]. Since our patient did not undergo steroid therapy and considering the alarmingly rapid growth of the number of lipomas, we investigated the possibility that her chemotherapy regimen might be the culprit. A quick scan of the literature revealed no reports of lipomatosis involving exposure to etoposide or ifosfamide, but a likely link to cisplatin, and a reported side effect of lipoma formation is fewer than 1 in 100 cases.

An extensive literature search using the terms "lipomatosis", "lipoma", "chemotherapy", "cisplatin" or "carboplatin", "platinum-based chemotherapy", and "germ cell tumor" was conducted using the "PubMed", "Web of Science", and "Scopus" databases in an attempt to answer the following question: how often are singular or multiple lipomas associated with the use of cisplatin or other platinum agents? Articles in the English literature published between 2000 to 2021 were included in our final search. The main steps of our literature scan are presented in Figure 4. We excluded articles irrelevant to our query and reports of cases where underlying genetic or metabolic conditions associated with a high incidence of lipomatous tumors were present (e.g., Cowden syndrome, Cushing's syndrome, Rb1 mutation, familial multiple lipomatosis, hamartoma tumor syndrome). A history of corticosteroid therapy was also an exclusion criterion due to its well-known association with the development of lipomatosis [23].

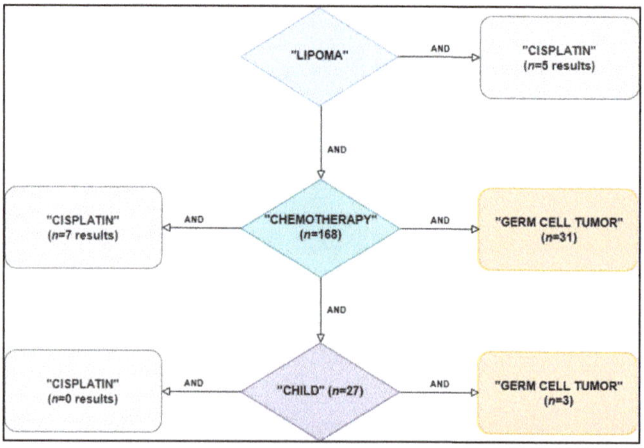

Figure 4. Flowchart of steps and keywords used for database search.

The only two cases found relevant to our search described intestinal and cutaneous lipomatosis occurring in two adult cancer survivors who underwent cisplatin therapy [24,25]. Cutaneous lipomatosis was described as the most prevalent benign mesenchymal tumor, without a significant comorbid problem, that forms in the subcutaneous tissues of the trunk or proximal extremities and manifests as distinct rubbery lumps. Although in one report the lipomas were discovered 12 years after cisplatin therapy, it is a known fact that circulating platinum levels remain high decades after the end of the regimen [26]. The first reported patient was diagnosed with testicular non-seminoma GCT treated with a right orchiectomy and a treatment scheme using bleomycin, etoposide, and cisplatin, followed by bilateral aortoiliac embolectomies during BEP treatment. The role of chemotherapy as a cause or aggravator of intestinal lipomatosis has not been conclusively proven [27,28]. Notably, the development and expansion of the lipomas seen in this patient's follow-up studies after chemotherapy treatment imply a causal involvement of the chemotherapy, but the potential of a congenital illness with a late presenting phenotype and limited penetrance persists. Multiple case reports demonstrate chemotherapy-induced cutaneous lipomatosis, although many of these patients had previously received steroids, which have a well-established relationship with acquired lipomatosis.

Significantly, the second case report described a patient with testicular cancer who was also treated with BEP chemotherapy and acquired cutaneous lipomatosis; it was hypothesized that chemotherapy might cause adipocyte proliferation via a mechanism that has yet to be determined [25]. Nonetheless, the patient had received a peripheral blood stem cell transplant, but there are no previous reports of multiple lipomatosis developing after such procedures. Regarding stem cell mobilization and transplant, there are no published reports of adipose cell proliferation as a well-known adverse effect. However, it would be beneficial to be able to show that chemo-mobilization agents may also drive adipocyte proliferation in addition to blood cells [29]. In fact, the stimulating effect of the Macrophage Colony-Stimulating Factor (MCSF) on adipocytes is well-established, although little is known about GCSF. In spite of this, recent research indicates a link between the increased release of GCSF and the expansion of cultured adipocyte sizes [30].

3.2. Germ Cell Tumors in Pediatric Oncology

The presented case represents a very rare occurrence in pediatric oncology, both by the mixed histology of the GCT and even more for the association of platinum-based chemotherapy with the development of systemic lipomatosis. Regarding the malignant diagnosis in this patient, it is important to acknowledge the rarity of a mixed GCT in pediatric oncology [31,32]. Germ cell tumors are a group of rare and highly heterogeneous neoplasms with respect to their site of presentation, histology, age of onset, and molecular signature [33,34]. Their incidence follows a bimodal curve, particularly high in children under four years of age and another peak later in adolescence. The highest incidence of GCT is seen in infants up to one month, where they account for 30–45% of neonatal tumors, although only 5% are proven malignant [33]. While GCT represents 3.6% of all cancers before puberty, their incidence spikes up to 12% in the 15–19-year age group that defined our presented case. In Europe, as well as in the USA and Asia, the incidence is significantly higher in males versus females, with 64 cases per million for males and 4 per million in female patients, such as our finding [35].

In pediatric patients, a quarter of all malignant GCT presents two or more histological types, thus classified as mixed germ cell tumors [36,37]. In females, the ovary is the most common site for GCT, with an incidence of 1 in 400 thousand in Europe, thus it is a rare cancer by RARECARE definition with an incidence that falls below 6 in 100 thousand cases [35]. GCTs make up 25% of all diagnosed ovarian masses and 3-5% of malignant ovarian cancers, as presented in this particular case. They are the most frequent ovarian malignancies in women under 25 years of age [38,39]. Although mixed GCT_S of the ovary are exceptionally rare (less than 1% of ovarian GCTs), they are the most common non-seminomatous tumors in adolescent girls, having the lowest 5-year relative survival rate in the 10–19 age group [37,40].

As presented in our rare case, surgical intervention was fundamental in the diagnosis and management of the large tumor, and therefore it is often the first step of the treatment in the majority of large-volume tumors. A laparotomy is usually performed, given the size and solid nature of these tumors. Since germ cell tumors occur mostly in the younger population, when possible, fertility-sparing interventions are performed in which the unaffected ovary and fallopian tube are preserved [41–44]. In cases of locally extended tumors or recurrent tumors, pre-surgical chemotherapy is preferred in order to reduce the tumor size and ensure better chances of a second step, complete carcinogenic surgical resection [45]. For GCTs, chemotherapy often involves the use of etoposide, ifosfamide, or cisplatin, a platinum-based drug that was used for the 16-year-old girl presented in our case [46]. However, when a clear diagnosis cannot be made before surgery, or the volume of the tumor creates a life-threatening circumstance, surgery will be performed before the chemotherapy management.

3.3. Platinum-Based Chemotherapy in Pediatric-Oncology

While in adults any immature teratoma classified higher than stage 1, based on the FIGO system, and a lower differentiation than Grade 1A, would be treated with chemotherapy without considering prior surgical management, in pediatric patients the use of chemotherapy has been challenged, with authors suggesting that complete resection of the tumor alone is sufficient, sometimes regardless of the grading, with no significant differences in overall 4–5 year survival of these patients compared to those who receive platinum-based chemotherapy [26,47–49]. It is worth noting, however, that literature on refractory and relapsed germ cell tumors in the pediatric population is scarce. Tumors that have previously been treated by surgery alone continue to have an excellent prognosis; however, in cases formerly treated with antineoplastic drugs, survival rates drop quite significantly. The main prognostic factor in such situations seems to come down to whether or not complete resection is possible. In order to achieve this, platinum-based chemotherapy is recommended beforehand [45].

Although chemotherapy treatment guidelines are clear in adult patients, they become uncertain in the pediatric population when involving ovarian GCTs of mixed histology due to the very low incidence of these cases. The use of cisplatin-based chemotherapy has been a breakthrough, not only in the treatment of gynecological cancers, but also in neuroblastomas, hepatoblastomas, medulloblastomas, and other solid tumors, drastically improving survival rates, and several studies have proven its superiority over carboplatin. This observation has led to the introduction of cisplatin in most chemotherapy schemes (the exception being the British protocol, which uses carboplatin) in combination with etoposide, ifosfamide, bleomycin, or vinblastin [50]. The downside of using platinum-based drugs is that blood serum platinum concentration remains high nearly 20 years after receiving cytotoxic therapy [51]. Platinum agents and long-term exposure to circulating platinum have been associated with long-term sequelae such as paresthesia, lipid disorder, hypertension, and gonadal dysfunction, as well as an increased risk of second malignancies and delayed cardiotoxicity [52].

The downside of using platinum-based chemotherapy is the risk of developing secondary tumors. This theory was substantiated by research that discovered a dose-dependent impact of chemotherapy on GCTs in 96 chemotherapy-treated individuals. In larger cohort studies, a dose-dependent link between platinum-based chemotherapy and GCTs eradication has not been studied [53]. In addition, in contrast to our case involving a 16-year old female patient with an ovarian GCT, the documented occurrences of contralateral tumoral recurrence were exclusively in male individuals. Moreover, all patients were young adults, with no reports being available on the recurrence of GCT due to platinum-based chemotherapy. Whether or not such a correlation exists is clinically significant since a growing number of patients with testicular GCTs may receive lower doses of platinum-based chemotherapy as an adjuvant treatment with one or two rounds of chemotherapy for high-risk early-stage illness gains popularity.

3.4. Platinum-Based Chemotherapy and Lipomatosis

Even though the underlying process is uncertain, cisplatin-based treatment might be a causative factor, although without significant proof. Cisplatin has an alkylating-like property, binding DNA and inducing apoptosis. Even if the growth of fat cells created by an antimitotic agent may seem counterintuitive, it is vital to recognize the alterations in the body caused by cisplatin. In reality, an increased body mass index and metabolic syndrome are potential side effects that may be attributed to hormonal changes brought on by chemotherapy, such as a testosterone deficit.

Quite the contrary, it has been demonstrated that cisplatin inhibits the migration and differentiation potential of human adipose-derived mesenchymal stem cells, and it can be assumed that the same mechanism is applicable to peripheral blood stem cells [54]. Adipose stem cells show promise in the repair of cytotoxic tissue injury in vitro and in vivo animal studies [55,56]. This is due to their chemotropism for sites of cell injury and inflammation, as well as their potential to differentiate into various cell lines under the right stimuli [57]. Chang et al. showed that adipose stem cells from subcutaneous adipose tissue collected from a patient who underwent cisplatin therapy four months prior displayed a significantly reduced differentiation capacity compared to those sampled from a healthy donor.

Our patient showed no signs of preexisting lipomas or abnormal growth in the affected area on the MRI performed before and after surgery prior to the start of platinum-based chemotherapy. The first signs of the abnormal growth could be seen less than a month after the last platinum agent-based cycle (cisplatin), the slight increase in size (~1–2 mm) was confirmed six months later, and surgery became necessary 18 months after the last PEI cycle (Figure 3C,D). Although the time correlation of cisplatin therapy with the onset of lipomatosis in our patient seems to be strong, it cannot be excluded that other factors might have played a significant role in this condition. Obesity may have been associated with an increased release of adipocyte-produced steroidogenic enzymes, leading to increased paracrine estrogen release [58]. Moreover, estrogen concentrations during hormone replacement therapy after bilateral adnexectomy may have differed from those occurring physiologically. In light of this and given the fact that only two cases of lipomatosis after CDDP therapy have been reported, the effect of estrogens seems more significant.

While the occurrence of lipomas after non-steroid chemotherapy regiments has not been frequently reported, chemotherapy-associated fatty depositions and liver injury is a well-known phenomenon that can manifest itself in a wide range of patterns from transient transaminitis, cholestasis, and venous occlusion to acute liver failure or neoplastic transformation. Cisplatin, one of the antineoplastic agents that our patient has received, has been noted to temporarily elevate liver enzymes, but it is rarely linked to severe cases of hepatotoxicity and fatty infiltration of the liver. However, another platinum-based agent, namely oxaliplatin, has been recently associated with the development of hepatosteatosis in adults who underwent hepatic resections due to colorectal or hepatic metastases [59]. One study that aimed to evaluate whether such lesions were evident in the pediatric population as well failed to find severe histopathological modifications, such as steatosis, fibrosis, or necrosis [15]. The reason why such findings appear in adult patients and not pediatric patients remains unknown, but it could be explained by the different drug combinations used in the pediatric population compared to the adult population, as well as by the shorter period of time elapsed between the end of neoadjuvant chemotherapy cycle and surgery in pediatric patients.

In the case of our patient, it would be unwise to disregard the nutritional factor that might have played a role in the development of the fatty liver. In the midst of the obesity epidemic, which has unfortunately not spared children, non-alcoholic fatty liver disease (NAFLD) has become the number one cause of chronic hepatic disorders in both children and adults [60]. Moreover, a statistically significant relationship between liver steatosis and hepatic lipomas was found with 50% of them also correlating with steatosis, and two-thirds of the lipomas were associated with obesity and hypercholesterolemia [61]. However, our patient had an increased BMI prior to the surgery and chemotherapy; therefore, obesity alone might not necessarily be the only factor involved in the development of lipomas.

Another factor that should be taken into account when discussing the etiology of the lipomatosis that our patient developed is the hormonal replacement therapy (HRT) administered after surgery. Although the process is not yet known, estrogen seems to promote the subcutaneous accumulation of adipose tissue by inhibiting lipolysis in the subcutaneous deposits only. That way, intra-abdominal fat deposits, which are typically seen in the male population and post-menopausal women, are shifted predominantly toward the subcutaneous regions. After menopause, either physiological or medically induced, a redistribution of fat is observed, with fat starting to accumulate in the visceral area. HRT, among other treatments, prevents both visceral fat distribution and additional weight gain [62,63]. Given that the HRT received by our patient was carefully monitored and deemed adequate, we consider it unlikely to have played a role in the development of lipomatosis.

The two phenomena described as "Chemotherapeutic retroconversion" and "growing teratoma syndrome" could explain the lipomatosis that appeared in our patient after chemotherapy [64,65]. Mature teratomas are germ cell tumors composed of different cell types, with the potential to mimic any tissue type of the body. According to the previously mentioned concepts, chemotherapy either contributes to the maturation of the immature cells in the immature teratomas or induces apoptosis in the immature cells, allowing the mature and slow-dividing cells to flourish. These theories were tested in a Japanese descriptive study published in 2021 [65] by analyzing the response of embryonic stem cells and induced pluripotent stem cell-derived immature teratomas to cisplatin in mice. The study showed that, after being treated with cisplatin, the tumors increased their maturation score, and on histological examination, mature tissues were found. Also, after treatment with cisplatin, the tumors exhibited immunostaining a lower immaturity index. Thus, the study concluded that cisplatin not only preferably destroys immature cells but also induces maturation in the teratoma tissue.

4. Conclusions

Despite the existing evidence, lack of genetic analysis of the patients and experimental methods to determine causality leave the possible association between the use of platinum-based agents and the development of lipomatosis uncertain. The underlying mechanism by which cisplatin or other platinum agents might affect adipocytes and favor the development of lipomatous tumors is yet to be understood. This case report and literature review provides further important evidence into a possible very rare side effect of cisplatin, although it requires further research and in vitro studies. Until more data is available, we suggest that our fellow clinicians consider previous treatment with platinum agents as a possible cause in cases of rapid development and growth of lipomas when nutritional, hormonal, or familial predisposition do not warrant the clinical manifestations.

Author Contributions: Conceptualization, E.B. and A.G.B.; methodology, E.B. and A.G.B.; validation, A.B. and I.C.M.; formal analysis, R.A.N. and B.F.; investigation, R.A.N. and B.F.; resources, A.I.B. and F.G.H.; data curation, A.I.B. and F.G.H.; writing—original draft preparation, I.D.H. and S.T.A.; writing—review and editing, M.R., O.S., A.A. (Abduljabar Adi) and A.A. (Adnan Anjary); visualization, A.B.; supervision, I.C.M. All authors have read and agreed to the published version of the manuscript.

Funding: This research received no external funding.

Institutional Review Board Statement: The study was conducted according to the guidelines of the Declaration of Helsinki, and approved by the Ethics Committee of the "Louis Turcanu" Emergency Clinical Hospital for Children (protocol code 16915/12.11.2020).

Informed Consent Statement: Written informed consent has been obtained from the patient to publish this paper.

Acknowledgments: The authors would like to give thanks to their colleagues from the Departments of Pediatric Oncology and Radiology and Imagistic Medicine of the "Louis Turcanu" Emergency Clinical Hospital for Children, for their assistance with data collection.

Conflicts of Interest: The authors declare no conflict of interest.

References

1. Hashizume, N.; Aiko, T.; Fukahori, S.; Ishii, S.; Saikusa, N.; Koga, Y.; Higashidate, N.; Sakamoto, S.; Tsuruhisa, S.; Nakahara, H.; et al. Benign mesenteric lipomatous tumor in a child: A case report and literature review. *Surg. Case Rep.* **2020**, *6*, 243. [CrossRef] [PubMed]
2. Putra, J.; Al-Ibraheemi, A. Adipocytic tumors in Children: A contemporary review. *Semin. Diagn. Pathol.* **2019**, *36*, 95–104. [CrossRef]

3. Schiltz, D.; Mueller, K.; Ortner, C.; Tschernitz, S.; Anker, A.; Felthaus, O.; Schreml, J.; Koller, M.; Prantl, L.; Schreml, S. Multiple Symmetric Lipomatosis: A Cross-Sectional Study to Investigate Clinical Features and Patients' Quality of Life. *Symmetry* **2021**, *13*, 1823. [CrossRef]
4. Al Ghazal, P.; Grönemeyer, L.-L.; Schön, M.P. Lipomatosen. *JDDG J. Dtsch. Dermatol. Ges.* **2018**, *16*, 313–328. [CrossRef] [PubMed]
5. Lemaitre, M.; Aubert, S.; Chevalier, B.; Jannin, A.; Bourry, J.; Prévost, G.; Lefebvre, H.; Vantyghem, M.-C. Rare Forms of Lipomatosis: Dercum's Disease and Roch-Leri Mesosomatous Lipomatosis. *J. Clin. Med.* **2021**, *10*, 1292. [CrossRef]
6. Sollier, C.; Capel, E.; Aguilhon, C.; Smirnov, V.; Auclair, M.; Douillard, C.; Ladsous, M.; Defoort-Dhellemmes, S.; Gorwood, J.; Braud, L.; et al. LIPE-related lipodystrophic syndrome: Clinical features and disease modeling using adipose stem cells. *Eur. J. Endocrinol.* **2021**, *184*, 155–168. [CrossRef] [PubMed]
7. Gudjoncik, A.; De Carvalho, A.; Loris, L.; Chavent, A.; Cercueil, J.-P.; Jazayeri, S.; Cottin, Y. Postoperative mediastinal lipomatosis. *Arch. Cardiovasc. Dis.* **2008**, *101*, 593–594. [CrossRef] [PubMed]
8. Humblot, S.; Weber, J.; Korganow, A.-S.; Hammann, B.; Pasquali, J.; Martin, T. Les lipomatoses induites par la corticothérapie. *Rev. Méd. Interne* **1997**, *18*, 396–401. [CrossRef] [PubMed]
9. Dasari, S.; Tchounwou, P.B. Cisplatin in cancer therapy: Molecular mechanisms of action. *Eur. J. Pharmacol.* **2014**, *740*, 364–378. [CrossRef]
10. Sharmin, S.; Rahaman, M.; Martorell, M.; Sastre-Serra, J.; Sharifi-Rad, J.; Butnariu, M.; Bagiu, I.C.; Bagiu, R.V.; Islam, M.T. Cytotoxicity of synthetic derivatives against breast cancer and multi-drug resistant breast cancer cell lines: A literature-based perspective study. *Cancer Cell Int.* **2021**, *21*, 612. [CrossRef]
11. Brown, A.; Kumar, S.; Tchounwou, P.B. Cisplatin-Based Chemotherapy of Human Cancers. *J. Cancer Sci. Ther.* **2019**, *11*, 97. [PubMed]
12. Iacob, R.; Manolescu, D.L.; Stoicescu, E.R.; Fabian, A.; Malita, D.; Oancea, C. Breast Cancer—How Can Imaging Help? *Healthcare* **2022**, *10*, 1159. [CrossRef] [PubMed]
13. Ranasinghe, R.; Mathai, M.L.; Zulli, A. Cisplatin for cancer therapy and overcoming chemoresistance. *Heliyon* **2022**, *8*, e10608. [CrossRef] [PubMed]
14. Tsuji, K.; Doyama, H. S-1 induced hepatic steatosis in patients with pancreatic cancer: Retrospective analysis. *World J. Gastrointest. Oncol.* **2017**, *9*, 314–318. [CrossRef] [PubMed]
15. Martin-Benitez, G.; Marti-Bonmati, L.; Barber, C.; Vila, R. Hepatic lipomas and steatosis: An association beyond chance. *Eur. J. Radiol.* **2012**, *81*, e491–e494. [CrossRef]
16. Kucharz, E.J.; Kopeć-Mędrek, M.; Kramza, J.; Chrzanowska, M.; Kotyla, P. Dercum's disease (adiposis dolorosa): A review of clinical presentation and management. *Reumatologia* **2019**, *57*, 281–287. [CrossRef]
17. Folescu, R.; Levai, C.M.; Grigoraş, M.L.; Arghirescu, T.S.; Talpoş, I.C.; Gîndac, C.M.; Zamfir, C.; Poroch, V.; Anghel, M.D. Expression and significance of Ki-67 in lung cancer. *Rom. J. Morphol. Embryol.* **2018**, *59*, 227–233. [PubMed]
18. Uxa, S.; Castillo-Binder, P.; Kohler, R.; Stangner, K.; Müller, G.A.; Engeland, K. Ki-67 gene expression. *Cell Death Differ.* **2021**, *28*, 3357–3370. [CrossRef]
19. Grigoraş, M.L.; Arghirescu, T.S.; Folescu, R.; Talpoş, I.C.; Gîndac, C.M.; Zamfir, C.; Cornianu, M.; Anghel, M.D.; Levai, C.M. Expression of E-cadherin in lung carcinoma, other than those with small cells (NSCLC). *Rom. J. Morphol. Embryol.* **2017**, *58*, 1317–1325.
20. Rosen, R.D.; Sapra, A. TNM Classification. In *StatPearls [Internet]*; StatPearls Publishing: Treasure Island, FL, USA, 2022.
21. Benedet, J.L.; Bender, H.; Jones, H., 3rd; Ngan, H.Y.; Pecorelli, S. FIGO staging classifications and clinical practice guidelines in the management of gynecologic cancers. FIGO Committee on Gynecologic Oncology. *Int. J. Gynaecol. Obstet.* **2000**, *70*, 209–262.
22. Rebelo, J.; Moreira, F.; Morgado, M.; Preto, A.S.; Madureira, A. Growing Teratoma Syndrome: A Rare Outcome. *Case Rep. Urol.* **2021**, *2021*, 8884787. [CrossRef]
23. Gurguş, D.; Grigoraş, M.L.; Motoc, A.G.M.; Zamfir, C.L.; Cornianu, M.; Faur, C.I.; Pop, D.L.; Folescu, R. Clinical relevance and accuracy of p63 and TTF-1 for better approach of small cell lung carcinoma versus poorly differentiated nonkeratinizing squamous cell carcinoma. *Rom. J. Morphol. Embryol.* **2019**, *60*, 139–143. [PubMed]
24. Hu, S.; Mojtahed, A.; Covington, A.; Thompson, W.; Volpicelli, N.; McCarthy, D. Intestinal Lipomatosis and Chemotherapy: A Growing Concern. *Am. J. Dig. Dis.* **2016**, *61*, 3151–3154. [CrossRef] [PubMed]
25. Bracaglia, R.; D'Ettorre, M.; Gentileschi, S.; Mingrone, G.; Tambasco, D. Multiple lipomatosis after stem cell trasplant and chemotherapy: A case report. *Eur. Rev. Med. Pharmacol. Sci.* **2014**, *18*, 413–415. [PubMed]
26. Gietema, J.A.; Meinardi, M.T.; Messerschmidt, J.; Gelevert, T.; Alt, F.; Uges, D.R.; Sleijfer, D.T. Circulating plasma platinum more than 10 years after cisplatin treatment for testicular cancer. *Lancet* **2000**, *355*, 1075–1076. [CrossRef] [PubMed]
27. Cronin, P.A.; Myers, E.; Redmond, H.P.; O'Reilly, S.; Kirwan, W.O. Lipomatosis: An unusual side-effect of cytotoxic chemotherapy? *Acta Derm. Venereol.* **2010**, *90*, 303–304. [CrossRef] [PubMed]
28. Gotoh, M.; Kitahara, T.; Sakuta, J.; Akahane, D.; Ohyashiki, K. Multiple lipoma with hyperlipidemia in a multiple myeloma patient treated with bortezomib/dexamethazone. *Leuk. Res.* **2010**, *34*, e120–e121. [CrossRef] [PubMed]
29. Levine, J.; Jensen, M.; Eberhardt, N.; O'Brien, T. Adipocyte macrophage colony-stimulating factor is a mediator of adipose tissue growth. *J. Clin. Investig.* **1998**, *101*, 1557–1564. [CrossRef]
30. Skurk, T.; Alberti-Huber, C.; Herder, C.; Hauner, H. Relationship between Adipocyte Size and Adipokine Expression and Secretion. *J. Clin. Endocrinol. Metab.* **2007**, *92*, 1023–1033. [CrossRef] [PubMed]

31. Jawass, M.A.; Alezzi, J.; Bin Gouth, H.S.; Bahwal, S.A.; Bamatraf, F.F.; Ba'Amer, A.A. Pattern of malignancies in children <15 years of age reported in Hadhramout Cancer Registry, Yemen between 2002 and 2014. *Saudi. Med. J.* **2016**, *37*, 513–520. [CrossRef]
32. Jinca, C.; Petrescu, C.A.M.; Boeriu, E.; Oprisoni, A.; Balint-Gib, L.; Baica, M.; Popa, C.; Andreescu, N.; Serban, M.; Ursu, E.; et al. The impact of immunological and biomolecular investigations on the outcome of children with acute lymphoblastic leukemia—Experience of IIIrd Paediatric Clinic Timisoara. *Rev. Romana Med. Lab.* **2018**, *26*, 77–85. [CrossRef]
33. Mosbech, C.H.; Rechnitzer, C.; Brok, J.S.; Rajpert-De Meyts, E.; Hoei-Hansen, C.E. Recent Advances in Understanding the Etiology and Pathogenesis of Pediatric Germ Cell Tumors. *J. Pediatr. Hematol. Oncol.* **2017**, *36*, 263–270. [CrossRef] [PubMed]
34. Trama, A.; Berrino, F. The Epidemiology of Malignant Germ Cell Tumors: The EUROCARE Study. In *Pathology and Biology of Human Germ Cell Tumors*, 1st ed.; Nogales, F.F., Jimenez, R.E., Eds.; Springer: Berlin/Heidelberg, Germany, 2017; pp. 11–21.
35. Steliarova-Foucher, E.; Colombet, M.; Ries, L.A.G.; Moreno, F.; Dolya, A.; Bray, F.; Hesseling, P.; Shin, H.Y.; Stiller, C.A.; IICC-3 contributors. International incidence of childhood cancer, 2001–2010: A population-based registry study. *Lancet Oncol.* **2017**, *18*, 719–731. [CrossRef] [PubMed]
36. Murray, M.J.; Nicholson, J.C.; Coleman, N. Biology of childhood germ cell tumours, focussing on the significance of microRNAs. *Andrology* **2015**, *3*, 129–139. [CrossRef]
37. Zambrano, E.; De Stefano, D.V.; Reyes-Múgica, M. Pediatric Germ Cell Tumors. In *Pathology and Biology of Human Germ Cell Tumors*, 1st ed.; Nogales, F.F., Jimenez, R.E., Eds.; Springer: Berlin/Heidelberg, Germany, 2017; pp. 381–391.
38. Ulbright, T.M. Germ cell tumors of the gonads: A selective review emphasizing problems in differential diagnosis, newly appreciated, and controversial issues. *Mod. Pathol.* **2005**, *18* (Suppl. S2), S61–S79. [CrossRef]
39. Jimenez, R.E.; Gupta, S.; Herrera-Hernandez, L.P.; Sebo, T.J. Testicular Germ Cell Tumors. In *Pathology and Biology of Human Germ Cell Tumors*, 1st ed.; Nogales, F.F., Jimenez, R.E., Eds.; Springer: Berlin/Heidelberg, Germany, 2017; pp. 294–296.
40. Smith, H.O.; Berwick, M.; Verschraegen, C.F.; Wiggins, C.; Lansing, L.; Muller, C.Y.; Qualls, C.R. Incidence and Survival Rates for Female Malignant Germ Cell Tumors. *Obstet. Gynecol.* **2006**, *107*, 1075–1085. [CrossRef]
41. You, W.; Dainty, L.A.; Rose, G.S.; Krivak, T.; McHale, M.T.; Olsen, C.H.; Elkas, J.C. Gynecologic Malignancies in Women Aged Less Than 25 Years. *Obstet. Gynecol.* **2005**, *105*, 1405–1409. [CrossRef]
42. Fonseca, A.; Frazier, A.L.; Shaikh, F. Germ Cell Tumors in Adolescents and Young Adults. *J. Oncol. Pract.* **2019**, *15*, 433–441. [CrossRef]
43. Triarico, S.; Capozza, M.A.; Mastrangelo, S.; Attinà, G.; Maurizi, P.; Ruggiero, A. Gynecological cancer among adolescents and young adults (AYA). *Ann. Transl. Med.* **2020**, *8*, 1–12. [CrossRef]
44. Berney, D.M.; Stoneham, S.; Arora, R.; Shamash, J.; Lockley, M. Ovarian germ cell tumour classification: Views from the testis. *Histopathology* **2020**, *76*, 25–36. [CrossRef]
45. Shaaban, A.; Rezvani, M.; Elsayes, K.M.; Baskin, H., Jr.; Mourad, A.; Foster, B.R.; Jarboe, E.A.; Menias, C.O. Ovarian Malignant Germ Cell Tumors: Cellular Classification and Clinical and Imaging Features. *RadioGraphics* **2014**, *34*, 777–801. [CrossRef] [PubMed]
46. Friedman, C.; Fenster, T. Laparoscopic Treatment of Mixed Malignant Ovarian Germ Cell Tumor in a 16-Year-Old Female Adolescent. *J. Pediatr. Adolesc. Gynecol.* **2016**, *29*, e91–e94. [CrossRef]
47. Faure-Conter, C.; Rocourt, N.; Sudour-Bonnange, H.; Vérité, C.; Martelli, H.; Patte, C.; Frappaz, D.; Orbach, D. Les tumeurs germinales de l'enfant. *Bull. Cancer* **2013**, *100*, 381–391. [CrossRef] [PubMed]
48. Luczak, J.; Bagłaj, M. Ovarian teratoma in children: A plea for collaborative clinical study. *J. Ovarian Res.* **2018**, *11*, 75. [CrossRef] [PubMed]
49. Kaur, B. Pathology of malignant ovarian germ cell tumours. *Diagn. Histopathol.* **2020**, *26*, 289–297. [CrossRef]
50. Boer, H.; Proost, J.H.; Nuver, J.; Bunskoek, S.; Gietema, J.Q.; Guebels, B.M.; Altena, R.; Zwart, N.; Oosting, S.F.; Vonk, J.M.; et al. Long-term exposure to circulating platinum is associated with late effects of treatment in testicular cancer survivors. *Ann. Oncol.* **2015**, *26*, 2305–2310. [CrossRef]
51. Blok, J.M.; Groot, H.J.; Huele, E.H.; De Wit, R.; Horenblas, S.; Nuver, J.; Groenewegen, G.; Bosch, J.R.; Witjes, J.A.; Tromp, J.M.; et al. Dose-Dependent Effect of Platinum-Based Chemotherapy on the Risk of Metachronous Contralateral Testicular Cancer. *J. Clin. Oncol.* **2021**, *39*, 319–327. [CrossRef]
52. Chang, Y.-H.; Liu, H.-W.; Chu, T.-Y.; Wen, Y.-T.; Tsai, R.-K.; Ding, D.-C. Cisplatin-Impaired Adipogenic Differentiation of Adipose Mesenchymal Stem Cells. *Cell Transplant.* **2017**, *26*, 1077–1087. [CrossRef]
53. Yao, W.; Hu, Q.; Ma, Y.; Xiong, W.; Wu, T.; Cao, J.; Wu, D. Human adipose-derived mesenchymal stem cells repair cisplatin-induced acute kidney injury through antiapoptotic pathways. *Exp. Ther. Med.* **2015**, *10*, 468–476. [CrossRef]
54. Meligy, F.Y.; Elgheed, A.T.A.; Alghareeb, S.M. Therapeutic effect of adipose-derived mesenchymal stem cells on Cisplatin induced testicular damage in adult male albino rat. *Ultrastruct. Pathol.* **2019**, *43*, 28–55. [CrossRef]
55. Spaeth, E.; Klopp, A.; Dembinski, J.; Andreeff, M.; Marini, F. Inflammation and tumor microenvironments: Defining the migratory itinerary of mesenchymal stem cells. *Gene Ther.* **2008**, *15*, 730–738. [CrossRef] [PubMed]
56. Floyd, J.; Mirza, I.; Sachs, B.; Perry, M.C. Hepatotoxicity of Chemotherapy. *Semin. Oncol.* **2006**, *33*, 50–67. [CrossRef]
57. Scuderi, M.G.; Magro, G.G.; Di Cataldo, A.; Pesce, A.; Scalora, L.; Vecchio, G.M.; Portale, R.; Di Benedetto, V.; Puleo, S. Evaluation of Neoadjuvant Chemotherapy Effects on Liver Parenchyma in Resected Pediatric Malignancies. *Pediatr. Hematol. Oncol.* **2013**, *30*, 392–399. [CrossRef] [PubMed]

58. Kiani, A.K.; Mor, M.; Bernini, A.; Fulcheri, E.; Michelini, S.; Herbst, K.L.; Buffelli, F.; Belgrado, J.P.; Kaftalli, J.; Stuppia, L.; et al. Steroid-converting enzymes in human adipose tissues and fat deposition with a focus on AKR1C enzymes. *Eur. Rev. Med. Pharmacol. Sci.* **2021**, *25* (Suppl. S1), 23–32. [CrossRef] [PubMed]
59. Nobili, V.; Alisi, A.; Valenti, L.; Miele, L.; Feldstein, A.E.; Alkhouri, N. NAFLD in children: New genes, new diagnostic modalities and new drugs. *Nat. Rev. Gastroenterol. Hepatol.* **2019**, *16*, 517–530. [CrossRef] [PubMed]
60. Mattsson, C.; Olsson, T. Estrogens and Glucocorticoid Hormones in Adipose Tissue Metabolism. *Curr. Med. Chem.* **2007**, *14*, 2918–2924. [CrossRef]
61. Pedersen, S.B.; Kristensen, K.; Hermann, P.A.; Katzenellenbogen, J.A.; Richelsen, B. Estrogen Controls Lipolysis by Up-Regulating α2A-Adrenergic Receptors Directly in Human Adipose Tissue through the Estrogen Receptor α. Implications for the Female Fat Distribution. *J. Clin. Endocrinol. Metab.* **2004**, *89*, 1869–1878. [CrossRef]
62. DiSaia, P.J.; Saltz, A.; Kagan, A.R.; Morrow, C.P. Chemotherapeutic retro-conversion of immature teratoma of the ovary. *Obstet. Gynecol.* **1977**, *49*, 346–350.
63. Logothetis, C.J.; Samuels, M.L.; Trindade, A.; Johnson, D.E. The growing teratoma syndrome. *Cancer* **1982**, *50*, 1629–1635. [CrossRef]
64. Amsalem, H. Growing teratoma syndrome vs. chemotherapeutic retroconversion Case report and review of the literature. *Gynecol. Oncol.* **2004**, *92*, 357–360. [CrossRef] [PubMed]
65. Kurata, A.; Takanashi, M.; Ohno, S.-I.; Fujita, K.; Kuroda, M. Cisplatin induces differentiation in teratomas derived from pluripotent stem cells. *Regen. Ther.* **2021**, *18*, 117–126. [CrossRef] [PubMed]

medicina

MDPI

Article

Multiparametric MRI Features of Breast Cancer Molecular Subtypes

Madalina Szep, Roxana Pintican *, Bianca Boca, Andra Perja, Magdalena Duma, Diana Feier, Bogdan Fetica, Dan Eniu, Sorin Marian Dudea and Angelica Chiorean

Department of Radiology and Medical Imaging, "Iuliu Hatieganu" University of Medicine and Pharmacy, 400374 Cluj-Napoca, Romania

* Correspondence: roxana.pintican@gmail.com

Abstract: *Background and Objectives*: Breast cancer (BC) molecular subtypes have unique incidence, survival and response to therapy. There are five BC subtypes described by immunohistochemistry: luminal A, luminal B HER2 positive and HER2 negative, triple negative (TNBC) and HER2-enriched. Multiparametric breast MRI (magnetic resonance imaging) provides morphological and functional characteristics of breast tumours and is nowadays recommended in the preoperative setting. *Aim*: To evaluate the multiparametric MRI features (T2-WI, ADC values and DCE) of breast tumours along with breast density and background parenchymal enhancement (BPE) features among different BC molecular subtypes. *Materials and Methods*: This was a retrospective study which included 344 patients. All underwent multiparametric breast MRI (T2WI, ADC and DCE sequences) and features were extracted according to the latest BIRADS lexicon. The inter-reader agreement was assessed using the intraclass coefficient (ICC) between the ROI of ADC obtained from the two breast imagers (experienced and moderately experienced). *Results*: The study population was divided as follows: 89 (26%) with luminal A, 39 (11.5%) luminal B HER2 positive, 168 (48.5%) luminal B HER2 negative, 41 (12%) triple negative (TNBC) and 7 (2%) with HER2 enriched. Luminal A tumours were associated with special histology type, smallest tumour size and persistent kinetic curve (all p-values < 0.05). Luminal B HER2 negative tumours were associated with lowest ADC value (0.77×10^{-3} mm^2/s^2), which predicts the BC molecular subtype with an accuracy of 0.583. TNBC were associated with asymmetric and moderate/marked BPE, round/oval masses with circumscribed margins and rim enhancement (all p-values < 0.05). HER2 enriched BC were associated with the largest tumour size (mean 37.28 mm, p-value = 0.02). *Conclusions*: BC molecular subtypes can be associated with T2WI, ADC and DCE MRI features. ADC can help predict the luminal B HER2 negative cases.

Keywords: molecular subtypes; breast cancer; multiparametric MRI; ADC; luminal

Citation: Szep, M.; Pintican, R.; Boca, B.; Perja, A.; Duma, M.; Feier, D.; Fetica, B.; Eniu, D.; Dudea, S.M.; Chiorean, A. Multiparametric MRI Features of Breast Cancer Molecular Subtypes. *Medicina* **2022**, *58*, 1716. https://doi.org/10.3390/medicina58121716

Academic Editors: Călin Căinap, Nicolae Crisan and Maria Rosaria De Miglio

Received: 21 September 2022
Accepted: 15 November 2022
Published: 23 November 2022

1. Introduction

The limitations of traditional histological classification led to the development of a new molecular classification of breast cancer (BC), in early 2000 [1]. The quantitative global gene expression profiling (GEP) technique identified four intrinsic subtypes: (1) estrogen receptor (ER) positive -> luminal A; (2) ER negative -> HER2-enriched (Erb-B2 overexpression); (3) basal-like and (4) normal breast-like. However, the GEP technique was not economical for daily practice. Nowadays, immunohistochemistry (IHC) procedures, using protein expression have been employed as a method for BC subtyping in clinical practice [2]. The IHC describes five surrogates BC subtypes: (1) luminal A; (2) luminal B HER2 negatives; (3) luminal B HER2 positive; (4) HER2 positives non luminal (corresponding to Erb-B2 overexpression in the intrinsic type classification) and (5) triple negative (the basal-like type of intrinsic subtyping).

The BC molecular subtypes have unique incidence, survival and response to therapy [3–5]. The ER, progesterone receptor (PR) and HER2 (human epidermal growth factor receptor (2) status provide prognostic and predictive information [6]. In this regard, the luminal A subtype is the most indolent BC and has the best prognosis; the TNBC type has a poor prognosis and is often associated with genetic mutations (BRCA 1) and HER2 positive subtypes benefit from targeted anti-HER2 therapy.

Breast MRI (magnetic resonance imaging) is recommended for the preoperative staging and subsequent choice of appropriate therapy in women with BC [7]. It provides morphological and functional characteristics of the breast tumours. Previously published work has independently investigated the role of T2-weighted image (WI) characteristics, apparent diffusion coefficient (ADC) or dynamic contrast enhancement (DCE) patterns in the prediction of BC molecular subtypes [8–11]. Few studies have analyzed combined multiparametric MRI findings (T2WI and ADC) for association with specific BC subtypes [12]. Recently, the distribution and level of breast parenchymal enhancement (BPE) has been reported to be associated with BC subtypes, allowing an additional risk stratification and targeted screening tests [10].

Our aim was to evaluate the multiparametric MRI features (T2-WI, ADC values and DCE) of breast tumours along with BPE features among different BC molecular subtypes.

2. Materials and Methods

2.1. Study Population

This was a retrospective study approved by the Ethics Committee of Medimages Review Board (NR10/15092022, from 15 September 2022), and the need for written consent was waived.

Inclusion criteria were patients with BC (regardless of the disease stage), preoperative breast MRI with T2WI, DWI/ADC, DCE sequences and pathology reports who presented to our clinic (Medimages Breast Center) between January 2018 and March 2022.

Exclusion criteria were patients with inadequate or incomplete MRI images, pathology and immunohistochemistry reports (Figure 1).

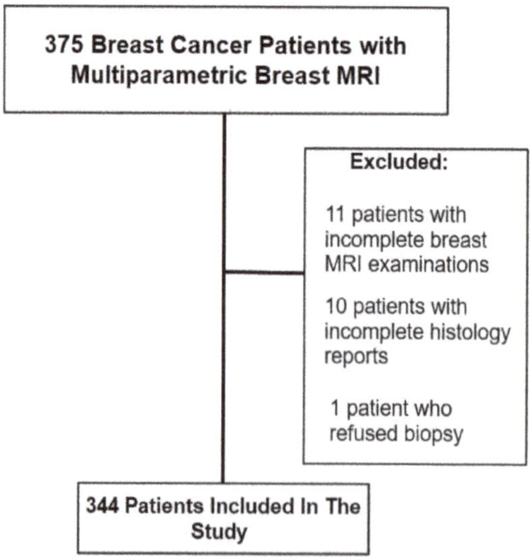

Figure 1. Study population.

2.2. MRI Acquisition and Features

All patients underwent 1.5 T MRI examinations performed in the prone position with a coil dedicated to breast imaging, using two MR machines (Siemens Magnetom Symphony TIM and Altea). The breast MRI protocol consisted of five sequences: (1) T1-WI turbo spin echo; (2) T2-WI turbo spin echo (TR = 5000 ms; TE = 120 ms, flip angle 90°; in-plane resolution 0.6 mm × 0.6 mm; 85 slices; slice thickness 2 mm; (3) T2 turbo inversion recovery magnitude (TIRM); (4) a T1WI vibe fs dynamic sequence (TR = 4.66 ms; TE = 2.3 ms; slice thickness 1.3 mm, with precontrast and five phases after the contrast administration (0.2 mL/kg, 3 mL/s); (5) DWI echo planar imaging with five b factors (0, 200, 400, 600, and 800 s/mm^2). The ADC maps were automatically calculated linearly by the method provided by the MRI vendor.

All morphological MRI features (breast density, BPE, T2-WI and DCE features) were reported by one radiologist (AC) with more than 15 years of experience in breast imaging, using the American College of Radiology BI-RADS lexicon (5th edition) [13]. Each breast with more than one breast lesion was included in the analysis only once; their highest assessment was used to guarantee statistical independence of each observation.

The BPE was recorded as symmetric or asymmetric, with a minimal, mild, moderate or marked level. The T2-WI features included shape and margins; enhancement patterns, distribution and intensity (including kinetic curves) were recorded for mass and non-mass separately.

ADC values were obtained by placing a standardized intratumoral ROI of 0.2 mm^2 in the darkest area corresponding to tumour enhancement. Breast parenchyma, tumour necrosis, haemorrhage and fat were excluded. Two ADC measurements were performed, first by the same experienced radiologist (AC) and second, by a moderately experienced breast radiologist with 5 years' experience (MS).

2.3. Pathology and Immunohistochemistry Data

Age and pathology data were reviewed from the medical records, including the histologic tumour type (no special type—NST or special types), and immunohistochemistry findings (ER, PR, HER2 status, ki-67% proliferation index).

BC molecular subtypes were assigned according to the latest St. Gallen International Expert Consensus (2013) classification system for IHC subtypes 2 (Table 1).

Table 1. Breast cancer molecular subtypes based on immunohistochemistry findings.

Luminal A	Luminal B	Luminal B-like	HER2+	Triple Negative (TN)
ER+	ER+	ER+	ER−	ER−
PR+	HER2−	HER2+	PR−	PR−
HER2−	High Ki-67/PR−	Any PR	HER2+	HER2−
Low Ki-67		Any Ki-67		

ER = estrogen receptor, PR = progesterone receptor, HER2 = human epidermal growth factor 2, Ki-67% = proliferation index with low Ki-67% < 20% and high ki-67% > 20%.

2.4. Statistical Analysis

Statistical analyses were performed using MedCalc software (version 19.2.6, Ostend, Belgium). The Mann–Whitney U test was used to compare the age, size and ADC values between one molecular BC subtype versus all the other subtypes (for example ADC value of Luminal A BCs versus ADC value of all the other BC molecular subtypes). To analyse associations between each BC molecular subtype, clinicopathological data and MRI findings, the Chi-square or Fisher's exact test were used. Using binary logistic analyses, we performed univariate and multivariate analyses to demonstrate the association between the presence of TNBC and different factors such as shape/margins/enhancement pattern. Results are expressed as unadjusted/adjusted odds ratio. ROC curve with AUC was calculated for statistically significant parameters. $p < 0.05$ was considered statistically significant.

The inter-reader agreement was assessed using the intraclass coefficient (ICC) between the ROI of ADC obtained from the two breast imagers. The following were selected as standards for strength of agreement: 0.01–0.20 = slight; 0.21–0.40 = fair; 0.41–0.60 = moderate; 0.61–0.80 = substantial; 0.81–0.99 = almost perfect; 1.0 = perfect.

3. Results

The study population consisted of 344 patients with breast cancer, with a 47.8 year mean age (ranging from 24 to 77 years old), having BC tumours with a mean size of 26 mm (ranging from 4.2 to 115 mm). The BC were as follows: 89 (26%) with luminal A, 39 (11.5%) luminal B HER2 positive, 168 (48.5%) luminal B HER2 negative, 41 (12%) triple negative (TNBC) and 7 (2%) with HER2 enriched.

3.1. Associations between 5 BC Molecular Subtypes and MRI Features

Patients with luminal B HER2 positive were associated with younger age compared to the other BC subtypes (44.5 mean age, p-value = 0.02), while patients with luminal A were associated with older age (50.2 mean age, p-value < 0.01).

The majority of the patients had a no-special-type (NST) BC histology type, with special subtypes (such as papillary or mucinous) associated with luminal A (p-value < 0.01).

There was no association between breast density and BC molecular subtypes (all $p > 0.05$).

The BPE was asymmetric in 18 (43.9%) patients with TNBC (p-value = 0.000), associated with moderate/marked level in 21 (51.21%) cases. There were no other statistically significant differences in BPE symmetry or level for the other BC molecular subtypes.

The smallest tumours were associated with luminal A (p-value < 0.01), and the largest with the HER2 enriched masses (p-value = 0.002).

For mass-appearing BC tumours, the shape was oval or round for 23 (56%, p-value = 0.01) of TNBC, with associated circumscribed margins (14.6%, p-value < 0.01) compared to the other BC subtypes (Figure 2).

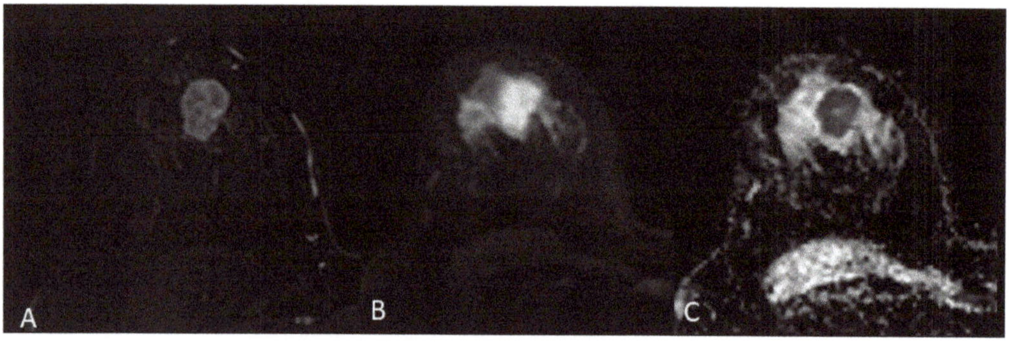

Figure 2. Triple negative breast cancer in a 33-year-old woman—subtracted T1-weighted contrast-enhanced image (**A**) shows a 3 cm oval (lobulated) mass with circumscribed margins, heterogeneous enhancement and restricted diffusion on DWI (**B**) and ADC (**C**).

A heterogeneous or rim enhancement was associated with TNBC (p-value < 0.01) in 26 (63.4%) and 15 (36.5%) cases, respectively. No BC mass was found to have internal dark septations (Figures 3 and 4).

For the non-mass enhancement appearance, there were no differences in the distribution, enhancement type or mean ADC values between BC molecular subtypes (all $p > 0.05$).

The majority of tumours had a "wash-out" (type 3) enhancement curve, but only luminal A tumours reached statistical significance. In addition, luminal A tumours were associated with a plateau (type 2) kinetic curve (p-value = 0.03) (Table 2).

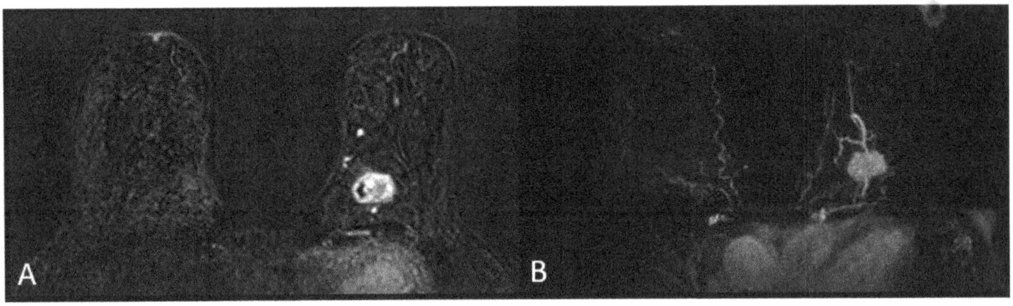

Figure 3. Triple negative breast cancer in a 60-year-old woman—heterogeneous enhancement on subtracted T1-weighted contrast-enhanced image (**A**) and maximum intensity projection (MIP, (**B**)).

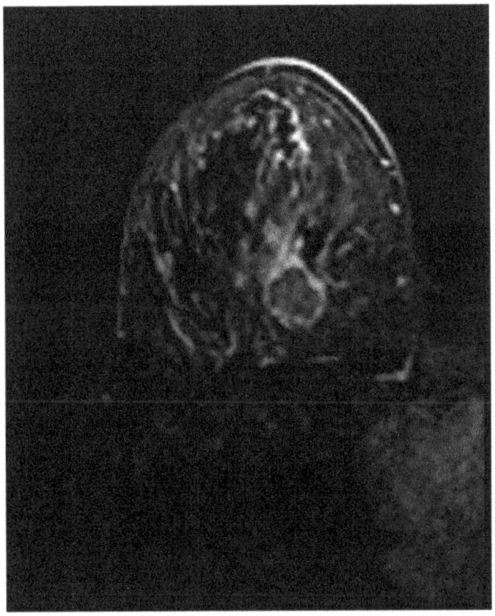

Figure 4. Triple negative breast cancer in a 32-year-old woman—rim enhancement on subtracted.

Table 2. Breast cancer molecular subtypes—association between pathology and MRI features.

Variable	Luminal A	Luminal B HER+	Luminal B HER−	Triple Negative	Her 2 Positive
Age mean	50.2	44.5	48.2	45	44.4
p-value	<0.01	0.02	0.57	0.04	0.36
Pathology					
NST	58	33	140	36	0
Other	31	6	28	5	7
p-value	**<0.01**	0.41	0.09	0.16	0.35
Breast density					
A	6	1	6	3	0
B	17	6	39	9	2
C	45	21	78	16	3
D	21	11	45	13	2
p-value	0.55	0.67	0.68	0.60	0.90

Table 2. *Cont.*

Variable	Luminal A	Luminal B HER+	Luminal B HER−	Triple Negative	Her 2 Positive
BPE					
Symmetric	79	34	141	23	5
Asymmetric	10	5	27	18	2
p-value	0.05	0.36	0.35	**<0.01**	0.61
Level					
Minimal	22	5	36	13	1
Mild	19	6	35	7	0
Moderate	43	23	85	13	5
Marked	5	5	12	8	1
p-value	0.56	0.29	0.60	**<0.01**	0.46
Mass					
Size, mean	21.3	29.9	27.7	25.1	37.28
p-value	**<0.01**	0.13	0.09	0.59	**0.02**
Shape					
Oval	24	11	45	21	2
Round	18	5	17	2	2
Irregular	40	16	81	12	3
Lobulated	3	6	15	6	0
p-value	0.02	0.50	0.25	**<0.01**	0.58
Margins					
Circumscribed	0	1	4	6	0
Irregular	60	27	104	31	5
Spiculated	25	10	50	4	2
p-value	0.13	0.94	0.24	**<0.01**	0.88
Enhancement					
Homogeneous	2	0	2	0	0
Heterogeneous	75	33	138	26	5
Rim	8	5	18	15	2
p-value	0.17	0.73	0.28	**<0.01**	0.55
ADC mean	0.80	0.80	0.77	0.82	0.80
Range	0.20–1.21	0.34–1.47	0.35–1.71	0.75–0.91	0.32–1.2
p-value	0.16	0.54	**<0.01**	0.12	0.87
Non-mass					NA *
Enhancement Distribution					
Focal	1	0	4	0	
Lineal	2	1	4	2	
Segmental	3	0	7	0	
Regional	2	1	4	0	
M. regions	1	1	0	0	
Diffuse	1	0	0	0	
p-value	0.66	0.29	0.28	0.31	
Type					
Homogeneous	6	0	1	1	
Heterogenous	4	3	7	1	
Clumped	0	0	9	0	
Cluster ring	0	0	1	0	
p-value	0.82	0.51	0.37	0.99	
ADC mean	0.88	0.86	0.86	NA *	
Range	0.74–1.02	1 case	0.60–1.29		
p-value	0.48	0.81	0.58		
Kinetic curves					
Persistent (1)	6	2	5	0	0
Plateau (2)	22	7	30	9	3
Wash-out (3)	61	30	133	32	4
p-value	0.10	0.82	0.30	0.39	0.31

Table 2. *Cont.*

Variable	Luminal A	Luminal B HER+	Luminal B HER−	Triple Negative	Her 2 Positive
ADC mean (mass + non-mass)	0.80	0.81	0.77	0.81	0.80
Range	0.20–1.21	0.34–1.47	0.35–1.71	0.34–1.47	0.75–0.91
p-value	0.12	0.59	**<0.01**	0.15	0.88
Total	89	39	168	41	7

* NA—impossibility to perform statistical tests; no dark septations were observed for the mass enhancement type; size is displayed in mm; range = minimum and maximum values.

The univariate and multivariate analysis remained statistically significant for TNBC (Table 3).

Table 3. Odds ratio for TNBC on univariate and multivariate analysis.

Variable	Odds Ratio (95%CI)	*p*-Value
Univariate analysis		
Shape – oval + round	2.16 (1.03–4.52)	0.04
Shape – lobulated	2.91 (0.99–8.51)	0.05
Margins – circumscribed	9.70 (2.81–33.45)	< 0.01
Enhancement – homogenous	0.00	0.99
Multivariate analysis		
Shape – oval + round	1.43 (0.64–3.19)	0.39
Shape – lobulated	2.64 (0.88–7.92)	0.08
Margins – circumscribed	9.12 (2.19–37.87)	< 0.01
Enhancement – homogenous	0.00	0.99

The lowest mean ADC values were obtained for luminal B HER2 negative tumours (0.77, *p*-value < 0.01). For a cut-off value of 0.88×10^{-3} mm^2/s^2, there was an AUC = 0.583 in predicting luminal B HER2 negative cases (*p*-value < 0.01), with a sensitivity of 79.17 (95% CI 72.2–85.0), specificity of 36.36 (95% CI 29.3–43.9), 54.3 positive predictive value and 64.6 negative predictive value (Figures 5 and 6).

We obtained almost perfect agreement between the ADC ROIs of the two readers (intraclass correlation = 0.85–0.91, weighted kappa = 0.78, standard error = 0.03, 95% CI 0.717–0.843).

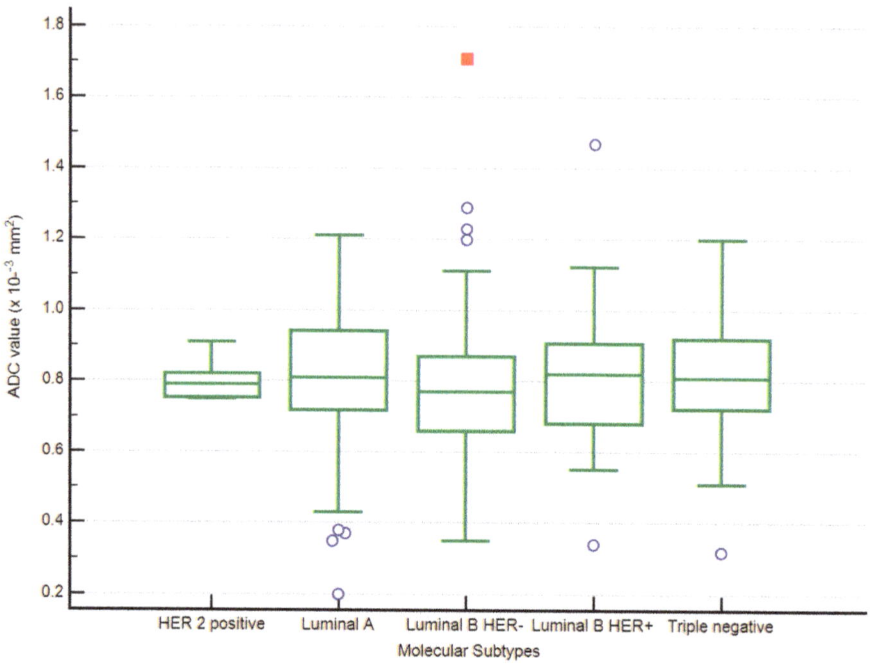

Figure 5. The distribution of ADC values in luminal B HER2 negative and other subtypes of BC.

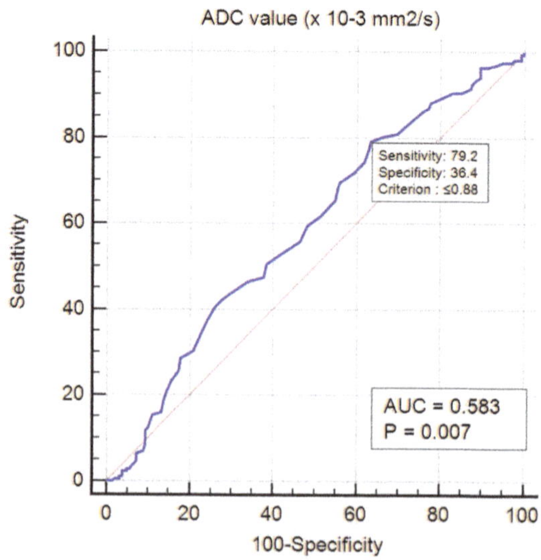

Figure 6. ROC curve for predicting luminal B HER2 negative BC.

3.2. *Associations between ER/PR Positive and Negative BC and MRI Features*

We further divided the entire study population into two groups according to positive or negative immunohistochemistry ER/PR status.

There were no differences for BC pathology type or breast density between the two groups (all *p*-values > 0.05).

Asymmetric BPE was associated with ER/PR negative group (p-value = 0.000), with no differences regarding the BPE level (p-value = 0.26).

Oval or round masses with circumscribed margins were associated with ER/PR negative group, while irregular masses with irregular or spiculated margins were associated with ER/PR positive tumours (p-values = 0.02 and < 0.01). All BC masses tested negative for ER/PR were associated with heterogeneous or rim enhancement (p-value < 0.01), while ER/PR positive tumours displayed rim enhancement only in 11% of the cases.

There were no differences regarding the ADC values or non-mass characteristics (distribution of enhancement, enhancement type) between ER/PR negative and positive tumours (all $p > 0.05$).

The persistent (type 1) kinetic curve was observed only in ER/PR positive tumours (4.6%) without reaching statistical significance (p-Value = 0.65) (Table 4).

Table 4. ER/PR positive and ER/PR negative cancers in association with MRI features.

Variable	ER/PR Positive	ER/PR Negative	*p*-Value
Pathology			
NST	231	43	0.06
Other	65	5	
Density			
A	13	3	
B	62	11	0.67
C	144	19	
D	77	15	
BPE			
			<0.01
Symmetric	254	28	
Asymmetric	42	20	
Minimal	63	14	
Mild	60	7	0.26
Moderate	151	18	
Marked	22	9	
Mass			
Size	26.08	26.95	0.14
Shape			
Oval	80	23	
Round	40	4	**0.02**
Irregular	137	15	
Lobulated	24	6	
Margins			
Circumscribed	5	6	**<0.01**
Irregular	191	36	
Spiculated	85	6	
Enhancement			
Homogeneous	4	0	**<0.01**
Heterogeneous	246	31	
Rim	31	17	
ADC mean	0.82	0.78	0.16

Table 4. *Cont.*

Variable	ER/PR Positive	ER/PR Negative	*p*-Value
Non-mass			
Enhancement Distribution			
Focal	5	0	0.31
Lineal	7	2	
Segmental	10	0	
Regional	7	0	
M. regions	2	0	
Diffuse	1	0	
Enhancement			
Homogeneous	16	1	0.95
Heterogenous	13	1	
Clumped	2	0	
Cluster ring	1	0	
Kinetic curves			
Persistent	13	0	0.26
Plateau	59	12	
Wash-out	224	36	
Total	281	48	

4. Discussion

In the current study, we found that T2WI, ADC values and DCE-based MRI features of BC differ between BC molecular subtypes.

In the univariate analysis, we found associations between shape, margins and enhancement pattern and TNBC. Lesions with oval/round margins or lobulated margins have more than two times higher odds of being TNBC when compared with the ones with an irregular shape. Circumscribed margins increased the odds of a lesion to be TNBC by 9.7 times. Moreover, the odds of TNBC status were approximately 4.3-fold greater in lesions with rim enhancement compared with the ones with heterogenous pattern. In the multivariate analysis, circumscribed margins and rim enhancement remained independently associated with TNBC.

As for the association of MRI features and ER/PR BC, our univariate analysis showed that lesions with irregular shapes are more likely to be ER/PR positive. Likewise, lesions with circumscribed margins have higher odds of being ER/PR positive. However, in the multivariate analysis, only circumscribed margins proved to be independently associated with ER/PR positive status, having an adjusted OR = 6.57 with a p-value < 0.01.

Few studies have included all five BC molecular subtypes in their analysis, and most papers have focused on differentiating ER/PR positive from ER/PR negative tumours to achieve statistical significance. In addition, the largest study population had 187 patients, and most authors reported association between BC molecular subtypes and one or two MRI based features (either T2WI features, or DWI/ADC or DCE features) [8–11]. We included 344 patients divided first into five BC molecular subtypes, and then into ER/PR positive and negative groups and analysed T2WI, ADC values and DCE features (enhancement pattern and kinetic curves) together.

Our results were consistent with other studies, which reported mass lesion type ($p < 0.001$), and smooth margins for TNBC [14–16]. Additionally, rim enhancement was reported in 76.8–80% of cases with TNBC, and has been proposed as a prognostic MRI feature for identifying TNBC [17,18]. In our study population, 15 cases (36.58%) presented with rim enhancement and reached statistical significance (p-value < 0.01).

It has been reported that ER/PR positive tumours (including luminal A, luminal B) are not associated with specific MRI features [18,19]. We found that irregular shaped masses with non-circumscribed margins were associated with the ER/PR positive group (all p-values < 0.05), in agreement with authors who reported that spiculated margins on ultrasound or mammography are associated with luminal tumours [20,21].

The persistent enhancement pattern (type I) has been associated in two studies with TNBC cases 14, 18, including in the young female population (< 30 years old). In our study, no TNBC mass had persistent kinetic enhancement, and in addition, rim enhancement was significantly associated with TNBC, consistent with Dogan et al. [15]. This could be explained by the heterogeneous group of TNBC, including special histology such as medullary carcinoma [22]. Another explanation of this heterogeneity is familial BC related to specific genetic alterations, which includes some cases of TNBC associated with benign imaging features including kinetics [23,24].

HER2 enriched tumours were found to have a kinetic curve pattern of more frequent "wash-out" or fast early rapid enhancement on MRI [25]. Our study did not reach statistical significance due to the small number of patients (only seven). However, we observed that this is associated with the ER/PR negative group, which includes the HER2 enriched patients.

There are conflicting results regarding the ADC/DWI, with authors reporting higher values in TNBC, and authors finding no association between ADC and BC molecular subtypes [19,26,27]. We found that luminal B HER2 negative was associated with the lowest ADC value (0.77×10^{-3} mm^2/s^2), which further predicts the BC molecular subtype with an accuracy of 0.583. Furthermore, we obtained almost perfect agreement between the two readers ADC ROIs, which supports the repeatability and reproducibility of this fast and easy method compared to the more technical ADC histogram making.

BPE may play a role as an imaging bridge to molecular BC subtypes allowing an additional risk stratification in breast MRI. Luminal B HER2 negative tumours may predominate in mild BPE, and TNBC in patients with marked BPE [28–30]. We found that asymmetric and moderate/marked BPE was associated with TNBC, and further with ER/PR negative group. However, no significant association was present between moderate/marked BPE and HER2 status or basal tumours (TNBC subtype) [2] and further studies are needed to make a conclusion.

Recent papers use radiomics-based MRI and machine learning in order to distinguish different BC molecular subtypes [31–34]. Leithner et al. reported 81–89% accuracy in distinguishing luminal A from luminal B, luminal B from triple-negative, luminal B from all others, and HER2-enriched from all others [35]. However, these techniques are time-consuming and the lack of standardized protocols remains a critical issue preventing entry into clinical practice.

This study had some limitations. First, this was a retrospective study, with one experienced breast imager reader that was aware of the presence of BC. Second, it was a single-centre study, which may weaken the statistical power of the results. Third, we did not include some features such as high T2WI signal, or DWI patterns (homogeneous or heterogeneous), as they are more prone to inter-reader variability. Fourth, even though this is the largest study on BC molecular subtypes, a larger prospective study needs to be conducted to validate the present results.

As future directions, where breast MRI will be increasingly used preoperatively, it could help change or adapt the patient management by suggesting a particular molecular type and may even indicate further examination (such as genetic testing for TNBC).

5. Conclusions

We found that BC molecular subtypes can be associated with multiparametric MRI features, especially TNBC, and that ER/PR positive tumours differ from ER/PR negative cancers. ADC can help predict the luminal B HER2 negative cases and ADC values obtained by ROI method is still a reliable method.

Author Contributions: Conceptualization: M.S., R.P., B.B., M.D., D.F., B.F., D.E., S.M.D. and A.C.; Methodology: M.S., R.P., B.B., S.M.D. and A.C.; Validation: R.P., B.B., D.F., S.M.D. and A.C.; Formal analysis, M.S., R.P., B.B., M.D., D.F., D.E. and A.C.; Investigation, M.S., A.P., M.D., D.F., B.F., D.E. and A.C.; Resources, M.S., A.P., D.F., B.F., D.E. and A.C.; Data curation, M.S., R.P., B.B., A.P., M.D., D.F., S.M.D. and A.C.; Writing—original draft, M.S., R.P., B.B., A.P., M.D., D.F., B.F., D.E., S.M.D. and A.C.; Supervision, R.P., M.D., S.M.D. and A.C.; Project administration, R.P., B.B., M.D., B.F., D.E., S.M.D. and A.C. All authors have read and agreed to the published version of the manuscript.

Funding: This research received no external funding.

Institutional Review Board Statement: The study was conducted in accordance with the Declaration of Helsinki, and approved by the Ethics Committee of Medimages Breast Center (NR10/15092022, approval date: 15 September 2022).

Informed Consent Statement: Informed consent was obtained from all subjects involved in the study.

Conflicts of Interest: The authors declare no conflict of interest.

References

1. Perou, C.M.; Sørlie, T.; Eisen, M.B.; van de Rijn, M.; Jeffrey, S.S.; Jeffrey, C.A.; Pollack, J.R.; Ross, D.T.; Johnsen, H.; Akslen, L.A.; et al. Molecular portraits of human breast tumours. *Nature* **2000**, *406*, 747–752. [CrossRef] [PubMed]
2. Goldhirsch, A.; Winer, E.P.; Coates, A.S.; Gelber, R.D.; Piccart-Gebhart, N.; Thürlimann, B.; Senn, H.J. Personalizing the treatment of women with earyely breast cancer: Highlights of the St Gallen International expert consensus on the primary therapy of early breast cancer 2013. *Ann. Oncol.* **2013**, *24*, 2206–2223. [CrossRef] [PubMed]
3. Rouzier, R.; Perou, C.M.; Symmans, W.F.; Ibrahim, N.; Cristofanilli, M.; Anderson, K.; Hess, K.R.; Stec, J.; Ayers, M.; Wagner, P.; et al. Breast cancer molecular subtypes respond differently to preoperative chemotherapy. *Clin. Cancer Res.* **2005**, *11*, 5678–5686. [CrossRef] [PubMed]
4. Anderson, W.F.; Rosenberg, P.S.; Prat, A.; Perou, C.M.; Sherman, M.E. How many etiological subtypes of breast cancer: Two, three, four, or more? *J. Natl. Cancer Inst.* **2014**, *106*, dju165. [CrossRef] [PubMed]
5. Anderson, K.N.; Schwab, R.B.; Martinez, M.E. Reproductive risk factors and breast cancer subtypes: A review of the literature. *Breast Cancer Res. Treat.* **2014**, *144*, 1–10. [CrossRef]
6. Rakha, E.A.; Reis-Filho, J.S.; Ellis, I.O. Combinatorial biomarker expression in breast cancer. *Breast Cancer Res. Treat.* **2010**, *120*, 293–308. [CrossRef] [PubMed]
7. Mann, R.M.; Kuhl, C.K.; Kinkel, K.; Boetes, C. Breast MRI: Guidelines from the European Society of Breast Imaging. *Eur. Radiol.* **2008**, *18*, 1307–1318. [CrossRef]
8. Sutton, E.J.; Dashevsky, B.Z.; Oh, J.H.; Veeraraghavan, H.; Apte, A.P.; Thakur, S.; Morris, E.; Deasy, J. Breast cancer molecular subtype classifier that incorporates MRI features. *J. Magn. Reson. Imaging* **2016**, *44*, 122–129. [CrossRef]
9. Kettunen, T.; Okuma, H.; Auvinen, P.; Sudah, M.; Tiainen, S.; Sutela, A.; Masarwah, A.; Tammi, M.; Tammi, R.; Oikari, S.; et al. Peritumoral ADC values in breast cancer: Region of interest selection, associations with hyaluronan intensity, and prognostic significance. *Eur. Radiol.* **2020**, *30*, 38–46. [CrossRef] [PubMed]
10. Cho, N. Imaging features of breast cancer molecular subtypes: State of the art. *J. Pathol. Transl. Med.* **2021**, *55*, 16–25. [CrossRef]
11. Uematsu, T.; Kasami, M.; Yuen, S. Triple-negative breast cancer: Correlation between MR imaging and pathologic findings. *Radiology* **2009**, *250*, 638–647. [CrossRef] [PubMed]
12. Yuen, S.; Monzawa, S.; Yanai, S.; Matsumoto, H.; Yata, Y.; Ichinose, Y.; Deai, T.; Hashimoto, T.; Tashiro, T.; Yamagami, K. The association between MRI findings and breast cancer subtypes: Focused on the combination patterns on diffusion-weighted and T2-weighted images. *Breast Cancer* **2020**, *27*, 1029–1037. [CrossRef] [PubMed]
13. D'Orsi, C.J.; Sickles, E.A.; Mendelson, E.B.; Morris, A.; Creech, E.W.; Butler, F.P.; Wiegmann, P.G.; Chatfield, B.M.; Meyer, W.L.; Wilcox, A.P. *ACR BI-RADS Atlas, Breast Imaging Reporting and Data System*, 5th ed.; American College of Radiology: Reston, VA, USA, 2013.
14. Yeh, R.H.; Yu, J.C.; Chu, C.H.; Ho, C.L.; Kao, H.W.; Liao, G.S.; Chen, H.W.; Kao, W.Y.; Yu, C.P.; Chao, T.Y. Distinct MR Imaging Features of Triple-Negative Breast Cancer with Brain Metastasis. J Neuroimaging. *J. Neuroimaging.* **2015**, *25*, 474–481. [CrossRef] [PubMed]
15. Dogan, B.E.; Turnbull, L.W. Imaging of triple-negative breast cancer. *Ann. Oncol.* **2012**, *23* (Suppl. 6), vi23–vi29. [CrossRef]
16. Bae, M.S.; Shin, S.U.; Ryu, H.S.; Han, W.; Im, S.A.; Park, I.A.; Noh, D.Y.; Moon, W.K. Pretreatment MR Imaging Features of Triple-Negative Breast Cancer: Association with Response to Neoadjuvant Chemotherapy and Recurrence-Free Survival. *Radiology* **2016**, *281*, 392–400. [CrossRef] [PubMed]
17. Teifke, A.; Behr, O.; Schmidt, M.; Victor, A.; Vomweg, T.W.; Thelen, M.; Lehr, H.-A. Dynamic MR imaging of breast lesions: Correlation with microvessel distribution pattern and histologic characteristics of prognosis. *Radiology* **2006**, *239*, 351–360. [CrossRef]
18. Huang, J.; Lin, Q.; Cui, C.; Fei, J.; Su, X.; Li, L.; Ma, J.; Zhang, M. Correlation between imaging features and molecular subtypes of breast cancer in young women ($\leq$30 years old). *Jpn J. Radiol.* **2020**, *38*, 1062–1074. [CrossRef]

19. Yetkin, D.I.; Akpınar, M.G.; Durhan, G.; Demirkazik, F.B. Comparison of clinical and magnetic resonance imaging findings of triple-negative breast cancer with non-triple-negative tumours. *Pol. J. Radiol.* **2021**, *86*, e269–e276. [CrossRef]
20. Irshad, A.; Leddy, R.; Pisano, E.; Baker, N.; Lewis, M.; Ackerman, S.; Campbell, A. Assessing the role of ultrasound in predicting the biological behavior of breast cancer. *AJR Am. J. Roentgenol.* **2013**, *200*, 284–290. [CrossRef]
21. Taneja, S.; Evans, A.J.; Rakha, E.A.; Green, A.R.; Ball, G.; Ellis, I.O. The mammographic correlations of a new immunohistochemical classifcation of invasive breast cancer. *Clin. Radiol.* **2008**, *11*, 1228–12355. [CrossRef]
22. Pintican, R.; Duma, M.; Chiorean, A.; Fetica, B.; Badan, M.; Bura, V.; Szep, M.; Feier, D.; Dudea, S. Mucinous versus medullary breast carcinoma: Mammography, ultrasound, and MRI findings. *Clin. Radiol.* **2020**, *75*, 483–496. [CrossRef] [PubMed]
23. Schrading, S.; Kuhl, C.K. Mammographic, US, and MR imaging phenotypes of familial breast cancer. *Radiology* **2008**, *246*, 58–70. [CrossRef] [PubMed]
24. Pintican, R.M.; Chiorean, A.; Duma, M.; Feier, D.; Szep, M.; Eniu, D.; Goidescu, I.; Dudea, S. Are Mutation Carrier Patients Different from Non-Carrier Patients? Genetic, Pathology, and US Features of Patients with Breast Cancer. *Cancers* **2022**, *14*, 2759. [CrossRef]
25. Elias, S.G.; Adams, A.; Wisner, D.J.; Esserman, L.J.; Veer, L.J.V.; Mali, W.P.; Gilhuijs, K.G.; Hylton, N.M. Imaging features of HER2 overexpression in breast cancer: A systematic review and meta-analysis. *Cancer Epidemiol Biomark. Prev.* **2014**, *23*, 1464–1483. [CrossRef]
26. Sharma, U.; Sah, R.G.; Agarwal, K.; Parshad, R.; Seenu, V.; Mathur, S.R.; Hari, S.; Jagannathan, N.R. Potential of Diffusion-Weighted Imaging in the Characterization of Malignant, Benign, and Healthy Breast Tissues and Molecular Subtypes of Breast Cancer. *Front. Oncol.* **2016**, *6*, 126. [CrossRef]
27. Kim, S.-Y.; Kim, E.-K.; Moon, H.J.; Yoon, J.H.; Koo, J.S.; Kim, S.G.; Kim, M.J. Association among T2 signal intensity, necrosis, ADC and Ki-67 in estrogen receptor-positive and HER2-negative invasive ductal carcinoma. *Magn Reson. Imaging* **2018**, *54*, 176–182. [CrossRef]
28. Dilorenzo, G.; Telegrafo, M.; La Forgia, D.; Ianora, A.A.S.; Moschetta, M. Breast MRI background parenchymal enhancement as an imaging bridge to molecular cancer sub-type. *Eur. J. Radiol.* **2019**, *113*, 148–152. [CrossRef]
29. Bignotti, B.; Signori, A.; Valdora, F.; Rossi, F.; Calabrese, M.; Durando, M.; Mariscotto, G.; Tagliafico, A. Evaluation of background parenchymal enhancement on breast MRI: A systematic review. *Br. J. Radiol.* **2017**, *90*, 20160542. [CrossRef]
30. Ha, R.; Mango, V.; Al-Khalili, R.; Mema, E.; Friedlander, L.; Desperito, E.; Wynn, R.T. Evaluation of association between degree of background parenchymal enhancement on MRI and breast cancer subtype. *Clin. Imaging* **2018**, *51*, 307–310. [CrossRef]
31. Xie, T.; Zhao, Q.; Fu, C.; Bai, Q.; Zhou, X.; Li, L.; Grimm, R.; Liu, L.; Gu, Y.; Peng, W. Differentiation of triple-negative breast cancer from other subtypes through whole-tumor histogram analysis on multiparametric MR imaging. *Eur. Radiol.* **2019**, *29*, 2535–2544. [CrossRef]
32. Song, L.; Lu, H.; Yin, J. Preliminary study on discriminating HER2 2+ amplification status of breast cancers based on texture features semi-automatically derived from pre-, post-contrast, and subtraction images of DCE-MRI. *PLoS ONE* **2020**, *15*, e0234800. [CrossRef] [PubMed]
33. Mazurowski, M.A.; Zhang, J.; Grimm, L.; Yoon, S.C.; Silber, J.I. Radiogenomic analysis of breast cancer: Luminal B molecular subtype is associated with enhancement dynamics at MR imaging. *Radiology* **2014**, *273*, 365–372. [CrossRef] [PubMed]
34. Grimm, L.J.; Zhang, J.; Mazurowski, M.A. Computational approach to radiogenomics of breast cancer: Luminal A and luminal B molecular subtypes are associated with imaging features on routine breast MRI extracted using computer vision algorithms. *J. Magn. Reason. Imaging* **2015**, *42*, 902–907. [CrossRef]
35. Leithner, D.; Horvat, J.V.; Marino, M.A.; Bernard-Davila, B.; Jochelson, M.S.; Ochoa-Albiztegui, R.E.; Martinez, D.; Morris, E.A.; Thakur, S.; Pinker, K. Radiomic signatures with contrast-enhanced magnetic resonance imaging for the assessment of breast cancer receptor status and molecular subtypes: Initial results. *Breast Cancer Res.* **2019**, *21*, 106. [CrossRef]

medicina

MDPI

Article

Characterization of Microscopic Multicellular Foci in Grossly Normal Renal Parenchyma of Von Hippel-Lindau Kidney

Nayef S. Al-Gharaibeh [1,2,*], Sharon B. Shively [3,4] and Alexander O. Vortmeyer [1]

1 Department of Pathology, Indiana University School of Medicine, 350 W 11th Street, Suite 4034, Indianapolis, IN 46202, USA
2 Department of Physiology and Biochemistry, Jordan University of Science and Technology, 22110 Al Ramtha, Jordan
3 Surgical Neurology Branch, National Institute of Neurological Disorders and Stroke, National Institutes of Health, Bethesda, MD 20894, USA
4 Department of Molecular Medicine, Institute for Biomedical Sciences, The George Washington University, Washington, DC 5246, USA
* Correspondence: nayef@just.edu.jo

Abstract: *Background and Objectives*: This study aims to describe the earliest renal lesions in patients with von Hippel-Lindau (VHL) disease, especially the multicellular microscopic pathologic events, to get information into the genesis of renal neoplasms in this condition. *Materials and Methods*: Multicellular events were identified, and 3dimensional reconstruction was performed in grossly normal kidney parenchyma from VHL disease patients by using H&E-stained slides previously prepared. *Results:* The lesions were measured and the volume of clusters was calculated. Immunohistochemistry was performed for downstream *HIF*-target protein carbonic anhydrase 9 (CAIX) as well as CD34 for assessment of angiogenesis. We divided lesions into four types according to lesion height/size. The number of lesions was markedly decreased from lesion 1 (smallest) to lesion 2, then from lesions 2 to 3, and again from lesion 3 to 4. Distribution was highly consistent in the four cases, and the same decrement pattern was seen in all blocks studied. The volumes of clusters were measured and divided into three categories according to their volume. The most frequent pathologic event in VHL kidneys was category 1 (smallest volume), then category 2, and then category 3. *Conclusion:* We demonstrate that tracking histologic and morphologic changes in 3 dimensions of multicellular microscopic pathologic events enabled us to confirm a protracted sequence of events from smaller to larger cellular amplification events in VHL kidney.

Keywords: 3dimensional reconstruction; von Hippel-Lindau disease; renal clear cells; kidney cancer; precursor structures

Citation: Al-Gharaibeh, N.S.; Shively, S.B.; Vortmeyer, A.O. Characterization of Microscopic Multicellular Foci in Grossly Normal Renal Parenchyma of Von Hippel-Lindau Kidney. *Medicina* **2022**, *58*, 1725. https://doi.org/10.3390/medicina58121725

Academic Editors: Nicolae Crisan and Călin Căinap

Received: 26 October 2022
Accepted: 16 November 2022
Published: 24 November 2022

1. Introduction

Von Hippel-Lindau (VHL) disease was first described in 1926 [1]. VHL disease is an autosomal dominant inherited familial tumor syndrome caused by germline mutation of the VHL tumor suppressor gene [2]. The VHL gene is a tumor suppressor gene on the short arm of chromosome 3 (3p25–26) [3]. The VHL protein (pVHL) is expressed in fetal and adult human tissues. The lack of this protein induces profound intracellular metabolic changes that closely resemble changes observed in oxidative stress [4]. Knudson proposed in 1971 a two-hit hypothesis for VHL disease. The first hit, germline mutation in the VHL gene, leads to the inactivation of a tumor suppressor allele. The second hit refers to somatic mutation, which occurs after the first hit and leads to the inactivation of the second allele [5]. Loss of VHL function induces negative regulation of the two alpha-subunits of hypoxia-inducible factor (HIF). HIF-regulated genes include vascular endothelial growth factor (VEGF), carbonic anhydrase IX (CAIX), erythropoietin (EPO), platelet-derived growth factor (PDGF), and the glucose-transporter 1 (GLUT-1) [3,6]. Cells with a loss of VHL-mediated HIF degradation express higher HIF alpha proteins, VEGFA, and CAIX. Loss of pVHL results in a lower breakdown and higher concentration of HIF, which leads to more increased secretion of pro-angiogenic HIF target proteins [2,6]. HIF is significant for tumor persistence because it stimulates angiogenesis, which results from increased levels of VEGF, PDGF-beta, or both. VEGF and PDGF-beta are necessary for the proliferation of endothelial cells and pericytes, respectively [7].

Patients affected by VHL disease develop specific types of heavily vascularized tumors in a highly selective subset of organs. Multiple tumors occur frequently. These include retinal, cerebellar, brainstem and spinal cord hemangioblastoma, microcystic cystadenoma, epididymal cystadenoma, pheochromocytoma and paraganglioma, endolymphatic sac tumor, as well as renal clear cell carcinoma [8–14].

Renal clear cell carcinoma (RCCC) shares histopathologic features with hemangioblastic central nervous system tumors and epididymal tumors. It is characterized by the proliferation of clear cells with abundant reactive vascularization [6,15]. Mandriota et al. showed that HIF activation could be identified in both early and late lesions in kidneys of VHL patients. In addition, there was an increase in vascularization around tubules containing CAIX-immunoreactive cells [9]. Loss of heterozygosity (LOH) on chromosome 3p has been reported in macroscopic RCCC of VHL and sporadic tumors [16–19].

Although clinical, radiological, and pathological data suggested that the precursors of RCCC in patients with VHL disease could be a benign or atypical renal lesion [20–22], the definitive transformation has not been well documented in the literature thus far.

Lubensky et al. showed that 25 of 26 clear-cell renal lesions lost the wild-type allele, and that was the first molecular evidence that LOH of the normal VHL disease gene occurs in benign and atypical clear-cell cysts. This may represent an early event in the development and progression of RCCC in VHL [13].

Characterization of precursor structures has provided more precise insight into the anatomic and cytologic origin of neoplastic processes associated with VHL disease. Precursor structures are far more numerous than frank tumors. VHL inactivation after the "second hit" is necessary but insufficient to develop mass-forming tumors [11,23]. Comparative molecular analysis of tumors and precursor structures may be helpful to identify a "third hit" promoting tumorigenesis from precursor material [11].

This study characterizes the multicellular microscopic pathologic events using a novel 3-dimensional algorithm consisting of methodological sampling and histological tracking. We tracked lesions in 3 dimensions to confirm their relationship to each other. This algorithm allows us to observe that microscopic clear cell proliferations follow a predictable pattern covering a morphological spectrum of histologic lesions from an isolated clear cell to potentially tumorigenic clear cell proliferation.

2. Materials and Methods

2.1. Ethics Statement and Sample Collection

Tissue has been obtained from seven kidneys from seven different patients obtained during post-mortem examination and used for this study. All tissues underwent fixation in 10% buffered formalin. Tissue collection included samples from four kidneys from patients with confirmed germline mutations of the VHL gene and established clinical diagnosis of VHL disease (designated as "VHL kidneys"). All VHL patients had additional VHL disease-associated tumors including hemangioblastomas, epididymal cystadenomas, endolymphatic sac tumors, and microcystic adenomas of the pancreas. Three patients were male and one female. The causes of death were pneumonia in two cases, metastatic renal cell carcinoma, and intracerebral hemorrhage.

The VHL patients' ages at the time of demise were between 50 and 65 years (average 58.5 years). Three control kidneys from patients with sporadic renal cell carcinoma, aged between 55 to 70 years (average 63.33 year), were also used for this study designated as "sporadic kidneys".

All kidneys were sectioned randomly into maximally 10 cm × 1.5 cm × 1.0 cm columnar cuboids. The obtained cuboids of each kidney were inspected, and cuboids showing any grossly visible tumor were rejected. Of the remaining cuboids, the cuboid with the least grossly visible pathologic changes was identified and selected. For each kidney under study, one cuboid was selected ($n = 7$). Cuboids were then sliced into 0.3 cm × 1.5 cm × 1.0 cm blocks and a maximum of 30 blocks per cuboid was created. All tissue blocks without grossly visible pathologic change were paraffin embedded. An initial section was taken from each block and stained with hematoxylin and eosin (H&E) for histologic evaluation and identification for 3D reconstruction.

2.2. Immunohistochemistry

All immunohistochemical studies were stained using the Dako Flex-hrp system and developed with DAB. CAIX (Cell Marque 379R-18) underwent antigen retrieval in a high pH buffer followed by 20 min incubation with the primary mouse monoclonal anti-human CAIX antibody. CD34, a mouse monoclonal from Agilent/Dako IR632, underwent antigen retrieval in low pH, then the antibody was incubated for 15 min, the linker 10 min, and the Flex-hrp for 10 min. HIF is a rabbit polyclonal antibody from Novus Biological (NB100-122). The Antigen retrieval was low pH. The antibody was diluted 1:50 and incubated 30 min. All immunohistochemical studies were developed with DAB for 10 min and counterstained with hematoxylin. All H&E and immunohistochemical slides were digitized and analyzed using ImageScope, version 12.1.0.5029. Vewport 12.1.3. Copyright © Aperio Technologies, Inc. 2003–2013.

2.3. 3Dimensional Analysis

We recently described an algorithm for tissue procurement, processing, and 3D analysis of tissue blocks of interest [23]. The histopathologic examination was performed on normal-appearing renal parenchyma with minimal gross pathology and relatively well-preserved histologic details. We investigated 10 serially sectioned VHL kidney blocks from 4 different patients and three serially sectioned control blocks (normal kidney tissue of patients with sporadic carcinoma). All the H&E-stained slides were scanned and digitized at 50-micron intervals. We investigated and characterized the lesions which revealed clear cell amplification. Since the microscopic clear cell proliferations follow a predictable pattern, all pathologic clear cell events in this study have been divided into the following types:

1-Chain: multiple adherent clear cells that fill more than 50% of a tubular circumference.

2-Cluster: Collection of clear cells expanding and filling a renal tubule.

3-Complex cluster: Multiple profiles of intratubular aggregates of clear cells next to each other, with possible growth outside the tubules.

2.4. First Quantitative Approach Applied

The quantification was confined to chains, clusters, and complex clusters. First, we annotated all lesions with the letter C with subsequent numbers (x) in each investigated block (C1; C2; C3, and so on). Subsequently, we followed those lesions into the following H&E-stained slides which are 50 microns deeper than the previous slide. If the lesion (C1) still showed in that second slide, the lesion retained the same annotation of (C1), but we annotated that lesion as (C1-2), and if the other lesion (C2) still showed in first, second and third H&E-stained slide, it was annotated as (C2-3) and so on (where the number after the dash designates the number of H&E-stained slides in which the lesion was detectable).

2.5. Second Quantitative Approach Applied

Knowing that the distance between H&E-stained slides was 50 microns we could calculate the lesion's height within the lumen of the renal tubule and divide them as follows:

Lesion 1 is equivalent to Cx-1 and is less than 50 microns in height in the tubular lumen.

Lesion 2 is equivalent to Cx-2 and is less than 100 microns in the tubular lumen.

Lesion 3 is equivalent to Cx-3 and is less than 150 microns in the tubular lumen.

Lesion 4 is equivalent to all lesions Cx-4 and exceeds 150 microns in height in the tubular lumen. Figure 1 illustrates the appearance of these lesions within the tissue block and how they would appear on different H&E-stained levels.

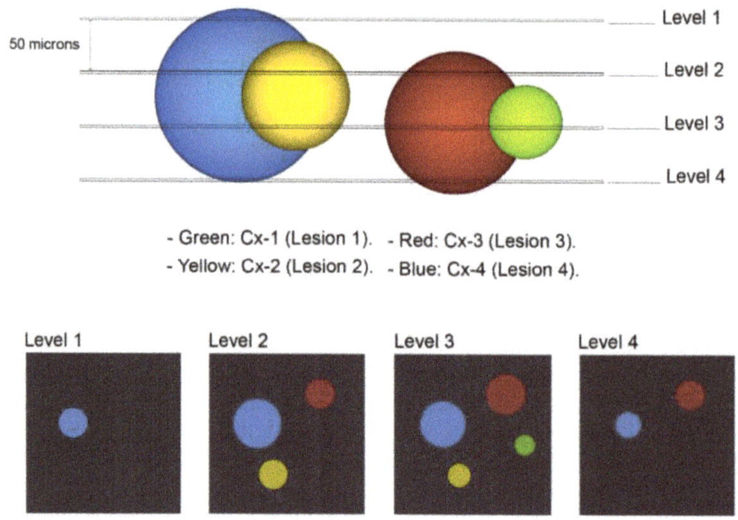

Figure 1. Diagram of VHL kidney block containing pathologic events of different sizes which are displayed in different colors. For the first quantitative approach, lesion annotation was performed, with retaining the annotation of a lesion that reoccurs in the next deeper level. The histologic presentation of different H&E-stained section levels into the block is shown (annotated as "Level 1", "Level 2", "Level 3" and "Level 4"). The second quantitative approach was concerned with identification of the exact number of levels lesions were present in: Green: Cx-1 (lesion 1): all chains or clusters which were seen on one H&E-stained slide only. Yellow: Cx-2 (lesion 2): all chains or clusters which were seen on two serially H&E-stained slides. Red: Cx-3 (lesion 3): all chains or clusters which were seen on three serially H&E- stained sections. Blue: Cx-4 (lesion 4): all chains or clusters which were seen on four serially H&E- stained sections and above.

2.6. Third Quantitative Approach Applied

We applied the following equations only to cluster lesions, and it represents a measurement of the volumes of lesions in cubic mm. The volume was calculated using a conical frustum as a model. The studied lesions were presumed to have a frustum shape. We measured the lesions by drawing a circle around the center of the lesion in a way that it approximated the area of the lesion. This circle annotation surface area was used as an equivalent to the lesion's surface area, and then we measured the radius of that circle. Subsequently, we measured the radius of the same lesion in the 50 microns deeper from the previous slide Figure 2A is an example of the radius measurement at three different H&E-stained levels that are 50 microns apart. Finally, we used the formula to calculate the Frustum volume [Volume of a conical frustum: $V = (1/3) \times \pi \times h \times [r1^2 + r2^2 + (r1 \times r2)]$, and we did add to the calculated frustum volumes two pyramid volumes, one on the top of the frustum and another on the bottom, with the assumption that all our lesions extend above and down by 25 µm, using the formula of pyramid volume ($V = (1/3) \times \pi \times h \times r^2$). The lesion 2 volume (Cx-2) = one frustum volume + 2 pyramids volumes. Lesion 3 (Cx-3) volume= 2 frustum volumes + 2 pyramids volumes. Figure 2B is a diagram explaining the volume measurement technique applied in this study.

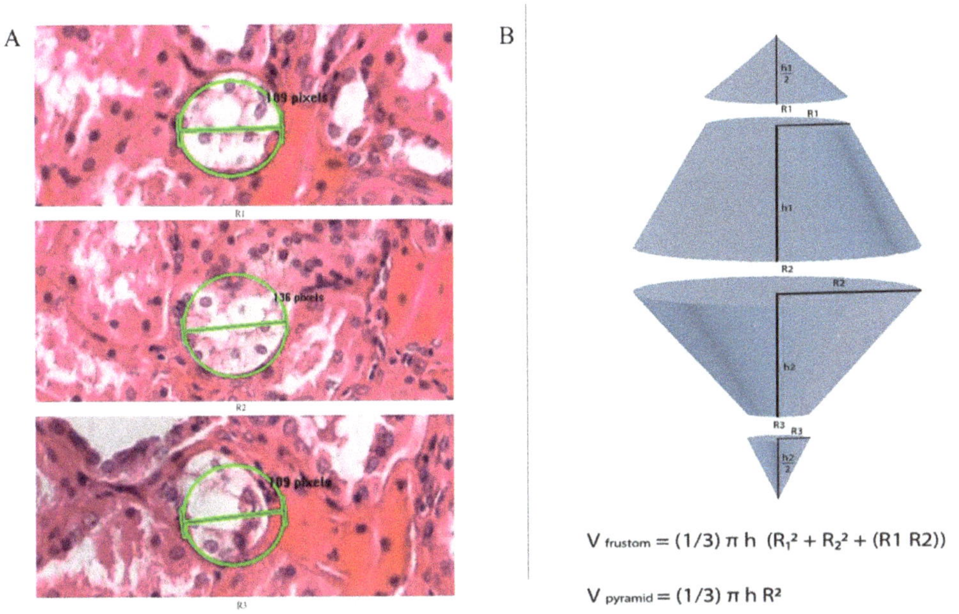

Figure 2. (**A**) Representation of H&E-stained slide illustrating diameter (2r) measurement for cluster lesion. First, we drew a circle over the lesion of approximately same area. Then we measured the radius of that circle which represents the lesion's radius, then measured the radius of this lesion in the tenth serial section, which is 50 microns from the previous slide. (**B**) Diagram illustrates the formula to measure the Frustum volume, $V = [(1/3) \times \pi \times h \times (r1^2 + r2^2 + (r1 \times r2)]$; we added the volume of two pyramids, one on the top and the other on the bottom of the cluster by using the formula of pyramid volume ($V = (1/3) \times \pi \times h \times r^2$). Volumes of lesion 2 (Cx-2) were measured by measuring the volume of one frustum R1 and R2, plus two pyramid volumes, where R1 and R2 are the radius of above and bottom pyramids respectively of 25 microns height. Volumes of lesion3 (Cx-3) were measured by calculating the volume of two frustums, plus two pyramid volumes, where R1 and R3 are the radius of above and bottom pyramids respectively of 25 microns height.

3. Results

The previous study [23] showed that most pathologic events in VHL patients were single clear cells, followed in lower frequency by chains, and the least frequent were complex or invasive clusters. In our control cases (normal kidney parenchyma of sporadic patients with resected renal cell carcinoma) only six to 12 isolated clear cells were identified. We did not detect any clear cell clusters in our three sporadic, serially sectioned control kidneys.

3.1. Lesions Account

After inspecting and annotating all the lesions, the total number of lesions visualized and annotated in the H&E-slides from 10 blocks of 4 VHL patients studied were 1194 chains, clusters, and complex clusters. The studied consecutive sections had a distance of 50 micrometer from each other. Those are subsequently subclassified as:

Lesion 1 (Cx-1): 958 chains or clusters (lesion present in one section only, not in any consecutive section).

Lesion 2 (Cx-2): 154 chains or clusters (lesion present in two consecutive sections only).

Lesion 3 (Cx-3): 53 chains or clusters (lesion present in three consecutive sections only).

Lesion 4 (Cx-4): 29 chains or clusters (lesion present in four consecutive sections only).

The total number of lesions studied in each block of all VHL patients is demonstrated in Table 1. The frequency pattern of lesions (the most frequent is lesion 1, then 2, 3, and the least are 4) is illustrated in Figure 3.

Table 1. Number of lesion repetition in blocks. The number in lower column represents the total number of all lesions annotated and analyzed (chains, clusters, and invasive clusters) in 10 blocks from 4 cases of VHL patients.

	Case 385			Case 393	Case 389			Case 386	
Block #7	**Block #24**	**Block #25**	**Block #13**	**Block #17**	**Block #1**	**Block #57**	**Block #12**	**Block #25**	**Block #26**
270	87	97	183	219	80	54	49	78	77

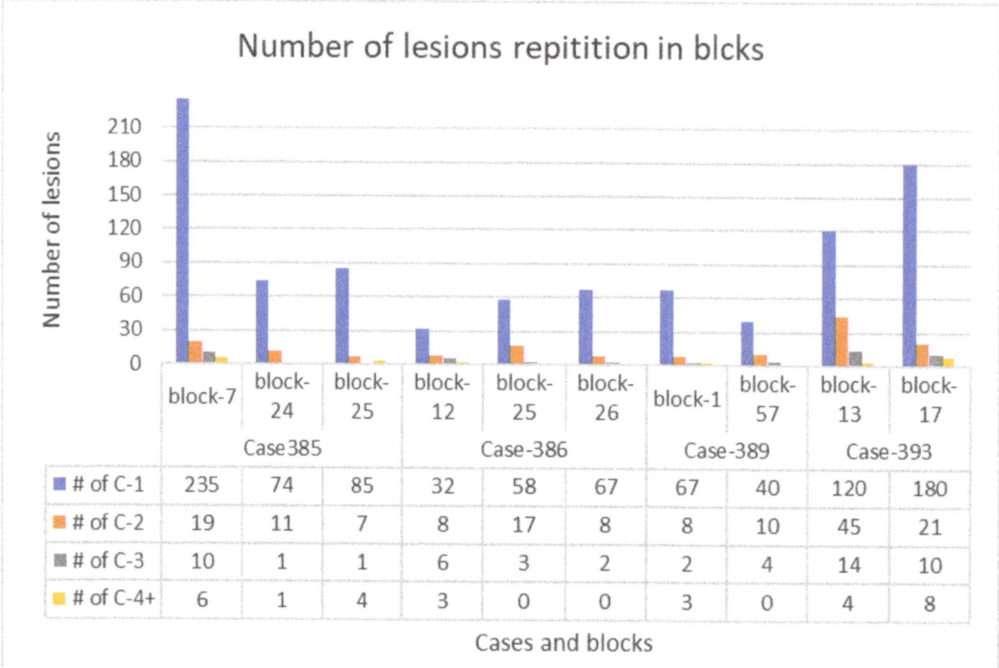

Figure 3. A lesion column chart showing the number of all types of lesions in each block from the four VHL patients. Blue columns—lesion 1 (Cx-1), orange columns—lesion 2 (Cx-2), gray columns—lesion 3 (Cx-3), and yellow columns—lesion 4 (Cx-4).

Table 2 summarizes the number of lesions and their percentages in the studied blocks.

Table 2. Prevalence of multicellular microscopic lesions. Prevalence of studied multicellular microscopic lesions in the normal cortex of kidney of patients with VHL disease.

Type of Lesions	Lesion 1	Lesion 2	Lesion 3	Lesion 4	Total
Number of lesions in all slides	958	154	53	29	1194
Percentage of each type from total	80.2%	12.9%	4.4%	2.5%	100%

3.2. Immunohistochemical Expression

Regarding the immunohistochemical studies, sixteen slides from seven blocks of four patients with VHL disease were stained with CAIX antibody. CAIX-positive cells were observed in all clear cell changes in all stained slides regardless of the size (small chains, big chains, clusters, and invasive clusters). Figure 4A shows representative images of the expression of CAIX in variably sized lesions. A CD34 stain was performed as an endothelial marker on sixteen slides from three blocks of two different VHL patients to detect the degree of vascularization. We observed a substantial increase in the extent of vascularization around tubules containing clear cells. The vessels were most evident around complex lesions, with a gradual decrease in the number of observed vessels in clusters followed Figure 4B. This marked increase in vascularization around clear cell lesions (CAIX-positive foci) shows that they promote angiogenesis.

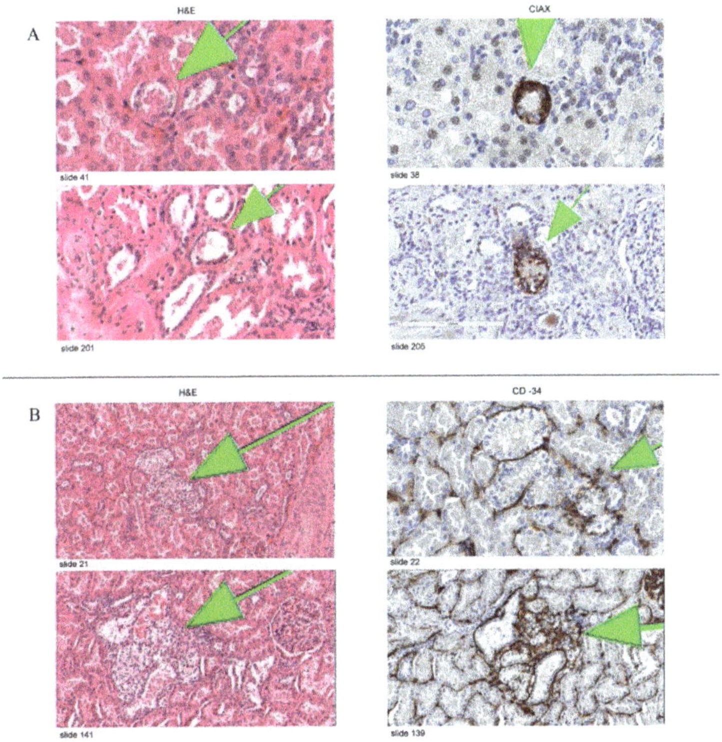

Figure 4. (**A**) Representative images of CAIX staining in VHL kidneys. The left column shows two lesions on H&E-stained slides, and the right column shows a serial section of the same lesion stained with anti-CAIX antibody. (**B**) Representative images of CD34 staining in VHL kidneys. Figure demonstrating a gradual increase in the density of the vasculature around the clear cell changes with the increment of the volume.

3.3. *Cluster Volumes*

We calculated the volumes of all clusters examined, which are 119 in total. We found that they range in volume from 0.000034–0.0671 mm^3. We subsequently subcategorized them into three groups from smallest volume to largest volume. In category one, there were 105 lesions (volume range 0.000034–0.00096 mm^3). Category two, there were 10 lesions (volume range 0.00131–0.00956 mm^3). Category three had four lesions (volume range 0.0102–0.0671 mm^3).

4. Discussion

Our study demonstrates that careful microscopic examination of grossly normal kidney tissue in VHL patients revealed numerous clear cell chains and clusters in comparison to grossly normal kidney tissue in patients with sporadic clear cell renal cell carcinoma. We investigated the multicellular foci in the grossly normal cortex of VHL kidneys and counted them in 10 blocks of 4 different VHL cases. We identified that the count of these pathological foci was sequentially decreased from lesion 1 to lesion 2, then from lesion 2 to 3, and finally, the lowest was lesion 4, in each case, and each block studied. Distribution was highly consistent in the four cases. The same sequential decrement was noticed in clusters volumes, too: the most frequent category 1 (105 clusters), category 2 (10 clusters), and the least was category 3 (4 clusters).

Given that we have cut through (i.e., exhausted the tissue) the tested blocks, H&E stained them, and performed a 3dimensional analysis of these structures by which approach we establish certainty that clear cell proliferations do not represent extensions of larger pathologic processes that were invisible on the original section. We primarily analyzed all H&E-stained sections for structures of interest, followed by examining expression of CAIX and CD34.

All samples from different multicellular clear cell foci examined were positive for CAIX, and it is well-established to detect activation of the HIF pathway. Expression of the *CAIX* gene is regulated by the VHL/HIF system via a HIF responsive element in the promoter, and CAIX constitutes an attractive marker for HIF activation because of the inducible response and the stability of the protein [24,25].

Mandriota et al. [9] confirmed that the foci of CAIX expression represented activation of the HIF pathway. They examined the expression of two other HIF-responsive genes encoding GLUT-1 [26] and VEGF [27]. Immunolabeling for GLUT-1 and in situ-hybridization for VEGF revealed high levels in overt CCRCC and collections of cells in the non-tumorous renal parenchyma that corresponded with the CAIX-expressing foci. As a marker for angiogenesis, CD34 expression was analyzed to observe a substantial increase in the extent of vascularization around tubules containing CAIX-positive cells.

One observation we had in this study after performing CD34 immunohistochemical stain is the proportional increase in the density of vasculature with the increment of the size of clear cell lesions from chains to clusters, to invasive foci. This observation supports the fact that these clear cell foci promote angiogenesis, which allows tumor growth and possible angiogenic invasion with continuous growth.

Given the findings in our studies, and since the microscopic clear cell proliferations follow a predictable pattern, we hypothesize that the renal neoplastic pathological events (i.e., atypical cysts and renal cancers) evolve from a variety of microscopic precursors (i.e., clear cells aggregates and clusters).

Mubarak et al. showed that the most frequently observed pathologic events in VHL kidney sections were either isolated clear cells within a renal tubule or two or more clear cells adherent to each other, likely representing changes secondary to an amplification event. More significant accumulations of clear cells along the tubular lining, as well as clustering of clear cells within renal tubules, appeared to be in the continuity of smaller preceding events [23].

Clear cell changes in renal tubules are not always neoplastic and can be seen secondary to metabolic or toxic etiologies (e.g., toxins, Alport nephropathy, osmotic tubulopathy secondary to radiographic contrast) [28–30]. In contrast to the clear cell changes seen in VHL patients, those changes are seen in cells displaying a brush border or tubular form of differentiation. Additionally, immunohistochemically the clear cells in VHL cases are consistently and reliably immunoreactive for anti-CAIX antibody [23,31].

In other studies in patients with VHL disease, it has been shown that careful microscopic examination of grossly normal tissue of the spinal cord and epididymis revealed precursor structures that are numerically more frequent than tumors [15,32]. Spinal cord tissue from patients with VHL disease presumed to be 'tumor-free' harbors microscopic foci of poorly differentiated cellular aggregates in nerve root tissue. Additionally, they showed that a small subset of VHL-deficient microscopic lesions extends beyond the nerve root to form early hemangioblastoma. Intraradicular precursors consist of scattered VHL-deficient cells with activation of HIF-2 alpha and HIF-dependent target proteins, including CAIX and VEGF and are associated with an extensive angiogenic response. Ultrastructural examination revealed the gradual transition from poorly differentiated VHL-deficient cells into vacuolated cells with a 'stromal' cell phenotype. The evolution of hemangioblastoma seems to involve multiple steps from a large pool of precursor lesions [33].

Shively et al. showed that in surgically resected nervous system hemangioblastomas of patients with VHL disease, HIF-1 alpha activation was associated with epithelioid structures, which were mainly seen in larger tumors. However, small tumors were composed mostly of poorly differentiated mesenchymal structures and did not show HIF-1 alpha activation. This suggests that the growth of nervous system tumors in VHL disease has a sequence of structural and molecular events [32].

The fact that the microscopic renal cystic and solid neoplasms containing only clear cell cytological features were found in patients with VHL disease [13] gave us a clue that those microscopic foci might play a role in the development of those neoplastic pathological lesions and a role in tumorigenesis.

Of particular interest in VHL disease has been the relationship between cysts and tumors. Additional studies may better clarify whether cysts and tumors develop via independent mechanisms or whether they are merely variations of an essentially uniform pathogenetic process. While consequences of dysregulation of VHL and HIF pathways provide numerous potential clues [34], our approach may be helpful in obtaining in-situ evidence in human tissues in the future.

Our morphological data showed different types of lesions which represent intriguing candidates for progression into cystic lesions and tumors in the kidney. The genetic and epigenetic mechanisms for these progressive events remain to be clarified. Walther et al. investigated normal-appearing renal tissues from VHL patients [22]. They detected an abundance of small clear cell tumorlets in VHL kidney; single H&E-stained sections from a multitude of paraffin blocks had been investigated to document and extrapolate the frequency of tumorlets, which was estimated as 600 per VHL kidney.

While previous studies by Mubarak et al. had been concerned with lesion quantification only, this study focuses on the larger simple and complex clear cell clusters, with more detailed information on their volumetric extent. Here we were able to not only show inverse correlation between lesion complexity and frequency, but—through 3D approach—between lesion size and frequency which we believe to represent the histomorphologic equivalent of linear "evolutionary tumorigenesis", recently postulated for VHL disease-associated clear cell carcinoma after whole genome analysis of a series of renal carcinomas [35].

Our findings demonstrate more than 1000 multicellular amplification events in less than one cubic cm of "normal" kidney tissue, the largest one measuring less than one thousandth of a cubic mm, undetectable by naked eye or radiologically. In the patients under study, VHL disease in human kidney therefore presents as wide-spread substitution of tubular cells with clear cells, only a small subset of which will undergo amplification into the next level of complexity. As we show here, this principle of protraction directly applies to lesion size.

Previously, this principle of protraction has been radiologically documented in far more than 1000 times larger, radiographically detectable lesions [36]. In conjunction with the "NIH rule" that tumors smaller than 3cm almost never metastasize (reviewed in [36]), this led to current therapeutic principles of pure observation of smaller nodules and nephron-sparing surgery at the time of surgical intervention.

Our findings may encourage diagnostic and academic pathologists to not only pay attention to primary tumor pathology in resection specimens, but also to adjacent normal kidney tissue; the quantity of our described precursor structures may give information on the "activity" of VHL pathology in these tissues, and may have prognostic implications for the likelihood of new tumor formation. Of particular current therapeutic interest is molecular targeting through immunotherapy [12], and resection specimens that are subjected to our approach may give new insight on immunotherapeutic efficacy on precursor structures. While the required serial sectioning of normal kidney tissue blocks is cumbersome, this activity can be automated by serial sectioning robots that have been recently developed [37]. Furthermore, advances in artificial intelligence development would allow to automatically scan for these relatively uniform precursor structures allowing for generation of quantitative data within reasonable time frames. Artificial intelligence analysis of 3D imaging data is rapidly developing; it has, e.g., shown encouraging results in limited-resection specimens of lung cancers [38], and we do not see any obstacles applying such technologies to analysis of "normal" tissues.

Finally, microdissection of precursor structures combined with targeted genetic analysis may give insight into earliest genetic changes leading to progression of precursors from single cell to complex cluster; 3D reconstruction could also be applied to serially sectioned frozen tissue blocks of normal VHL kidney allowing for mass spectroscopic analysis for earliest proteomic change after microdissection, and applicability of proteomics for diagnosis of primary VHL tumors has been previously demonstrated [39].

The main limitation of our study is the small number of cases tested. However, this study has examined sequential levels cut at 50 microns (from a total from 10 blocks of 4 VHL patients). Each slide was digitized and annotated accordingly. Each annotation was followed in each cross -section to accurately assess the frequency, sizes, and volumes of these clear cell chains and clusters occurring in the grossly normal kidneys of patients with VHL. Similar studies in the literature are lacking.

5. Conclusions

The study of morphological changes and histogenesis in VHL kidneys leads to a better understanding of the sequence of tumorigenesis. Additionally, this study has shown us the prevalence of these clear cell changes and sublocalization within the kidney parenchyma to be in close proximity to a certain anatomic/histologic structure in the kidney, which is leading us to a conclusion regarding the cell of origin in clear cell renal cell carcinoma [40]. Further studies are needed on the role these different microscopic and molecular events play in renal carcinogenesis.

Author Contributions: N.S.A.-G. participated in study execution, data interpretation, and manuscript writing. A.O.V. participated in study design, study execution, data interpretation, and manuscript writing. S.B.S. assisted in providing old data, additional IHC and drafting the new manuscript. All authors have read and agreed to the published version of the manuscript.

Funding: N.S. Al-Gharaibeh was in part funded by the Deanship of Research, Jordan University of Science of Technology (under the approval official number of 819/2020 and the application E-number of IRB approval 659/2019), and by the Department of Pathology, University of Indiana Medical School.

Institutional Review Board Statement: This study protocol was reviewed and approved by the ethical committee of the Deanship of Research at Jordan University of Science and Technology (659/2019).

Informed Consent Statement: Not applicable.

Data Availability Statement: All data supporting this study's findings are available with correspondents upon request.

Acknowledgments: Authors acknowledge Haitham Al-Gharaibeh for image development and editing assistance, and Ebaa Ababneh, for her review the data and participating in formatting text and references according to journal regulation.

Conflicts of Interest: The authors declare no conflict of interest with respect to research, authorship, and/or publication of this article.

Abbreviation and Annotations

VHL kidney	Kidney of von Hippel-Lindau disease patient
RCCC	Renal clear cell carcinoma
CAIX	Carbonic anhydrase IX
H&E	Hematoxylin and eosin
pVHL	VHL protein
VEGF	Vascular endothelial growth factor
EPO	Erythropoietin
PDGF	Platelet-derived growth factor
GLUT-1	Glucose-transporter 1
HIF	Hypoxia-inducible factor

References

1. Huntoon, K.; Oldfield, E.H.; Lonser, R.R. Dr. Arvid Lindau and discovery of von Hippel-Lindau disease. *J. Neurosurg. JNS* **2015**, *123*, 1093–1097. [CrossRef] [PubMed]
2. Haddad, N.M.; Cavallerano, J.D.; Silva, P.S. Von hippel-lindau disease: A genetic and clinical review. *Semin Ophthalmol.* **2013**, *28*, 377–386. [CrossRef] [PubMed]
3. Richards, F.M.; Maher, E.R.; Latif, F.; Phipps, M.E.; Tory, K.; Lush, M.; Crossey, P.A.; Oostra, B.; Enblad, P.; Gustavson, K.H. Detailed genetic mapping of the von Hippel-Lindau disease tumour suppressor gene. *J. Med. Genet.* **1993**, *30*, 104–107. [CrossRef]
4. Kaelin, W.G., Jr. Molecular basis of the VHL hereditary cancer syndrome. *Nat. Rev. Cancer.* **2002**, *2*, 673–682. [CrossRef] [PubMed]
5. Knudson Alfred, G. Mutation and Cancer: Statistical Study of Retinoblastoma. *Proc. Natl. Acad. Sci. USA* **1971**, *68*, 820–823. [CrossRef] [PubMed]
6. Latif, F.; Tory, K.; Gnarra, J.; Yao, M.; Duh, F.M.; Orcutt, M.L.; Stackhouse, T.; Kuzmin, I.; Modi, W.; Lerman, M.I. Identification of the von Hippel-Lindau disease tumor suppressor gene. *Science* **1993**, *260*, 1317–1320. [CrossRef] [PubMed]
7. Lonser, R.R.; Glenn, G.M.; Walther, M.; Chew, E.Y.; Libutti, S.K.; Linehan, W.M.; Oldfield, E.H. von Hippel-Lindau disease. *Lancet* **2003**, *361*, 2059–2567. [CrossRef]
8. Glasker, S.; Tran, M.G.; Shively, S.B.; Ikejiri, B.; Lonser, R.R.; Maxwell, P.H.; Zhuang, Z.; Oldfield, E.H.; Vortmeyer, A.O. Epididymal cystadenomas and epithelial tumourlets: Effects of VHL deficiency on the human epididymis. *J. Pathol.* **2006**, *210*, 32–41. [CrossRef]
9. Mandriota, S.J.; Turner, K.J.; Davies, D.R.; Murray, P.G.; Morgan, N.V.; Sowter, H.M.; Wykoff, C.C.; Maher, E.R.; Harris, A.L.; Ratcliffe, P.J.; et al. HIF activation identifies early lesions in VHL kidneys: Evidence for site-specific tumor suppressor function in the nephron. *Cancer Cell* **2002**, *1*, 459–468. [CrossRef]
10. Neumann, H.P.; Eggert, H.R.; Weigel, K.; Friedburg, H.; Wiestler, O.D.; Schollmeyer, P. Hemangioblastomas of the central nervous system. A 10-year study with special reference to von Hippel-Lindau syndrome. *J. Neurosurg.* **1989**, *70*, 24–30. [CrossRef]
11. Vortmeyer, A.O.; Yuan, Q.; Lee, Y.S.; Zhuang, Z.; Oldfield, E.H. Developmental effects of von Hippel-Lindau gene deficiency. *Ann. Neurol.* **2004**, *55*, 721–728. [CrossRef] [PubMed]
12. Glasker, S.; Vergauwen, E.; Koch, C.A.; Kutikov, A.; Vortmeyer, A.O. Von Hippel-Lindau disease: Current challenges and future prospects. *OncoTargets Ther.* **2020**, *13*, 5669–5690. [CrossRef] [PubMed]
13. Lubensky, I.A.; Pack, S.; Ault, D.; Vortmeyer, A.O.; Libutti, S.K.; Choyke, P.L.; Walther, M.M.; Linehan, W.M.; Zhuang, Z. Multiple neuroendocrine tumors of the pancreas in von Hippel-Lindau disease patients: Histopathological and molecular genetic analysis. *Am. J. Pathol.* **1998**, *153*, 223–231. [CrossRef] [PubMed]
14. Fearon, E.R.; Vogelstein, B. A genetic model for colorectal tumorigenesis. *Cell* **1990**, *61*, 759–767. [CrossRef] [PubMed]
15. Vortmeyer, A.O.; Alomari, A.K. Pathology of the Nervous System in Von Hippel-Lindau Disease. *J. Kidney Cancer VHL* **2015**, *2*, 114–129. [CrossRef] [PubMed]
16. Anglard, P.; Tory, K.; Brauch, H.; Weiss, G.H.; Latif, F.; Merino, M.J.; Lerman, M.I.; Zbar, B.; Linehan, W.M. Molecular analysis of genetic changes in the origin and development of renal cell carcinoma. *Cancer Res.* **1991**, *51*, 1071–1077. [PubMed]
17. Seizinger, B.R.; Rouleau, G.A.; Ozelius, L.J.; Lane, A.H.; Farmer, G.E.; Lamiell, J.M.; Haines, J.; Yuen, J.W.M.; Collins, D.; Majoor-Krakauer, D.; et al. Von Hippel-Lindau disease maps to the region of chromosome 3 associated with renal cell carcinoma. *Nature* **1988**, *332*, 268–269. [CrossRef] [PubMed]
18. Tory, K.; Brauch, H.; Linehan, M.; Barba, D.; Oldfield, E.; Filling-Katz, M.; Seizinger, B.; Nakamura, Y.; White, R.; Marshall, F.F.; et al. Specific genetic change in tumors associated with von Hippel-Lindau disease. *J. Natl. Cancer Inst.* **1989**, *81*, 1097–1101. [CrossRef]
19. Zbar, B.; Brauch, H.; Talmadge, C.; Linehan, M. Loss of alleles of loci on the short arm of chromosome 3 in renal cell carcinoma. *Nature* **1987**, *327*, 721–724. [CrossRef]
20. Choyke, P.L.; Glenn, G.M.; Walther, M.M.; Zbar, B.; Weiss, G.H.; Alexander, R.B.; Hayes, W.S.; Long, J.P.; Thakore, K.N.; Linehan, W.M. The natural history of renal lesions in von Hippel-Lindau disease: A serial CT study in 28 patients. *AJR Am. J. Roentgenol.* **1992**, *159*, 1229–1234. [CrossRef]
21. Solomon, D.; Schwartz, A. Renal pathology in von Hippel-Lindau disease. *Hum. Pathol.* **1988**, *19*, 1072–1079. [CrossRef] [PubMed]

22. Walther, M.M.; Lubensky, I.A.; Venzon, D.; Zbar, B.; Linehan, W.M. Prevalence of microscopic lesions in grossly normal renal parenchyma from patients with von Hippel-Lindau disease, sporadic renal cell carcinoma and no renal disease: Clinical implications. *J. Urol.* **1995**, *154*, 2010–2014, discussion 4–5. [CrossRef] [PubMed]
23. Mubarak, M.; Al-Gharaibeh, N.; Sommaruga, S.; Li, J.; Vortmeyer, A.O. Histological Tracking into the Third Dimension: Evolution of Early Tumorigenesis in VHL Kidney. *J. Kidney Cancer VHL* **2021**, *8*, 5–14. [CrossRef] [PubMed]
24. Ivanov, S.V.; Kuzmin, I.; Wei, M.H.; Pack, S.; Geil, L.; Johnson, B.E.; Stanbridge, E.J.; Lerman, M.I. Down-regulation of transmembrane carbonic anhydrases in renal cell carcinoma cell lines by wild-type von Hippel-Lindau transgenes. *Proc. Natl. Acad. Sci. USA* **1998**, *95*, 12596–12601. [CrossRef] [PubMed]
25. Wykoff, C.C.; Beasley, N.J.; Watson, P.H.; Turner, K.J.; Pastorek, J.; Sibtain, A.; Wilson, G.D.; Turley, H.; Talks, K.L.; Maxwell, P.H.; et al. Hypoxia-inducible expression of tumor-associated carbonic anhydrases. *Cancer Res.* **2000**, *60*, 7075–7083.
26. Ebert, B.L.; Firth, J.D.; Ratcliffe, P.J. Hypoxia and mitochondrial inhibitors regulate expression of glucose transporter-1 via distinct Cis-acting sequences. *J. Biol. Chem.* **1995**, *270*, 29083–29089. [CrossRef]
27. Iyer, N.V.; Kotch, L.E.; Agani, F.; Leung, S.W.; Laughner, E.; Wenger, R.H.; Gassmann, M.; Gearhart, J.D.; Lawler, A.M.; Aimee, Y.Y.; et al. Cellular and developmental control of O2 homeostasis by hypoxia-inducible factor 1 alpha. *Genes Dev.* **1998**, *12*, 149–162. [CrossRef]
28. Dickenmann, M.; Oettl, T.; Mihatsch, M.J. Osmotic nephrosis: Acute kidney injury with accumulation of proximal tubular lysosomes due to administration of exogenous solutes. *Am. J. Kidney Dis.* **2008**, *51*, 491–503. [CrossRef]
29. Moreau, J.F.; Droz, D.; Sabto, J.; Jungers, P.; Kleinknecht, D.; Hinglais, N.; Michel, J.R. Osmotic Nephrosis Induced by Water-Soluble Triiodinated Contrast Media in Man. A Retrospective Study of 47 Cases. *Radiology* **1975**, *115*, 329–336. [CrossRef]
30. Zhou, C.; Vink, R.; Byard, R.W. Hyperosmolarity Induces Armanni-Ebstein-like Renal Tubular Epithelial Swelling and Cytoplasmic Vacuolization. *J. Forensic. Sci.* **2017**, *62*, 229–232. [CrossRef]
31. Benej, M.; Pastorekova, S.; Pastorek, J. Carbonic anhydrase IX: Regulation and role in cancer. *Subcell Biochem.* **2014**, *75*, 199–219. [PubMed]
32. Shively, S.B.; Beltaifa, S.; Gehrs, B.; Duong, H.; Smith, J.; Edwards, N.A.; Lonser, R.R.; Raffeld, M.; Vortmeyer, A.O. Protracted haemangioblastic proliferation and differentiation in von Hippel-Lindau disease. *J. Pathol.* **2008**, *216*, 514–520. [CrossRef] [PubMed]
33. Vortmeyer, A.O.; Tran, M.G.; Zeng, W.; Gläsker, S.; Riley, C.; Tsokos, M.; Ikejiri, B.; Merrill, M.J.; Raffeld, M.; Zhuang, Z.; et al. Evolution of VHL tumourigenesis in nerve root tissue. *J. Pathol.* **2006**, *210*, 374–382. [CrossRef] [PubMed]
34. Frew, I.J.; Moch, H. A clearer view of the molecular complexity of clear cell renal cell carcinoma. *Ann. Rev. Pathol.* **2015**, *10*, 263–289. [CrossRef] [PubMed]
35. Mitchell, T.J.; Turajlic, S.; Rowan, A.; Nicol, D.; Farmery, J.H.R.; O'Brien, T.; Martincorena, I.; Tarpey, P.; Angelopoulos, N.; Yates, L.R.; et al. Timing the landmark events in the evolution of clear cell renal cell cancer: TRACERx Renal. *Cell* **2018**, *173*, 611–623. [CrossRef] [PubMed]
36. Meister, M.; Choyke, P.; Anderson, C.; Patel, U. Radiological evaluation, management, and surveillance of renal masses in von Hippel-Lindau disease. *Clin. Radiol.* **2009**, *64*, 589–600. [CrossRef] [PubMed]
37. Spowart, J.E. Automated serial sectioning for 3-D analysis of microstructures. *Scr. Mater.* **2006**, *55*, 5–10. [CrossRef]
38. Kudo, Y.; Shimada, Y.; Matsubayashi, J.; Kitamura, Y.; Makino, Y.; Maehara, S.; Hagiwara, M.; Park, J.; Yamada, T.; Takeuchi, S.; et al. Artificial intelligence analysis of three-dimensional imaging data derives factors associated with postoperative recurrence in patients with radiologically solid-predominant small-sized lung cancers. *Eur. J. Cardiothorac. Surg.* **2022**, *61*, 751–760. [CrossRef]
39. Vortmeyer, A.O.; Weil, R.J.; Zhuang, Z. Proteomic applications for differential diagnosis of histologically identical tumors. *Neurology* **2003**, *61*, 1626–1627. [CrossRef]
40. Al-Gharaibeh, N.S.; Temm, C.J.; Shively, S.B.; Vortmeyer, A.O. Kidney in VHL disease: Early cl ear cell proliferation occurs in the distal tubular system. *Oncol. Rep.* **2022**, *48*, 222.

medicina

MDPI

Article

Elevated Preoperative NMPR Predicts an Unfavorable Chance of Survival in Resectable Esophageal Squamous Cell Carcinoma

Meng-Ying Peng [1,2,†], Zhi-Gang Zuo [2,†], Feng-Jun Cao [2], Yuan-Dong Yu [2], Xiao-Jun Cai [2,*] and Guo-Xing Wan [2,3,*]

1 Graduate Student Training Base, Graduate School of Jinzhou Medical University, Jinzhou 121004, China
2 Department of Oncology, Renmin Hospital, Hubei University of Medicine, Shiyan 442000, China
3 Institute of Cancer, Renmin Hospital, Hubei University of Medicine, Shiyan 442000, China
* Correspondence: xiaojuncaishiyan@126.com (X.-J.C.); 15gxwan@stu.edu.cn (G.-X.W.)
† These authors contributed equally to this work.

Abstract: *Background and objectives*: Combined peripheral neutrophil–platelet indexes reflecting the systemic inflammatory status have been reported to predict the clinical outcome in patients with various types of cancer. However, the prognostic value of combined neutrophil–platelet indexes in operable esophageal squamous cell carcinoma (ESCC) remains unclear. The study introduced a novel combined neutrophil–meanplateletvolume–platelet ratio (NMPR) index and investigated its clinical and prognostic value in patients with operable ESCC receiving curative surgery. *Materials and Methods*: A retrospective analysis of the clinicopathologic data of 277 consecutive ESCC patients who received curative resection at Zhejiang Cancer Hospital in China between January 2007 and December 2010 was conducted (the training cohort). In addition, the clinicopathologic data of 101 resectable ESCC patients at Renmin Hospital of Hubei University of Medicine between December 2018 and June 2021 were collected (the external validation cohort). The optimal cutoff value of NMPR concerning overall survival (OS) in the training cohort was determined by X-tile software. Univariate and multivariate Cox regression analyses were used to evaluate the prognostic value of NMPR along with other variables in the training cohort, which was further validated with the same cutoff value in the external validation cohort. Significant predictors of OS were used to construct the nomogram, of which the discrimination and calibration was evaluated by concordance index (C-index) and calibration plots. *Results*: With a cutoff value of 16.62, the results from both the training and external validation cohorts supported the association of high NMPR (>16.62) with increased tumor length and advanced T stage but not with other variables. In the training cohort, a significant association between shorter OS and high NMPR ($p = 0.04$) as well as high CRP ($p < 0.001$), poor tumor differentiation ($p = 0.008$), advanced T stage ($p = 0.006$), advanced N stage ($p < 0.001$) and high CEA ($p = 0.007$) was revealed. Additionally, the high NMPR was verified to independently predict unfavorable OS ($p = 0.049$) in the external validation cohort. The C-index of the OS nomogram cooperating significant predictors in the training cohort was 0.71 and the calibration plots of the OS nomogram fitted well. *Conclusions*: The present study demonstrates that high NMPR is an independent predictor of unfavorable OS in resectable ESCC patients without neoadjuvant therapy.

Keywords: esophageal squamous cell carcinoma; neutrophil–mean-platelet-volume–platelet ratio; prognosis; biomarker

Citation: Peng, M.-Y.; Zuo, Z.-G.; Cao, F.-J.; Yu, Y.-D.; Cai, X.-J.; Wan, G.-X. Elevated Preoperative NMPR Predicts an Unfavorable Chance of Survival in Resectable Esophageal Squamous Cell Carcinoma. *Medicina* **2022**, *58*, 1808. https://doi.org/10.3390/medicina58121808

Academic Editors: Nicolae Crisan and Calin Cainap

Received: 22 October 2022
Accepted: 5 December 2022
Published: 8 December 2022

1. Introduction

Esophageal cancer (EC) ranks sixth in annual incidence and fourth in mortality in China [1]. Unlike in Western countries, esophageal squamous cell carcinoma (ESCC) is the predominant histopathological type accounting for approximately 90% of the EC cases in Asian populations [2]. Surgical resection with or without chemoradiotherapy remains the mainstay of curative treatment for operable ESCC [3]. Although great progress in perioperative techniques, staging methods, surgical and oncological management has been made, the prognosis of operable ESCC patients remains poor due to tumor recurrence [4]. Some clinicopathologic characteristics including performance status, nutrition status, tumor location, tumor differentiation, vessel invasion, TNM (tumor, node, metastasis) stage, extent of surgical resection, and response to chemoradiotherapy are firmly considered prognostic in operable ESCC [5–7]. However, disease recurrence or progression in ESCC patients is not solely determined by the clinicopathologic characteristics of the tumor, and host-related factors including the systemic inflammatory response may have a significant impact as well [8]. Therefore, accurate identification of prognostic factors is critical for the comprehensive evaluation of clinical outcomes and optimization of the treatment strategies.

Accumulating evidence demonstrates that systemic inflammatory status reflected by some peripheral blood parameters such as neutrophils, lymphocytes and platelets, play vital roles in the progression of various cancers [9]. Particularly, neutrophils and platelets, as important pro-tumor inflammatory cells in circulating system, can release many cytokines, chemokines and growth factors, and thus lead to tumor growth and metastasis [10]. Indeed, a variety of inflammatory indicators, such as the neutrophil-to-lymphocyte ratio (NLR), platelet-to-lymphocyte ratio (PLR) and mean-platelet-volume-to-platelet-count ratio (MPV/PC) have been identified as prognostic biomarkers in cancers, including ESCC [3,11,12]. Moreover, a recent study reported the improved ability of preoperative combined NLR and PLR score (CNPS) compared to NLR or PLR alone in predicting postoperative survival in early-stage gastric cancer [13]. Additionally, the CNPS has been found to independently predict the overall survival (OS) in ESCC patients treated with chemoradiotherapy [14,15]. More recently, the combined neutrophil–platelet score (NPS) was found to predict survival in a variety of common solid cancers, including gastroesophageal cancer [16]. However, the prognostic value of the combined neutrophil–platelet index in resectable ESCC patients remains to be elucidated. Furthermore, the previously defined neutrophil–platelet indexes only include the number of platelets while the increased release of biological active factors in a systemic inflammatory condition is mainly derived from the activated platelets, and the findings were not externally validated. We suppose that taking the neutrophil, platelet and MPV, which is recognized as an indicator of platelet activation, into account together may be a more viable means of constructing the neutrophil–platelet index for survival prediction.

Therefore, we introduce a novel combined neutrophil–MPV–platelet ratio (NMPR) index in the present study and evaluate the prognostic value of NMPR in resectable ESCC patients.

2. Materials and Methods

2.1. Study Population and Design

The study population was comprised of patients from the training cohort (ZJ cohort) and the external validation cohort (RM cohort) in China. The raw data of the training cohort was derived from the public dataset from a retrospective study conducted by Feng et al. [17], which could be obtained without further permission and ethic approval according to the public policy and ethic statement of the dataset. In the training cohort, a total of 277 resectable ESCC patients were recruited between January 2007 and December 2010 at Zhejiang Cancer Hospital. Another 101 patients with ESCC receiving radical resection at Renmin Hospital of Hubei University of Medicine from December 2018 and June 2021 were enrolled as the external validation cohort. All ESCC patients were pathologically confirmed and the data of blood examination were obtained preoperatively. The design and patient exclusion criteria have been previously reported [17]. Briefly, eligible patients were required with complete pretreatment laboratory data of interest and complete follow-up data, including OS or progression-free survival (PFS), but without preoperative treatment, any form of acute inflammatory diseases or infections, unstable or uncontrolled systemic diseases and perioperative distant metastases. The study was designed to evaluate the prognostic value of NMPR in the training cohort and verify it in the external validation cohort. Approval for the study was granted by the Ethics Committee of Renmin Hospital of Hubei University of Medicine with written informed consent from all participants.

2.2. Data Collection and Definition

Clinicopathologic data were collected, including age, gender, tumor location (upper, middle and lower), tumor differentiation (well, moderate and poor), tumor length, vessel invasion and tumor stage. Laboratory data including neutrophil count (10^9/L), mean platelet volume (MPV, fL), platelet count (PC, 10^9/L), C-reaction protein (CRP, mg/L) and carcinoembryonic antigen (CEA, ng/mL) levels were collected from the blood test results within one week prior to surgery. Furthermore, the survival data in the training cohort (ZJ cohort) were obtained from the dataset provided by Feng et al. [17]. The OS and PFS survival data in the external validation cohort (RM cohort) were obtained according to the follow-up as follows: once every 3–6 months for years 1–2, once every 6 months for years 3–5, and then once every year after year 5 with the assessment of physical examination, thoracic and abdominal computed tomography (CT), routine blood tests, hepatic and renal function tests, and blood tumor marker tests. In our laboratory (RM cohort), blood routine test was performed through automated blood cell counter (Sysmex XN-90000; Sysmex, Kobe, Japan), while CRP level was analyzed with a Hitachi Modular P800 (Hitachi High-tech Corporation, Tokyo, Japan) and CEA level was measured using a Cobas e 601 auto-analyzer (Roche Diagnostics International Ltd, Rotkreuz, Switzerland). In the ZJ cohort, the blood routine test, CRP and CEA levels were measured as previously described [17]. All tumor stage was determined according to the American Joint Committee on Cancer (AJCC-7) classification system and the histopathologic postoperative pTNM stage (tumor-node-metastasis) categories system (International Union against Cancer; UICC-7). NMPR was calculated as follows: NMPR = neutrophil $\times$ MPV/log10(PC).

2.3. Statistical Analyses

The optimal cutoff level for NMPR to predict OS in training cohort was determined using the X-tile software (version:3.6.1, Copyright Yale University 2003–2005), and this cut-off value was used to divide the study population into high-NMPR and low-NMPR groups. The differences of clinicopathologic parameters between groups were analyzed by the Chi-square test. Time-dependent receiver operating characteristics (ROC) curve was used to evaluate the prediction ability of NMPR, which indicated by the areas under the ROC curve (AUROC) value. The survival curve as well as cumulative survival rates and median OS and PFS was plotted and calculated by the Kaplan–Meier method. Significant prognostic parameters in univariate Cox regression analysis were selected for multivariate Cox regression analysis to identify independent prognostic factors and the corresponding hazard ratio (HR) and 95% confidence intervals (95% CIs) were obtained. All the statistical analyses were performed using SPSS 26.0 statistics software (IBM, Chicago, Illinois) and R-based packages (survival [version 3.4.0], survminer [version 0.4.9], survivalROC [version 1.0.3], pROC [version 1.18.0]). Moreover, the nomogram was constructed with the independent prognostic factors of OS in the training cohort and established idiomatically by the R-based rms package (version 6.3.0) according to the instruction. The performance of the nomogram was evaluated by Harrell's concordance index (C-index) and calibration plot. The methods for nomogram construction and evaluation were detailed as our previous study [18]. For all statistical analyses, a 2-side $p < 0.05$ was considered statistically significant.

3. Results

3.1. Clinicopathologic Characteristics

A total of 277 patients with resectable ESCC were incorporated into the present study as the training cohort and 101 resectable ESCC patients were enrolled as the external validation cohort (Table 1). The median follow-up time was 46.1 and 24.5 months for the test and validation cohort, respectively. The median age was 59 (36–80) in the training cohort and 62 (42–78) in the external validation cohort. Some baseline characteristics between the test and validation cohort were remarkably different except for gender, tumor differentiation and T stage (Table 1). More patients with age >60 years, tumor length <3.0 cm, lower location of tumor, N0 stage and earlier TNM stage were found in the validation cohort than that in test cohort (Table 1). Moreover, the preoperative CEA and CRP levels were not comparable due to the lack of relevant information in the validation cohort.

3.2. Associations between NMPR and Clinicopathologic Parameters

As shown in Figure 1, the optimal cutoff values of NMPR were estimated by X-tile software which divided the studied population into a high-NMPR group (NMPR >16.62) and low-NMPR group (NMPR $\leq$ 16.62). The associations between NMPR and clinicopathologic parameters are summarized in Table 1. In the training cohort, there were 155 (56.0%) patients in the low-NMPR group and 122 (44.0%) patients in the high-NMPR group. The NMPR was found to be significantly associated with tumor length (p = 0.004) and CRP level (p = 0.028), but marginally associated with T stage (p = 0.055), while there was no significant association of NMPR with other clinicopathologic parameters. In the external validation cohort, there were 60 (59.4%) patients in the low-NMPR group and 41 (40.6%) patients in the high-NMPR group with the same cutoff value to the training cohort. The NMPR was found to be significantly associated with tumor location (p = 0.042), tumor length (p = 0.013) and T stage (p = 0.002), but not associated with other clinicopathologic parameters. Collectively, the results from the two cohorts both supported a significant association of NMPR with tumor length and T stage.

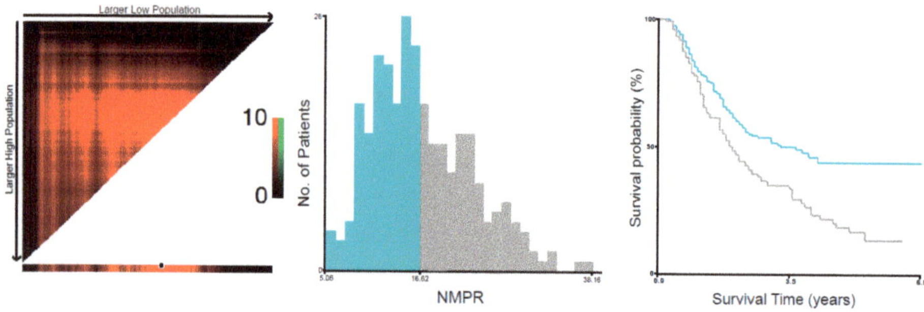

Figure 1. Determination of cut-off values of NMPR level in the training cohort and overall survival analyses. NMPR, neutrophil–mean-platelet-volume–platelet ratio.

3.3. Associations between NMPR and Prognosis in Patients with Resectable ESCC

Firstly, the ROC curve was used to evaluate the ability of NMPR with established cutoff value to predict overall survival in the training cohort. As shown in Figure 2, the AUC value was 0.57, 0.58, 0.64 and 0.65 at 24 months, 36 months, 60 months and 72 months, respectively, which suggested a possibility that the NMPR was more able to predict the long-term survival outcome in resectable ESCC. Additionally, the Kaplan–Meier cumulative survival curve is plotted as Figure 3A with a median OS of 22 months (95% CI:17.7–26.3) in the high-NMPR group and a median OS of 38 months (95% CI:22.2–53.8) in the low-NMPR group. Correspondingly, the 2-, 3-, 5-, and 6-year OS rate were 44%, 35%, 16% and 14% in the high-NMPR group and 57%, 50%, 44% and 44% in the low-NMPR group, respectively. Then, univariate Cox regression analysis revealed that high NMPR (p = 0.001), tumor length (p = 0.002), vessel invasion (p = 0.004), poor tumor differentiation (p = 0.023), advanced T stage (p < 0.001), advanced N stage (p < 0.001), high CEA (p = 0.009) and high CRP (p < 0.001) were significantly associated with poor OS (Table 2). The multivariate Cox regression analysis involving significant factors indicated by the univariate Cox regression analysis demonstrated that high NMPR (p = 0.04) along with high CRP (p < 0.001), poor tumor differentiation (p = 0.008), advanced T stage (p = 0.006), advanced N stage (p < 0.001) and high CEA (p = 0.007) might be independent predictors for unfavorable OS (Table 2). Only OS was included in the analysis due to the lack of PFS data in the original dataset.

Table 1. Baseline characteristics of the training cohort (ZJ Cohort) and external validation cohort (SY cohort) according to NMPR group.

Characteristics	Total	NMPR of ZJ Cohort		*p*-Value	Total	NMPR of SY Cohort		*p*-Value	*p*-Value
	(N = 277)	≤16.62 (n = 155)	>16.62 (n = 122)		(N = 101)	≤16.62 (n = 60)	>16.62 (n = 41)		
Age (years)				0.886				0.299	0.008
≤60	158 (57%)	89 (56%)	69 (44%)		42 (42%)	23 (55%)	19 (45%)		
>60	119 (43%)	66 (55%)	53 (45%)		59 (58%)	37 (63%)	22 (37%)		
Gender				0.126				0.631	0.899
Female	37 (13%)	25 (68%)	12 (32%)		14 (14%)	9 (64%)	5 (36%)		
Male	240 (87%)	130 (54%)	110 (46%)		87 (86%)	51 (59%)	36 (41%)		
Tumor length (cm)				0.004				0.013	0.001
≤3.0	74 (27%)	52 (70%)	22 (30%)		46 (46%)	33 (72%)	13 (28%)		
>3.0	203 (73%)	103 (51%)	100 (49%)		55 (54%)	27 (49%)	28 (51%)		
Tumor location				0.488				0.042	<0.001
Upper + Middle	145 (52%)	84 (58%)	61 (42%)		22 (22%)	17 (77%)	5 (23%)		
Lower	132 (48%)	71 (54%)	61 (46%)		79 (78%)	43 (54%)	36 (46%)		
Vessel invasion				0.474				0.719	<0.001
Negative	232 (84%)	132 (57%)	100 (43%)		58 (57%)	33 (57%)	25 (43%)		
Positive	45 (16%)	23 (51%)	22 (49%)		43 (43%)	27 (63%)	16 (37%)		
Differentiation				0.326				0.152	0.364
Well+ Moderate	223 (81%)	128 (57%)	95 (43%)		77 (76%)	48 (62%)	29 (38%)		
Poor	54 (19%)	27 (50%)	27 (50%)		24 (24%)	12 (50%)	12 (50%)		
T stage				0.055				0.002	0.491
T1 + T2	99 (36%)	63 (64%)	36 (36%)		40 (40%)	31 (76%)	9 (24%)		
T3 + T4	178 (64%)	92 (52%)	86 (48%)		61 (60%)	29 (48%)	32 (52%)		
N stage				0.616				0.338	<0.001
N0	150 (54%)	86 (57%)	64 (43%)		77 (76%)	47 (61%)	30 (39%)		
N+	127 (46%)	69 (54%)	58 (46%)		24 (24%)	13 (54%)	11 (46%)		
TNM stage				0.337				0.312	0.004
I + II	161 (58%)	94 (58%)	67 (42%)		75 (74%)	46 (61%)	29 (39%)		
III	116 (42%)	61 (53%)	55 (47%)		26 (26%)	14 (54%)	12 (46%)		
CEA (ng/mL)				0.926					
≤5.0	239 (86%)	134 (56%)	105 (44%)						
>5.0	38 (14%)	21 (55%)	17 (45%)						
CRP (mg/L)				0.028					
≤10.0	198 (71%)	119 (60%)	79 (40%)						
>10.0	79 (29%)	36 (46%)	43 (54%)						

NMPR, neutrophil–mean-platelet-volume–platelet ratio; TNM, tumor node metastasis; CEA, carcinoembryonic antigen; CRP, C-reactive protein.

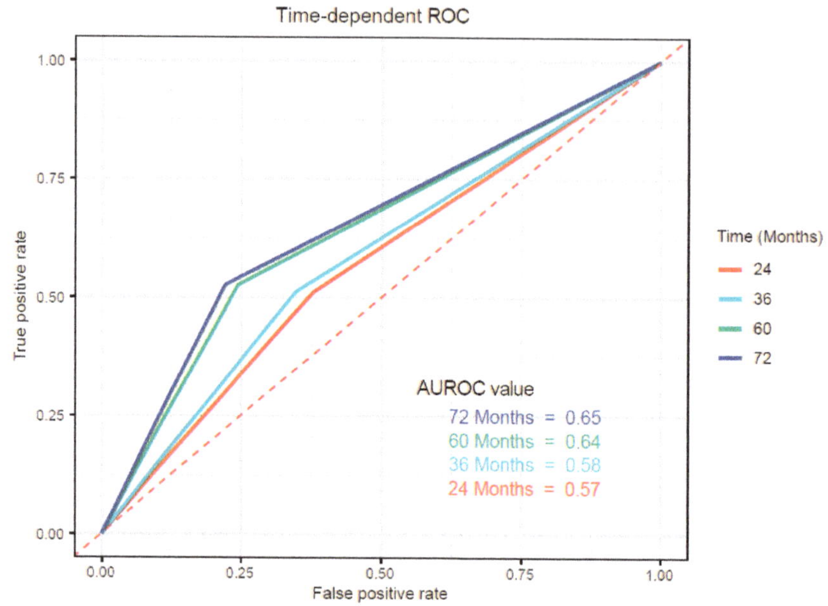

Figure 2. Predictive ability of NMPR in resectable esophageal squamous cell carcinoma by ROC curves in 24 months, 36 months, 60 months and 72 months in the training cohort. NMPR, neutrophil–mean-platelet-volume–platelet ratio.

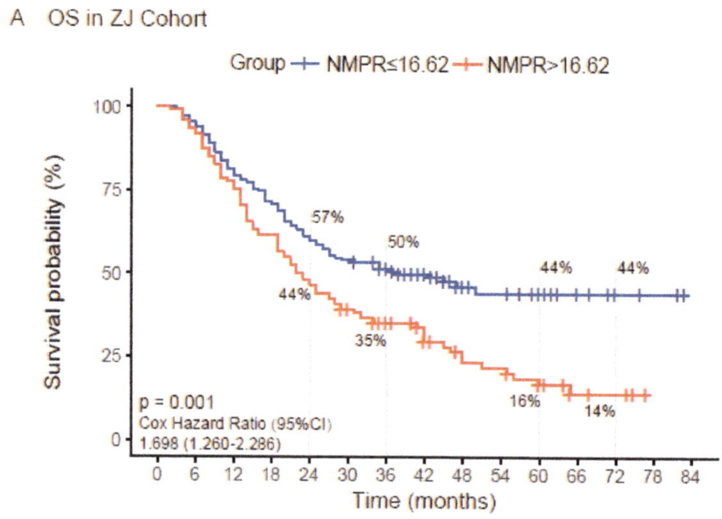

Number at risk

NMPR≤16.62	155	148	126	111	95	84	71	55	25	20	15	7	4	2	0
NMPR>16.62	122	114	95	75	59	47	36	30	17	14	11	4	3	0	0

Figure 3. *Cont.*

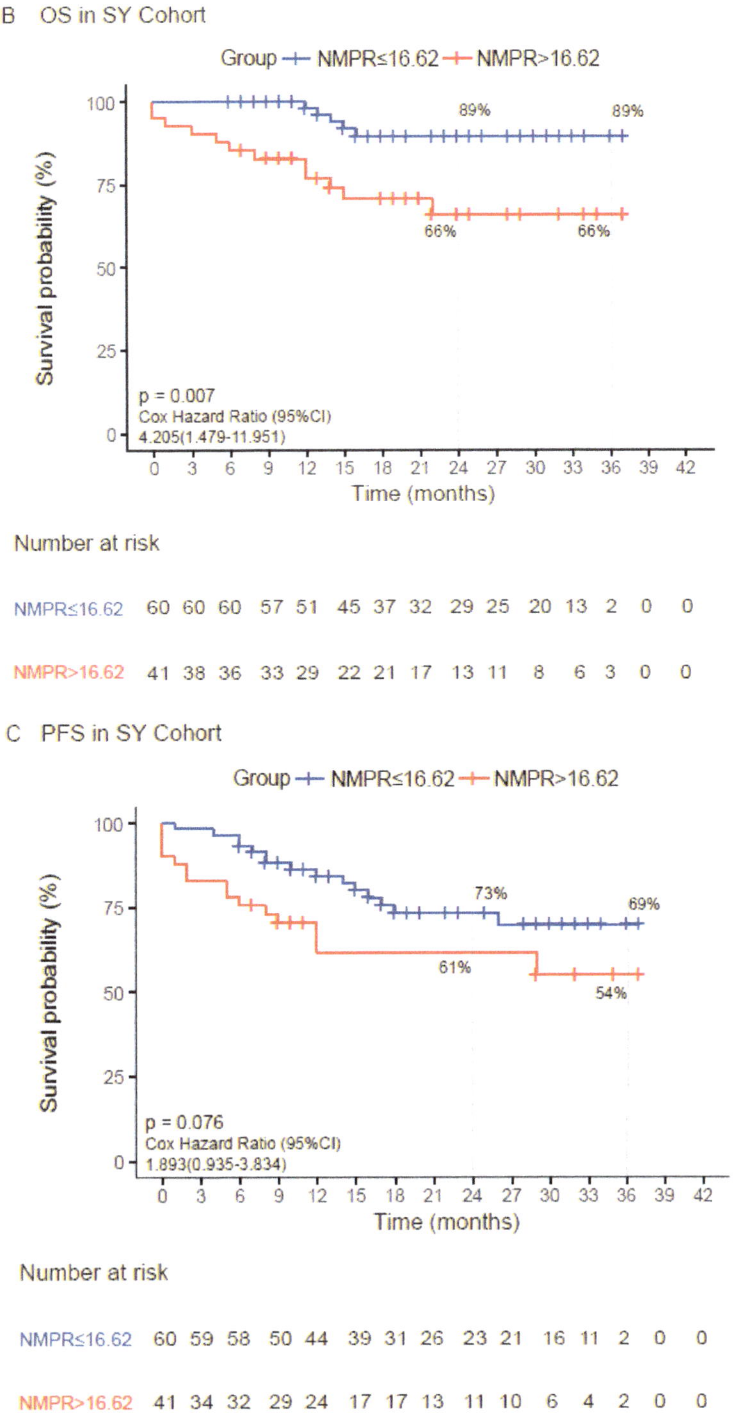

Figure 3. Kaplan–Meier curves for PFS or OS according to NMPR in the training cohort (**A**) and the external validation cohort (**B**,**C**). PFS, progression-free survival; OS, overall survival; NMPR, neutrophil–mean-platelet-volume–platelet ratio.

Table 2. Univariate and multivariate Cox regression analyses for overall survival in the training cohort (ZJ Cohort).

Variable	Univariate		*p*-Value	Multivariate		*p*-Value
	HR	95% CI		HR	95% CI	
Age (≤60 y/>60 y)	1.088	0.806–1.468	0.583			
Gender (Female/Male)	1.392	0.854–2.268	0.185			
Tumor length (≤3 cm/>3 cm)	1.767	1.224–2.553	0.002	1.106	0.734–1.667	0.630
CRP (≤10 ng/mL/>10 ng/mL)	2.258	1.652–3.083	<0.001	2.383	1.718–3.307	<0.001
Tumor location (Upper + Middle/Lower)	1.168	0.868–1.573	0.305			
Vessel invasion (Negative/Positive)	1.725	1.194–2.492	0.004	0.986	0.669–1.453	0.942
Differentiation (Well + Moderate/Poor)	1.515	1.060–2.165	0.023	1.674	1.143–2.452	0.008
T stage (T1-2/T3-4)	2.170	1.541–3.057	<0.001	1.718	1.168–2.528	0.006
N stage (N0/N+)	2.626	1.932–3.569	<0.001	2.257	1.632–3.123	<0.001
CEA (≤5 mg/L/>5 mg/L)	1.707	1.114–2.546	0.009	1.752	1.167–2.630	0.007
Adjuvant therapy (No/Yes)	1.291	0.944–1.766	0.110			
NMPR (≤16.62/>16.62)	1.698	1.260–2.286	0.001	1.380	1.014–1.878	0.040

NMPR, neutrophil-mean-platelet-volume–platelet ratio; HR, hazard ratio; CI, confidence interval; CRP, C-reactive protein; CEA, carcinoembryonic antigen. The former of the group in every variable was set as reference in Cox regression analyses.

3.4. Validation of the Prognostic Value of NMPR in an External Cohort

The prognostic value of NMPR suggested by the training cohort was further validated with the same cutoff value in an external cohort. Likewise, the ROC curve was used to evaluate the ability of NMPR with established cutoff value to predict overall survival in the validation cohort. As shown in Figure S1, the AUC value was 0.68 and 0.53 at 24 months and 36 months, respectively. The AUC value was not determined for 60 months and 72 months due to a shorter follow-up period. For survival analysis, as shown in Figure 3B, the median OS and PFS were both not reached in the high-NMPR group and low-NMPR group. The 2- and 3-year OS rate were both 66% in high-NMPR group and were both 89% in low-NMPR group, respectively. Likewise, the 2- and 3-year PFS rate were 61% and 54% in the high-NMPR group and 73% and 69% in the low-NMPR group, respectively (Figure 3C). Similarly, the Cox regression analysis was performed to verify the prognostic value of NMPR. Notably, the CRP and CEA levels were not included in the Cox regression analysis due to the lack of relevant information in the validation cohort. As a result, the univariate Cox regression analysis revealed that high NMPR ($p = 0.007$), tumor length ($p = 0.034$), advanced T stage ($p = 0.013$) and advanced N stage ($p = 0.019$) were significantly associated with poor OS (Table 3). The multivariate Cox regression analysis revealed that only high NMPR ($p = 0.049$), and no other parameters, was independently associated with poor OS. Similarly, the univariate Cox regression analysis regarding PFS showed that the vessel invasion ($p < 0.001$), tumor length ($p = 0.012$), advanced T stage ($p = 0.001$) and advanced N stage ($p < 0.001$) were significantly associated with shorter PFS, while high NMPR ($p = 0.076$) was marginally associated with shorter PFS (Table 4). Correspondingly, the multivariate Cox regression analysis qualified that vessel invasion ($p = 0.025$) might independently predict PFS, although advanced T stage ($p = 0.072$) was marginally associated with shorter PFS (Table 4). Collectively, the results verified that high NMPR was a potential independent predictor of poor OS.

3.5. Development of NMPR-Based Nomogram for OS

In light of the limited sample size and lack of some information, the nomogram was only developed in the training cohort. The independent predictors including NMPR, CRP, CEA, T stage, N stage and tumor differentiation were used to establish nomogram (Figure 4A). The evaluation revealed that the predictive accuracy of the nomogram is acceptable with a C-index of 0.71 (95% CI; 0.676–0.748). The calibration plots of the nomograms also showed a good coherence between the predictions and actual values in the probability of 2- and 3-year OS (Figure 4B).

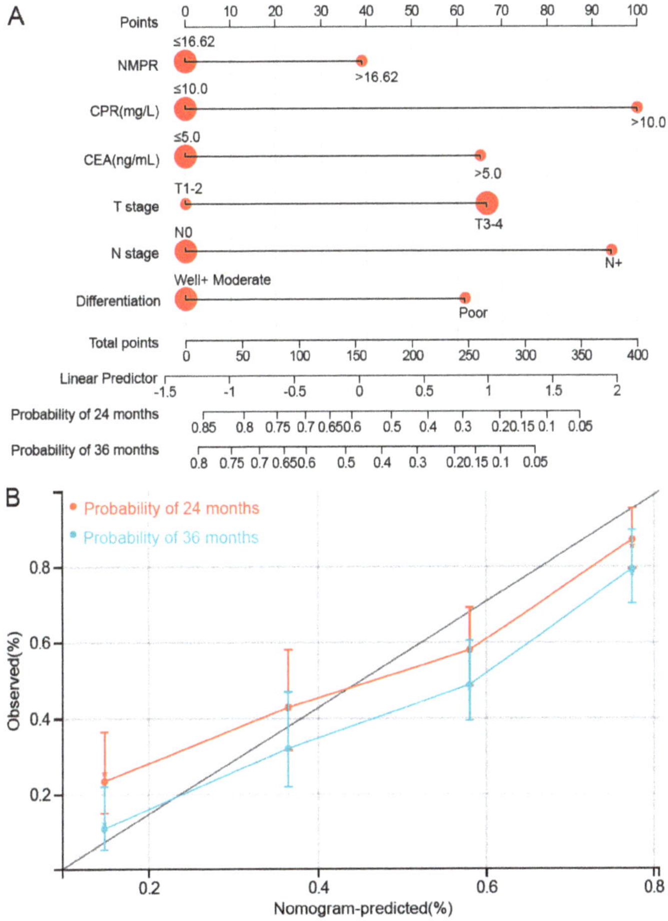

Figure 4. The nomogram (**A**) and calibration plots (**B**) for OS prediction in the training cohort. OS, overall survival; NMPR, neutrophil–mean-platelet-volume–platelet ratio; CRP, C-reaction protein; CEA, carcinoembryonic antigen.

Table 3. Univariate and multivariate Cox regression analyses for overall survival in the external validation cohort (SY Cohort).

Variable	Univariate		*p*-Value	Multivariate		*p*-Value
	HR	95% CI		HR	95% CI	
Age (≤60 y/>60 y)	1.869	0.657–5.311	0.241			
Gender (Female/Male)	2.591	0.344–19.545	0.356			
Tumor length (≤3 cm/>3 cm)	3.362	1.092–10.345	0.034	1.148	0.303–4.350	0.840
Tumor location (Upper + Middle/Lower)	2.313	0.528–10.125	0.266			
Vessel invasion (Negative/Positive)	2.417	0.894–6.537	0.082			
Differentiation (Well + Moderate/Poor)	1.009	0.329–3.097	0.987			
T stage (T1-2/T3-4)	6.490	1.477–28.510	0.013	3.226	0.588–18.141	0.176
N stage (N0/N+)	3.141	1.211–8.151	0.019	2.165	0.785–5.972	0.136
NMPR (≤16.62/>16.62)	4.205	1.479–11.951	0.007	3.029	1.007–9.112	0.049

NMPR, neutrophil–mean-platelet-volume–platelet ratio; HR, hazard ratio; CI, confidence interval; The former of the group in every variable was set as reference in Cox regression analyses.

Table 4. Univariate and multivariate Cox regression analyses for progress-free survival in the external validation cohort (SY Cohort).

Variable	Univariate		*p*-Value	Multivariate		*p*-Value
	HR	95% CI		HR	95% CI	
Age (≤60 y/>60 y)	1.701	0.800–3.616	0.168			
Gender (Female/Male)	5.510	0.751–40.433	0.093			
Tumor length (≤3 cm/>3 cm)	2.748	1.254–6.019	0.012	1.406	0.592–3.337	0.440
Tumor location (Upper + Middle/Lower)	1.048	0.451–2.434	0.913			
Vessel invasion (Negative/Positive)	4.689	2.094–10.499	<0.001	2.788	1.134–6.852	0.025
Differentiation (Well + Moderate/Poor)	1.738	0.814–3.709	0.153			
T stage (T1-2/T3-4)	5.164	1.959–13.612	0.001	2.709	0.915–8.023	0.072
N stage (N0/N+)	3.775	1.851–7.702	<0.001	1.637	0.915–8.023	0.224
NMPR (≤16.62/>16.62)	1.893	0.935–3.834	0.076			

NMPR, neutrophil–mean-platelet-volume–platelet ratio; HR, hazard ratio; CI, confidence interval; The former of the group in every variable was set as reference in Cox regression analyses.

4. Discussion

For decades, the close link between cancer and inflammation has been widely accepted. An inflammatory tumor environment has proven detrimental by promoting the proliferation and survival of malignant cells as well as angiogenesis and metastasis in the development of ESCC [19]. Testing leukocytes obtained from peripheral blood is an inexpensive, highly reliable, and reproducible method to examine the status of systemic inflammatory responses, by which numerous valuable indicators such as NLR, PLR and RDW have been confirmed as prognostic, even treatment-guided, in a variety of cancers including ESCC [3,20–22]. In the present study, the first of its kind to the best of our knowledge, we investigated and validated the prognostic value of NMPR, demonstrating an adverse impact of high NMPR on OS in patients with resectable ESCC. Moreover, we developed a nomogram incorporating NMPR, T stage, N stage, tumor differentiation, CEA and CRP to predict the OS with reasonable discriminations and calibrations.

Accumulating evidence shows that the systemic inflammatory response is associated with clinical outcome in multiple types of cancers [23]. Among the peripheral blood cells, neutrophils and platelets are important elements involved in systemic inflammation and the main contributors affecting patient outcomes. Mechanically, the neutrophil population increased under inflammatory conditions, such as in a tumor-bearing body, secreting a large amount of nitric oxide, arginase, cytokines and reactive oxygen species, resulting in disorders of T cell activation and stimulating the production of vascular endothelial growth factor (VEGF) to accelerate tumor neovascularization [24–26]. Activated platelets were also a critical source of growth factors and cytokines, especially transforming growth factor β and VEGF to support multiple malignant biological behaviors of cancer cells [3], and platelet count was often found increased in patients with solid tumors [27,28]. Similarly, another platelet indices, MPV, which was usually recognized as a hallmark for platelet activation, was used to reflect the function status of platelets under inflammatory conditions, and MPV/PC was classically used as a marker of platelet activation in peripheral blood [29]. Thus, clinically, elevated levels of neutrophil, PC or MPV were useful indicators of systemic inflammation in cancer patients, which were also linked to a poor prognosis in a variety of cancers including ESCC.

Recently, studies showed that platelet activation triggers platelet–neutrophil interaction and modulates the immune system, leading to the progression of inflammation condition in several inflammatory diseases [30–32]. Similar interaction between platelets and neutrophils was also found in cancer development and progression. The direct and indirect interplay between platelet and neutrophil was able to regulate several aspects of tumor-associated pathology and influence the tumor immune microenvironment in cancer at various stages [33]. Further mechanistic investigation demonstrated that platelets might augment rolling and firm neutrophil adhesion via the crosstalk between platelets and neutrophils initiated by the platelet surface marker P-selectin (CD62P) and neutrophil surface marker P-selectin-binding glycoprotein 1 (PSGL-1) [34,35]. Subsequently, platelet–neutrophil interaction fostered mutual activation and neutrophil extracellular traps and triggered the release of platelet granule contents and the generation of lipid mediators, which affected the functions of various effector immune cells including T-helper, CD8+ T and natural killer cells in the tumor environment via various cytokines, chemokines and prostaglandins signaling such as Toll-like receptors, interleukin-1 (IL-1), IL-17, CCL2, IL-8, PD-L1, IFN-γ, and TNF-α, etc., facilitating immune evasion of cancer cells and tumor development and dissemination [33]. Therefore, the measurement of combined neutrophil and platelet indices values may reflect the pro-tumor inflammatory status more accurately in cancer patients.

In the present study, we developed a novel inflammation-based NMPR score comprised of neutrophil and MPV to Log (PC) ratio, of which the prognostic value has yet to be examined in cancer patients. Our study investigated the effect of preoperative NMPR level on the survival of ESCC patients, demonstrating a significant association of high NMPR with shorter OS. Moreover, the multivariate Cox regression analysis suggested that NMPR might be an independent prognostic indicator for ESCC, which was further validated in an external cohort to add the reliability of the results. However, a borderline but not significant association of NMPR with PFS was revealed (p = 0.07), which may be due to the limited sample size in the validation cohort or heterogeneity in neoadjuvant therapy or surgery schedule. Indeed, in several previous studies, a part of NMPR, MPV/PC ratio has been reported to predict the clinical outcome of resectable, locally advanced and apatinib-treated advanced ESCC, respectively [11,17,36]. However, the NMPR seemed to improve the prediction efficiency for long survival with higher AUROC value compared to MPV/PC in the previous study (0.64 vs. 0.60) [17], suggesting the superior value of NMPR in survival prediction of resectable ESCC. Similarly, the prognostic value of another method of combined neutrophil and platelet indices expressed as neutrophil–platelet score (NPS, neutrophils $\leq 7.5 \times 10^9$/L and platelets $\leq 400 \times 10^9$/L defined as score of 0; neutrophils $\geq 7.5 \times 10^9$/L or platelets ≥ 400 x10^9/L defined as score of 1; neutrophils > 7.5×10^9/L or platelets > 400×10^9/L defined as score of 2) was investigated in locally advanced ESCC treated with chemoradiotherapy as well as other cancers [14,16]. Consistent with our findings, high NPS was found to be associated with poor survival in a variety of common cancers including gastroesophageal cancer, implying the high value of combined neutrophil and platelet indices in esophageal cancer patients. However, the main difference between NMPR and NPS was whether the MPV was included or not. As mentioned previously, the systemic pro-tumor inflammatory status may be reflected by neutrophils, platelets and their interaction, and the activation of platelets is a key step in the release of a large amount of inflammatory factors and the initiation of inflammation process to favor the tumor, which is likely indicated by MPV. Biologically, it is, therefore, more plausible to include MPV when combining neutrophils and platelets for the evaluation of systemic inflammatory conditions in cancer patients. Additionally, the relationship between NMPR and clinicopathologic parameters was also investigated in the study. We found that high NMPR was significantly associated with several unfavorable factors including longer tumor length, advanced T stage and high CRP level, supporting a potential association of NMPR with increased inflammation induced by cancer progression. Furthermore, with the ZJ test cohort, this study identified several other prognostic factors including tumor differentiation, T stage, N stage, CEA and CRP, which were well-established in previous studies regarding esophageal cancer [2,17,37–39]. However, their prognostic value was not validated in the external RM validation cohort, which may be explained by several points: limited sample size, short follow-up, heterogeneity of baseline characteristics between test and validation cohorts, missing data such as CEA and CRP in RM cohort, difference in adjuvant therapy and late-line treatment. Using the ZJ cohort with a larger sample size, we further constructed a nomogram with clinical parameters of T stage, N stage, tumor differentiation, and preoperative NMPR, CRP and CEA levels. The nomogram was evaluated and considered reliable with an acceptable predictive accuracy by exhibiting an acceptable calibration curve, based on looking at the nomograph prediction probability versus actual probability and presenting a concordance index (c-index) of 0.71. Our result, however, needed further validation.

It is noteworthy that our study is the first attempt to evaluate the prognosis significance of inflammation-based index NMPR in patients with ESCC. The NMPR is simple and comprised of components of a blood test with low cost, which is of great value for the patients in developing countries and low-income territories. Still, there are several potential limitations in the current study. First, the retrospective study has an inherent selection bias. Although the prognostic value of NMPR was revealed, the multivariate Cox regression analysis in the external validation cohort was not adjusted by several important factors, such as CEA and CRP levels, due to the missing data which were not routinely examined preoperatively. Likewise, data asymmetry of the baseline characteristics between the high and low NMPR group and between the test and validation cohorts might affect the result to some extent as well, which was indicated by the significant difference in survival outcome between the test and validation cohort (Figure 3). Further, considering that blood parameters were dynamic, blood samples were not obtained at an accordant timepoint and without repeated test, which might introduce an irreconcilable bias and negate the utility of the test. Then, different equipment among different medical institutions may result in inconsistency in the results of blood parameter tests, which may affect the applicability of our findings. Lastly, the nomogram was constructed using several significant prognostic parameters in the test cohort, which was not validated by an external cohort. The reliability and generalization are therefore limited. Finally, there might be insufficient power to validate the significance of NMPR with a short follow-up period in the cohort with relatively small sample size, considering the lack of a significant association of NMPR with PFS. Our findings should be interpreted with caution in light of the defects mentioned above.

5. Conclusions

In conclusion, the present study revealed that preoperative elevated NMPR was associated with unfavorable clinicopathologic features in resectable ESCC patients. The results of our study emphasize the importance of combined neutrophil and platelet indices by demonstrating that preoperative elevated NMPR was independently associated with poor survival outcome in patients with resectable ESCC. Due to its ease of accessibility and low cost, large-scale prospective validation studies are warranted to confirm these results to promote its clinical application.

Supplementary Materials: The following are available online at https://www.mdpi.com/article/10.3390/medicina58121808/s1, Figure S1: Predictive ability of NMPR in resectable esophageal squamous cell carcinoma by ROC curves in 24 months and 36 months in the validation cohort. NMPR, neutrophil mean platelet volume platelet ratio.

Author Contributions: M.-Y.P.: Writing—original draft. Z.-G.Z.: Methodology, Investigation. G.-X.W.: Writing—Review and editing. F.-J.C.: Data curation. Y.-D.Y.: Data curation. M.-Y.P.: Data curation. M.-Y.P.: Software. Z.-G.Z.: Software. Z.-G.Z.: Funding acquisition. G.-X.W.: Interpretation of results. X.-J.C.: Supervision. G.-X.W.: Supervision. X.-J.C. and G.-X.W.: Conceptualization. All authors have read and agreed to the published version of the manuscript.

Funding: This study was supported by the Guiding project of scientific technology research and development of Shiyan (21Y59).

Institutional Review Board Statement: The study was conducted according to the guidelines of the Declaration of Helsinki, and approved by the local Ethics Committee of Renmin Hospital of Hubei University of Medicine (IRB Study ID: syrmyy2022-033, 27 June 2022).

Informed Consent Statement: Informed consent was obtained from all subjects involved in the study.

Data Availability Statement: The dataset of ZJ cohort was public and derived from a previous publication by Feng et al. (https://peerj.com/articles/7246/#supplemental-information, accessed on 27 June 2022). The dataset in RM cohort was available from the corresponding author on reasonable request.

Acknowledgments: The authors appreciate Jifeng Feng, Chen Sheng, Qiang Zhao and Pengcheng Chen for their spirit of scientific sharing.

Conflicts of Interest: The authors declare no conflict of interest.

References

1. Zheng, R.; Zhang, S.; Zeng, H.; Wang, S.; Sun, K.; Chen, R.; Li, L.; Wei, W.; He, J. Cancer incidence and mortality in China, 2016. *J. Natl. Cancer Cent.* **2022**, *2*, 1–9.
2. Hu, J.; Kuang, P.; Chen, D.; Chen, Y.; Wen, Z. Prognostic significance of serum carcinoembryonic antigen and squamous cell carcinoma antigen in patients with esophageal squamous cell carcinoma undergoing radical esophagectomy. *Transl. Cancer Res.* **2020**, *9*, 2460–2471. [CrossRef] [PubMed]
3. Yodying, H.; Matsuda, A.; Miyashita, M.; Matsumoto, S.; Sakurazawa, N.; Yamada, M.; Uchida, E. Prognostic Significance of Neutrophil-to-Lymphocyte Ratio and Platelet-to-Lymphocyte Ratio in Oncologic Outcomes of Esophageal Cancer: A Systematic Review and Meta-analysis. *Ann. Surg. Oncol.* **2016**, *23*, 646–654. [CrossRef] [PubMed]
4. He, S.; Xu, J.; Liu, X.; Zhen, Y. Advances and challenges in the treatment of esophageal cancer. *Acta Pharm. Sin. B* **2021**, *11*, 3379–3392. [CrossRef] [PubMed]
5. Lagergren, J.; Smyth, E.; Cunningham, D.; Lagergren, P. Oesophageal cancer. *Lancet* **2017**, *390*, 2383–2396. [CrossRef]
6. Smyth, E.C.; Lagergren, J.; Fitzgerald, R.C.; Lordick, F.; Shah, M.A.; Lagergren, P.; Cunningham, D. Oesophageal cancer. *Nat. Rev. Dis. Primers* **2017**, *3*, 17048. [CrossRef]
7. Li, D.L.; Zhang, L.; Yan, H.J.; Zheng, Y.B.; Guo, X.G.; Tang, S.J.; Hu, H.Y.; Yan, H.; Qin, C.; Zhang, J.; et al. Machine learning models predict lymph node metastasis in patients with stage T1-T2 esophageal squamous cell carcinoma. *Front Oncol.* **2022**, *12*, 986358. [CrossRef]
8. Xie, X.; Luo, K.J.; Hu, Y.; Wang, J.Y.; Chen, J. Prognostic value of preoperative platelet-lymphocyte and neutrophil-lymphocyte ratio in patients undergoing surgery for esophageal squamous cell cancer. *Dis. Esophagus* **2016**, *29*, 79–85. [CrossRef]
9. Ravindranathan, D.; Master, V.A.; Bilen, M.A. Inflammatory Markers in Cancer Immunotherapy. *Biology* **2021**, *10*, 325. [CrossRef]
10. Huang, Z.; Liu, Y.; Yang, C.; Li, X.; Pan, C.; Rao, J.; Li, N.; Liao, W.; Lin, L. Combined neutrophil/platelet/lymphocyte/differentiation score predicts chemosensitivity in advanced gastric cancer. *BMC Cancer* **2018**, *18*, 515. [CrossRef]
11. Sun, S.Y.; Zhao, B.Q.; Wang, J.; Mo, Z.X.; Zhao, Y.N.; Wang, Y.; He, J. The clinical implications of mean platelet volume and mean platelet volume/platelet count ratio in locally advanced esophageal squamous cell carcinoma. *Dis. Esophagus* **2018**, *31*, dox125. [CrossRef] [PubMed]
12. Dupre, A.; Malik, H.Z. Inflammation and cancer: What a surgical oncologist should know. *Eur. J. Surg. Oncol* **2018**, *44*, 566–570. [CrossRef] [PubMed]
13. Sun, X.; Liu, X.; Liu, J.; Chen, S.; Xu, D.; Li, W.; Zhan, Y.; Li, Y.; Chen, Y.; Zhou, Z. Preoperative neutrophil-to-lymphocyte ratio plus platelet-to-lymphocyte ratio in predicting survival for patients with stage I-II gastric cancer. *Chin. J. Cancer* **2016**, *35*, 57. [CrossRef] [PubMed]
14. Miao, C.; Zhu, S.; Pan, H.; Cao, X.; Yuan, S.; Hu, X. Combined neutrophil-platelet score and hemoglobin level predict survival in esophageal squamous cell carcinoma patients treated with chemoradiotherapy. *Oncotarget* **2017**, *8*, 87971–87979. [CrossRef]
15. Zhu, S.; Miao, C.W.; Wang, Z.T.; Peng, L.; Li, B. Sensitivity value of hematological markers in patients receiving chemoradiotherapy for esophageal squamous cell carcinoma. *OncoTargets Ther.* **2016**, *9*, 6187–6193. [CrossRef]
16. Watt, D.G.; Proctor, M.J.; Park, J.H.; Horgan, P.G.; McMillan, D.C. The Neutrophil-Platelet Score (NPS) Predicts Survival in Primary Operable Colorectal Cancer and a Variety of Common Cancers. *PLoS ONE* **2015**, *10*, e142159. [CrossRef]
17. Feng, J.F.; Sheng, C.; Zhao, Q.; Chen, P. Prognostic value of mean platelet volume/platelet count ratio in patients with resectable esophageal squamous cell carcinoma: A retrospective study. *PeerJ* **2019**, *7*, e7246. [CrossRef]
18. Di, Q.S.; Xu, T.; Song, Y.; Zuo, Z.G.; Cao, F.J.; Yu, X.J.; Tang, J.Y.; Zhang, W.; Li, C.; Wan, G.X.; et al. High C-Reactive Protein to Albumin Ratio Predicts Inferior Clinical Outcomes in Extranodal Natural Killer T-Cell Lymphoma. *Dose Response* **2020**, *18*, 710600640. [CrossRef]
19. Sharma, T.; Gupta, A.; Chauhan, R.; Bhat, A.A.; Nisar, S.; Hashem, S.; Akhtar, S.; Ahmad, A.; Haris, M.; Singh, M.; et al. Cross-talk between the microbiome and chronic inflammation in esophageal cancer: Potential driver of oncogenesis. *Cancer Metastasis Rev.* **2022**, *41*, 281–299. [CrossRef]
20. Li, X.; Zhang, S.; Lu, J.; Li, C.; Li, N. The prognostic value of systemic immune-inflammation index in surgical esophageal cancer patients: An updated meta-analysis. *Front. Surg.* **2022**, *9*, 922595. [CrossRef]
21. Diakos, C.I.; Charles, K.A.; McMillan, D.C.; Clarke, S.J. Cancer-related inflammation and treatment effectiveness. *Lancet Oncol.* **2014**, *15*, e493–e503. [CrossRef] [PubMed]

22. Han, F.; Liu, Y.; Cheng, S.; Sun, Z.; Sheng, C.; Sun, X.; Shang, X.; Tian, W.; Wang, X.; Li, J.; et al. Diagnosis and survival values of neutrophil-lymphocyte ratio (NLR) and red blood cell distribution width (RDW) in esophageal cancer. *Clin. Chim. Acta* **2019**, *488*, 150–158. [CrossRef] [PubMed]
23. He, C.; Zhang, Y.; Cai, Z.; Lin, X. The prognostic and predictive value of the combination of the neutrophil-to-lymphocyte ratio and the platelet-to-lymphocyte ratio in patients with hepatocellular carcinoma who receive transarterial chemoembolization therapy. *Cancer Manag. Res.* **2019**, *11*, 1391–1400. [CrossRef] [PubMed]
24. Tazzyman, S.; Niaz, H.; Murdoch, C. Neutrophil-mediated tumour angiogenesis: Subversion of immune responses to promote tumour growth. *Semin. Cancer Biol.* **2013**, *23*, 149–158. [CrossRef]
25. Huang, H.; Liu, Q.; Zhu, L.; Zhang, Y.; Lu, X.; Wu, Y.; Liu, L. Prognostic Value of Preoperative Systemic Immune-Inflammation Index in Patients with Cervical Cancer. *Sci. Rep.* **2019**, *9*, 3284. [CrossRef]
26. Xiong, S.; Dong, L.; Cheng, L. Neutrophils in cancer carcinogenesis and metastasis. *J. Hematol. Oncol.* **2021**, *14*, 173. [CrossRef]
27. Zhou, Q.; Huang, F.; He, Z.; Zuo, M.Z. Clinicopathological and prognostic significance of platelet count in patients with ovarian cancer. *Climacteric* **2018**, *21*, 60–68. [CrossRef]
28. Ishii, Y.; Hamashima, T.; Yamamoto, S.; Sasahara, M. Pathogenetic significance and possibility as a therapeutic target of platelet derived growth factor. *Pathol. Int.* **2017**, *67*, 235–246. [CrossRef]
29. Kamath, S.; Blann, A.D.; Lip, G.Y. Platelet activation: Assessment and quantification. *Eur. Heart J.* **2001**, *22*, 1561–1571. [CrossRef]
30. Tokoro, T.; Makino, I.; Harada, S.; Okamoto, K.; Nakanuma, S.; Sakai, S.; Kinoshita, J.; Nakamura, K.; Miyashita, T.; Tajima, H.; et al. Interactions Between Neutrophils and Platelets in the Progression of Acute Pancreatitis. *Pancreas* **2020**, *49*, 830–836. [CrossRef]
31. Liang, X.; Xiu, C.; Liu, M.; Lin, C.; Chen, H.; Bao, R.; Yang, S.; Yu, J. Platelet-neutrophil interaction aggravates vascular in fl ammation and promotes the progression of atherosclerosis by activating the TLR4/NF-kappaB pathway. *J. Cell Biochem.* **2019**, *120*, 5612–5619. [CrossRef] [PubMed]
32. Lisman, T. Platelet-neutrophil interactions as drivers of inflammatory and thrombotic disease. *Cell Tissue Res.* **2018**, *371*, 567–576. [CrossRef] [PubMed]
33. Stoiber, D.; Assinger, A. Platelet-Leukocyte Interplay in Cancer Development and Progression. *Cells* **2020**, *9*, 855. [CrossRef] [PubMed]
34. Abd, E.H.; Abdel, M.I.; El, H.R.; Meshaal, S.; Shimila, I.; Radwan, E.R. Potential Role of Neutrophil-Platelet Interaction in Increased Susceptibility to Infection of Patients with Down Syndrome. *Lab. Med.* **2022**, *53*, 405–411.
35. Sreeramkumar, V.; Adrover, J.M.; Ballesteros, I.; Cuartero, M.I.; Rossaint, J.; Bilbao, I.; Nacher, M.; Pitaval, C.; Radovanovic, I.; Fukui, Y.; et al. Neutrophils scan for activated platelets to initiate inflammation. *Science* **2014**, *346*, 1234–1238. [CrossRef] [PubMed]
36. Su, R.; Zhu, J.; Wu, S.; Luo, H.; He, Y. Prognostic Significance of Platelet (PLT) and Platelet to Mean Platelet Volume (PLT/MPV) Ratio During Apatinib Second-Line or Late-Line Treatment in Advanced Esophageal Squamous Cell Carcinoma Patients. *Technol. Cancer Res. Treat.* **2022**, *21*, 2091152434. [CrossRef] [PubMed]
37. Guo, X.W.; Zhou, J.Y.; Jiang, W.; Ji, L.; Liu, Y.C.; Yin, X.X. The Combination of Preoperative Nutritional Risk Screening-2002 and Neutrophil-to-Lymphocyte Ratio is a Useful Prognostic Marker in Patients with Esophageal Squamous Cell Carcinoma. *Nutr. Cancer* **2021**, *73*, 588–595. [CrossRef] [PubMed]
38. Huang, Y.; Liu, J.S.; Feng, J.F. The combination of preoperative serum C-reactive protein and carcinoembryonic antigen is a useful prognostic factor in patients with esophageal squamous cell carcinoma: A combined ROC analysis. *OncoTargets Ther.* **2015**, *8*, 795–803. [CrossRef]
39. Zhou, X.; Li, Y.; Wang, W.; Wang, S.; Hou, J.; Zhang, A.; Lv, B.; Gao, C.; Yan, Z.; Pang, D.; et al. Regulation of Hippo/YAP signaling and Esophageal Squamous Carcinoma progression by an E3 ubiquitin ligase PARK2. *Theranostics* **2020**, *10*, 9443–9457. [CrossRef]

medicina

MDPI

Case Report

Adrenal Gland Primary Neuroblastoma in an Adult Patient: A Case Report and Literature Review

Teodora Telecan [1,2,†], Iulia Andras [1,2,†], Maria Raluca Bungardean [3], Diana Muntean [4], Claudia Militaru [5], Ion Perciuleac [2], Andrei Maga [2] and Nicolae Crisan [1,2,*]

1 Department of Urology, University of Medicine and Pharmacy Iuliu Hatieganu, 400012 Cluj-Napoca, Romania
2 Department of Urology, Municipal Clinical Hospital, 400139 Cluj-Napoca, Romania
3 Department of Pathology, University of Medicine and Pharmacy Iuliu Hatieganu, 400012 Cluj-Napoca, Romania
4 Department of Pathology, Municipal Clinical Hospital, 400139 Cluj-Napoca, Romania
5 Department of Pharmacology, Toxicology and Clinical Pharmacology, University of Medicine and Pharmacy Iuliu Hatieganu, 400012 Cluj-Napoca, Romania
* Correspondence: drnicolaecrisan@gmail.com
† These authors contributed equally to this work.

Abstract: Neuroblastoma (NB) is an undifferentiated malignant tumor of the sympathetic ganglia, occurring in children under 5 years of age. However, it is a rare histology in adult patients, occurring once per every 10 million patients per year. We present the case of a 68-year-old male patient presented to our department for right lumbar pain, asthenia, loss of weight and altered general status. The contrast-enhanced abdominal computer tomography revealed bilateral adrenal tumoral masses of 149 mm and 82 mm on the right and left sides, respectively, with invasion of the surrounding organs. The patient underwent right 3D laparoscopic adrenalectomy and right radical nephrectomy. The pathological result concluded that the excised tumor was a neuroblastoma of the adrenal gland. The patient followed adjuvant oncological treatment; however, due to disease progression, he passed away 22 months after the surgery. To our knowledge, less than 100 cases of adrenal NB in adult patients have been published, the eldest case being diagnosed at 75 years of age; meanwhile, the largest reported tumor measured 200 mm, and was excised through open surgery. Minimally invasive techniques have been limited so far to smaller, organ-confined diseases, thus making the present case the largest adrenal NB removed entirely laparoscopically. Neuroblastoma in the adult population is a rare finding, with worse prognosis compared to pediatric patients. The available literature does not provide enough data for standardized, multimodal management, as the patients are treated following adapted pediatric protocols, thus reinforcing the need for international, multidisciplinary boards for rare tumors.

Keywords: adrenal gland; 3D laparoscopic adrenalectomy; neuroblastoma

Citation: Telecan, T.; Andras, I.; Bungardean, M.R.; Muntean, D.; Militaru, C.; Perciuleac, I.; Maga, A.; Crisan, N. Adrenal Gland Primary Neuroblastoma in an Adult Patient: A Case Report and Literature Review. *Medicina* **2023**, *59*, 33. https://doi.org/10.3390/medicina59010033

Academic Editor: Dah-Shyong Yu

Received: 10 November 2022
Revised: 17 December 2022
Accepted: 20 December 2022
Published: 23 December 2022

1. Introduction

Neuroblastoma (NB) is an undifferentiated tumor of the peripheric sympathetic nervous system, being most often located in the paravertebral sympathetic ganglia (chest and retroperitoneum), as well as in the medulla of the adrenal gland [1]. It is the most common extracranial malignancy in young children, occurring once per every 7000 live births [2]. The median age at diagnosis is 19 months, while 90% of NB are diagnosed before 5 years of age [3]. Patients are stratified as low-, intermediate- or high-risk, depending on their age at diagnosis, stage, MYCN proto-oncogene status and tumor cell ploidy [4]. Depending on the risk category, treatment options range from active surveillance approaches and radical surgery for low- and intermediate-risk cases—with overall survival at 5 years being as high as 90%—whilst high-risk patients must undergo neo-adjuvant chemotherapy and surgery, followed by myeloablative and radiation therapy. For this category of patients, the 5-year survival rate does not exceed 50% [5,6].

However, in the adult population, it is a rare occurrence, being estimated as 1 in 10 million adults/year [7]. It is considered that adults have worse prognosis than children, and overall survival rates at 5 years for patients older than 20 years are estimated as 36.3% [3]. Moreover, ATRX translocation is more frequent than MYCN proto-oncogene mutation in this subgroup, being present in 11% of cases, as well as ALK (up to 14% in high-risk NB) and TERT rearrangements (23% of cases) [8,9].

We hereby present the case of an adult patient diagnosed incidentally with NB of the adrenal gland, along with a literature review of similar cases.

2. Case Presentation

This is the case of a 68-year-old male patient, who was referred to our department in September 2020 for right lumbar pain, asthenia, loss of weight and altered general status. His past medical history revealed hypertension and peripheral venous insufficiency of the lower limbs, associated with a venous ulcer of the right calf.

Upon examination, a mass could be palpated in the right upper quadrant. The abdominal ultrasound showed a parenchymatous tumor at the upper pole of the right kidney with a diameter of approximately 140 mm.

The patient underwent contrast-enhanced, thoraco-abdominal, and pelvic computer tomography (CECT-TAP), which confirmed a 149 × 106 mm tumoral mass with inhomogeneous structure and necrotic content, located in the right suprarenal region and extending into the upper renal pole, while compressing the right hepatic lobe and adhering at the level of the right diaphragmatic crus. The radiological characteristics indicated its adrenal origin, rather than an upper pole renal carcinoma. A similar mass of 82 × 63 mm was identified contralaterally, extending into the posterior gastric wall, spleen, pancreas, and upper left renal pole. No signs of lymph node involvement or distant metastases have been detected (Figure 1).

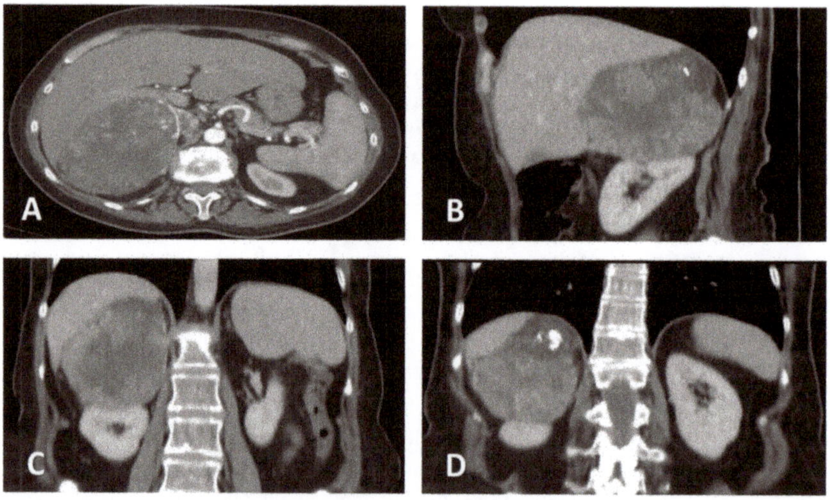

Figure 1. Contrast-enhanced abdominal computer tomography. (**A**) Transversal plane, showing the close contact of the tumor with the right hepatic lobe. (**B**) Sagittal plane. (**C**) Coronal plane, showing possible hepatic and renal invasion, as well as central necrosis. (**D**) Coronal plane, illustrating upper tumoral calcification and suspected invasion of the diaphragm.

A full endocrinological evaluation was conducted, showing an ACTH-dependent Cushing Syndrome, with an ectopic secretory pattern (Table 1). Pituitary MRI did not reveal any abnormalities of the gland.

Table 1. Hormonal profile, suggestive of ectopic ACTH secretion.

Parameters	Determined Values	Reference Values
24 h urine free cortisol	187.5 µg/24 h	<50 µg/24 h
Serum ACTH	286.2 pg/mL	46–52 pg/mL
Metanephrines	5.7 µmole	<7.1 µmole
Vanillylmandelic acid	38.3 µmole	<50 µmole
Homovanillic acid	46.9 µmole	<82.4 µmole
Aldosterone	14 µg/24 h	3–25 µg/24 h
Seric sodium	139 mEq/L	135–145 mEq/L
Seric potassium	4.4 mEq/L	3.5–5.2 mEq/L

The case was analyzed by a multidisciplinary uro-oncological board and the surgical indication for right adrenalectomy was established based on the size, the radiologic description and hormonal profile. We proposed a 3D laparoscopic approach. The preoperative blood work revealed moderate anemia (hemoglobin 9.8 g/dL) and urinary tract infection, treated with a full course of antibiotics adapted according to the provided antibiogram.

2.1. Surgical Technique

The patient was placed in left lateral decubitus, with the surgical table bent at a 45-degree angle. The intervention was performed using a 5 trocar transperitoneal approach (Figure 2) and the Karl Storz 3D laparoscopy tower (Tuttlingen, Germany).

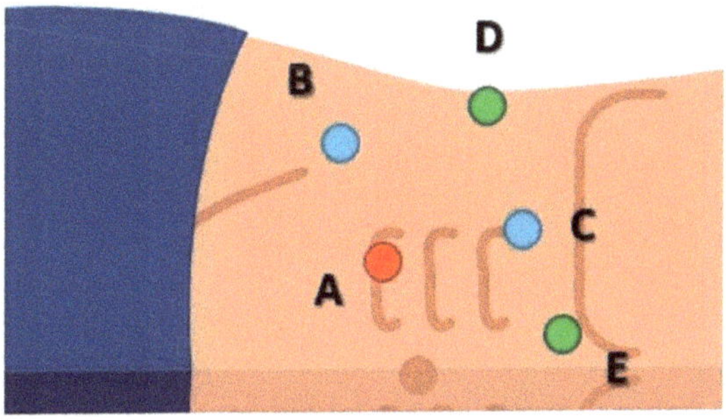

Figure 2. Trocar placement. (**A**) 10 mm optic trocar, placed paraumbilically, at the right border of the rectus abdominis muscle. Two 10 to 12 mm working trocars were inserted for the main surgeon. (**B**) Working trocar for the left hand, placed on the imaginary line drawn between the umbilicus and the anterior iliac spine, 2 cm cranially from the latter. (**C**) Working trocar for the right hand, placed on the line that links the umbilicus with the costal margin, 2 cm below the latter. Next, two trocars of 5 mm each were placed for the assistant surgeon. (**D**) Trocar placed on the mid-axillary line, used for suction. (**E**) Epigastric trocar, used for liver retraction.

The ascending colon and duodenum were medialized, exposing the inferior vena cava (IVC). The tumoral mass was identified in the right lumbar area, being adherent to the upper pole of the right kidney, diaphragm, and visceral side of the liver (Figure 3), thus adrenalectomy with radical nephrectomy was decided on.

The renal pedicle comprised of 1 artery and 2 veins was identified, clipped, and divided, as well as the central adrenal vein. The circumferential dissection of the surgical specimen was difficult due to the tumoral invasion of the liver and IVC, but uneventful. The surgical specimen was retrieved through a modified Gibson incision. A peritoneal drainage was placed. The operative time was 210 min and was carried out entirely laparoscopically, with minimum blood loss.

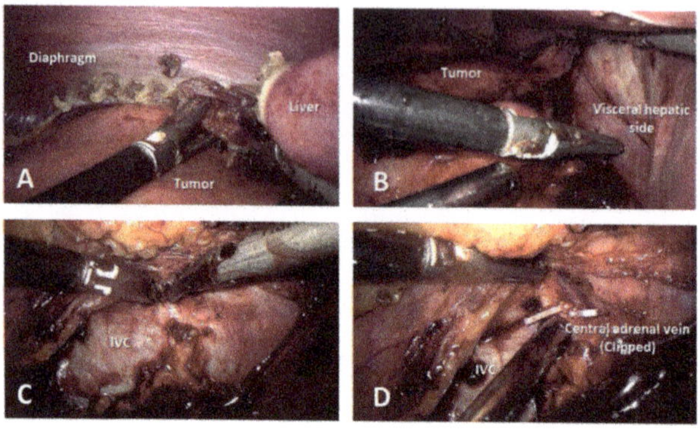

Figure 3. Intraoperative images. (**A**) Tumoral invasion of the diaphragm. (**B**) Infiltration of the visceral side of the liver. (**C**) Difficult isolation of inferior vena cava. (**D**) Central adrenal vein clipped and divided.

The patient was mobilized on the first post operative day (POD). The drainage was removed on the third POD and the patient was discharged on the seventh POD.

2.2. Pathological Report

Macroscopically, the tumoral mass measured 155 × 110 × 70 mm and appeared as a grey mass of hard consistency, polylobate, with multiple areas of necrosis identified on the transversal section.

From a microscopical standpoint, a proliferation of small polyhedric cells was observed in standard hematoxylin—eosin stain, with nuclear atypia and up to 3 mitosis per high power field. The intercellular matrix was fibrillar and poorly represented. Tumoral emboli could be identified in lymphatic and venous vessels, as well as perineural infiltration (Figure 4). Approximatively 20% of the tumoral mass was necrotic tissue. The surgical margins were negative.

Further, immunohistochemical stains were carried out in order to rule out possible differential diagnostics (Table 2). Cytogenetic analysis revealed the absence of MYCN and ALK translocation, while ATRX mutation was present. Taking all of these findings into consideration, as well as the high Ki—67 index (over 80%), the pathologist concluded that the tumor was an undifferentiated neuroblastoma with poor histological prognosis according to the International Neuroblastoma Pathological Classification (INPC), considered Stage 3, L2.

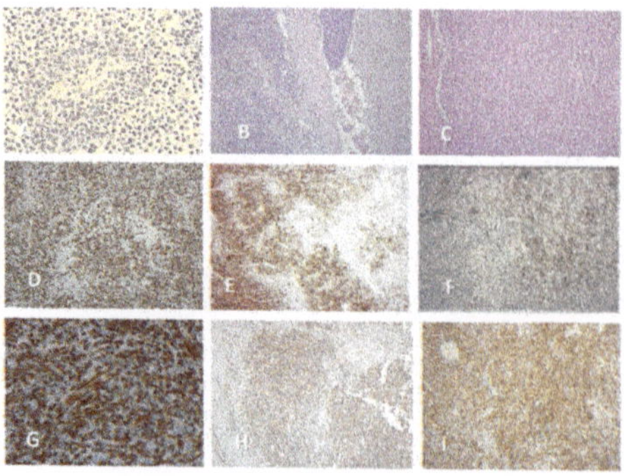

Figure 4. Microscopic aspects of the adrenal mass (magnification size). (**A**) Rosette-like arrangement of tumoral cells, around a blood vessel (×40). (**B**) Venous tumoral emboli (×10). (**C**) Hematoxylin—eosin stain (×10). (**D**) Ki—67 stain (×20). (**E**) Positive CD 10 stain (parcelar pattern) (×10). (**F**) Positive

CD 56 stain (parcelar pattern) (×10). (**G**) Positive WT 1 stain (×100). (**H**) Positive Vimentin stain (×10). (**I**) Positive CD 99 stain (×40).

Table 2. Immunohistochemical stains, cytogenetic evaluation (*) and differential diagnosis.

Positive Markers	Negative Markers	Excluded Tumors
CD 10 CD 56 CD 99 WT 1 Vimentin ATRX (*)	CD 45 CD 4 CD 8 CD 20 CD 34 CD 31 MelanA Bcl 2 Tdl EMA TTF 1 Chromogranin A NSE S100 PSAP CKAE1/AE3 PAX 8 Inhibin Desmin ALK (*) MYCN (*)	- Melanoma, pulmonary, renal, prostate, germ cells tumors metastases - Adrenal carcinoma - Lymphoma - Neuroendocrine tumors - Primitive Neuroectodermal Tumors (PNET) - Mesonephric carcinoma - Small cells desmoplastic tumors - Sarcoma - Ganglioneuroblastoma

2.3. Follow-Up

The patient was referred to the oncologist for adjuvant treatment, starting 2 months after the surgery. He received three cycles of carboplatin and etoposide combined therapy, over the course of 4 months. Due to the rarity of the diagnosis in his age group, an artificial intelligence-based decision support tool was employed (Oncompass™ Gmbh., Schindellegi, Switzerland) that indicated pembrolizumab as being suitable in this case, based on the immunologic profile identified at immunohistochemical analysis. The patient underwent three cycles of pembrolizumab (200 mg intravenously, 21 days apart). However, 9 months after the surgery, the CECT-TAP revealed tumoral progression (left adrenal mass) and local recurrence at the level of the right tumoral bed (right hepatic lobe, right psoas muscle), as well as para-aortic and interaortocaval lymph node metastases. Subsequently, the patient received third line palliative treatment, comprised of three cycles of docetaxel. As the disease progressed, fourth line treatment was initiated, including a combination of doxorubicin, cyclophosphamide, and vincristine, between November 2021 and March 2022. Eventually, the patient needed fifth line therapy, consisting of weekly gemcitabine administration; however, he passed away in July 2022, 22 months after the surgery, due to multiple organ insufficiency caused by the metastases.

3. Discussions

Neuroblastoma is an endodermal tumor, having the possibility of developing in any region of the body, along the autonomous nervous system plexi [10]. The adrenal gland is the most common location, followed by the paravertebral ganglia. To date, under 100 cases over 10 years of age have been published in the literature, even less if we restrict the searching criteria to the adult population. A summary of the published cases can be found in Table 3.

Table 3. Cases of adrenal neuroblastoma in adult patients published to date.

No.	Author, Year	Age of the Patient (Years)	Maximum Diameter (mm)	Side	Excreted Hormones	Treatment	Details	Pathological Diagnosis	Complications and Follow-Up
1.	Suzuki et al., 1985 [11]	13	120	Right	Vanilmandelic acid Metanephrines	Open surgery	Right adrenalectomy Lymphadenectomy	Homer-Wright rosettes	Uneventful recovery. Systemic chemotherapy (Vincristine, Doxorubicin, Cyclophosphamide). Alive at 12 months follow-up.
2.	Custodio et al., 1999 [12]	32	140 Tumoral thrombus invading the IVC and extending into the right atrium	Right	None	Open surgery	Right adrenalectomy, nephrectomy and right hepatic lobectomy; IVC thrombectomy and right atriotomy	Synaptophysin (+) Neurofilament (+) No CYT available	Chyloperitoneum and recurrent pleural effusions. Systemic chemotherapy and regional radiotherapy 3 months after surgery. No data regarding disease recurrence available
3.	Genc et al., 2003 [13]	30	66	Left	None	Open surgery	Left adrenalectomy	Homer-Wright rosettes No IHC performed No CYT available	Uneventful recovery. Systemic chemotherapy, disease free at the 12 months follow-up.
4.	Genc et al., 2005 [14]	52	70	Left	Vanilmandelic acid Metanephrines	Open surgery	Left adrenalectomy	No IHC performed No CYT available	Systemic chemotherapy (Cisplatin, Vincristin, Ifosfamide). Multiple liver metastases. Deceased 10 months after the surgery.
5.	Schalk et al., 2005 [15]	51	170 Retroperitoneal, mediastinal and cervical metastases	Right	Cathecolamines	Open surgery	Right adrenalectomy and nephrectomy, and lymphadenectomy. Positive surgical margins	Ki-67 (40%) MYCN (−)	Systemic chemotherapy. Tumoral progression. Deceased 9 months after surgery.
6.	Gupta et al., 2013 [16]	47	N/A	Right	None	Open surgery	Right adrenalectomy and nephrectomy	No IHC performed No CYT available	Tumoral progression after 9 months. Systemic chemotherapy (Cisplatin and Etoposide as first line, then Irinotecan and Carboplatin as second line) and regional radiotherapy (30 Gy/10 fractions/2 weeks)).

Table 3. *Cont.*

No.	Author, Year	Age of the Patient (Years)	Maximum Diameter (mm)	Side	Excreted Hormones	Treatment	Details	Pathological Diagnosis	Complications and Follow-Up
7.	Kurokawa et al., 2016 [17]	62	70	Left	None	Open surgery	Left adrenalectomy and splenectomy	NSE (+) Vimentin (+) Synaptophysin (+) S-100 (+) UCHL-1 (−) L26 (−) No CYT available	Tumoral recurrence after 4 months. Radiotherapy (50 Gy).
8.	Casteleti et al., 2017 [18]	11	N/A	Right	None	Open surgery	Right adrenalectomy	MYCN (+)	Recurrence: 5.6 years Survival: 11.37 years Status: alive
		19		Left			Left adrenalectomy		Recurrence: N/A Survival: 3.15 years Status: dead
		11		Left			Left adrenalectomy		Recurrence: 9.3 years Survival: 10 years Status: dead
9.	Manjunath et al., 2018 [19]	15	241 Liver invasion Retroperitoneal adenopaties	Right	Metanephrines	Neoadjuvant chemotherapy (Cisplatin, Etoposide, Doxorubicin, Cyclophos-phamide + Open surgery	Right adrenalectomy with upper right renal pole resection and retroperitoneal lymphadenec-tomy; omental metastasis resection	Horner-Wright rosettes NSE (+) Synaptophysin (+) LCA (−) WT1 (−) CD99 (−)	Uneventful recovery. Regional external radiotherapy. Alive at 6 months follow-up.
10.	Majumder et al., 2018 [20]	23	156 Liver invasion and metastases	Right	None	FNA + Systemic chemotherapy	FNA from one of the liver metastases	Horner-Wright rosettes	Poorly tolerated chemotherapy (Completed 3 cycles).

Table 3. *Cont.*

No.	Author, Year	Age of the Patient (Years)	Maximum Diameter (mm)	Side	Excreted Hormones	Treatment	Details	Pathological Diagnosis	Complications and Follow-Up
11.	McCarthy et al., 2019 [21]	11	105 Multiple bone metastases	Left	Catecholamines Chromogranin A	Open surgery	Left adrenalectomy	MYCN (−) ARTX (+)	Systemic chemotherapy (First line: Cyclophosphamide, Topotecan, Doxorubicin, Etoposide, Vincristine, Cisplatin; Second line: Irinotecan/Temozolomide puls Dinutuximab) Deceased 21 months from diagnosis.
		13	67	Left	None	Open surgery	Left adrenalectomy	MYCN (−) ALK (+)	Disease recurrence (left renal fossa mess, para-aortic adenopathy, bone metastases) Systemic chemotherapy (Topotecan, Cytoxan, Cisplatin, Etoposide, Doxorubicin, Vincristine) plus immunotherapy (Dinutuximab) and radiation therapy. Deceased 23 months from diagnosis.
12.	Ramsingh et al., 2019 [22]	22	80	Left	None	Laparoscopic surgery	Left adrenalectomy	Synaptophysin (+) Chromogranin A (+) PGP9.5 (+) CD56 (+) Neurofilament (+) Tyrosine hydroxylase (+) NSE (+) NB84A (+) Ki-67 (10%)	Uneventful recovery. Platinum-based chemotherapy.

Table 3. *Cont.*

No.	Author, Year	Age of the Patient (Years)	Maximum Diameter (mm)	Side	Excreted Hormones	Treatment	Details	Pathological Diagnosis	Complications and Follow-Up
13.	Zhang et al., 2019 [23]	75	45	Left	Aldosterone	Laparoscopic surgery	Left adrenal-sparing tumorectomy	CD56 (+) Synaptophysin (+) Vimentin (+) Ki-67 (>30%) CD99 (−) EMA (−) MyoD1 (−) HHF35 (−) Chromogranin A (−) S-100 (−) No CYT available	Patient refused adjuvant oncological treatment. Multiple brain and lung metastases. Deceased 22 months after the surgery.
14.	Thapa et al., 2020 [24]	35	200	Left	None	Open surgery	Left adrenalectomy and nephrectomy, paraaortic lymph node dissection	Homer-Wright rosettes Low Ki-67 No CYT available	Uneventful recovery. Patient refused adjuvant oncological treatment. No signs of recurrence at 69 months after the surgery.
15.	Xu et al., 2021 [25]	40	53	Left	None	Laparoscopic surgery	Left adrenalectomy	Synaptophysin (+) Chromogranin A (+) CD56 (+) CD99 (+) S-100 (partially +) Ki-67 (80%)	Patient refused adjuvant oncological treatment. Multiple lung metastases. Deceased 36 months after the surgery.
16.	Guzman Gomez et al., 2022 [26]	24	80	Right	Cathecolamines	Open surgery	Right adrenalectomy and retroperitoneal lymphadenectomy	N/A	Local recurrence: paravertebral retroperitoneal mass; retrocrural, retrocaval and para-aortic adenopathies. Systemic chemotherapy (Cisplatin, Etoposide).

CYT = cytogenetic analysis; IHC = immunohistochemical stain; Ki-67 = proliferation index; MYCN = avian myelocytomatosis viral oncogene neuroblastoma derived homolog; NSE = neuron-specific enolase; EMA = epithelial membrane antigen; FNA = fine needle aspiration.

To our knowledge, the presented case is one of the eldest in current literature. Zhang et al. [23] reported the case of an even older female patient (75-year-old), who underwent laparoscopic adrenal sparing surgery. The patient refused adjuvant therapy and died due to cancer progression three years later. Furthermore, the current report presents the largest neoplasia of this histology managed completely with a minimally-invasive approach. Schalk et al. [15] and Thapa et al. [24] published cases with masses of 170 mm and 200 mm, respectively, but managed in an open manner. Regarding the laparoscopic approach, it has been taken into consideration for smaller, less invasive tumors. The largest adrenalectomy, that was carried out in a totally laparoscopic fashion, was reported by Ramsingh et al. [22], with the tumor measuring 80 mm.

Although most NB are initially diagnosed as non-secreting incidentalomas, some are prone to secreting hormones that alter the hemodynamic balance of the patient, most often catecholamines, thus mimicking a pheochromocytoma. Alternatively, NB can present as the cause of a Conn's Syndrome. Although atypical, this is considered to be secondary to renal artery invasion or obstruction, due to large tumors that plunge onto the renal pedicle, increasing the renin production and inducing hyperaldosteronism [27]. The endocrinological evaluation revealed, in this case, an ectopic secretion of ACTH, confirmed by the normal aspect of the pituitary gland on MRI, together with an increased level of ACTH. It is a rare occurrence in adrenal NB, explained by the lack of cellular differentiation characterizing this type of histology, and is much rarer when it comes to ganglioneuroblastomas or ganglioneuromas [28].

Pathological diagnosis represents a challenge, even in the era of cytogenetic and molecular testing. MYCN gene amplification is characteristic for neuroblastomas, encountered in 20–25% of pediatric cases and associated with worse prognosis [29]. However, it is suggested that in adult-onset neuroblastoma, MYCN amplification is less common, while other somatic mutations such as ATRX tend to be more frequent [3]. This was consistent with our patient's phenotype, as well as with other reported cases in the literature [15]. In terms of histological prognostic factors, the proliferation index (Ki-67) is suggested to indicate the oncological outcomes more accurately than MYCN amplification. Patients having an index over 25% register a survival rate of 34% at three years, compared to 76% in cases with Ki-67 below the cut-off value [30].

To date, there is no consensus regarding the treatment protocol in adult NB. Tumors staged L1 (localized and without invasion of vital nearby structures) and L2 (as long as negative surgical margins are assured) have the possibility of being managed through a multimodal scheme, comprised of surgery, radiation and chemotherapy (most frequently adapted from pediatric protocols), associated with radioactive iodine meta-iodobenzylguanidine (I131 MIBG) therapy [31]. According to the European Society of Endocrinology [32], percutaneous biopsy was ruled out, as the tumor was considered amendable for surgical removal, and hence no prior histological confirmation was needed. Additionally, the sensitivity of the procedure ranges between 50% and 70% [33,34], while the complication rate is as high as 11% [34]. Although locally-advanced, totally 3D laparoscopic adrenalectomy was feasible in our center, as the surgical team has over 10 years of experience in complex minimally-invasive procedures. Therefore, the presented case received surgical treatment for the largest tumoral mass, as well as multiline immuno- and chemotherapy.

Since no standardized protocol is available, personalized oncological assessment has been attempted, however it did not prevent the disease progression. Further chemotherapy protocols have been conducted according to Pediatric Neuroblastoma Protocols, including docetaxel, cyclophosphamide, and doxorubicin, with dose adjustment [35]. Similar approaches have been previously attempted by Genc et al. [14], Schalk et al. [15] and Gupta et al. [16]. However, the reported cases presented distant metastases under systemic chemotherapy and succumbed to the disease at 10 and 9 months after radical surgery, respectively, while the latter underwent additional regional radiotherapy. Tumors characterized by rare histology subtypes lack standardized therapeutic approaches, hence in such scenarios, AI-driven decision support tools may have an important role. In the present case, the main therapeutic targets and their alterations have been identified through cytogenetic analysis. Based on the tumoral genotype, the algorithm synthesizes a hypothetic prognostic model that predicts the expected survival if the patient receives certain immunotherapy molecules, thus ranking each agent and assigning the most suitable one. The process is known as digital therapy planning.

The overall survival reported in the literature is heterogenous, due to the scarcity of cases in the adult population. For the full multimodal protocol, the estimated overall survival at 5 years for stage I and II masses is 83%, while for stages III and IV it drops to 28% [31]. In our case, the patient survived 22 months after the initial diagnosis, similar to some reported cases that had refused adjuvant therapy [23,25]. This finding reinforces the need for international, multidisciplinary tumor boards, in order to reach a consensus regarding the best clinical practice in rare cancers and to increase patients' survival rates.

The main limitation that we encountered in the management of the presented case was the fact that the final diagnosis was made by ruling out other small blue round cell tumors. Other limitations were the lack of access to I131 MIBG imaging and therapy, as well as the lack of widely applied therapeutic protocols. Regarding the review process, few cases have been published, thus making it difficult to draw solid conclusions regarding incidence, treatment protocols—and how they were adapted from the pediatric population—as well as survival rates.

4. Conclusions

Adrenal neuroblastoma is a rare occurrence in the adult and elderly population, with heterogenous and atypical clinical presentation, and worse outcome than those diagnosed in early childhood.

Author Contributions: Conceptualization, T.T., I.A. and N.C.; validation, M.R.B. and D.M.; investigation, T.T.; resources, I.P., A.M. and C.M.; writing—original draft preparation, T.T., M.R.B. and C.M.; writing—review and editing, I.A. and D.M.; supervision, N.C. All authors have read and agreed to the published version of the manuscript.

Funding: This research received no external funding.

Institutional Review Board Statement: The case report was conducted in accordance with the Declaration of Helsinki and approved by the Municipal Clinical Hospital's Ethical Committee (protocol code: 28/2022, approved on 6 August 2022).

Informed Consent Statement: Written informed consent has been obtained from the patient's relatives.

Conflicts of Interest: The authors declare no conflict of interest.

References

1. Smith, L.; Minter, S.; O'Brien, P.; Kraveka, J.M.; Medina, A.M.; Lazarchick, J. Neuroblastoma in an adult: Case presentation and literature review. *Ann. Clin. Lab. Sci.* **2013**, *43*, 81–84. [PubMed]
2. Ward, E.; DeSantis, C.; Robbins, A.; Kohler, B.; Jemal, A. Childhood and adolescent cancer statistics, 2014. *CA Cancer J. Clin.* **2014**, *64*, 83–103. [CrossRef] [PubMed]
3. Suzuki, M.; Kushner, B.H.; Kramer, K.; Basu, E.M.; Roberts, S.S.; Hammond, W.J.; LaQuaglia, M.P.; Wolden, S.L. Treatment and outcome of adult-onset neuroblastoma. *Int. J. Cancer* **2018**, *143*, 1249–1258. [CrossRef] [PubMed]
4. Sokol, E.; Desai, A.V. The Evolution of Risk Classification for Neuroblastoma. *Children* **2019**, *6*, 27. [CrossRef] [PubMed]
5. Irwin, M.S.; Naranjo, A.; Zhang, F.F.; Cohn, S.L.; London, W.B.; Gastier-Foster, J.M.; Ramirez, N.C.; Pfau, R.; Reshmi, S.; Wagner, E.; et al. Revised Neuroblastoma Risk Classification System: A Report From the Children's Oncology Group. *J. Clin. Oncol.* **2021**, *39*, 3229–3241. [CrossRef]
6. Maris, J.M. Recent advances in neuroblastoma. *N. Engl. J. Med.* **2010**, *362*, 2202–2211. [CrossRef]
7. Rogowitz, E.; Babiker, H.M.; Kanaan, M.; Millius, R.A.; Ringenberg, Q.S.; Bishop, M. Neuroblastoma of the elderly, an oncologist's nightmare: Case presentation, literature review and SEER database analysis. *Exp. Hematol. Oncol.* **2014**, *3*, 20. [CrossRef]
8. Trigg, R.M.; Turner, S.D. ALK in Neuroblastoma: Biological and Therapeutic Implications. *Cancers* **2018**, *10*, 113. [CrossRef]

9. Valentijn, L.J.; Koster, J.; Zwijnenburg, D.A.; Hasselt, N.E.; van Sluis, P.; Volckmann, R.; van Noesel, M.M.; George, R.E.; Tytgat, G.A.; Molenaar, J.J.; et al. TERT rearrangements are frequent in neuroblastoma and identify aggressive tumors. *Nat. Genet.* **2015**, *47*, 1411–1414. [CrossRef]
10. Papaioannou, G.; McHugh, K. Neuroblastoma in childhood: Review and radiological findings. *Cancer Imaging*. **2005**, *30*, 116–127. [CrossRef]
11. Suzuki, H.; Honzumi, M.; Funada, M.; Tomiyama, H. Metachronous bilateral adrenal neuroblastoma. *Cancer* **1985**, *56*, 1490–1492. [CrossRef]
12. Custodio, C.M.; Semelka, R.C.; Balci, N.C.; Mitchell, K.M.; Freeman, J.A. Adrenal neuroblastoma in an adult with tumor thrombus in the inferior vena cava. *J. Magn. Reson. Imaging*. **1999**, *9*, 621–623. [CrossRef]
13. Genc, F.A.; Aksoy, M.; Kapran, Y.; Tunca, F.; Tanakol, R.; Basar, Y.; Terzioglu, T. Adrenal neuroblastoma in an adult: Report of a case. *Surg. Today* **2003**, *33*, 879–881. [CrossRef]
14. Genc, H.; Haciyanli, M.; Haciyanli, S.G.; Gelal, F.; Avci Uçarsoy, A.; Dag, F. An adult adrenal neuroblastoma: A case report. *Acta Chir. Belg.* **2005**, *105*, 673–676. [CrossRef]
15. Schalk, E.; Mohren, M.; Koenigsmann, M.; Buhtz, P.; Franke, A.; Jentsch-Ullrich, K. Metastatic adrenal neuroblastoma in an adult. *Oncol. Res. Treat.* **2005**, *28*, 353–355. [CrossRef]
16. Gupta, P.; Maiti, A.; Aich, R.K.; Deb, A.R. Adrenal neuroblastoma in an adult. *J. Cancer Res. Ther.* **2013**, *9*, 96–98.
17. Kurokawa, S.; Mizuno, K.; Nakane, A.; Moritoki, Y.; Nishio, H.; Kamisawa, H.; Kubota, Y.; Okada, A.; Kawai, N.; Hayashi, Y.; et al. Adrenal Neuroblastoma in an Adult: Effect of Radiotherapy on Local Progression after Surgical Removal. *Case Rep. Urol.* **2016**, *2016*, 2657632. [CrossRef]
18. Casteleti, P.; Senerchia, A.A.; Caran, E. Neuroblastoma in Patients Over 10 Years Old: A 21-Year Experience at a Single Institution. *Austin Pediatr.* **2017**, *4*, 1052.
19. Manjunath, S.; Mouli, S.; Singh, K.R.; Rana, C.; Ramakant, P.; Mishra, A. Neuroblastoma in Late Adolescence: Case Report and Review of Literature. *World J. Endoc. Surg.* **2018**, *10*, 157–162.
20. Majumder, S.N.; Amin, M.R.; Mallick, P.K.; Nath, R.; Islam, M.R. Adrenal Neuroblastoma: A Rare Entity in Adulthood. *Rep. Cancer Treat* **2018**, *2*, 1–5.
21. McCarthy, L.C.; Chastain, K.; Flatt, T.G.; Taboada, E.; Garola, R.; Herriges, J.; Cooley, L.D.; Ahmed, A.A. Neuroblastoma in Adolescents and Children Older than 10 Years: Unusual Clinicopathologic and Biologic Features. *J. Pediatr. Hematol. Oncol.* **2019**, *41*, 586–595. [CrossRef] [PubMed]
22. Ramsingh, J.; Casey, H.; Watson, C. Adult neuroblastoma: A rare diagnosis of an adrenal mass. *BMJ Case Rep.* **2019**, *12*, e228730. [CrossRef] [PubMed]
23. Zhang, H.; Feng, Z. Adrenal neuroblastoma in an elderly adult: A case report and review of the literature. *J. Med. Case Rep.* **2019**, *13*, 284. [CrossRef] [PubMed]
24. Thapa, B.B.; Yadav, S.; Pant, S.; Rajkarnikar, P.; Mandal, P. Surgical Management of Giant L2 Adrenal Neuroblastoma in Adult Male. *Case Rep. Urol.* **2020**, *2020*, 8890223. [CrossRef] [PubMed]
25. Xu, S.; Zhang, W.; Che, B.; Zhang, J.; He, J.; Liu, D.; Chen, P.; Mu, Y.; Chen, K.; Tang, K. Adult adrenal neuroblastoma: A case report. *Mol. Clin. Oncol.* **2021**, *15*, 225. [CrossRef]
26. Guzman Gómez, G.E.; Urbano, M.A.; Martínez, V. Adrenal Neuroblastoma Producing Catecholamines Diagnosed in Adults: Case Report. *Case Rep. Oncol.* **2022**, *15*, 682–686. [CrossRef]
27. Moon, S.B. Childhood neuroblastoma masquerading as pheochromocytoma: Case report. *Int. Med. Case Rep. J.* **2016**, *9*, 65–67. [CrossRef]
28. Corcuff, J.B.; Deminiere, C.; Trouillas, J.; Puel, O.; Perel, Y.; Barat, P. Ectopic Cushing's Syndrome due to an Adrenal Ganglioneuroma. *Horm. Res. Paediatr.* **2010**, *73*, 405–408. [CrossRef]
29. Mossé, Y.P.; Deyell, R.J.; Berthold, F.; Nagakawara, A.; Ambros, P.F.; Monclair, T.; Cohn, S.L.; Pearson, A.D.; London, W.B.; Matthay, K.K. Neuroblastoma in older children, adolescents and young adults: A report from the International Neuroblastoma Risk Group project. *Pediatr. Blood Cancer* **2014**, *61*, 627–635. [CrossRef]
30. Krams, M.; Hero, B.; Berthold, F.; Parwaresch, R.; Harms, D.; Rudolph, P. Proliferation marker KI-S5 discriminates between favorable and adverse prognosis in advanced stages of neuroblastoma with and without MYCN amplification. *Cancer* **2002**, *94*, 854–861. [CrossRef]
31. Podda, M.G.; Luksch, R.; Polastri, D.; Gandola, L.; Piva, L.; Collini, P.; Cefalo, G.; Terenziani, M.; Ferrari, A.; Casanova, M.; et al. Neuroblastoma in patients over 12 years old: A 20-year experience at the Istituto Nazionale Tumori of Milan. *Tumori J.* **2010**, *96*, 684–689. [CrossRef]
32. Fassnacht, M.; Dekkers, O.M.; Else, T.; Baudin, E.; Berruti, A.; de Krijger, R.; Haak, H.R.; Mihai, R.; Assie, G.; Terzolo, M. European Society of Endocrinology Clinical Practice Guidelines on the management of adrenocortical carcinoma in adults, in collaboration with the European Network for the Study of Adrenal Tumors. *Eur. J. Endocrinol.* **2018**, *179*, G1–G46. [CrossRef]
33. Mazzaglia, P.J.; Monchik, J.M. Limited Value of Adrenal Biopsy in the Evaluation of Adrenal Neoplasm: A Decade of Experience. *Arch Surg.* **2009**, *144*, 465–470. [CrossRef]

34. Williams, A.R.; Hammer, G.D.; Else, T. Transcutaneous biopsy of adrenocortical carcinoma is rarely helpful in diagnosis, potentially harmful, but does not affect patient outcome. *Eur. J. Endocrinol.* **2014**, *170*, 829–835. [CrossRef]
35. Twist, C.J.; Schmidt, M.L.; Naranjo, A.; London, W.B.; Tenney, S.C.; Marachelian, A.; Shimada, H.; Collins, M.H.; Esiashvili, N.; Adkins, E.S.; et al. Maintaining Outstanding Outcomes Using Response- and Biology-Based Therapy for Intermediate-Risk Neuroblastoma: A Report From the Children's Oncology Group Study ANBL0531. *J. Clin. Oncol.* **2019**, *37*, 3243–3255. [CrossRef]

medicina

MDPI

Article

Associations between Body Mass Index and Prostate Cancer: The Impact on Progression-Free Survival

Dorel Popovici [1], Cristian Stanisav [2], Marius Pricop [3,*], Radu Dragomir [4], Sorin Saftescu [1,*] and Daniel Ciurescu [5]

1 Department of Oncology, Faculty of Medicine, Victor Babeş University of Medicine and Pharmacy Timisoara, Eftimie Murgu Square 2, 300041 Timisoara, Romania
2 Departments of Radiology, Victor Babeş University of Medicine and Pharmacy Timisoara, 300041 Timisoara, Romania
3 Department of Oral and Maxillo-Facial Surgery, Faculty of Dental Medicine, Victor Babeș University of Medicine and Pharmacy Timisoara, Eftimie Murgu Square 2, 300041 Timisoara, Romania
4 Departments of Obstetrics and Gynecology, Victor Babeş University of Medicine and Pharmacy Timisoara, 300041 Timisoara, Romania
5 Department of Medical and Surgical Specialties, Faculty of Medicine, Transylvania University of Braşov, 500019 Braşov, Romania
* Correspondence: pricop.marius@umft.ro (M.P.); sorin.saftescu@umft.ro (S.S.)

Abstract: *Background and objectives:* This study aimed to evaluate the impact of body mass index on PCa outcomes in our institution and also to find if there are statistically significant differences between the variables. *Materials and Methods:* A retrospective chart review was performed to extract information about all male patients with prostate cancer between 1 February 2015, and 25 October 2022, and with information about age, weight, height, follow-up, and PSA. We identified a group of 728 patients, of which a total of 219 patients resulted after the inclusion and exclusion criteria were applied. The primary endpoint was progression-free survival, which was defined as the length of time that the patient lives with the disease, but no relapses occur, and this group included 105 patients. In this case, 114 patients had a biological, local or metastatic relapse and were included in the progression group. *Results:* Our study suggests that prostate cancer incidence rises with age (72 ± 7.81 years) in men with a normal BMI, but the diagnostic age tends to drop in those with higher BMIs, i.e., overweight, and obese in the age range of 69.47 ± 6.31 years, respectively, 69.1 ± 7.51 years. A statistically significant difference was observed in the progression group of de novo metastases versus the absent metastases group at diagnostic ($p = 0.04$). The progression group with metastases present (n = 70) at diagnostic had a shorter time to progression, compared to the absent metastases group (n = 44), 18.04 ± 11.37 months, respectively, 23.95 ± 16.39 months. Also, PSA levels tend to diminish with increasing BMI classification, but no statistically significant difference was observed. *Conclusions:* The median diagnostic age decreases with increasing BMI category. Overweight and obese patients are more likely to have an advanced or metastatic prostate cancer at diagnosis. The progression group with metastatic disease at diagnostic had a shorter time to progression, compared to the absent metastases group. Regarding prostate serum antigen, the levels tend to become lower in the higher BMI groups, possibly leading to a late diagnosis.

Keywords: prostate cancer; obesity; overweight; body mass index; age; metastases; PSA

Citation: Popovici, D.; Stanisav, C.; Pricop, M.; Dragomir, R.; Saftescu, S.; Ciurescu, D. Associations between Body Mass Index and Prostate Cancer: The Impact on Progression-Free Survival. *Medicina* **2023**, *59*, 289. https://doi.org/10.3390/medicina59020289

Academic Editors: Giuseppe Lucarelli and Dah-Shyong Yu

Received: 12 December 2022
Revised: 30 December 2022
Accepted: 26 January 2023
Published: 2 February 2023

1. Introduction

Globally, according to a study in 2016, 39% of adults aged 18 years old and older were overweight (39% of men and 40% of women) and about 13% obese (11% of men and 15% of women) [1]. Obesity is a major public concern around the world because it is also a disease in its own right and is now considered to be a cause of at least 13 types of cancer [2]. Each year, it is estimated that 19 million new cancer cases are diagnosed worldwide—around 10 million cases are in men and 9.2 million in women, with a mortality rate of 54%, and 43%, respectively [3]. The biological pathway between obesity and cancer is currently being investigated, but not completely understood. Sakers et al. suggest that the negative health effects of obesity come from physiologic stimuli that induce alterations in adipose tissue metabolism, structure, and phenotype. Simply explained, adipocytes lose their plasticity causing a diminished or aberrant response to signaling and promoting the pathological outcome [4]. Obese people suffer from significant metabolic and endocrinological abnormalities that lead to enhanced insulin and insulin-growth factor signaling, dysregulation of sex hormone metabolism, and adipose tissue-derived inflammation [2,5]. Experimental animal models have shown that obesity leads to cancers of the mammary gland, colon, skin, and prostate [6].

According to Globocan, in 2020, prostate cancer (PCa) was among the most diagnosed cancers worldwide, with around 1.4 million men affected by the disease (7.8%), being surpassed by breast (2.2 million), lung (2.2 million) and colorectal cancer (1.9 million). Globally, prostate cancer was the second most diagnosed cancer in male subjects (15.1%), following lung cancer (15.4%), and taking the fifth place in the mortality rate [3].

In Europe, the estimated number of new cases of PCa was ~470,000 (20% of the male total), which makes it the most frequent cancer diagnosed in men. Additionally, the cumulative risk of being diagnosed with prostate cancer before the age of 75 is 8.2% (1 in 12 men), while the risk of PCa death before the age of 75 is 1% (1 in 103 men) [7].

In Romania, 8055 new PCa cases were diagnosed in 2020, representing 8.15% among all cancers in men aged 45+, and taking second place, after lung cancer [3,7]. The proportion of men diagnosed with PCa before the age of 60 is 1.2%, with a mortality rate of 0.2%, and after the age of 60 is 14.2%, with a mortality rate of 7.5% [3].

Studies suggest that nonmodifiable risk factors, besides age, include several others such as family history of cancer, height, lower testosterone level, type 2 diabetes, higher serum glucose, and high insulin levels. Accounted as significant modifiable risk factors are overweight and obesity, high intake of red meat, fat, dairy, and eggs, consumption of fish, and soy foods, tobacco smoking, and alcohol consumption [8].

Over time, PCa incidence and mortality were significantly different during the past years, worldwide, and they seem tightly correlated to the use of prostate-specific antigen (PSA) measurement in the male population [9]. Incidence rates for PCa are estimated to rise by +71.6% worldwide, followed by a rise of +97.1% in mortality rate, between 2020 and 2040. The highest incidence will be registered in Africa (+106.8%), Asia (+94.1%), Latin America and the Caribbean, (+81.5%), and Oceania (+47.7%), followed by the lowest incidence rates in Europe (+27.6%) and Northern America (+23.5). The mortality rate will also rise significantly on all continents, with Asia (+112.7%) leading the charts, followed by Africa (112.3%), Latin America and Caribbean (+110.4%), Oceania (+92.5%), Northern America (+80.7%) and Europe (+53.2%). In Romania, the incidence is estimated to rise by +21.5%, followed by a mortality rate of +31.2% [3].

The current study aimed to evaluate the impact of body mass index (BMI) on PCa outcomes in our institution and also to find if there are statistically significant differences between the variables.

2. Materials and Methods

2.1. Criteria

We performed a retrospective chart review to extract information about all male patients with prostate cancer seen in our electronic health record system between 1 February 2015 and 25 October 2022, with information about age, weight, height, follow-up, and PSA. We identified 728 patients, of which we excluded 509 patients due to missing information on height, weight; PSA measured in other medical laboratories; low body mass index; association of other malignancies and patients missing from follow-up visits. This resulted in a total of 219 patients who were included in the final analysis. The primary endpoint was progression-free survival (PFS), which was defined as the length of time that the patient lives with the disease, but no relapses occur, and this group included 105 patients. In this case, 114 patients had a biological, local or metastatic relapse and were included in the progression group (Figure 1). The body mass index (BMI) of each patient was calculated using weight and height, documented in the patient medical history. This analysis is partly based on self-declaration of weight and height, which might be underestimated by the patients, which might lead to potential deviations. Data about the radical prostatectomy, orchiectomy, lymphadenectomy and pathological staging is limited because the surgeries and the histopathological exam were performed in other hospitals. This study was approved by the committee board members of OncoHelp Association Timisoara.

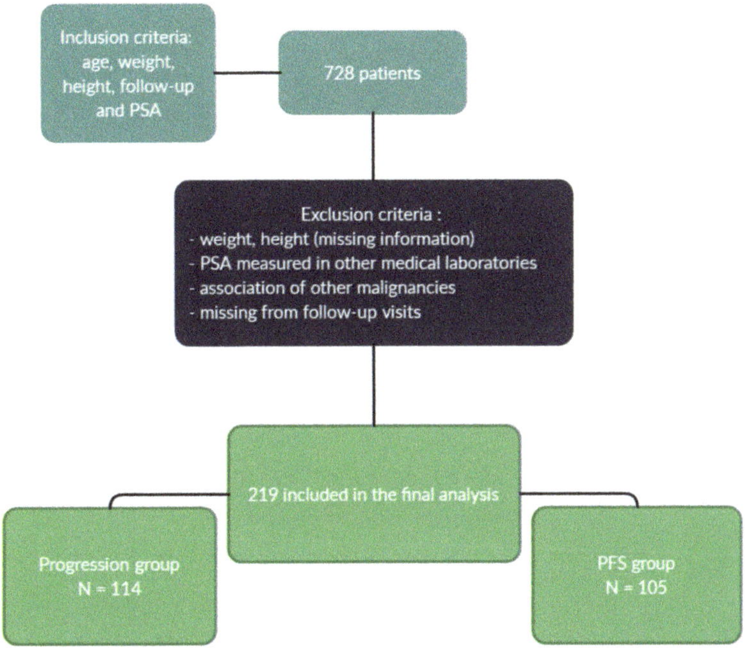

Figure 1. Inclusion and exclusion criteria.

2.2. *Statistical Analysis*

Numeric variables were expressed as mean (±SD) and discrete outcomes as absolute and relative (%) frequencies. We created three groups according to the values of BMI. Group comparability was assessed by comparing baseline follow-up duration between groups. Normality and heteroskedasticity of continuous data were assessed with Shapiro-Wilk and Levene's test, respectively. Continuous outcomes were compared with ANOVA, Welch ANOVA, or Kruskal-Wallis tests according to data distribution. Discrete outcomes were compared with chi-squared or Fisher's exact test accordingly. The alpha risk was set to 5% and two-tailed tests were used. The difference between ages according to modalities of BMI was assessed with the ANOVA.

If the null hypothesis of the ANOVA test was rejected, post-hoc pairwise analyses were performed with Tukey's HSD test. The alpha risk was set to 5% ($\alpha = 0.05$). Statistical analysis was performed with EasyMedStat—version 3.20; www.easymedstat.com (accessed on 21 October 2022).

3. Results

3.1. *Patient Population*

The clinical characteristics of patients are shown in Table 1. Of the 219 patients, 28% (62) were categorized as normal weight (NW) (18.5–24.9 kg/m^2), 43% (94) as overweight (OW) (25–29.9 kg/m^2), and 29% (63) as obese (OB) (30+ kg/m^2).

Table 1. Patient clinical characteristics.

Variable	(18.0–24.9) N = 62 (28%)	(25.0–29.9) N = 94 (43%)	(30.0–42.0) N = 63 (29%)	p-Value
Age	72.68 (± 7.81) 95% CI: [70.69; 74.66] Range: (53.0; 87.0)	69.47 (± 6.31) 95% CI: [68.18; 70.76] Range: (54.0; 86.0) RR 1.42 OR 2.44	69.1 (± 7.51) 95% CI: [67.2; 70.99] Range: (51.0; 82.0) RR 1.48 OR 2.63	0.007
cTNM Stage				
2a	0 (0.0%)	6 (6.38%)	3 (4.76%)	
2b	3 (4.84%)	5 (5.32%)	6 (9.52%)	
3a	6 (9.68%)	8 (8.51%)	3 (4.76%)	
3b	13 (20.97%)	19 (20.21%)	11 (17.46%)	0.567
3c	0 (0.0%)	3 (3.19%)	1 (1.59%)	
4a	9 (14.52%)	12 (12.77%)	10 (15.87%)	
4b	31 (50.0%)	41 (43.62%)	29 (46.03%)	
Metastases De novo	32 (51.61%)	44 (46.81%)	29 (46.03%)	0.788
Recurrent	12 (19.35%)	24 (25.53%)	13 (20.63%)	0.614
Gleason score				
≤3 + 4=7	15 (27.80%)	24 (25.50%)	14 (22.22%)	0.905
>4 + 3=7	46 (72.19%)	71 (74.50%)	49 (77.77%)	
GnRH agonist	36 (58.06%)	56 (59.57%)	46 (73.02%)	0.147
Bisphosphonates	13 (20.96%)	10 (10.64%)	7 (11.11%)	0.422
Radical prostatectomy	15 (24.19%)	25 (26.61%)	15 (23.81%)	0.907
Orchiectomy	2 (3.23%)	4 (4.32%)	3 (5.0%)	0.914

GnRH—gonadotropin-releasing hormone.

3.2. BMI and Age

We identified a statistically significant difference ($p < 0.05$) regarding age in the BMI categorized groups. The median age for the NW group was 72 ± 7.81 years vs. 69.47 ± 6.31 years in the OW group and 69.1 ± 7.51 years in the OB group (Table 1, Figure 2). The difference between ages according to modalities of BMI was assessed with ANOVA. The null hypothesis was rejected ($p = 0.007$), so we performed a post-hoc Tukey test to explore the differences between the means of all three groups and observed a significant difference between the NW and OW groups ($p = 0.023$) and NW and OB groups ($p = 0.0087$; Table 2). There was no significant difference ($p = 0.9$) between the OW and OB groups.

Table 2. Tukey HSD test results.

Pairwise Comparisons		$HSD_{0.05} = 2.84$ $HSD_{0.01} = 3.54$	$Q_{0.5} = 3.34$ $Q_{0.5} = 4.12$
NW–OW	$G_1 = 72.69$ $G_2 = 69.49$	3.19	Q = 3.76 ($p = 0.023$)
NW–OB	$G_1 = 72.69$ $G_3 = 69.10$	3.59	Q = 4.23 ($p = 0.0087$)
OW–OB	$G_2 = 69.49$ $G_3 = 69.10$	0.4	Q = 0.47 ($p = 0.94$)

NW—normal weight; OW—overweight; OB—obese; HSD—honestly significant difference.

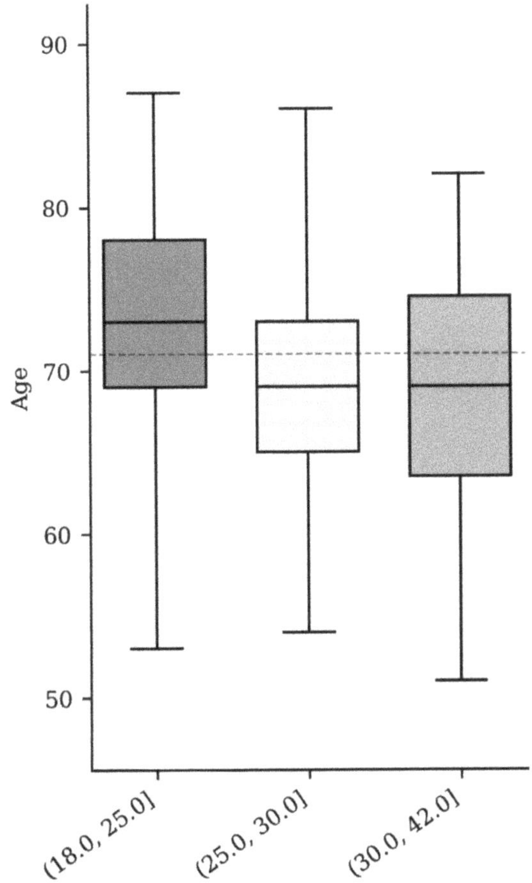

Figure 2. BMI based on age.

3.3. BMI and cTNM Stage

Group stages in correlation with BMI were represented as follows: stage 2–4.84% NW, 11.70% OW, and 14.28% OB: stage 3–30.65% NW, 31.91% OW and 23.81% OB and stage 4–64.52% NW, 56.39% OW and 61.90%. No statistically significant difference was found between the groups ($p = 0.567$).

3.4. BMI and De Novo Metastases

De novo metastasis rates were 51.61%, 46.81%, and 46.03% in patients categorized as NW, OW, and OB, respectively. No statistically significant difference was found ($p = 0.788$).

3.5. BMI and Recurrent Metastases

Recurrent metastasis rates were 19.35%, 25.53%, and 20.63% in patients for which BMI was NW, OW and OB. No statistically significant difference was found between the groups ($p = 0.614$).

3.6. BMI and Gleason Score

The Gleason score $\geq 4 + 3 = 7$ was most seen between the groups, with 46 (72.19%) in the NW group, 71 (74.50%) in the OW group and 49 (77.77%) in the OB group. No statistically significant difference was found between the groups, overall ($p = 0.905$).

3.7. BMI and GnRH Agonists

Leuprorelin and Triptorelin were the most used agonists among PCa patients (rates of 58.06%, 59.57%, and 73.02% in the NW, OW and OB groups). No statistically significant difference was found between GnRH agonists ($p = 0.147$).

3.8. BMI and Bisphosphonates

The use rates of zoledronic acid were 20.96%, 10.64%, and 11.11% within the NW, OW, and OB groups, respectively, with no statistically significant differences ($p = 0.422$).

3.9. BMI and Radical Prostatectomy or Orchiectomy

Approximately $\frac{1}{4}$ of the patients from each BMI category group underwent radical prostatectomy (24.19%, 26.61%, and 23.81% for the NW, OW, and OB groups), while orchiectomy was applied in much lower rates (3.23%, 4.32%, and 5.0), with no statistically significant difference between therapeutic approaches within the groups ($p = 0.907$ and $p = 0.914$).

3.10. BMI and PSA

The median PSA for the NW group was 123.45 ± 366.58 vs. 48.67 ± 132.6 in the OW group and 54.23 ± 156.6 in the OB group, with no statistically significant difference ($p = 0.1$).

3.11. Progression Group According to De Novo Metastases

Overall, 114 events that define progression of PCa were included in this analysis. 70 patients (61%) in the group of events were recorded with de novo metastases and 44 (39%) were without. The median progression time for those patients who presented with de novo metastasis was 15.91 months, while 20.17 months was for those without de novo metastases. The Mann-Whitney test was used to compare the progression group median according to de novo metastasis. There was a statistically significant difference in the progression group between the patients with present and absent metastases ($p = 0.04$, Table 3).

Table 3. Progression group and de novo metastasis.

De Novo Metastasis	Present	Absent
Mean ± SD	18.04 ± 11.37 (months)	23.95 ± 16.39 (months)
Median	15.91	20.17
Min–Max	0.0–55.76	1.12–84.43
N	70	44

3.12. Progression Group According to Recurrent Metastases

From the same group of 114 patients that experienced events, 49 patients (43%) in the group presented with recurrent metastases and 65 patients (57%) did not have recurrent metastases. The median progression time was 16.8 and 17.03 months for patients with, and without recurrent metastasis, respectively. The Mann-Whitney test was used to compare the median progression time according to recurrent metastasis. There was no statistically significant difference ($p = 0.859$, Table 4).

Table 4. Progression group and recurrent metastasis.

Recurrent Metastasis	Present	Absent
Mean ± SD	20.51 ± 13.17 (months)	20.18 ± 14.3 (months)
Median	16.8	17.03
Min–Max	1.12–59.7	0.0–84.43
N	49	65

3.13. Progression Group According to BMI

Finally, from the same group of 114 patients who presented with progression of the disease, 32 (28%) were NW, 46 (31%) were OW, and 36 (41%) were OB. The median time of progression was 15.42, 19.56, and 17.01 months for the patients categorized as NW, OW, and OB, respectively. The Kruskal-Wallis one-way analysis of variance was used to compare the median time in the progression group according to BMI. There was no statistically significant difference between NW and the other two groups ($p = 0.4$, Table 5). We also performed a Mann-Whitney test to compare in the progression group the median time between NW and OW, NW and OB, and OW and OB groups. There was no statistically difference between NW and OW ($p = 0.3$), NW and OB ($p = 0.1$), and OW and OB ($p = 0.2$).

Table 5. Progression group and body mass index.

BMI	(18.5–24.9 kg/m^2)	(25.0–29.9 kg/m^2)	(30.0–42.0 kg/m^2)
Mean ± SD	15.76 ± 11.64 (months)	21.24 ± 11.36 (months)	23.22 ± 17.23 (months)
Median	15.42	19.56	17.01
Min–Max	1.12–55.76	0.0–51.91	4.31–84.43
N	32	46	36

We used the Kaplan-Meier method to estimate the progression of PCa as an end-point, from the diagnostic date until the date of the last consultation in the progression group. The log-rank non-parametric test for comparison of progression distributions was used to compare recurrence differences between the NW, OW, and OB groups. There was no difference between the recurrence distributions ($p = 0.09$, Figure 3).

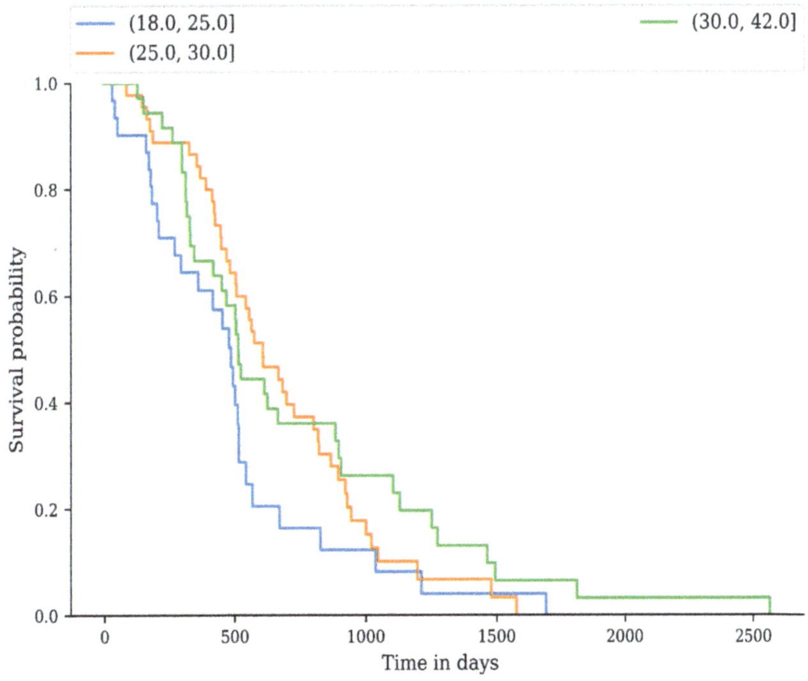

Figure 3. Progression of prostate cancer based on BMI.

4. Discussions

Prostate cancer remains a global health burden and cases will continue to rise, but pharmacological and technological advances are considerably improving and should enable future precision and improved clinical outcomes [10]. A proof-of-concept for future precision was recently demonstrated in some studies, where the teams used a Transient Receptor Potential Melastatin 8 channel (TRPM8) agonist, an ion channel in the plasma membrane, encapsulated into a Lipid NanoCapsule in order to inhibit PCa cell migration. The use of TRPM8 has a great impact in castration-resistant prostate cancer, which is considered to be the most aggressive form of PCa, because it is usually resistant to androgen deprivation therapy [11,12]. However, it is clear that over time a high index of body mass will influence our general physical well-being, along with age and other factors, and can lead to negative results in cases of pathologies, such as cancers.

Our study suggests that prostate cancer incidence rises with age (72 ± 7.81 years) in men with a normal BMI, but the diagnostic age tends to drop in those with higher BMIs, i.e., overweight, and obese in the age range of 69.47 ± 6.31 years, respectively, 69.1 ± 7.51 years. A single-center retrospective study from 2022 investigated the relationship between BMI and prostate cancer risk in 1079 Italian men, which also revealed that overweight and obese men were diagnosed at younger ages, compared to normal-weight patients. Furthermore, excessive fat accumulation contributes to a more favorable tumor microenvironment onset and growth [13].

We also observed, as well mentioned in the study cited above, that PSA tends to diminish with increasing BMI classification, but no statistically significant difference was observed. Both factors, higher BMI and lower PSA, can lead to a high-grade PCa, avoiding an early diagnostic, also seen in this study, where almost half of OW and OB patients had a stage IV PCa diagnostic.

A study from 2017 suggests that long-term weight gain is associated with an increased risk of high-grade PCa among never smokers and among men who were overweight or obese at age 21 [14]. Furthermore, OB men tend to have higher levels of insulin, insulin-like growth factor-1, and lower level of androgens and adiponectin, suggesting that inflammatory and hormonal pathways are involved. Prolonged hyperinsulinemia raises the bioavailability of IGF-1, which has been shown to promote proliferation and inhibit apoptosis in normal prostate and tumor cells in vitro, increasing the risk of PCa [14–18]. Adiponectin is secreted by the adipose tissue and is inversely related to the degree of adiposity [19]. This protein hormone promotes apoptosis and inhibits proliferation and angiogenesis, and higher concentrations have been shown to decrease the risk of high-grade prostate cancer [14,19] Opposed to adiponectin, leptin concentrations are directly related to adiposity and the biological effects are to stimulate cell proliferation and promote angiogenesis [20].

The immune response could also play an important role in the progression of PCa. In a recent study by Fujita et al., they researched the relationship between high-fat diet (HFD) induced inflammation and tumor progression PCa in mice and found that local inflammation of the prostate is one of the most important factors for the progression of PCa in obese and HFD-fed mice in early and late stages. Interestingly, the number of B cells, T cells, macrophages, and mast cells and the ratio of CD8/CD4 T cells were not changed by the HFD, but the number of myeloid-derived suppressor cells and the M2/M1 macrophage ratio were significantly increased in the HFD-fed mice compared with the control group. Using celecoxib, a cyclooxygenase 2 inhibitor, the promotion of tumor growth by the HFD was canceled, which suggests that inflammation plays a specific role in tumor progression caused by HFD. All these hormonal and immune changes in obese people lead to chronic inflammation and play an important role in the development and progression of PCa [21].

We observed a statistically significant difference in the progression group of de novo metastases versus the absent metastases group at diagnostic (p = 0.04). The progression group with metastases present (n = 70) at diagnostic had a shorter progression time, compared to the absent metastases group (n = 44), 18.04 ± 11.37 months, respectively,

23.95 ± 16.39 months. No other statistically significant difference was found between the progression and recurrent metastasis group ($p = 0.859$) or BMI ($p = 0.4$).

Our study has its limitations. Since the cohort study has not had enough participants, it is possible that the results could not meet the criteria for significance. For example, the progression of PCa in the BMI groups was not statistically significant between the NW group and the OW, or OB group, but looking at the Kaplan-Meier curve, the NW line has a more obvious decrease at some point, compared to the OW and OB groups. Some studies suggest that there is an inverse association in the progression of the disease between obese and normal-weight patients, where normal-weight patients have a faster relapse and obesity has a protective effect from the PCa recurrence [22,23]. This inverse association is called the obesity paradox. A study conducted by Martini et al. and another study by Schiffmann et al., revealed that obese patients treated with docetaxel and prednisone for metastatic castration-resistant prostate cancer benefited of a protective factor against overall mortality and death, respectively, the second team mentioned that increased BMI was associated with a decreased risk of metastases after radical prostatectomy [24,25]. We propose that urologists should be attentive to radical prostatectomy procedures in overweight and obese patients in order to avoid positive surgical margins. Marenco et al. suggests the fluorescent confocal microscopy as a novel technique that could be used for real-time diagnosis of PCa and also for the evaluation of surgical margins during radical prostatectomy. An advantage of fluorescent confocal microscopy, compared to other intraoperative histological evaluations, could be the rapid application to whole tissue sections [26]. Regarding lymph node dissection, we think that a more aggressive approach is currently suited for the patients with increased BMI, but we believe that further studies need to be corelated with the molecular mechanisms underlying PCa migration, in order to enable a better clinical, surgical and pharmacological management. Such molecular mechanisms were studied recently in vivo and revealed that the upregulation of HGK (a component of Mitogen-Activated Protein Kinase Kinase Kinase Kinase 4), Culin 4B (a scaffold protein with oncogenic activity), overexpression of Human Homebox B9 (a key transcription factor that promotes metastases) and low levels of Receptor tyrosine kinase-like receptor (a noncanonical Wnt receptor) play a major role in the aggressive behavior of PCa [27–30]. Our data about the radical prostatectomy, orchiectomy, lymphadenectomy and pathological staging is limited because the surgeries and the histopathological exam were performed in other hospitals.

5. Conclusions

Our results suggest that the median diagnostic age decreases with increasing BMI category. Furthermore, overweight, and obese patients are more likely to have an advanced or metastatic prostate cancer at diagnosis. In the metastatic group we observed that the progression of the disease has a shorter interval, leading to a faster relapse. The levels of prostate serum antigen tend to become lower in the higher BMI groups, possibly leading to a late diagnosis. Further studies are needed to determine if there is an inverse association between progression of prostate cancer in normal weight and overweight or obese patients, involving excessive fat as a protective mechanism against prostate cancer. We want to address a further question. Does the subcutaneous fat exert a protective effect on the development and recurrence of prostate cancer or other pathways are involved?

Author Contributions: Conceptualization, D.P. and D.C.; methodology, D.P., R.D., S.S. and M.P.; validation, D.P., R.D., S.S. and D.C.; formal analysis, C.S.; investigation, C.S.; resources, D.P., R.D. and S.S.; data curation, D.P., C.S., R.D., S.S. and D.C.; writing—original draft preparation, D.P.; writing—review and editing, D.P., R.D. and S.S.; visualization, D.P., C.S., R.D., S.S. and M.P.; supervision, D.P. and D.C.; project administration, D.P. All authors provided critical feedback and helped shape the data, analysis, and redacted the conclusions of the study. All the authors have read and approved the final version of the manuscript for publication. All authors have read and agreed to the published version of the manuscript.

Funding: This research received no external funding.

Institutional Review Board Statement: Not applicable.

Informed Consent Statement: All patients gave their informed consent for the procedure. The study protocol was conducted according to the principles of the Declaration of Helsinki after the approval of our institution's Ethical Committee. All patients provided written informed consent for the study participation and data collection.

Data Availability Statement: The data generated or analyzed during this study are included in this published article or are available from the corresponding author on reasonable request.

Conflicts of Interest: The authors declare that they have no competing interest.

References

1. Ritchie, H.; Roser, M. Obesity. Our World in Data. 2017. Available online: https://ourworldindata.org/obesity (accessed on 12 October 2022).
2. Lauby-Secretan, B.; Scoccianti, C.; Loomis, D.; Grosse, Y.; Bianchini, F.; Straif, K. Body Fatness and Cancer—Viewpoint of the IARC Working Group. *N. Engl. J. Med.* **2016**, *375*, 794–798. [CrossRef]
3. Sung, H.; Ferlay, J.; Siegel, R.L.; Laversanne, M.; Soerjomataram, I.; Jemal, A.; Bray, F. Global Cancer Statistics 2020: GLOBOCAN Estimates of Incidence and Mortality Worldwide for 36 Cancers in 185 Countries. *CA Cancer J. Clin.* **2021**, *71*, 209–249. [CrossRef] [PubMed]
4. Sakers, A.; De Siqueira, M.K.; Seale, P.; Villanueva, C.J. Adipose-Tissue Plasticity in Health and Disease. *Cell* **2022**, *185*, 419–446. [CrossRef]
5. Renehan, A.; Zwahlen, M.; Egger, M. Adiposity and cancer risk: New mechanistic insights from epidemiology. *Nat. Rev. Cancer* **2015**, *15*, 484–498. [CrossRef] [PubMed]
6. Ray, A.; Cleary, M.P. Animal Models to Study the Interplay Between Cancer and Obesity. In *Adipose Tissue and Cancer*; Kolonin, M.G., Ed.; Springer: New York, NY, USA, 2013; pp. 99–119. [CrossRef]
7. Dyba, T.; Randi, G.; Bray, F.; Martos, C.; Giusti, F.; Nicholson, N.; Gavin, A.; Flego, M.; Neamtiu, L.; Dimitrova, N.; et al. The European Cancer Burden in 2020: Incidence and Mortality Estimates for 40 Countries and 25 Major Cancers. *Eur. J. Cancer* **2021**, *157*, 308–347. [CrossRef] [PubMed]
8. Chung, B.H.; Horie, S.; Chiong, E. The Incidence, Mortality, and Risk Factors of Prostate Cancer in Asian Men. *Prostate Int.* **2019**, *7*, 1–8. [CrossRef]
9. Rawla, P. Epidemiology of Prostate Cancer. *World J. Oncol.* **2019**, *10*, 63–89. [CrossRef]
10. Rebello, R.J.; Oing, C.; Knudsen, K.E.; Loeb, S.; Johnson, D.C.; Reiter, R.E.; Gillessen, S.; Van der Kwast, T.; Bristow, R.G. Prostate Cancer. *Nat. Rev. Dis. Prim.* **2021**, *7*, 9. [CrossRef]
11. Grolez, G.P.; Hammadi, M.; Barras, A.; Gordienko, D.; Slomianny, C.; Völkel, P.; Angrand, P.O.; Pinault, M.; Guimaraes, C.; Potier-Cartereau, M.; et al. Encapsulation of a TRPM8 Agonist, WS12, in Lipid Nanocapsules Potentiates PC3 Prostate Cancer Cell Migration Inhibition through Channel Activation. *Sci. Rep.* **2019**, *9*, 7926. [CrossRef]
12. Di Sarno, V.; Giovannelli, P.; Medina-Peris, A.; Ciaglia, T.; Di Donato, M.; Musella, S.; Lauro, G.; Vestuto, V.; Smaldone, G.; Di Matteo, F.; et al. New TRPM8 Blockers Exert Anticancer Activity over Castration-Resistant Prostate Cancer Models. *Eur. J. Med. Chem.* **2022**, *238*, 114435. [CrossRef]
13. Baio, R.; Napodano, G.; Caruana, C.; Molisso, G.; Di Mauro, U.; Intilla, O.; Pane, U.; D'Angelo, C.; Francavilla, A.B.; Guarnaccia, C.; et al. Association between Obesity and Frequency of High-grade Prostate Cancer on Biopsy in Men: A Single-center Retrospective Study. *Mol. Clin. Oncol.* **2022**, *17*, 127. [CrossRef] [PubMed]
14. Lavalette, C.; Duverger, E.C.; Artaud, F.; Rébillard, X.; Lamy, P.; Trétarre, B.; Cénée, S.; Menegaux, F. Body Mass Index Trajectories and Prostate Cancer Risk: Results from the EPICAP Study. *Cancer Med.* **2020**, *9*, 6421–6429. [CrossRef] [PubMed]
15. Dickerman, B.A.; Ahearn, T.U.; Giovannucci, E.; Stampfer, M.J.; Nguyen, P.L.; Mucci, L.A.; Wilson, K.M. Weight Change, Obesity and Risk of Prostate Cancer Progression among Men with Clinically Localized Prostate Cancer: Weight Change, Obesity and Prostate Cancer Risk. *Int. J. Cancer* **2017**, *141*, 933–944. [CrossRef] [PubMed]
16. Chan, J.M.; Stampfer, M.J.; Ma, J.; Gann, P.; Gaziano, J.M.; Pollak, M.; Giovannucci, E. Insulin-Like Growth Factor-I (IGF-I) and IGF Binding Protein-3 as Predictors of Advanced-Stage Prostate Cancer. *Cancer Spectr. Knowl. Environ.* **2002**, *94*, 1099–1106. [CrossRef]
17. Cao, Y.; Nimptsch, K.; Shui, I.M.; Platz, E.A.; Wu, K.; Pollak, M.N.; Kenfield, S.A.; Stampfer, M.J.; Giovannucci, E.L. Prediagnostic Plasma IGFBP-1, IGF-1 and Risk of Prostate Cancer: Prediagnostic IGFBP-1, IGF-1 and Prostate Cancer. *Int. J. Cancer* **2015**, *136*, 2418–2426. [CrossRef]
18. Roddam, A.W.; Allen, N.E.; Appleby, P.; Key, T.J.; Ferrucci, L.; Carter, H.B.; Metter, E.J.; Chen, C.; Weiss, N.S.; Fitzpatrick, A.; et al. Insulin-like Growth Factors, Their Binding Proteins, and Prostate Cancer Risk: Analysis of Individual Patient Data from 12 Prospective Studies. *Ann. Intern. Med.* **2008**, *149*, 461. [CrossRef]
19. Li, H.; Stampfer, M.J.; Mucci, L.; Rifai, N.; Qiu, W.; Kurth, T.; Ma, J. A 25-Year Prospective Study of Plasma Adiponectin and Leptin Concentrations and Prostate Cancer Risk and Survival. *Clin. Chem.* **2010**, *56*, 34–43. [CrossRef]

20. Cantarutti, A.; Bonn, S.E.; Adami, H.-O.; Grönberg, H.; Bellocco, R.; Bälter, K. Body Mass Index and Mortality in Men with Prostate Cancer: BMI and Mortality in Men With Prostate Cancer. *Prostate* **2015**, *75*, 1129–1136. [CrossRef]
21. Fujita, K.; Hayashi, T.; Matsushita, M.; Uemura, M.; Nonomura, N. Obesity, Inflammation, and Prostate Cancer. *J. Clin. Med.* **2019**, *8*, 201. [CrossRef]
22. Porter, M.P.; Stanford, J.L. Obesity and the Risk of Prostate Cancer. *Prostate* **2005**, *62*, 316–321. [CrossRef]
23. Giovannucci, E.; Michaud, D. The Role of Obesity and Related Metabolic Disturbances in Cancers of the Colon, Prostate, and Pancreas. *Gastroenterology* **2007**, *132*, 2208–2225. [CrossRef] [PubMed]
24. Schiffmann, J.; Karakiewicz, P.I.; Rink, M.; Manka, L.; Salomon, G.; Tilki, D.; Budäus, L.; Pompe, R.; Leyh-Bannurah, S.-R.; Haese, A.; et al. Obesity Paradox in Prostate Cancer: Increased Body Mass Index Was Associated with Decreased Risk of Metastases after Surgery in 13,667 Patients. *World J. Urol.* **2018**, *36*, 1067–1072. [CrossRef] [PubMed]
25. Martini, A.; Shah, Q.N.; Waingankar, N.; Sfakianos, J.P.; Tsao, C.-K.; Necchi, A.; Montorsi, F.; Gallagher, E.J.; Galsky, M.D. The Obesity Paradox in Metastatic Castration-Resistant Prostate Cancer. *Prostate Cancer Prostatic Dis.* **2022**, *25*, 472–478. [CrossRef] [PubMed]
26. Marenco, J.; Calatrava, A.; Casanova, J.; Claps, F.; Mascaros, J.; Wong, A.; Barrios, M.; Martin, I.; Rubio, J. Evaluation of Fluorescent Confocal Microscopy for Intraoperative Analysis of Prostate Biopsy Cores. *Eur. Urol. Focus* **2021**, *7*, 1254–1259. [CrossRef]
27. Garcia-Garcia, S.; Rodrigo-Faus, M.; Fonseca, N.; Manzano, S.; Győrffy, B.; Ocaña, A.; Bragado, P.; Porras, A.; Gutierrez-Uzquiza, A. HGK Promotes Metastatic Dissemination in Prostate Cancer. *Sci. Rep.* **2021**, *11*, 12287. [CrossRef]
28. Qi, M.; Hu, J.; Cui, Y.; Jiao, M.; Feng, T.; Li, X.; Pang, Y.; Chen, X.; Qin, R.; Su, P.; et al. CUL4B Promotes Prostate Cancer Progression by Forming Positive Feedback Loop with SOX4. *Oncogenesis* **2019**, *8*, 23. [CrossRef]
29. Sui, Y.; Hu, W.; Zhang, W.; Li, D.; Zhu, H.; You, Q.; Zhu, R.; Yi, Q.; Tang, T.; Gao, L.; et al. Insights into Homeobox B9: A Propeller for Metastasis in Dormant Prostate Cancer Progenitor Cells. *Br. J. Cancer* **2021**, *125*, 1003–1015. [CrossRef]
30. Tseng, J.-C.; Huang, S.-H.; Lin, C.-Y.; Wang, B.-J.; Huang, S.-F.; Shen, Y.-Y.; Chuu, C.-P. ROR2 Suppresses Metastasis of Prostate Cancer via Regulation of MiR-199a-5p–PIAS3–AKT2 Signaling Axis. *Cell Death Dis.* **2020**, *11*, 376. [CrossRef]

medicina

MDPI

Review

Ovarian Cancer—Insights into Platinum Resistance and Overcoming It

Andrei Havasi [1,*], Simona Sorana Cainap [2,3], Ana Teodora Havasi [4] and Calin Cainap [1,5]

1 Department of Oncology, "Iuliu Hatieganu" University of Medicine and Pharmacy, 400012 Cluj-Napoca, Romania
2 Pediatric Clinic No. 2, Department of Pediatric CNardiology, Emergency County Hospital for Children, 400177 Cluj-Napoca, Romania
3 Department of Mother and Child, "Iuliu Hatieganu" University of Medicine and Pharmacy, 400013 Cluj-Napoca, Romania
4 Department of Gastroenterology, County Hospital Oradea, 410169 Oradea, Romania
5 Department of Medical Oncology, The Oncology Institute "Prof. Dr. Ion Chiricuta", 400015 Cluj-Napoca, Romania
* Correspondence: havasi.andrei@gmail.com

Abstract: Ovarian cancer is the most lethal gynecologic malignancy. Platinum-based chemotherapy is the backbone of treatment for ovarian cancer, and although the majority of patients initially have a platinum-sensitive disease, through multiple recurrences, they will acquire resistance. Platinum-resistant recurrent ovarian cancer has a poor prognosis and few treatment options with limited efficacy. Resistance to platinum compounds is a complex process involving multiple mechanisms pertaining not only to the tumoral cell but also to the tumoral microenvironment. In this review, we discuss the molecular mechanism involved in ovarian cancer cells' resistance to platinum-based chemotherapy, focusing on the alteration of drug influx and efflux pathways, DNA repair, the dysregulation of epigenetic modulation, and the involvement of the tumoral microenvironment in the acquisition of the platinum-resistant phenotype. Furthermore, we review promising alternative treatment approaches that may improve these patients' poor prognosis, discussing current strategies, novel combinations, and therapeutic agents.

Keywords: ovarian cancer; platinum resistance; platinum resistance mechanisms; overcoming platinum resistance

Citation: Havasi, A.; Cainap, S.S.; Havasi, A.T.; Cainap, C. Ovarian Cancer—Insights into Platinum Resistance and Overcoming It. *Medicina* **2023**, *59*, 544. https://doi.org/10.3390/medicina59030544

Academic Editor: Paolo Zola

Received: 1 February 2023
Revised: 26 February 2023
Accepted: 7 March 2023
Published: 10 March 2023

1. Introduction

Ovarian cancer is a significant cause of morbidity and mortality worldwide, responsible for 300,000 new cases each year and almost as many deaths [1]. Often diagnosed in an advanced stage, it is the most lethal gynecological cancer, with a 5-year survival rate of 26–42%, depending on the initial stage [2]; however, more than 40% of stage III/IV patients die within the first year and 25% within the first 90 days following diagnosis [3]. Ovarian tumors may arise from epithelial, stromal, or germ cells, where over 90% of malignant ovarian tumors arise from epithelial cells [4]. A heterogeneous disease, epithelial ovarian cancer (EOC) comprises several histological subtypes: high-grade serous ovarian cancer (HGSOC) (70–80%), endometrioid (10%), clear cell (10%), mucinous (3%), and low-grade serous (<5%) [5].

Most high-grade serous tumors are sporadic; however, up to 15% of patients with ovarian cancer have a genetic predisposition. BRCA1 and BRCA2 mutations are responsible for most hereditary epithelial ovarian cancer, while Lynch syndrome is associated with clear-cell and endometrioid tumors [6]. Patients with a hereditary predisposition have a younger age at presentation, a history of other cancers, and positive family history. Traditionally patients that fit this profile are considered for genetic testing; however, up to 44% of BRCA carriers have no family history, and the highest annual risk for ovarian cancers for BRCA carriers is for those between 50 and 69 years old [7,8]. Almost all

BRCA-mutated cancers are high-grade serous ovarian cancers [9]. However, high-grade serous tumors are not limited to the BRCA mutation. TP53 mutations are present in almost 97% of tumors, and other homologous recombination repair mutations, including EMSY, RAD51, ATM, BARD1, BRIP1, ATR, PALB2, RB1, and CDKN2A, have been identified in 11% of patients [10].

Surgery with complete cytoreduction is the standard of care in the management of EOC. In patients in which resection with no gross residual disease can be achieved, surgery can be offered upfront, followed by adjuvant systemic therapy. For patients with stage III and IV, bulky, extensive disease, neoadjuvant chemotherapy followed by interval debulking surgery and adjuvant chemotherapy was associated with higher R0 resection rates and better survival outcomes than primary debulking surgery [11]. Surgery continues to play an essential role in the management of EOC, even in recurrent disease. Secondary debulking surgery in platinum-sensitive recurrent EOC has been linked to a 5-month progression-free survival (PFS) improvement. This survival benefit was more significant in patients where R0 resection had been achieved [12].

Systemic treatment is essential in the management of ovarian cancer. The EORTC-ACTION [13] and ICON1 [14] trials established almost 20 years ago the superior outcomes associated with the use of platinum and taxane-based chemotherapy in ovarian cancer compared with monotherapy and observation alone. Patient selection for adjuvant chemotherapy depends on histological subtype and tumor grade. In very early-stage IA tumors, observation is recommended for low-grade serous, grade 1,2 endometrioid, and grade 1,2 mucinous ovarian cancer. Nevertheless, there remains a question regarding the benefit of adjuvant treatment for clear-cell carcinoma stage I A, B, and C1; stage IB and IC; grade 1,2 endometrioid; and stage IB/C low-grade serous [15]. Regarding the number of chemotherapy administrations, the standard number of cycles remains six; however, the GOG157 trial reported similar outcomes when using three or six of paclitaxel and carboplatin [16].

The combination of paclitaxel and carboplatin is the standard of care treatment for patients with advanced disease [17]. Other various dose-dense administration schedules and the association of intraperitoneal chemotherapy have been investigated in numerous trials but with conflicting results [18]. Despite adjuvant treatment and optional surgery, patients with advanced ovarian cancer often present high recurrence rates. Therefore, several trials have investigated the use of maintenance therapies; paclitaxel marginally improved progression-free survival (PFS) [19], and maintenance with bevacizumab proved to provide benefits only in PFS terms, except for a subgroup of poor-prognosis patients where a trend toward improved overall survival (OS) could be observed [20]. Additionally, for BRCA-mutated advanced ovarian cancer, a PARP inhibitor (PARPi) can also be considered as maintenance after first-line treatment [21].

The primary challenge in the treatment of cancer is treatment resistance. Unfortunately, ovarian cancer is no exception to treatment resistance, with particular importance being the resistance to platinum compounds. Traditionally platinum resistance was defined on the basis of the duration of the response to platinum-containing chemotherapy. Patients who initially respond to platinum-based chemotherapy and relapse 6 months or longer after the initial treatment were classified as platinum sensitive, while patients who relapse within under 6 months after platinum-based chemotherapy were considered platinum resistant. Within the platinum-resistant group, a subgroup of patients presents with the worst prognosis: platinum-refractory ovarian cancer, with a disease that progresses during or within 1 month of platinum-containing first-line chemotherapy [22]. However, the current classification of platinum resistance, which is based on the 6-month platinum-free interval, has several shortcomings: the use of bevacizumab or bevacizumab with olaparib as maintenance therapy for patients responding to first-line platinum-based chemotherapy significantly prolonged PFS, rendering the evaluation of platinum response difficult [23,24]; platinum rechallenge in patients with platinum-free intervals longer than 6 months results in response rates of only 47–66% [25,26], while a platinum-free interval shorter than

6 months does not exclude a benefit from the addition of platinum to chemotherapy [27,28]. Furthermore, the studies used to define platinum resistance determined recurrence on the basis of clinical and radiological evidence of disease or clinical symptoms. These studies took place before the widespread use of CA125 to detect recurrence, and additionally, the imaging techniques available were inferior to currently used high-resolution CT, MRI, or PET-CT [22].

Most patients generally respond well to platinum-based chemotherapy, with only 20% of HGSOC presenting from the beginning with the platinum-resistant disease. However, the majority of initially platinum-sensitive patients will develop secondary platinum resistance following multiple recurrences with progressively shorter progression-free survival [15]. Therefore eventually, platinum resistance influences the prognosis of every ovarian cancer patient, representing one of the key prognostic factors influencing overall survival. In the following review, we will discuss the underlying mechanisms of platinum resistance, available biomarkers, and possibilities of overcoming resistance.

2. Molecular Mechanisms of Platinum Resistance in High-Grade Ovarian Cancer

Platinum compounds exert their cytotoxic anticancer effects mainly by forming covalent bonds to the DNA, thus generating DNA crosslinks and inhibiting DNA replication, eventually leading to cell death. The mechanisms of platinum resistance are multifactorial and comprise genetic and epigenetic alterations as well as immune and environmental factors frequently involving more than one mechanism of resistance [29]. Figure 1 summarizes the main mechanisms in the development of platinum resistance.

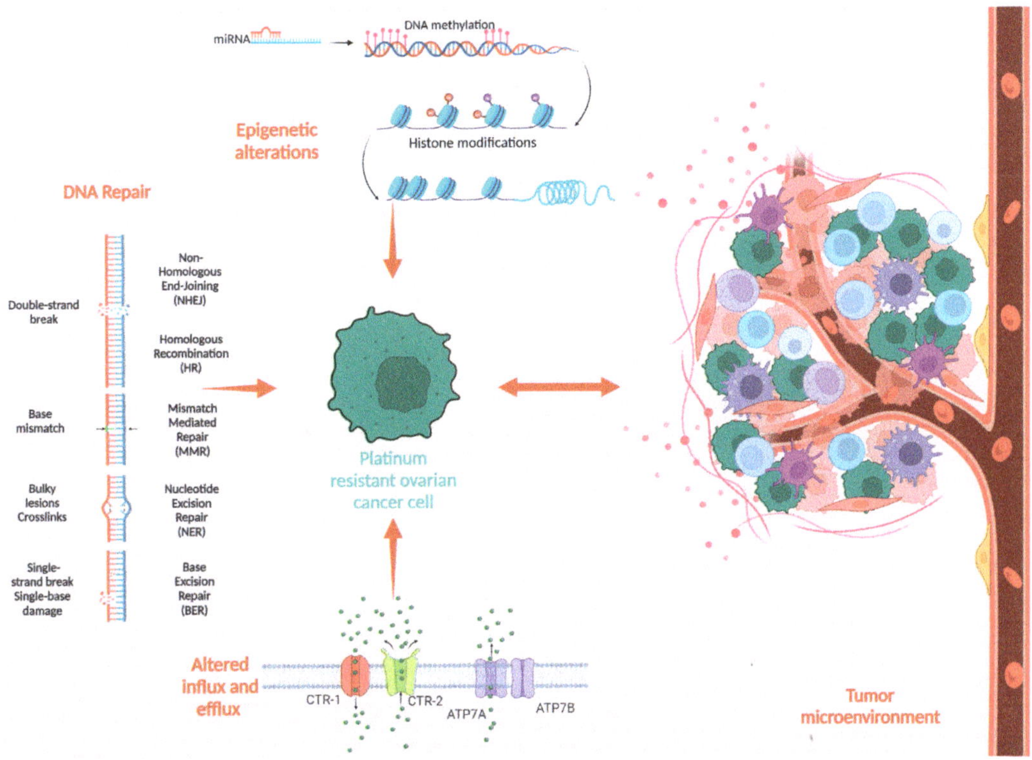

Figure 1. Schematic overview of the mechanisms of platinum resistance in ovarian cancer.

2.1. Alteration of Drug Influx and Efflux Pathways

One of the most-agreed-upon mechanisms of platinum resistance is the dysregulation of drug influx and efflux pathways that modulate the transport of platinum salts in the cancer cell. As a result, platinum-resistant cell lines display a reduction in cisplatin concentration, varying from 20% to 70% [30]. The copper transporter 1 (CTR-1), a transmembrane influx transporter involved in copper homeostasis, also plays a crucial role in the intracellular uptake of platinum salts. The knockout of CTR-1 in mouse cell lines led to platinum resistance via decreased intracellular platinum concentrations [31]; similarly, the overexpression of CTR-1 led to increased sensitivity to platinum in ovarian cell lines [32]. Additionally, Song et al. [33] demonstrated that the upregulation of CTR-1 expression in cisplatin-resistant small cell lung cancer cell lines restored platinum sensitivity. Ishida et al. correlated tumoral CTR-1 mRNA levels with a response to platinum-based chemotherapy in 15 patients with stage III or IV HGSOC who underwent cytoreductive surgery. Patients with platinum-sensitive disease expressed significantly higher levels of CTR-1 mRNA compared with platinum-resistant or refractory disease. These results were further validated by using clinical and array-based data from the Cancer Genome Atlas, on a subset of 91 stage III and IV HGSOC patients who underwent surgery followed by platinum-based adjuvant chemotherapy. Patients with high CTR-1 expression had significantly prolonged disease-free survival compared with those expressing low levels of CTR-1 [34]. Organic cation transporters (OCTs) are part of the solute carrier family and are involved in the cellular uptake of platinum derivates. Furthermore, low OCT6 expression has been associated with platinum resistance in human lung cancer cell lines [35].

The copper transporter 2 (CTR-2) is also involved in regulating cellular platinum levels; however, it acts as a platinum efflux transporter. Higher CTR-2 expression was linked to platinum resistance in ovarian cancer cell lines [36]. The copper exporters ATP7A and ATP7B are also involved in platinum efflux and subsequent resistance. ATP7A is responsible for the intracytoplasmic sequestration of platinum derivates blocking their access to the nucleus, while ATP7B facilitates drug efflux via the secretory pathway. The overexpression of both ATP7A and ATP7B has been associated with platinum resistance, whereas blocking their activity restores platinum sensitivity [37–39]. The altered expression of multidrug resistance proteins (MRPs) has been linked to multidrug resistance and worse outcomes in multiple cancers. Arts et al. [40] found that increased MRP2 and MRP4 expression was linked to platinum resistance and poor outcomes in ovarian cancer. Similarly, three other reports have associated high MRP2 levels and resistance to platinum-based chemotherapy in various cancers, including ovarian cancer [41–43].

2.2. DNA Repair

DNA is the main target of platinum-based anticancer drugs, and the cell's ability to recognize and repair drug-induced DNA damage can influence its sensitivity or resistance to platinum chemotherapy. The primary mechanism through which platinum chemotherapy exerts its cytotoxic effects is the formation of DNA monoadducts that evolve through covalent binding to DNA crosslinks that can occur either on the same DNA strand or on the opposite strands, generating interstrand crosslinks that block DNA synthesis and transcription if they are not repaired. The DNA damage response (DDR) mechanism is activated in the presence of DNA lesions. DDR consists of several signaling pathways responsible for enforcing cell-cycle arrest and, depending on the severity of DNA damage, either DNA repair or the activation of apoptosis for cells presenting with unrepairable DNA lesions [44]. Six major DNA repair pathways have been described: mismatch repair (MMR), base excision repair (BER), nucleotide excision repair (NER), homologous recombination (HR), nonhomologous end joining (NHEJ), and Fanconi anemia (FA). An intertwined activation of these pathways is responsible for repairing DNA lesions and preventing the development of various pathologies, including cancer [45,46].

The same pathways are also accountable for preventing the accumulation of DNA lesions secondary to platinum-based chemotherapy, and their variation may promote platinum sensitivity or resistance. The upregulation of DNA repair proteins may lead to removing platinum adducts and repairing tumoral DNA, decreasing treatment efficacy. Most platinum-resistant tumors display the upregulation of DNA damage repair proteins such as BRCA 1/2, mismatch repair proteins MSH1 and MSH2, excision repair cross-complementing (ERCC) proteins, RAD51, and Fanconi anemia complementation group D2 [29,47]. BRCA1/2-mutated HGSOC have increased sensitivity to DNA-damaging agents such as PARPis and platinum agents and have an improved overall response to platinum therapy [7,48]. CDK12, a kinase involved in the HR pathway, is mutated in 3% of ovarian cancer patients. Preclinical data have associated low CDK12 expression with higher susceptibility to cisplatin and PARPis [49]. Replication protein A (RPA) recognizes single-stranded DNA lesions interfering with the replication fork and acts as an activation platform for DNA damage repair via NER. RPA-deficient ovarian cancer cells cannot efficiently repair cisplatin-induced DNA lesions via NER and display increased platinum sensitivity [50]. NER alterations are present in 8% of HGSOC and are associated with increased sensitivity to platinum chemotherapy, similar to BRCA1/2-mutated patients [51]. ERCC1, a NER-associated protein, is one of the most promising biomarkers for platinum sensitivity in these patients. Low ERCC1 expression was associated with platinum sensitivity [52–54], but these findings were inconsistent across multiple studies, where some reported an absent or negative correlation between ERCC1 and a response to platinum [47].

2.3. *Epigenetic Alterations*

Epigenetic processes influence gene expression without changing the DNA sequence. They are essential in ensuring normal genome functioning and ensuring altered epigenetic regulation results in the development of various pathologies, including cancer. Three key processes are involved in the epigenetic regulation of HGSOC: DNA methylation, histone modification, and microRNAs (miRs).

2.3.1. DNA Methylation

DNA methylation modulates gene expression via DNA methyltransferase enzymes that catalyze the addition of a methyl group or an ethyl group onto the fifth carbon of a cytosine ring to form methylcytosine. DNA methylation frequently occurs in areas known as CpG islands, often located in the promoter region of genes. Increased cytosine methylation in the promoter region is known as hypermethylation and decreases gene expression by inhibiting transcription factors and RNA polymerase from binding DNA and undergoing transcription [55]. The role of DNA methylation in ovarian cancer chemoresistance has been extensively studied. Lum et al. [56] analyzed DNA methylation in 36 HGSOC samples segregated on the basis of platinum sensitivity. They identified 749 probes corresponding to 296 genes that were significantly differently methylated in platinum-sensitive samples and in platinum-resistant samples; furthermore, they observed that hypermethylation was more often present in platinum-resistant samples than in platinum-sensitive ones. Two other reports found the same association between hypermethylation and platinum resistance [57,58]; however, these findings are inconsistent across studies. Lund et al. [59] found that the majority (1251 of 1488) of the differentially methylated sites were hypomethylated in cisplatin-resistant samples. A pathway analysis of the 452 hypermethylated genes associated with platinum resistance, by Cardenas et al. [58], found the epithelial–mesenchymal transition (EMT) pathway to be the most influenced by aberrant methylation in the development of the chemoresistant phenotype. MSX1 encodes a member of the muscle segment homeobox gene family and can influence EMT in ovarian cancer. The hypomethylation of MSX1 leads to decreased MSX1 expression, which is associated with cisplatin resistance in ovarian cancer cell lines, while MSX1 overexpression sensitizes cells to cisplatin [60]. LAMA3 (laminin alpha 3), a component of the cell base membrane, plays an important role in cell adhesion, migration, and embryo differentiation. Reduced LAMA3 expression has been associated with EMT in various tumors, including ovarian cancer. Feng et al. demonstrated that the hypermethylation of LAMA3 was responsible for the reduced expression and that decreased LAMA3 levels were correlated with chemoresistance and poor outcomes [61]. The SOX9, ZIC1, and TWIST genes involved in EMT were also associated with a hypermethylated status in platinum-resistant ovarian cancer [56]. The aberrant methylation of genes involved in the wingless/integrated (Wnt) signaling pathway was also associated with platinum resistance in HGSOC. FZD1, FZD10, and GSK3B were found to be differentially methylated in both platinum-resistant samples and platinum-sensitive ones [62]. Wang et al. found that DNA methylation via the PI3K-Akt pathway is associated with low BRCA1 expression in ovarian cancer cell lines, and BRCA1 demethylation was associated with the development of platinum resistance [63].

2.3.2. Histone Modifications

Histone modifications, regulated by histone-modifying enzymes, directly affect gene expression by altering the chromatin structure. Histones are susceptible to several changes, including acetylation, methylation, phosphorylation, ubiquitination, glycosylation, sumoylation, ADP-ribosylation, and carbonylation. However, histone acetylation is of particular importance as it has been associated with ovarian cancer pathogenesis [64]. Histone acetyltransferase (HAT) enzymes add acetyl groups to the histone surface, enabling RNA polymerase II interaction and favoring gene expression. Meanwhile, histone deacetylase (HDAC) enzymes remove acetyl groups from histones and increase chromatin compaction, thus restricting RNA polymerase II access with subsequently decreased gene expression [62]. Cacan et al. [65] demonstrated HDAC1 involvement in cisplatin resistance in ovarian cancer cells. The suppression of HDAC1 and DNA methyltransferase activity in platinum-resistant ovarian cancer cells restored cisplatin-mediated cell deaths through the upregulation of RGS10, an essential regulator of cell survival and chemoresistance. Liu et al. [66] demonstrated that HDAC1 knockdown in cisplatin-resistant cell lines suppressed proliferation and increased apoptosis and chemosensitivity through the downregulation of the c-Myc oncogene and the upregulation of miR-34a. Furthermore, cisplatin treatment in platinum-sensitive cells increased HDAC1 and c-Myc expression while inactivating miR-34a, leading cells to acquire chemoresistance to cisplatin.

2.3.3. MicroRNAs

MicroRNAs are small 19–25-nucleotides-long single-stranded noncoding RNAs in the post-translational regulation of gene expression. Multiple miRs have altered expression in HGSOC and are associated with carcinogenesis, progression, metastasis, and drug resistance [67]. MiR-mediated platinum resistance arises through multiple mechanisms influenced by microRNA dysregulation. MiR-130a was found to be involved in platinum resistance occurrence by altering cellular cisplatin uptake. The overexpression of miR-130a was associated with platinum resistance by targeting the SOX9/miR-130a/CTR1 axis [68]. Cisplatin resistance can also be secondary to increased cellular drug efflux. The ATP7A and ATP7B transporters are associated with platinum chemotherapy resistance and are influenced by miRs expression. MiR-139 dysregulation influences ATP7A and ATP7B expression with secondary platinum resistance. Platinum-resistant cell lines presented low levels of MiR-139 and high ATP7A and ATP7B expression. MiR-139 overexpression enhanced the suppressive effect of cisplatin on resistant cell lines. Furthermore, there is an inverse correlation between miR-139 and ATP7A/B expression [69]. MiR-15a and miR-16 are also involved in ATP7B regulation and platinum resistance. MiR-15a and miR-16 transfection in cisplatin-resistant cell lines and murine models have restored cisplatin sensitivity by inhibiting ATP7B expression [70]. MiR also influences MRP2-associated resistance. The upregulation of miR-490-3p and downregulation of miR-411 was associated with increased cisplatin sensitivity via the inhibition of MRP2 expression ovarian cell lines [71,72]. MiR-514 downregulation was associated with advanced stages of and poor outcomes in ovarian cancer. MiR-514 also increases cisplatin chemosensitivity by targeting ATP-binding cassette subfamily members ABCA1, ABCA10, and ABCF2 [73].

MicroRNA modulation influences pathways involved in the process of DNA repair and the secondary platinum resistance induced by their activation. One study demonstrated that miR-211 expression enhanced platinum sensitivity in ovarian cancer cells by targeting DDR. MiR-211 facilitated platinum-induced DNA damage by targeting DDR effector genes, including POLH, TDP1, ATRX, MRPS11, and ERCC6L2 [74]. Enhanced nucleotide excision repair is associated with resistance to platinum chemotherapy. ERCC1, an essential effector of the NER pathway, is characterized as a potential biomarker for platinum resistance and is a direct target of miR-30a-3p. Increased miR-30a-3p restored cisplatin sensitivity by targeting ERCC1 and ATP7A [75]. NER-pathway-induced resistance is also influenced by miR-770-5p expression. Downregulated in cisplatin-resistant cell lines, miR-770-5p overexpression restored cisplatin sensitivity by directly targeting ERCC2, an effector of the NER pathway [76]. Zhu et al. [77] also demonstrated miR-770-5p involvement in cisplatin resistance; the long noncoding RNA nuclear paraspeckle assembly transcript 1 (NEAT1) has been shown to enable treatment resistance by inhibiting miR-770-5p and upregulating PARP1 expression, a promoter of platinum resistance. MiR-9 inhibits homologous recombination-associated resistance by targeting BRCA1. Patients with high MiR-9 expression have better chemotherapy responses and increased platinum sensitivity. MiR-9 levels were inversely correlated with BRCA1 expression and treatment with miR-9-sensitized BRCA1-proficient cell lines to cisplatin [78]. MiR-506 and miR-152 can increase platinum sensitivity by targeting RAD51 and suppressing HR [79,80]. Choi et al. demonstrated that miR-622 could be responsible for platinum and PARPi resistance in BRCA1-mutated tumors by restoring HR-mediated double-stranded break repair [81]. MiR-146a, miR-148a, and miR-545 are linked to improved outcomes in ovarian cancer patients by targeting BRCA1/2 expression [82]. In contrast, miR-493-5p expression promotes platinum and PARPi resistance in BRCA2-mutated ovarian carcinoma by reducing nucleases and other factors involved in maintaining genomic stability, thus resulting in relatively stable replication forks, diminished single-strand annealing, and increased R-loop formation [83]. The epigenetic mechanisms of resistance are also influenced by microRNA modulation. Liu et al. demonstrated that the upregulation of miR-200b and miR-200c restored cisplatin cytotoxicity by directly targeting the DNA methyltransferases (DNMT) responsible for DNA methylation, often associated with treatment resistance [84]. Low levels of miR-30a-5p and miR-30c-5p are associated with cisplatin resistance through DNMT upregulation and subsequent hypermethylation. DNMT1 is a direct target of miR-30a-5p and miR-30c-5p, and the overexpression of miR-30a-5p and miR-30c-5p-inhibited DNMT1 promoted cisplatin sensitivity and partially reversed EMT in ovarian cancer cell lines [85]. MiR-152 and miR-185 were also found to be downregulated in platinum-resistant ovarian cell lines, and their upregulation reversed cisplatin sensitivity, increased apoptosis, and inhibited proliferation by targeting DNMT1 [86].

Robust data suggest an EMT association with platinum resistance in ovarian cancer. MicroRNAs mediate platinum resistance or sensitivity by regulating EMT [87]. MiR-186 downregulation was associated with EMT and chemoresistance by targeting Twist1 in ovarian cancer cell lines [88]. MiR-363 low expression was also linked to chemoresistance and carcinogenesis via Snail-induced EMT [89]. Zhan et al. demonstrated that miR-1294 is downregulated in cisplatin-resistant ovarian cancer cell lines and that the overexpression of miR-1294 prevented platinum resistance by directly targeting IGF1R and inhibiting EMT [90]. MiR-20a promotes a cisplatin-resistant phenotype in ovarian cancer cells by activating EMT [91]. High oncogenic miR-205-5p and miR-216a levels were linked to platinum resistance in ovarian cancer cell lines by targeting the PTEN/Akt pathway [92,93]. MiR-483-3p and miR-224-5p conferred platinum resistance by suppressing protein kinase C family members [94–96]. MiR-1180 was associated with bone-marrow-derived mesenchymal stem-cell-induced platinum resistance in HGSOC cells. Mir-1180 overexpression leads to Wnt signaling and secondary glycolysis-induced chemoresistance [97]. A high expression of the platinum-refractory phenotype miR-98-5p directly targets Dicer1 and suppresses its activity, causing global miR downregulation; additionally, it inhibits cyclin-dependent kinase inhibitor 1A (CDKN1A), a promoter of cisplatin sensitivity [80,98]. Table 1 summarizes microRNA involvement in ovarian cancer platinum resistance.

Table 1. MicroRNA involvement in HGSOC platinum resistance.

MicroRNA	Target Gene	Effect on Cisplatin Response	Ref.
miR-130a	SOX9/miR-130a/CTR1 axis	Resistance	[68]
miR-411	MRP2		[72]
miR-622	Ku70, Ku80		[81]
miR-20a	EMT		[91]
miR-205-5p	PTEN		[92]
miR-216a	STAT3/miR-216a/PTEN axis		[93]
miRr-483-3p	PKC-alpha		[95]
miR-224-5p	PKC-delta		[96]
mir-1180	Wnt		[97]
miR-98-5p	Dicer1, CDKN1A		[98]
miR-493-5p	MRE11, CHD4, EXO1, RNASEH2A, FEN1, SSRP1		[83]
miR-139	ATP7A/B	Sensitivity	[69]
miR-15amiR-16	ATP7B		[70]
miR-490-3p	MRP2		[71]
miR-514	ABCA1, ABCA10, ABCF2		[73]
miR-211	POLH, TDP1, ATRX, MRPS11, ERCC6L2		[74]
miR-30a-3p	ERCC1		[75]
miR-770-5p	ERCC2, NEAT1		[76,77]
miR-9	BRCA1		[78]
miR-506	RAD51		[79]
miR-152	RAD51, DNMT		[80,86]
miR-146a, miR-148a, miR-545	BRCA1/2		[83]
miR-200b, miR-200c	DNMT		[84]
miR-30a-5p, miR-30c-5p	DNMT		[85]
miR-185	DNMT		[86]
miR-186	Twist1		[88]
miR-363	Snail-induced EMT		[89]
miR-1294	IGF1R		[90]

2.4. Tumoral Microenvironment

Ovarian cancer arises in a unique tumoral microenvironment (TME) that plays a crucial part in the natural history of the disease. The TME comprises stromal cells, immune cells, endothelial cells, adipocytes, bone-marrow-derived cells, lymphocytes, and the extracellular matrix (ECM), which play essential roles in supporting tumor progression through signaling molecules that promote cell growth, differentiation, and invasiveness. Unlike the cells of other epithelial tumors, ovarian cancer cells detach from their origin in the ovary and the fallopian tube and adhere to the mesothelial layers of the peritoneum, covering the abdominal organs and invading the submesothelial layers. In addition, ovarian cancer cells can survive in the ascitic fluid, which acts as a medium wherein tumor cells disseminate throughout the entire abdominal cavity. Alongside ovarian tumor cells, the ascitic fluid also includes multiple types of nontumorigenic cells regulated by soluble factors and extracellular vesicles that promote tumor growth and metastasis [99–102].

The extracellular matrix consists of glycosaminoglycans, proteoglycans, hyaluronan, collagen, fibronectin, vitronectin, elastin, laminin, and other glycoproteins that sustain tissue integrity but also regulate cell migration, growth, and protein synthesis [102,103]. In ovarian cancer, the ECM signaling is dysregulated through the activation of cancer-associated fibroblasts (CAFs) and tumor-associated macrophages (TAMs), which leads to excessive ECM remodeling associated with tumor progression but also treatment resistance through the activation of multiple signaling pathways [99]. Osterman et al. demonstrated the role of ECM in promoting platinum resistance in ovarian cancer, in which ECM inhibits focal adhesion kinase (FAK), a cytosolic tyrosine kinase activated by matrix and integrin receptors that controls cell motility. High FAK expression is associated with ovarian cancer cells resistant to platinum chemotherapy. Combining FAK inhibition with platinum chemotherapy overcame this resistance and increased apoptosis [104]. Cell-adhesion-mediated drug resistance (CAM-DR) enables cells to rapidly evade cytotoxic stress by interacting with the elements of the ECM. CAM-DR markers CD44, basigin (CD147), HE4, integrin α5, and β1 were elevated in chemoresistant HGSOC patients and were associated with poor outcomes [105]. Growing ovarian cancer cells in collagen type 1 decreased their platinum sensitivity by activating CAM-DR via integrin β1. Integrin β1 knockdown restored platinum sensitivity in platinum-sensitive ovarian cell lines but not in platinum-resistant ones, suggesting CAM-DR activation via integrin β1 as an initial mechanism of resistance in ovarian cancer [106]. Proteomic profiling of chemoresistant HGSOC revealed the overexpression of 10 ECM-associated proteins, specifically decorin, versican, CD147, fibulin-1, extracellular matrix protein 1, biglycan, fibronectin 1, dermatopontin, alpha-cardiac actin, and an EGF-containing fibulin-like extracellular matrix protein 1 [107]. Additionally, carboplatin treatment increased hyaluronan expression in ovarian cancer cells, leading to chemoresistance by the upregulation of the membrane ATP-binding cassette transporter proteins (ABCB3, ABCC1, ABCC2, and ABCC3) in CD44-expressing ovarian cells. Treatment with hyaluronan oligomers restored platinum sensitivity in chemoresistant cells [108].

Ovarian cancer cell and mesothelial cell crosstalk promotes tumor adhesion and invasion, but ovarian-cancer-associated mesothelial cells also induce chemoresistance through the ATP-binding cassette transporter protein induction of the fibronectin 1/Akt signaling pathway [109]. Cancer-associated fibroblasts (CAFs) occur in the TME secondary to inflammation and hypoxia. They promote tumor growth, proliferation, and metastasis; inhibit immune regulation; and modulate cell metabolism but are also involved in treatment resistance [110]. CAFs can obstruct chemotherapy transport to the cancer cell by creating physical barriers and microvascular compression. Additionally, they can mediate resistance by secreting cysteine and glutathione, thus reducing the intracellular concentration of cisplatin via competition to DNA binding sites and platinum efflux through an ATP-dependent glutathione S-conjugate export pump [111]. Functional studies have revealed that CAFs and cancer-associated adipocytes (CAAs) are also able to transfer miR-21 to the ovarian cancer cell, where it inhibits apoptosis and confers chemoresistance by the downregulation of APAF1 [112]. CAAs represent essential elements of the ovarian cancer milieu, promoting metastasis and chemoresistance. Lipidomic analysis found that CAAs were responsible for the secretion of arachidonic acid, a chemoprotective lipid mediator that acts directly on the ovarian tumor cell and inhibits cisplatin-induced apoptosis through Akt pathway activation [113]. Tumor-associated macrophages (TAMs) were also found to promote chemoresistance. Hypoxic TAMs were responsible for the exosomal transfer of miR-223 to the ovarian cancer cells that promote drug resistance by activating the PTEN-PI3K/AKT pathway [114].

3. Overcoming Platinum Resistance in Ovarian Cancer

Platinum resistance is one of the most important prognostic factors in ovarian cancer and one of the main factors driving HGSOC mortality. Therefore, overcoming platinum resistance is considered one of the most significant challenges in ovarian cancer. The current management of platinum-resistant disease involves treatment with nonplatinum chemotherapy, such as paclitaxel, pegylated liposomal doxorubicin, or topotecan alone or in association with the antiangiogenic agent bevacizumab, which improved PFS compared to chemotherapy alone [115]. Alternative treatment strategies may include gemcitabine or etoposide. Nevertheless, platinum rechallenge can also be an option even for platinum-resistant disease. Various studies demonstrated longer PFS and higher response rates for platinum-based associations compared with monotherapy, especially in patients with a platinum-free interval longer than 3 months. However, new biomarkers that may enable the selection of patients that benefit from this strategy are needed [27,116–118].

PARP inhibitors make up a class of drugs that inhibits the activity of an alternate DNA repair pathway. Single-strand DNA breaks are detected by the PARP family of proteins that initiate DNA repair through the BER pathway. PARPis block the activity of PARP1, leading to the accumulation of single-strand DNA breaks and, eventually, double-stranded DNA breaks, which only a functional HR pathway can repair. Therefore, PARPis exploit HR deficiency to promote cancer cell death [119]. Although platinum and PAPRis share a common mechanism of resistance, specifically through the reactivation of the HR pathway, PARPis are an option worth exploring in the management of platinum-resistant disease. Kaufman et al. [120] reported an objective response rate of 31.1% and stable disease in 40.4% of the platinum-resistant BRCA-mutated ovarian cancer patients treated with olaparib. A similar response rate of 33.5% was reported by Fong et al. They demonstrated a clear connection between the platinum response and the clinical benefit of olaparib in BRCA-mutated ovarian cancer. Further, 61.5% of the platinum-sensitive patients responded (partial or complete response) according to the RECIS or GCIG criteria, compared with 41.7% in the platinum-resistant group. Platinum-refractory patients had the lowest response rates; there were no radiologic responders, and only one patient had stable disease lasting for more than four cycles [121]. Similar response rates were reported for rucaparib, niraparib, and veliparib administration in the setting of platinum-resistant HGSOC [121–124].

Recently, combinational therapy with PARPi has gained attention. The association between PARPis and antiangiogenic agents was investigated across several clinical trials. Niraparib and the antiangiogenic tyrosine kinase inhibitor (TKI) anlotinib demonstrated promising objective response rates (ORR): 50% with a PFS of 9.2 months in platinum-resistant ovarian cancer patients [125]. A combined treatment of olaparib and bevacizumab resulted in a superior response and 3-year survival compared with bevacizumab and albumin-bound paclitaxel [126]. However, the association of cediranib and olaparib failed to achieve superior outcomes in platinum-resistant disease compared with chemotherapy [127,128].

Ataxia telangiectasia and RAD3-related protein kinase (ATR)/checkpoint kinase 1 (CHK1) have attracted significant attention as possible targets for anticancer therapy because of their role in regulating cell-cycle checkpoints. The ATR/CHK1 pathway acts as a sensor detecting single-stranded DNA breaks that lead to cell-cycle arrest. Combined ATR and PARP inhibition has been evaluated across multiple studies; despite promising preclinical data, the phase 2 CAPIRI trial failed to demonstrate a clinical benefit in platinum-resistant epithelial ovarian cancer [129,130]. Prexasertib, a CHK1 inhibitor, was also evaluated in BRCA wild-type HGSOC, where the majority (79%) of the patients had platinum-resistant or refractory disease. Prexasertib showed clinical activity, where 33% of the patients had a partial response (PR) and 29% had stable disease (SD) [131]. WEE-1 inhibitors target the WEE-1 kinase, a G2 cell-cycle checkpoint regulator, resulting in increased apoptosis secondary to the accumulation of irreparable genetic lesions [132]. The WEE-1 inhibitor AZD1775 was evaluated in a phase 2 trial and demonstrated clinical activity, with a 43% ORR and 5.3-month PFS in p53-mutated platinum-resistant or refractory ovarian cancer patients [132]. BET inhibitors bind the bromodomains of BET proteins interfering with BRCA1 and RAD51 expression. BET inhibition in ovarian cancer cell lines resulted in HR deficiency, thus providing an argument for combined BET and PARP inhibition. Olaparib combined with various BET acted synergistically, increasing either treatment's efficacy alone, irrespective of HR status. Additionally, the association of BET inhibition with cisplatin chemotherapy exhibited the same synergistic activity. The coadministration of cisplatin and BET inhibitors increased ovarian cancer cells' sensitivity to cisplatin even in resistant cell lines [133,134].

Epigenetic dysregulation is involved in the acquisition of the platinum-resistant phenotype through multiple mechanisms; thus, epigenetic modulators have been investigated as potential therapies to reverse platinum resistance and resensitize tumors to platinum salts. DNMT inhibitors showed modest clinical activity in monotherapy, but combined treatment may enhance sensitivity to platinum compounds. When combined with carboplatin, the DNMT inhibitor guadecitabine showed a superior 6-month PFS compared with physicians' choice treatment: 37% vs. 11% [135]. Similarly, combining carboplatin with low-dose decitabine resulted in a clinical benefit rate of 70%, with an ORR of 35% and a median PFS of 309 days [136]. Hypermethylation has been associated with an immunosuppressive tumoral milieu by silencing tumoral antigen expression and downregulating programmed death ligand (PDL) expression [137,138]. On the basis of these findings, it was hypothesized that the association of epigenetic therapy and immune checkpoint inhibitors (ICIs) could boost ovarian cancer tumoral immunogenicity and increase ICI efficiency [139]. Chen et al. [140] evaluated the hypomethylating agent guadecitabine in association with pembrolizumab in 35 platinum-resistant ovarian cancer patients, where 8.6% of the patients had PR and 22.9% SD, resulting in a clinical benefit rate of 31.4%, with a median response duration of 6.8 months. The association of the CC-486 hypomethylating agent and durvalumab was also investigated in a phase II basket trial that included platinum-resistant ovarian cancer; however, the association failed to achieve any clinical activity [141].

HDAC inhibitors were also evaluated; however, they failed to demonstrate consistent efficacy across studies [142,143]. An association between avelumab and entinostat, a class I selective HDAC inhibitor, was also assessed in pretreated ovarian cancer patients but failed to improve PFS compared with avelumab alone [144]. The association between HDAC inhibitors and DNMT inhibitors was also evaluated to determine their synergistic activity [145]. Falchook et al. [146] investigated the association of azacytidine and valproic acid in restoring carboplatin sensitivity in a phase 1 trial, with a clinical benefit rate of 18.8% but with high toxicity, where 81% of the patients reported grade ≥ 3 adverse events, including fatigue, neutropenia, and vomiting. Preclinical models evaluated the association between immunotherapy and the combination of DNMT1 with an enhancer of zeste homologue 2 (EZH2) inhibition in ovarian cancer cells. EZH2-mediated histone H3 lysine 27 trimethylation and DNMT1-mediated DNA methylation were shown to repress the production of the T helper 1 type of chemokines: CXC-motif chemokine 9 (CXCL9) and CXCL10. Combined EZH2 and DNMT1 inhibition increased effector T-cell tumor infiltration, inhibited tumor progression, and improved the therapeutic efficacy of PDL-1 blockade [147].

An immune checkpoint blockade aims to restore T-cell function and reverse tumor-associated immune-evasion mechanisms, with the aim of producing a sustained T-cell-mediated antitumoral response. Unfortunately, despite promising results in various solid tumors, checkpoint inhibition has failed to provide a significant benefit in ovarian cancer. Immune checkpoint inhibitor monotherapy with nivolumab, pembrolizumab, avelumab, or atezolizumab showed a favorable toxicity profile but was unable to provide substantial clinical benefit, with an ORR of 6–22% [148].

The disappointing efficacy of ICI monotherapy represented the rationale for investigating ICI-combined treatment strategies. One promising combination is ICIs and PARPi because HR-deficient tumors display high PD-1 expression, and the accumulation of double-stranded DNA breaks enables the buildup of neoantigens [149]. The efficacy of the niraparib and pembrolizumab combination was assessed in recurrent platinum-resistant ovarian cancer patients. An ORR of 18% with a disease control rate of 65% was observed irrespective of platinum sensitivity, BRCA, or HR status [150]. Lampert et al. evaluated a durvalumab and olaparib combination in recurrent ovarian cancer. Most patients were platinum resistant (86%) and had the BRCA wild type (77%). Although the disease control rate was 71%, the clinical activity was modest, with an ORR of 14% [151].

The association between antiangiogenic therapy and ICI was also investigated. Hypoxia and VEGF dysregulation promote an immunosuppressive microenvironment by shifting the T helper 1 antitumoral response to a T helper 2 protumorigenic response; antigen presentation by dendritic cells is also inhibited; and VEGF itself has immunosuppressive properties [152]. Liu et al. [153] assessed the efficacy of a nivolumab and bevacizumab combination in relapsed ovarian cancer. The ORR was 40% in platinum-sensitive and 16.7% in platinum-resistant disease. The median PFS was 7.7 months in the platinum-resistant subgroup and 12.1 months in the platinum-sensitive one. Bevacizumab was also evaluated in combination with pembrolizumab and cyclophosphamide in recurrent ovarian cancer. Patients with platinum-resistant disease had an ORR of 43.3%, where 93.3% of patients exhibited a clinical benefit and had a 5.5-month median duration of response [154].

Chemotherapy was shown to induce an immunogenic antitumoral response and promote a proinflammatory tumoral microenvironment through the release of inflammatory signals from dying tumor cells [155]. This rationale was the basis for investigating the safety and efficacy of chemotherapy and ICI association in multiple solid tumors, including ovarian cancer. The JAVELIN Ovarian 200 trial evaluated compared avelumab and pegylated liposomal doxorubicin (PLD) monotherapy to the avelumab and PLD combination in 566 platinum-resistant ovarian cancer patients. Neither avelumab monotherapy nor the combination of avelumab and PLD improved PFS or OS compared with PLD monotherapy. However, there was a higher ORR for the combo in the PDL1-positive group compared with the PDL1-negative one: 18.5% vs. 3.4%. This ORR also translated into a survival advantage for the PDL1-positive patient subgroup [156]. The association between PLD and pembrolizumab was also evaluated in 23 platinum-resistant ovarian cancer patients, where 52.2% of patients achieved a clinical benefit from the combinational treatment, with an ORR of 26.1% and a favorable toxicity profile. There was no significant correlation between PDL1 expression and an objective response [157].

Copper transporter dysregulation has been validated as a critical mechanism of platinum resistance in HGSOC, and this is the basis for targeting copper homeostasis as a mechanism to resensitize ovarian cancer cells to platinum compounds. Using cisplatin-resistant ovarian cancer cell lines, Liang et al. [158] demonstrated that cisplatin resistance is associated with the decreased expression of the high-affinity copper transporter 1 (hCTR1). Furthermore, they revealed that copper chelators resensitize cells to cisplatin by enhancing hCTR1 expression. Following this preclinical data, the association between carboplatin and the copper-lowering agent trientine was evaluated in platinum-resistant patients. The association was well tolerated and displayed antitumor activity, especially in patients with lowered ceruloplasmin and copper levels, but the response rates remained low, warranting improvement [159]. The association between trientine carboplatin and PLD was also assessed in a dose escalation study involving patients with relapsed epithelial ovarian, tubal, and peritoneal cancers. The combination was well tolerated and safe, rendering a clinical benefit rate of 33.3% in the platinum-resistant group and 50% in the partially platinum-sensitive group [160]. Tranilast (an analog of tryptophan metabolite and an inhibitor of histamine release) and telmisartan (an angiotensin II receptor antagonist) were shown to facilitate platinum compound delivery to the nucleus by targeting ATP7B expression and trafficking in platinum-resistant IGROV-CP20 ovarian cancer cell lines. Amphotericin B was also shown to promote cisplatin toxicity by inhibiting ATP7B expression but with an inferior safety profile compared with tranilast and telmisartan [161]. Theaflavin-3,3′-digallate (TF3), a black tea polyphenol, was also shown to enhance ovarian cancer cells' sensitivity to cisplatin. TF3 increased the intracellular accumulation of cisplatin and enhanced platinum DNA damage by decreasing glutathione levels and upregulating CTR1 levels [162].

4. Conclusions

Alongside surgery, chemotherapy is the cornerstone of treatment in advanced ovarian cancer, and platinum-based combinations continue to be the most effective first-line treatment for these patients. Despite the initial efficacy, most patients will present recurrent disease. Rechallenge with platinum-based chemotherapy is the treatment of choice for patients with platinum-sensitive disease, defined as a platinum-free interval longer than 6 months. These patients usually respond to platinum rechallenge and have a better prognosis than those who are platinum resistant. Therefore, we can safely consider platinum resistance as one of the most important prognostic factors in ovarian cancer.

Although classically defined on the basis of the 6-month platinum-free cutoff interval, the concept of platinum resistance is an everchanging concept owing to the widespread availability of CA125, high-resolution and functional imaging that enables early recurrence detection, and the changes in maintenance therapy now that bevacizumab and PARPis have managed to prolong PFS, thus delaying recurrence. Nevertheless, resistance to platinum-based cytotoxic agents is a complex concept resulting from an interplay between mechanisms. Tumor cells can modify the intracellular concentration of chemotherapy by changing the expression of cellular influx and efflux transporters. Changes in the DNA repair pathways are involved in platinum resistance, but they can also represent targetable therapeutic opportunities. Recent data have revealed that the dysregulation of epigenetic control processes with aberrant miR expression, histone acetylation, and DNA methylation modulates these resistance mechanisms' expression. The tumoral microenvironment is essential in ovarian cancer carcinogenesis and progression but can also promote treatment resistance through EMT and various adaptative signals modulated by the stromal cells of the tumoral milieu. Despite the substantial progress in understanding the underlying mechanisms of platinum resistance, more research is necessary to understand their interplay and contribution to achieving the resistant phenotype.

Several treatment strategies have been evaluated in the setting of platinum-resistant ovarian cancer, and treatment associations involving PARP inhibition, antiangiogenic agents, immune checkpoint inhibitors, and chemotherapy have shown promising results. However, more research is necessary to identify biomarkers that enable better patient stratification.

Author Contributions: Conceptualization, A.H. and C.C.; methodology, A.H., S.S.C. and A.T.H.; formal analysis, A.H.; investigation, A.H. and S.S.C.; data curation, A.H. and A.T.H.; writing—original draft preparation, A.H., S.S.C. and A.T.H.; writing—review and editing, S.S.C.; visualization, A.H.; supervision, C.C.; project administration, C.C.; funding acquisition, C.C. All authors have read and agreed to the published version of the manuscript.

Funding: This work was supported by the grant partnership for the transfer of knowledge in biogenomics applications in oncology and related fields—BIOGENONCO; the project was cofinanced by FEDR through the Competitiveness Operational Programme 2014–2020, contract no. 10/01.09.2016, code ID P_40_318, MySMIS 105774.

Institutional Review Board Statement: Not applicable.

Informed Consent Statement: Not applicable.

Data Availability Statement: Data sharing not applicable.

Conflicts of Interest: The authors declare no conflict of interest.

References

1. Sung, H.; Ferlay, J.; Siegel, R.L.; Laversanne, M.; Soerjomataram, I.; Jemal, A.; Bray, F. Global Cancer Statistics 2020: GLOBOCAN Estimates of Incidence and Mortality Worldwide for 36 Cancers in 185 Countries. *CA Cancer J. Clin.* **2021**, *71*, 209–249. [CrossRef] [PubMed]
2. Torre, L.A.; Trabert, B.; DeSantis, C.E.; Miller, K.D.; Samimi, G.; Runowicz, C.D.; Gaudet, M.M.; Jemal, A.; Siegel, R.L. Ovarian Cancer Statistics, 2018. *CA Cancer J. Clin.* **2018**, *68*, 284–296. [CrossRef] [PubMed]
3. Urban, R.R.; He, H.; Alfonso, R.; Hardesty, M.M.; Gray, H.J.; Goff, B.A. Ovarian Cancer Outcomes: Predictors of Early Death. *Gynecol. Oncol.* **2016**, *140*, 474–480. [CrossRef]
4. Sankaranarayanan, R.; Ferlay, J. Worldwide Burden of Gynaecological Cancer: The Size of the Problem. *Best Pract. Res. Clin. Obstet. Gynaecol.* **2006**, *20*, 207–225. [CrossRef] [PubMed]
5. McCluggage, W.G. Morphological Subtypes of Ovarian Carcinoma: A Review with Emphasis on New Developments and Pathogenesis. *Pathology* **2011**, *43*, 420–432. [CrossRef] [PubMed]
6. Reid, B.M.; Permuth, J.B.; Sellers, T.A. Epidemiology of Ovarian Cancer: A Review. *Cancer Biol. Med.* **2017**, *14*, 9–32. [CrossRef]
7. Alsop, K.; Fereday, S.; Meldrum, C.; DeFazio, A.; Emmanuel, C.; George, J.; Dobrovic, A.; Birrer, M.J.; Webb, P.M.; Stewart, C.; et al. BRCA Mutation Frequency and Patterns of Treatment Response in BRCA Mutation–Positive Women With Ovarian Cancer: A Report From the Australian Ovarian Cancer Study Group. *J. Clin. Oncol.* **2012**, *30*, 2654–2663. [CrossRef]
8. Finch, A.P.M.; Lubinski, J.; Møller, P.; Singer, C.F.; Karlan, B.; Senter, L.; Rosen, B.; Maehle, L.; Ghadirian, P.; Cybulski, C.; et al. Impact of Oophorectomy on Cancer Incidence and Mortality in Women With a BRCA1 or BRCA2 Mutation. *J. Clin. Oncol.* **2014**, *32*, 1547–1553. [CrossRef]
9. Jayson, G.C.; Kohn, E.C.; Kitchener, H.C.; Ledermann, J.A. Ovarian Cancer. *Lancet* **2014**, *384*, 1376–1388. [CrossRef]
10. Lheureux, S.; Braunstein, M.; Oza, A.M. Epithelial Ovarian Cancer: Evolution of Management in the Era of Precision Medicine. *CA Cancer J. Clin.* **2019**, *69*, 280–304. [CrossRef]
11. Kuroki, L.; Guntupalli, S.R. Treatment of Epithelial Ovarian Cancer. *BMJ* **2020**, *371*, m3773. [CrossRef]
12. Du Bois, A.; Sehouli, J.; Vergote, I.; Ferron, G.; Reuss, A.; Meier, W.; Greggi, S.; Jensen, P.T.; Selle, F.; Guyon, F.; et al. Randomized Phase III Study to Evaluate the Impact of Secondary Cytoreductive Surgery in Recurrent Ovarian Cancer: Final Analysis of AGO DESKTOP III/ENGOT-Ov20. *J. Clin. Oncol.* **2020**, *38*, 6000. [CrossRef]
13. Trimbos, J.B.; Vergote, I.; Bolis, G.; Vermorken, J.B.; Mangioni, C.; Madronal, C.; Franchi, M.; Tateo, S.; Zanetta, G.; Scarfone, G.; et al. Impact of Adjuvant Chemotherapy and Surgical Staging in Early-Stage Ovarian Carcinoma: European Organisation for Research and Treatment of Cancer-Adjuvant ChemoTherapy in Ovarian Neoplasm Trial. *J. Natl. Cancer Inst.* **2003**, *95*, 113–125. [CrossRef]
14. Colombo, N.; Guthrie, D.; Chiari, S.; Parmar, M.; Qian, W.; Swart, A.M.; Torri, V.; Williams, C.; Lissoni, A.; Bonazzi, C. International Collaborative Ovarian Neoplasm Trial 1: A Randomized Trial of Adjuvant Chemotherapy in Women With Early-Stage Ovarian Cancer. *J. Natl. Cancer Inst.* **2003**, *95*, 125–132. [CrossRef]
15. Colombo, N.; Sessa, C.; du Bois, A.; Ledermann, J.; McCluggage, W.G.; McNeish, I.; Morice, P.; Pignata, S.; Ray-Coquard, I.; Vergote, I.; et al. ESMO–ESGO Consensus Conference Recommendations on Ovarian Cancer: Pathology and Molecular Biology, Early and Advanced Stages, Borderline Tumours and Recurrent Disease. *Ann. Oncol.* **2019**, *30*, 672–705. [CrossRef]
16. Bell, J.; Brady, M.F.; Young, R.C.; Lage, J.; Walker, J.L.; Look, K.Y.; Rose, G.S.; Spirtos, N.M. Randomized Phase III Trial of Three versus Six Cycles of Adjuvant Carboplatin and Paclitaxel in Early Stage Epithelial Ovarian Carcinoma: A Gynecologic Oncology Group Study. *Gynecol. Oncol.* **2006**, *102*, 432–439. [CrossRef]
17. Karam, A.; Ledermann, J.A.; Kim, J.-W.; Sehouli, J.; Lu, K.; Gourley, C.; Katsumata, N.; Burger, R.A.; Nam, B.-H.; Bacon, M.; et al. Fifth Ovarian Cancer Consensus Conference of the Gynecologic Cancer InterGroup: First-Line Interventions. *Ann. Oncol.* **2017**, *28*, 711–717. [CrossRef]
18. Lheureux, S.; Gourley, C.; Vergote, I.; Oza, A.M. Epithelial Ovarian Cancer. *Lancet* **2019**, *393*, 1240–1253. [CrossRef]
19. Markman, M.; Liu, P.Y.; Moon, J.; Monk, B.J.; Copeland, L.; Wilczynski, S.; Alberts, D. Impact on Survival of 12 versus 3 Monthly Cycles of Paclitaxel (175 Mg/M2) Administered to Patients with Advanced Ovarian Cancer Who Attained a Complete Response to Primary Platinum-Paclitaxel: Follow-up of a Southwest Oncology Group and Gynecologic Oncol. *Gynecol. Oncol.* **2009**, *114*, 195–198. [CrossRef]
20. Oza, A.M.; Cook, A.D.; Pfisterer, J.; Embleton, A.; Ledermann, J.A.; Pujade-Lauraine, E.; Kristensen, G.; Carey, M.S.; Beale, P.; Cervantes, A.; et al. Standard Chemotherapy with or without Bevacizumab for Women with Newly Diagnosed Ovarian Cancer (ICON7): Overall Survival Results of a Phase 3 Randomised Trial. *Lancet Oncol.* **2015**, *16*, 928–936. [CrossRef]
21. Moore, K.; Colombo, N.; Scambia, G.; Kim, B.-G.; Oaknin, A.; Friedlander, M.; Lisyanskaya, A.; Floquet, A.; Leary, A.; Sonke, G.S.; et al. Maintenance Olaparib in Patients with Newly Diagnosed Advanced Ovarian Cancer. *N. Engl. J. Med.* **2018**, *379*, 2495–2505. [CrossRef]
22. Davis, A.; Tinker, A.V.; Friedlander, M. "Platinum Resistant" Ovarian Cancer: What Is It, Who to Treat and How to Measure Benefit? *Gynecol. Oncol.* **2014**, *133*, 624–631. [CrossRef]
23. Ray-Coquard, I.; Pautier, P.; Pignata, S.; Pérol, D.; González-Martín, A.; Berger, R.; Fujiwara, K.; Vergote, I.; Colombo, N.; Mäenpää, J.; et al. Olaparib plus Bevacizumab as First-Line Maintenance in Ovarian Cancer. *N. Engl. J. Med.* **2019**, *381*, 2416–2428. [CrossRef]
24. Haunschild, C.E.; Tewari, K.S. Bevacizumab Use in the Frontline, Maintenance and Recurrent Settings for Ovarian Cancer. *Future Oncol.* **2020**, *16*, 225–246. [CrossRef]

25. Pfisterer, J.; Plante, M.; Vergote, I.; du Bois, A.; Hirte, H.; Lacave, A.J.; Wagner, U.; Stähle, A.; Stuart, G.; Kimmig, R.; et al. Gemcitabine Plus Carboplatin Compared With Carboplatin in Patients With Platinum-Sensitive Recurrent Ovarian Cancer: An Intergroup Trial of the AGO-OVAR, the NCIC CTG, and the EORTC GCG. *J. Clin. Oncol.* **2006**, *24*, 4699–4707. [CrossRef]
26. Icon and Ago Collaborators. Paclitaxel plus Platinum-Based Chemotherapy versus Conventional Platinum-Based Chemotherapy in Women with Relapsed Ovarian Cancer: The ICON4/AGO-OVAR-2.2 Trial. *Lancet* **2003**, *361*, 2099–2106. [CrossRef]
27. Lindemann, K.; Gao, B.; Mapagu, C.; Fereday, S.; Emmanuel, C.; Alsop, K.; Traficante, N.; Harnett, P.R.; Bowtell, D.D.L.; DeFazio, A. Response Rates to Second-Line Platinum-Based Therapy in Ovarian Cancer Patients Challenge the Clinical Definition of Platinum Resistance. *Gynecol. Oncol.* **2018**, *150*, 239–246. [CrossRef]
28. Le Saux, O.; Ray-Coquard, I.; Labidi-Galy, S.I. Challenges for Immunotherapy for the Treatment of Platinum Resistant Ovarian Cancer. *Semin. Cancer Biol.* **2021**, *77*, 127–143. [CrossRef]
29. Galluzzi, L.; Senovilla, L.; Vitale, I.; Michels, J.; Martins, I.; Kepp, O.; Castedo, M.; Kroemer, G. Molecular Mechanisms of Cisplatin Resistance. *Oncogene* **2012**, *31*, 1869–1883. [CrossRef]
30. Siddik, Z.H. Cisplatin: Mode of Cytotoxic Action and Molecular Basis of Resistance. *Oncogene* **2003**, *22*, 7265–7279. [CrossRef]
31. Lee, J.; Petris, M.J.; Thiele, D.J. Characterization of Mouse Embryonic Cells Deficient in the Ctr1 High Affinity Copper Transporter. Identification of a Ctr1-Independent Copper Transport System. *J. Biol. Chem.* **2002**, *277*, 40253–40259. [CrossRef]
32. Holzer, A.K.; Samimi, G.; Katano, K.; Naerdemann, W.; Lin, X.; Safaei, R.; Howell, S.B. The Copper Influx Transporter Human Copper Transport Protein 1 Regulates the Uptake of Cisplatin in Human Ovarian Carcinoma Cells. *Mol. Pharmacol.* **2004**, *66*, 817–823. [CrossRef]
33. Song, I.-S.; Savaraj, N.; Siddik, Z.H.; Liu, P.; Wei, Y.; Wu, C.J.; Kuo, M.T. Role of Human Copper Transporter Ctr1 in the Transport of Platinum-Based Antitumor Agents in Cisplatin-Sensitive and Cisplatin-Resistant Cells. *Mol. Cancer Ther.* **2004**, *3*, 1543–1549. [CrossRef]
34. Ishida, S.; McCormick, F.; Smith-McCune, K.; Hanahan, D. Enhancing Tumor-Specific Uptake of the Anticancer Drug Cisplatin with a Copper Chelator. *Cancer Cell* **2010**, *17*, 574–583. [CrossRef]
35. Oguri, T.; Kunii, E.; Fukuda, S.; Sone, K.; Uemura, T.; Takakuwa, O.; Kanemitsu, Y.; Ohkubo, H.; Takemura, M.; Maeno, K.; et al. Organic Cation Transporter 6 Directly Confers Resistance to Anticancer Platinum Drugs. *Biomed. Rep.* **2016**, *5*, 639–643. [CrossRef]
36. Kilari, D.; Guancial, E.; Kim, E.S. Role of Copper Transporters in Platinum Resistance. *World J. Clin. Oncol.* **2016**, *7*, 106–113. [CrossRef]
37. Yoshizawa, K.; Nozaki, S.; Kitahara, H.; Ohara, T.; Kato, K.; Kawashiri, S.; Yamamoto, E. Copper Efflux Transporter (ATP7B) Contributes to the Acquisition of Cisplatin-Resistance in Human Oral Squamous Cell Lines. *Oncol. Rep.* **2007**, *18*, 987–991. [CrossRef]
38. Li, Y.-Q.; Yin, J.-Y.; Liu, Z.-Q.; Li, X.-P. Copper Efflux Transporters ATP7A and ATP7B: Novel Biomarkers for Platinum Drug Resistance and Targets for Therapy. *IUBMB Life* **2018**, *70*, 183–191. [CrossRef]
39. Chisholm, C.L.; Wang, H.; Wong, A.H.-H.; Vazquez-Ortiz, G.; Chen, W.; Xu, X.; Deng, C.-X. Ammonium Tetrathiomolybdate Treatment Targets the Copper Transporter ATP7A and Enhances Sensitivity of Breast Cancer to Cisplatin. *Oncotarget* **2016**, *7*, 84439–84452. [CrossRef]
40. Arts, H.J.; Katsaros, D.; de Vries, E.G.; Massobrio, M.; Genta, F.; Danese, S.; Arisio, R.; Scheper, R.J.; Kool, M.; Scheffer, G.L.; et al. Drug Resistance-Associated Markers P-Glycoprotein, Multidrug Resistance-Associated Protein 1, Multidrug Resistance-Associated Protein 2, and Lung Resistance Protein as Prognostic Factors in Ovarian Carcinoma. *Clin. Cancer Res. Off. J. Am. Assoc. Cancer Res.* **1999**, *5*, 2798–2805.
41. Januchowski, R.; Sterzyńska, K.; Zaorska, K.; Sosińska, P.; Klejewski, A.; Brązert, M.; Nowicki, M.; Zabel, M. Analysis of MDR Genes Expression and Cross-Resistance in Eight Drug Resistant Ovarian Cancer Cell Lines. *J. Ovarian Res.* **2016**, *9*, 65. [CrossRef] [PubMed]
42. Korita, P.V.; Wakai, T.; Shirai, Y.; Matsuda, Y.; Sakata, J.; Takamura, M.; Yano, M.; Sanpei, A.; Aoyagi, Y.; Hatakeyama, K.; et al. Multidrug Resistance-Associated Protein 2 Determines the Efficacy of Cisplatin in Patients with Hepatocellular Carcinoma. *Oncol. Rep.* **2010**, *23*, 965–972. [CrossRef]
43. Yamasaki, M.; Makino, T.; Masuzawa, T.; Kurokawa, Y.; Miyata, H.; Takiguchi, S.; Nakajima, K.; Fujiwara, Y.; Matsuura, N.; Mori, M.; et al. Role of Multidrug Resistance Protein 2 (MRP2) in Chemoresistance and Clinical Outcome in Oesophageal Squamous Cell Carcinoma. *Br. J. Cancer* **2011**, *104*, 707–713. [CrossRef]
44. Pilié, P.G.; Tang, C.; Mills, G.B.; Yap, T.A. State-of-the-Art Strategies for Targeting the DNA Damage Response in Cancer. *Nat. Rev. Clin. Oncol.* **2019**, *16*, 81–104. [CrossRef]
45. Curtin, N.J. DNA Repair Dysregulation from Cancer Driver to Therapeutic Target. *Nat. Rev. Cancer* **2012**, *12*, 801–817. [CrossRef]
46. Lord, C.J.; Ashworth, A. The DNA Damage Response and Cancer Therapy. *Nature* **2012**, *481*, 287–294. [CrossRef]
47. Damia, G.; Broggini, M. Platinum Resistance in Ovarian Cancer: Role of DNA Repair. *Cancers* **2019**, *11*, 119. [CrossRef]
48. Brown, J.S.; O'Carrigan, B.; Jackson, S.P.; Yap, T.A. Targeting DNA Repair in Cancer: Beyond PARP Inhibitors. *Cancer Discov.* **2017**, *7*, 20–37. [CrossRef]
49. Chilà, R.; Guffanti, F.; Damia, G. Role and Therapeutic Potential of CDK12 in Human Cancers. *Cancer Treat. Rev.* **2016**, *50*, 83–88. [CrossRef]
50. Bélanger, F.; Fortier, E.; Dubé, M.; Lemay, J.-F.; Buisson, R.; Masson, J.-Y.; Elsherbiny, A.; Costantino, S.; Carmona, E.; Mes-Masson, A.-M.; et al. Replication Protein A Availability during DNA Replication Stress Is a Major Determinant of Cisplatin Resistance in Ovarian Cancer Cells. *Cancer Res.* **2018**, *78*, 5561–5573. [CrossRef]
51. Ceccaldi, R.; O'Connor, K.W.; Mouw, K.W.; Li, A.Y.; Matulonis, U.A.; D'Andrea, A.D.; Konstantinopoulos, P.A. A Unique Subset of Epithelial Ovarian Cancers with Platinum Sensitivity and PARP Inhibitor Resistance. *Cancer Res.* **2015**, *75*, 628–634. [CrossRef]

52. Kuhlmann, J.D.; Wimberger, P.; Bankfalvi, A.; Keller, T.; Schöler, S.; Aktas, B.; Buderath, P.; Hauch, S.; Otterbach, F.; Kimmig, R.; et al. ERCC1-Positive Circulating Tumor Cells in the Blood of Ovarian Cancer Patients as a Predictive Biomarker for Platinum Resistance. *Clin. Chem.* **2014**, *60*, 1282–1289. [CrossRef]
53. Avraam, K.; Pavlakis, K.; Papadimitriou, C.; Vrekoussis, T.; Panoskaltsis, T.; Messini, I.; Patsouris, E. The Prognostic and Predictive Value of ERCC-1, P53, Bcl-2 and Bax in Epithelial Ovarian Cancer. *Eur. J. Gynaecol. Oncol.* **2011**, *32*, 516–520.
54. Zhao, M.; Li, S.; Zhou, L.; Shen, Q.; Zhu, H.; Zhu, X. Prognostic Values of Excision Repair Cross-Complementing Genes MRNA Expression in Ovarian Cancer Patients. *Life Sci.* **2018**, *194*, 34–39. [CrossRef]
55. Koukoura, O.; Spandidos, D.A.; Daponte, A.; Sifakis, S. DNA Methylation Profiles in Ovarian Cancer: Implication in Diagnosis and Therapy. *Mol. Med. Rep.* **2014**, *10*, 3–9. [CrossRef]
56. Lum, E.; Vigliotti, M.; Banerjee, N.; Cutter, N.; Wrzeszczynski, K.O.; Khan, S.; Kamalakaran, S.; Levine, D.A.; Dimitrova, N.; Lucito, R. Loss of DOK2 Induces Carboplatin Resistance in Ovarian Cancer via Suppression of Apoptosis. *Gynecol. Oncol.* **2013**, *130*, 369–376. [CrossRef]
57. Chan, D.W.; Lam, W.-Y.; Chen, F.; Yung, M.M.H.; Chan, Y.-S.; Chan, W.-S.; He, F.; Liu, S.S.; Chan, K.K.L.; Li, B.; et al. Genome-Wide DNA Methylome Analysis Identifies Methylation Signatures Associated with Survival and Drug Resistance of Ovarian Cancers. *Clin. Epigenet.* **2021**, *13*, 142. [CrossRef]
58. Cardenas, H.; Fang, F.; Jiang, G.; Perkins, S.M.; Zhang, C.; Emerson, R.E.; Hutchins, G.; Keer, H.N.; Liu, Y.; Matei, D.; et al. Methylomic Signatures of High Grade Serous Ovarian Cancer. *Epigenetics* **2021**, *16*, 1201–1216. [CrossRef]
59. Lund, R.J.; Huhtinen, K.; Salmi, J.; Rantala, J.; Nguyen, E.V.; Moulder, R.; Goodlett, D.R.; Lahesmaa, R.; Carpén, O. DNA Methylation and Transcriptome Changes Associated with Cisplatin Resistance in Ovarian Cancer. *Sci. Rep.* **2017**, *7*, 1469. [CrossRef]
60. Bonito, N.A.; Borley, J.; Wilhelm-Benartzi, C.S.; Ghaem-Maghami, S.; Brown, R. Epigenetic Regulation of the Homeobox Gene MSX1 Associates with Platinum-Resistant Disease in High-Grade Serous Epithelial Ovarian Cancer. *Clin. Cancer Res. Off. J. Am. Assoc. Cancer Res.* **2016**, *22*, 3097–3104. [CrossRef]
61. Feng, L.-Y.; Yan, B.-B.; Huang, Y.-Z.; Li, L. Abnormal Methylation Characteristics Predict Chemoresistance and Poor Prognosis in Advanced High-Grade Serous Ovarian Cancer. *Clin. Epigenet.* **2021**, *13*, 141. [CrossRef]
62. Matthews, B.G.; Bowden, N.A.; Wong-Brown, M.W. Epigenetic Mechanisms and Therapeutic Targets in Chemoresistant High-Grade Serous Ovarian Cancer. *Cancers* **2021**, *13*, 5993. [CrossRef]
63. Wang, Y.-Q.; Zhang, J.-R.; Li, S.-D.; He, Y.-Y.; Yang, Y.-X.; Liu, X.-L.; Wan, X.-P. Aberrant Methylation of Breast and Ovarian Cancer Susceptibility Gene 1 in Chemosensitive Human Ovarian Cancer Cells Does Not Involve the Phosphatidylinositol 3′-Kinase-Akt Pathway. *Cancer Sci.* **2010**, *101*, 1618–1623. [CrossRef] [PubMed]
64. Yang, Q.; Yang, Y.; Zhou, N.; Tang, K.; Lau, W.B.; Lau, B.; Wang, W.; Xu, L.; Yang, Z.; Huang, S.; et al. Epigenetics in Ovarian Cancer: Premise, Properties, and Perspectives. *Mol. Cancer* **2018**, *17*, 109. [CrossRef] [PubMed]
65. Cacan, E.; Ali, M.W.; Boyd, N.H.; Hooks, S.B.; Greer, S.F. Inhibition of HDAC1 and DNMT1 Modulate RGS10 Expression and Decrease Ovarian Cancer Chemoresistance. *PLoS ONE* **2014**, *9*, e87455. [CrossRef]
66. Liu, X.; Yu, Y.; Zhang, J.; Lu, C.; Wang, L.; Liu, P.; Song, H. HDAC1 Silencing in Ovarian Cancer Enhances the Chemotherapy Response. *Cell. Physiol. Biochem. Int. J. Exp. Cell. Physiol. Biochem. Pharmacol.* **2018**, *48*, 1505–1518. [CrossRef]
67. Ghafouri-Fard, S.; Shoorei, H.; Taheri, M. MiRNA Profile in Ovarian Cancer. *Exp. Mol. Pathol.* **2020**, *113*, 104381. [CrossRef]
68. Feng, C.; Ma, F.; Hu, C.; Ma, J.-A.; Wang, J.; Zhang, Y.; Wu, F.; Hou, T.; Jiang, S.; Wang, Y.; et al. SOX9/MiR-130a/CTR1 Axis Modulates DDP-Resistance of Cervical Cancer Cell. *Cell Cycle* **2018**, *17*, 448–458. [CrossRef]
69. Xiao, F.; Li, Y.; Wan, Y.; Xue, M. MircroRNA-139 Sensitizes Ovarian Cancer Cell to Cisplatin-Based Chemotherapy through Regulation of ATP7A/B. *Cancer Chemother. Pharmacol.* **2018**, *81*, 935–947. [CrossRef]
70. Dwivedi, S.K.D.; Mustafi, S.B.; Mangala, L.S.; Jiang, D.; Pradeep, S.; Rodriguez-Aguayo, C.; Ling, H.; Ivan, C.; Mukherjee, P.; Calin, G.A.; et al. Therapeutic Evaluation of MicroRNA-15a and MicroRNA-16 in Ovarian Cancer. *Oncotarget* **2016**, *7*, 15093–15104. [CrossRef]
71. Tian, J.; Xu, Y.-Y.; Li, L.; Hao, Q. MiR-490-3p Sensitizes Ovarian Cancer Cells to Cisplatin by Directly Targeting ABCC2. *Am. J. Transl. Res.* **2017**, *9*, 1127–1138. [PubMed]
72. Chen, F.D.; Chen, H.H.; Ke, S.C.; Zheng, L.R.; Zheng, X.Y. SLC27A2 Regulates MiR-411 to Affect Chemo-Resistance in Ovarian Cancer. *Neoplasma* **2018**, *65*, 915–924. [CrossRef] [PubMed]
73. Xiao, S.; Zhang, M.; Liu, C.; Wang, D. MiR-514 Attenuates Proliferation and Increases Chemoresistance by Targeting ATP Binding Cassette Subfamily in Ovarian Cancer. *Mol. Genet. Genom.* **2018**, *293*, 1159–1167. [CrossRef] [PubMed]
74. Wang, T.; Hao, D.; Yang, S.; Ma, J.; Yang, W.; Zhu, Y.; Weng, M.; An, X.; Wang, X.; Li, Y.; et al. MiR-211 Facilitates Platinum Chemosensitivity by Blocking the DNA Damage Response (DDR) in Ovarian Cancer. *Cell Death Dis.* **2019**, *10*, 495. [CrossRef]
75. Gong, T.-T.; Liu, X.-D.; Zhan, Z.-P.; Wu, Q.-J. Sulforaphane Enhances the Cisplatin Sensitivity through Regulating DNA Repair and Accumulation of Intracellular Cisplatin in Ovarian Cancer Cells. *Exp. Cell Res.* **2020**, *393*, 112061. [CrossRef]
76. Zhao, H.; Yu, X.; Ding, Y.; Zhao, J.; Wang, G.; Wu, X.; Jiang, J.; Peng, C.; Guo, G.Z.; Cui, S. MiR-770-5p Inhibits Cisplatin Chemoresistance in Human Ovarian Cancer by Targeting ERCC2. *Oncotarget* **2016**, *7*, 53254–53268. [CrossRef]
77. Zhu, M.; Yang, L.; Wang, X. NEAT1 Knockdown Suppresses the Cisplatin Resistance in Ovarian Cancer by Regulating MiR-770-5p/PARP1 Axis. *Cancer Manag. Res.* **2020**, *12*, 7277–7289. [CrossRef]
78. Sun, C.; Li, N.; Yang, Z.; Zhou, B.; He, Y.; Weng, D.; Fang, Y.; Wu, P.; Chen, P.; Yang, X.; et al. MiR-9 Regulation of BRCA1 and Ovarian Cancer Sensitivity to Cisplatin and PARP Inhibition. *J. Natl. Cancer Inst.* **2013**, *105*, 1750–1758. [CrossRef]

79. Liu, G.; Xue, F.; Zhang, W. MiR-506: A Regulator of Chemo-Sensitivity through Suppression of the RAD51-Homologous Recombination Axis. *Chin. J. Cancer* **2015**, *34*, 485–487. [CrossRef]
80. Wang, Y.; Bao, W.; Liu, Y.; Wang, S.; Xu, S.; Li, X.; Li, Y.; Wu, S. MiR-98-5p Contributes to Cisplatin Resistance in Epithelial Ovarian Cancer by Suppressing MiR-152 Biogenesis via Targeting Dicer1. *Cell Death Dis.* **2018**, *9*, 447. [CrossRef]
81. Choi, Y.E.; Meghani, K.; Brault, M.-E.; Leclerc, L.; He, Y.J.; Day, T.A.; Elias, K.M.; Drapkin, R.; Weinstock, D.M.; Dao, F.; et al. Platinum and PARP Inhibitor Resistance Due to Overexpression of MicroRNA-622 in BRCA1-Mutant Ovarian Cancer. *Cell Rep.* **2016**, *14*, 429–439. [CrossRef] [PubMed]
82. Gu, Y.; Zhang, M.; Peng, F.; Fang, L.; Zhang, Y.; Liang, H.; Zhou, W.; Ao, L.; Guo, Z. The BRCA1/2-Directed MiRNA Signature Predicts a Good Prognosis in Ovarian Cancer Patients with Wild-Type BRCA1/2. *Oncotarget* **2015**, *6*, 2397–2406. [CrossRef] [PubMed]
83. Meghani, K.; Fuchs, W.; Detappe, A.; Drané, P.; Gogola, E.; Rottenberg, S.; Jonkers, J.; Matulonis, U.; Swisher, E.M.; Konstantinopoulos, P.A.; et al. Multifaceted Impact of MicroRNA 493-5p on Genome-Stabilizing Pathways Induces Platinum and PARP Inhibitor Resistance in BRCA2-Mutated Carcinomas. *Cell Rep.* **2018**, *23*, 100–111. [CrossRef] [PubMed]
84. Liu, J.; Zhang, X.; Huang, Y.; Zhang, Q.; Zhou, J.; Zhang, X.; Wang, X. MiR-200b and MiR-200c Co-Contribute to the Cisplatin Sensitivity of Ovarian Cancer Cells by Targeting DNA Methyltransferases. *Oncol. Lett.* **2019**, *17*, 1453–1460. [CrossRef] [PubMed]
85. Han, X.; Zhen, S.; Ye, Z.; Lu, J.; Wang, L.; Li, P.; Li, J.; Zheng, X.; Li, H.; Chen, W.; et al. A Feedback Loop Between MiR-30a/c-5p and DNMT1 Mediates Cisplatin Resistance in Ovarian Cancer Cells. *Cell. Physiol. Biochem. Int. J. Exp. Cell. Physiol. Biochem. Pharmacol.* **2017**, *41*, 973–986. [CrossRef]
86. Xiang, Y.; Ma, N.; Wang, D.; Zhang, Y.; Zhou, J.; Wu, G.; Zhao, R.; Huang, H.; Wang, X.; Qiao, Y.; et al. MiR-152 and MiR-185 Co-Contribute to Ovarian Cancer Cells Cisplatin Sensitivity by Targeting DNMT1 Directly: A Novel Epigenetic Therapy Independent of Decitabine. *Oncogene* **2014**, *33*, 378–386. [CrossRef]
87. Zou, X.; Zhao, Y.; Liang, X.; Wang, H.; Zhu, Y.; Shao, Q. Double Insurance for OC: MiRNA-Mediated Platinum Resistance and Immune Escape. *Front. Immunol.* **2021**, *12*, 641937. [CrossRef]
88. Zhu, X.; Shen, H.; Yin, X.; Long, L.; Xie, C.; Liu, Y.; Hui, L.; Lin, X.; Fang, Y.; Cao, Y.; et al. MiR-186 Regulation of Twist1 and Ovarian Cancer Sensitivity to Cisplatin. *Oncogene* **2016**, *35*, 323–332. [CrossRef]
89. Cao, L.; Wan, Q.; Li, F.; Tang, C.-E. MiR-363 Inhibits Cisplatin Chemoresistance of Epithelial Ovarian Cancer by Regulating Snail-Induced Epithelial-Mesenchymal Transition. *BMB Rep.* **2018**, *51*, 456–461. [CrossRef]
90. Zhang, Y.; Huang, S.; Guo, Y.; Li, L. MiR-1294 Confers Cisplatin Resistance in Ovarian Cancer Cells by Targeting IGF1R. *Biomed. Pharmacother.* **2018**, *106*, 1357–1363. [CrossRef]
91. Liu, Y.; Han, S.; Li, Y.; Liu, Y.; Zhang, D.; Li, Y.; Zhang, J. MicroRNA-20a Contributes to Cisplatin-Resistance and Migration of OVCAR3 Ovarian Cancer Cell Line. *Oncol. Lett.* **2017**, *14*, 1780–1786. [CrossRef] [PubMed]
92. Shi, X.; Xiao, L.; Mao, X.; He, J.; Ding, Y.; Huang, J.; Peng, C.; Xu, Z. MiR-205-5p Mediated Downregulation of PTEN Contributes to Cisplatin Resistance in C13K Human Ovarian Cancer Cells. *Front. Genet.* **2018**, *9*, 555. [CrossRef]
93. Jin, P.; Liu, Y.; Wang, R. STAT3 Regulated MiR-216a Promotes Ovarian Cancer Proliferation and Cisplatin Resistance. *Biosci. Rep.* **2018**, *38*, BSR20180547. [CrossRef]
94. Lan, H.; Yuan, J.; Zeng, D.; Liu, C.; Guo, X.; Yong, J.; Zeng, X.; Xiao, S. The Emerging Role of Non-Coding RNAs in Drug Resistance of Ovarian Cancer. *Front. Genet.* **2021**, *12*, 693259. [CrossRef] [PubMed]
95. Arrighetti, N.; Cossa, G.; De Cecco, L.; Stucchi, S.; Carenini, N.; Corna, E.; Gandellini, P.; Zaffaroni, N.; Perego, P.; Gatti, L. PKC-Alpha Modulation by MiR-483-3p in Platinum-Resistant Ovarian Carcinoma Cells. *Toxicol. Appl. Pharmacol.* **2016**, *310*, 9–19. [CrossRef]
96. Zhao, H.; Bi, T.; Qu, Z.; Jiang, J.; Cui, S.; Wang, Y. Expression of MiR-224-5p Is Associated with the Original Cisplatin Resistance of Ovarian Papillary Serous Carcinoma. *Oncol. Rep.* **2014**, *32*, 1003–1012. [CrossRef] [PubMed]
97. Gu, Z.-W.; He, Y.-F.; Wang, W.-J.; Tian, Q.; Di, W. MiR-1180 from Bone Marrow-Derived Mesenchymal Stem Cells Induces Glycolysis and Chemoresistance in Ovarian Cancer Cells by Upregulating the Wnt Signaling Pathway. *J. Zhejiang Univ. Sci. B* **2019**, *20*, 219–237. [CrossRef]
98. Guo, H.; Ha, C.; Dong, H.; Yang, Z.; Ma, Y.; Ding, Y. Cancer-Associated Fibroblast-Derived Exosomal MicroRNA-98-5p Promotes Cisplatin Resistance in Ovarian Cancer by Targeting CDKN1A. *Cancer Cell Int.* **2019**, *19*, 347. [CrossRef]
99. Ghoneum, A.; Almousa, S.; Warren, B.; Abdulfattah, A.Y.; Shu, J.; Abouelfadl, H.; Gonzalez, D.; Livingston, C.; Said, N. Exploring the Clinical Value of Tumor Microenvironment in Platinum-Resistant Ovarian Cancer. *Semin. Cancer Biol.* **2021**, *77*, 83–98. [CrossRef]
100. Pogge von Strandmann, E.; Reinartz, S.; Wager, U.; Müller, R. Tumor-Host Cell Interactions in Ovarian Cancer: Pathways to Therapy Failure. *Trends Cancer* **2017**, *3*, 137–148. [CrossRef]
101. Tan, D.S.P.; Agarwal, R.; Kaye, S.B. Mechanisms of Transcoelomic Metastasis in Ovarian Cancer. *Lancet. Oncol.* **2006**, *7*, 925–934. [CrossRef] [PubMed]
102. Ahmed, N.; Stenvers, K.L. Getting to Know Ovarian Cancer Ascites: Opportunities for Targeted Therapy-Based Translational Research. *Front. Oncol.* **2013**, *3*, 256. [CrossRef] [PubMed]
103. Cox, T.R.; Erler, J.T. Remodeling and Homeostasis of the Extracellular Matrix: Implications for Fibrotic Diseases and Cancer. *Dis. Model. Mech.* **2011**, *4*, 165–178. [CrossRef]
104. Diaz Osterman, C.J.; Ozmadenci, D.; Kleinschmidt, E.G.; Taylor, K.N.; Barrie, A.M.; Jiang, S.; Bean, L.M.; Sulzmaier, F.J.; Jean, C.; Tancioni, I.; et al. FAK Activity Sustains Intrinsic and Acquired Ovarian Cancer Resistance to Platinum Chemotherapy. *Elife* **2019**, *8*, e47327. [CrossRef] [PubMed]

105. Zhu, L.-C.; Gao, J.; Hu, Z.-H.; Schwab, C.L.; Zhuang, H.-Y.; Tan, M.-Z.; Yan, L.-M.; Liu, J.-J.; Zhang, D.-Y.; Lin, B. Membranous Expressions of Lewis y and CAM-DR-Related Markers Are Independent Factors of Chemotherapy Resistance and Poor Prognosis in Epithelial Ovarian Cancer. *Am. J. Cancer Res.* **2015**, *5*, 830–843.
106. Wantoch von Rekowski, K.; König, P.; Henze, S.; Schlesinger, M.; Zawierucha, P.; Januchowski, R.; Bendas, G. The Impact of Integrin-Mediated Matrix Adhesion on Cisplatin Resistance of W1 Ovarian Cancer Cells. *Biomolecules* **2019**, *9*, 788. [CrossRef]
107. Pan, S.; Cheng, L.; White, J.T.; Lu, W.; Utleg, A.G.; Yan, X.; Urban, N.D.; Drescher, C.W.; Hood, L.; Lin, B. Quantitative Proteomics Analysis Integrated with Microarray Data Reveals That Extracellular Matrix Proteins, Catenins, and P53 Binding Protein 1 Are Important for Chemotherapy Response in Ovarian Cancers. *Omics J. Integr. Biol.* **2009**, *13*, 345–354. [CrossRef]
108. Ricciardelli, C.; Ween, M.P.; Lokman, N.A.; Tan, I.A.; Pyragius, C.E.; Oehler, M.K. Chemotherapy-Induced Hyaluronan Production: A Novel Chemoresistance Mechanism in Ovarian Cancer. *BMC Cancer* **2013**, *13*, 476. [CrossRef]
109. Yoshihara, M.; Kajiyama, H.; Yokoi, A.; Sugiyama, M.; Koya, Y.; Yamakita, Y.; Liu, W.; Nakamura, K.; Moriyama, Y.; Yasui, H.; et al. Ovarian Cancer-Associated Mesothelial Cells Induce Acquired Platinum-Resistance in Peritoneal Metastasis via the FN1/Akt Signaling Pathway. *Int. J. Cancer* **2020**, *146*, 2268–2280. [CrossRef]
110. Zhang, M.; Chen, Z.; Wang, Y.; Zhao, H.; Du, Y. The Role of Cancer-Associated Fibroblasts in Ovarian Cancer. *Cancers* **2022**, *14*, 2637. [CrossRef]
111. Wang, W.; Kryczek, I.; Dostál, L.; Lin, H.; Tan, L.; Zhao, L.; Lu, F.; Wei, S.; Maj, T.; Peng, D.; et al. Effector T Cells Abrogate Stroma-Mediated Chemoresistance in Ovarian Cancer. *Cell* **2016**, *165*, 1092–1105. [CrossRef] [PubMed]
112. Au Yeung, C.L.; Co, N.-N.; Tsuruga, T.; Yeung, T.-L.; Kwan, S.-Y.; Leung, C.S.; Li, Y.; Lu, E.S.; Kwan, K.; Wong, K.-K.; et al. Exosomal Transfer of Stroma-Derived MiR21 Confers Paclitaxel Resistance in Ovarian Cancer Cells through Targeting APAF1. *Nat. Commun.* **2016**, *7*, 11150. [CrossRef] [PubMed]
113. Yang, J.; Zaman, M.M.; Vlasakov, I.; Roy, R.; Huang, L.; Martin, C.R.; Freedman, S.D.; Serhan, C.N.; Moses, M.A. Adipocytes Promote Ovarian Cancer Chemoresistance. *Sci. Rep.* **2019**, *9*, 13316. [CrossRef]
114. Zhu, X.; Shen, H.; Yin, X.; Yang, M.; Wei, H.; Chen, Q.; Feng, F.; Liu, Y.; Xu, W.; Li, Y. Macrophages Derived Exosomes Deliver MiR-223 to Epithelial Ovarian Cancer Cells to Elicit a Chemoresistant Phenotype. *J. Exp. Clin. Cancer Res.* **2019**, *38*, 81. [CrossRef]
115. Pujade-Lauraine, E.; Hilpert, F.; Weber, B.; Reuss, A.; Poveda, A.; Kristensen, G.; Sorio, R.; Vergote, I.; Witteveen, P.; Bamias, A.; et al. Bevacizumab Combined with Chemotherapy for Platinum-Resistant Recurrent Ovarian Cancer: The AURELIA Open-Label Randomized Phase III Trial. *J. Clin. Oncol. Off. J. Am. Soc. Clin. Oncol.* **2014**, *32*, 1302–1308. [CrossRef] [PubMed]
116. Tatsuki, S.; Shoji, T.; Abe, M.; Tomabechi, H.; Takatori, E.; Kaido, Y.; Nagasawa, T.; Kagabu, M.; Aida, T.; Baba, T. Efficacy and Safety of Platinum Rechallenge in Patients With Platinum-Resistant Ovarian, Fallopian Tube or Primary Peritoneal Cancer: A Multicenter Retrospective Study. *Anticancer Res.* **2022**, *42*, 4603–4610. [CrossRef] [PubMed]
117. Nasu, H.; Nishio, S.; Park, J.; Yoshimitsu, T.; Matsukuma, K.; Tasaki, K.; Katsuda, T.; Terada, A.; Tsuda, N.; Ushijima, K. Platinum Rechallenge Treatment Using Gemcitabine plus Carboplatin with or without Bevacizumab for Platinum-Resistant Ovarian Cancer. *Int. J. Clin. Oncol.* **2022**, *27*, 790–801. [CrossRef]
118. Taylor, S.E.; Beck, T.L.; Krivak, T.C.; Zorn, K.K.; Kelley, J.L.; Edwards, R.P. Oxaliplatin Salvage for Recurrent Ovarian Cancer: A Single Institution's Experience in Patient Populations with Platinum Resistant Disease or a History of Platinum Hypersensitivity. *Gynecol. Oncol.* **2014**, *134*, 68–72. [CrossRef]
119. Murai, J.; Huang, S.N.; Das, B.B.; Renaud, A.; Zhang, Y.; Doroshow, J.H.; Ji, J.; Takeda, S.; Pommier, Y. Trapping of PARP1 and PARP2 by Clinical PARP Inhibitors. *Cancer Res.* **2012**, *72*, 5588–5599. [CrossRef]
120. Kaufman, B.; Shapira-Frommer, R.; Schmutzler, R.K.; Audeh, M.W.; Friedlander, M.; Balmaña, J.; Mitchell, G.; Fried, G.; Stemmer, S.M.; Hubert, A.; et al. Olaparib Monotherapy in Patients With Advanced Cancer and a Germline BRCA1/2 Mutation. *J. Clin. Oncol.* **2014**, *33*, 244–250. [CrossRef]
121. Fong, P.C.; Yap, T.A.; Boss, D.S.; Carden, C.P.; Mergui-Roelvink, M.; Gourley, C.; De Greve, J.; Lubinski, J.; Shanley, S.; Messiou, C.; et al. Poly(ADP)-Ribose Polymerase Inhibition: Frequent Durable Responses in BRCA Carrier Ovarian Cancer Correlating With Platinum-Free Interval. *J. Clin. Oncol.* **2010**, *28*, 2512–2519. [CrossRef] [PubMed]
122. Coleman, R.L.; Sill, M.W.; Bell-McGuinn, K.; Aghajanian, C.; Gray, H.J.; Tewari, K.S.; Rubin, S.C.; Rutherford, T.J.; Chan, J.K.; Chen, A.; et al. A Phase II Evaluation of the Potent, Highly Selective PARP Inhibitor Veliparib in the Treatment of Persistent or Recurrent Epithelial Ovarian, Fallopian Tube, or Primary Peritoneal Cancer in Patients Who Carry a Germline BRCA1 or BRCA2 Mutation—An NRG. *Gynecol. Oncol.* **2015**, *137*, 386–391. [CrossRef] [PubMed]
123. Moore, K.N.; Secord, A.A.; Geller, M.A.; Miller, D.S.; Cloven, N.; Fleming, G.F.; Wahner Hendrickson, A.E.; Azodi, M.; DiSilvestro, P.; Oza, A.M.; et al. Niraparib Monotherapy for Late-Line Treatment of Ovarian Cancer (QUADRA): A Multicentre, Open-Label, Single-Arm, Phase 2 Trial. *Lancet. Oncol.* **2019**, *20*, 636–648. [CrossRef] [PubMed]
124. Oza, A.M.; Tinker, A.V.; Oaknin, A.; Shapira-Frommer, R.; McNeish, I.A.; Swisher, E.M.; Ray-Coquard, I.; Bell-McGuinn, K.; Coleman, R.L.; O'Malley, D.M.; et al. Antitumor Activity and Safety of the PARP Inhibitor Rucaparib in Patients with High-Grade Ovarian Carcinoma and a Germline or Somatic BRCA1 or BRCA2 Mutation: Integrated Analysis of Data from Study 10 and ARIEL2. *Gynecol. Oncol.* **2017**, *147*, 267–275. [CrossRef]
125. Liu, G.; Feng, Y.; Li, J.; Deng, T.; Yin, A.; Yan, L.; Zheng, M.; Xiong, Y.; Li, J.; Huang, Y.; et al. A Novel Combination of Niraparib and Anlotinib in Platinum-Resistant Ovarian Cancer: Efficacy and Safety Results from the Phase II, Multi-Center ANNIE Study. *EClinicalMedicine* **2022**, *54*, 101767. [CrossRef]

126. Sun, L.; Liu, C.; Li, Y. Effect of PARP Inhibitor Combined with Bevacizumab on Platinum-Resistant Recurrent Ovarian Epithelial Carcinoma. *Comput. Math. Methods Med.* **2022**, *2022*, 4600145. [CrossRef]
127. Lee, J.-M.; Moore, R.G.; Ghamande, S.; Park, M.S.; Diaz, J.P.; Chapman, J.; Kendrick, J.; Slomovitz, B.M.; Tewari, K.S.; Lowe, E.S.; et al. Cediranib in Combination with Olaparib in Patients without a Germline BRCA1/2 Mutation and with Recurrent Platinum-Resistant Ovarian Cancer: Phase IIb CONCERTO Trial. *Clin. Cancer Res. Off. J. Am. Assoc. Cancer Res.* **2022**, *28*, 4186–4193. [CrossRef]
128. Colombo, N.; Tomao, F.; Benedetti Panici, P.; Nicoletto, M.O.; Tognon, G.; Bologna, A.; Lissoni, A.A.; DeCensi, A.; Lapresa, M.; Mancari, R.; et al. Randomized Phase II Trial of Weekly Paclitaxel vs. Cediranib-Olaparib (Continuous or Intermittent Schedule) in Platinum-Resistant High-Grade Epithelial Ovarian Cancer. *Gynecol. Oncol.* **2022**, *164*, 505–513. [CrossRef]
129. Li, S.; Wang, T.; Fei, X.; Zhang, M. ATR Inhibitors in Platinum-Resistant Ovarian Cancer. *Cancers* **2022**, *14*, 5902. [CrossRef]
130. Shah, P.D.; Wethington, S.L.; Pagan, C.; Latif, N.; Tanyi, J.; Martin, L.P.; Morgan, M.; Burger, R.A.; Haggerty, A.; Zarrin, H.; et al. Combination ATR and PARP Inhibitor (CAPRI): A Phase 2 Study of Ceralasertib plus Olaparib in Patients with Recurrent, Platinum-Resistant Epithelial Ovarian Cancer. *Gynecol. Oncol.* **2021**, *163*, 246–253. [CrossRef]
131. Lee, J.-M.; Nair, J.; Zimmer, A.; Lipkowitz, S.; Annunziata, C.M.; Merino, M.J.; Swisher, E.M.; Harrell, M.I.; Trepel, J.B.; Lee, M.-J.; et al. Prexasertib, a Cell Cycle Checkpoint Kinase 1 and 2 Inhibitor, in BRCA Wild-Type Recurrent High-Grade Serous Ovarian Cancer: A First-in-Class Proof-of-Concept Phase 2 Study. *Lancet Oncol.* **2018**, *19*, 207–215. [CrossRef] [PubMed]
132. Do, K.; Doroshow, J.H.; Kummar, S. Wee1 Kinase as a Target for Cancer Therapy. *Cell Cycle* **2013**, *12*, 3159–3164. [CrossRef] [PubMed]
133. Andrikopoulou, A.; Liontos, M.; Koutsoukos, K.; Dimopoulos, M.-A.; Zagouri, F. Clinical Perspectives of BET Inhibition in Ovarian Cancer. *Cell. Oncol. (Dordr)* **2021**, *44*, 237–249. [CrossRef]
134. McMullen, M.; Karakasis, K.; Madariaga, A.; Oza, A.M. Overcoming Platinum and PARP-Inhibitor Resistance in Ovarian Cancer. *Cancers* **2020**, *12*, 1607. [CrossRef]
135. Oza, A.M.; Matulonis, U.A.; Alvarez Secord, A.; Nemunaitis, J.; Roman, L.D.; Blagden, S.P.; Banerjee, S.; McGuire, W.P.; Ghamande, S.; Birrer, M.J.; et al. A Randomized Phase II Trial of Epigenetic Priming with Guadecitabine and Carboplatin in Platinum-Resistant, Recurrent Ovarian Cancer. *Clin. Cancer Res. Off. J. Am. Assoc. Cancer Res.* **2020**, *26*, 1009–1016. [CrossRef]
136. Matei, D.; Shen, C.; Fang, F.; Schilder, J.; Li, M.; Arnold, A.; Zeng, Y.; Pilrose, J.M.; Kulesavage, C.; Balch, C.; et al. A Phase II Study of Decitabine and Carboplatin in Recurrent Platinum (Pt)-Resistant Ovarian Cancer (OC). *J. Clin. Oncol.* **2011**, *29*, 5011. [CrossRef]
137. Jones, P.A.; Baylin, S.B. The Fundamental Role of Epigenetic Events in Cancer. *Nat. Rev. Genet.* **2002**, *3*, 415–428. [CrossRef]
138. Yang, H.; Bueso-Ramos, C.; DiNardo, C.; Estecio, M.R.; Davanlou, M.; Geng, Q.-R.; Fang, Z.; Nguyen, M.; Pierce, S.; Wei, Y.; et al. Expression of PD-L1, PD-L2, PD-1 and CTLA4 in Myelodysplastic Syndromes Is Enhanced by Treatment with Hypomethylating Agents. *Leukemia* **2014**, *28*, 1280–1288. [CrossRef]
139. Chiappinelli, K.B.; Zahnow, C.A.; Ahuja, N.; Baylin, S.B. Combining Epigenetic and Immunotherapy to Combat Cancer. *Cancer Res.* **2016**, *76*, 1683–1689. [CrossRef]
140. Chen, S.; Xie, P.; Cowan, M.; Huang, H.; Cardenas, H.; Keathley, R.; Tanner, E.J.; Fleming, G.F.; Moroney, J.W.; Pant, A.; et al. Epigenetic Priming Enhances Antitumor Immunity in Platinum-Resistant Ovarian Cancer. *J. Clin. Investig.* **2022**, *132*, 14. [CrossRef]
141. Taylor, K.; Loo Yau, H.; Chakravarthy, A.; Wang, B.; Shen, S.Y.; Ettayebi, I.; Ishak, C.A.; Bedard, P.L.; Abdul Razak, A.R.; Hansen, A.; et al. An Open-Label, Phase II Multicohort Study of an Oral Hypomethylating Agent CC-486 and Durvalumab in Advanced Solid Tumors. *J. Immunother. Cancer* **2020**, *8*, e000883. [CrossRef] [PubMed]
142. Dizon, D.S.; Damstrup, L.; Finkler, N.J.; Lassen, U.; Celano, P.; Glasspool, R.; Crowley, E.; Lichenstein, H.S.; Knoblach, P.; Penson, R.T. Phase II Activity of Belinostat (PXD-101), Carboplatin, and Paclitaxel in Women with Previously Treated Ovarian Cancer. *Int. J. Gynecol. Cancer Off. J. Int. Gynecol. Cancer Soc.* **2012**, *22*, 979–986. [CrossRef] [PubMed]
143. Dizon, D.S.; Blessing, J.A.; Penson, R.T.; Drake, R.D.; Walker, J.L.; Johnston, C.M.; Disilvestro, P.A.; Fader, A.N. A Phase II Evaluation of Belinostat and Carboplatin in the Treatment of Recurrent or Persistent Platinum-Resistant Ovarian, Fallopian Tube, or Primary Peritoneal Carcinoma: A Gynecologic Oncology Group Study. *Gynecol. Oncol.* **2012**, *125*, 367–371. [CrossRef] [PubMed]
144. Cadoo, K.A.; Meyers, M.L.; Burger, R.A.; Armstrong, D.K.; Penson, R.T.; Gordon, M.S.; Fleming, G.F.; Moroney, J.W.; Hamilton, E.P.; Duska, L.R.; et al. A Phase II Randomized Study of Avelumab plus Entinostat versus Avelumab plus Placebo in Patients (Pts) with Advanced Epithelial Ovarian Cancer (EOC). *J. Clin. Oncol.* **2019**, *37*, 5511. [CrossRef]
145. Jones, P.A.; Issa, J.-P.J.; Baylin, S. Targeting the Cancer Epigenome for Therapy. *Nat. Rev. Genet.* **2016**, *17*, 630–641. [CrossRef] [PubMed]
146. Falchook, G.S.; Fu, S.; Naing, A.; Hong, D.S.; Hu, W.; Moulder, S.; Wheler, J.J.; Sood, A.K.; Bustinza-Linares, E.; Parkhurst, K.L.; et al. Methylation and Histone Deacetylase Inhibition in Combination with Platinum Treatment in Patients with Advanced Malignancies. *Investig. New Drugs* **2013**, *31*, 1192–1200. [CrossRef] [PubMed]
147. Peng, D.; Kryczek, I.; Nagarsheth, N.; Zhao, L.; Wei, S.; Wang, W.; Sun, Y.; Zhao, E.; Vatan, L.; Szeliga, W.; et al. Epigenetic Silencing of TH1-Type Chemokines Shapes Tumour Immunity and Immunotherapy. *Nature* **2015**, *527*, 249–253. [CrossRef]
148. Pirš, B.; Škof, E.; Smrkolj, V.; Smrkolj, Š. Overview of Immune Checkpoint Inhibitors in Gynecological Cancer Treatment. *Cancers* **2022**, *14*, 631. [CrossRef]
149. Strickland, K.C.; Howitt, B.E.; Shukla, S.A.; Rodig, S.; Ritterhouse, L.L.; Liu, J.F.; Garber, J.E.; Chowdhury, D.; Wu, C.J.; D'Andrea, A.D.; et al. Association and Prognostic Significance of BRCA1/2-Mutation Status with Neoantigen Load, Number of Tumor-Infiltrating Lymphocytes and Expression of PD-1/PD-L1 in High Grade Serous Ovarian Cancer. *Oncotarget* **2016**, *7*, 13587–13598. [CrossRef]

150. Konstantinopoulos, P.A.; Waggoner, S.; Vidal, G.A.; Mita, M.; Moroney, J.W.; Holloway, R.; Van Le, L.; Sachdev, J.C.; Chapman-Davis, E.; Colon-Otero, G.; et al. Single-Arm Phases 1 and 2 Trial of Niraparib in Combination With Pembrolizumab in Patients With Recurrent Platinum-Resistant Ovarian Carcinoma. *JAMA Oncol.* **2019**, *5*, 1141–1149. [CrossRef]
151. Lampert, E.J.; Zimmer, A.; Padget, M.; Cimino-Mathews, A.; Nair, J.R.; Liu, Y.; Swisher, E.M.; Hodge, J.W.; Nixon, A.B.; Nichols, E.; et al. Combination of PARP Inhibitor Olaparib, and PD-L1 Inhibitor Durvalumab, in Recurrent Ovarian Cancer: A Proof-of-Concept Phase II Study. *Clin. Cancer Res. Off. J. Am. Assoc. Cancer Res.* **2020**, *26*, 4268–4279. [CrossRef] [PubMed]
152. Leary, A.; Tan, D.; Ledermann, J. Immune Checkpoint Inhibitors in Ovarian Cancer: Where Do We Stand? *Ther. Adv. Med. Oncol.* **2021**, *13*, 17588359211039900. [CrossRef] [PubMed]
153. Liu, J.F.; Herold, C.; Gray, K.P.; Penson, R.T.; Horowitz, N.; Konstantinopoulos, P.A.; Castro, C.M.; Hill, S.J.; Curtis, J.; Luo, W.; et al. Assessment of Combined Nivolumab and Bevacizumab in Relapsed Ovarian Cancer: A Phase 2 Clinical Trial. *JAMA Oncol.* **2019**, *5*, 1731–1738. [CrossRef] [PubMed]
154. Zsiros, E.; Lynam, S.; Attwood, K.M.; Wang, C.; Chilakapati, S.; Gomez, E.C.; Liu, S.; Akers, S.; Lele, S.; Frederick, P.J.; et al. Efficacy and Safety of Pembrolizumab in Combination With Bevacizumab and Oral Metronomic Cyclophosphamide in the Treatment of Recurrent Ovarian Cancer: A Phase 2 Nonrandomized Clinical Trial. *JAMA Oncol.* **2021**, *7*, 78–85. [CrossRef]
155. Ocadlikova, D.; Lecciso, M.; Isidori, A.; Loscocco, F.; Visani, G.; Amadori, S.; Cavo, M.; Curti, A. Chemotherapy-Induced Tumor Cell Death at the Crossroads Between Immunogenicity and Immunotolerance: Focus on Acute Myeloid Leukemia. *Front. Oncol.* **2019**, *9*, 1004. [CrossRef] [PubMed]
156. Pujade-Lauraine, E.; Fujiwara, K.; Ledermann, J.A.; Oza, A.M.; Kristeleit, R.; Ray-Coquard, I.-L.; Richardson, G.E.; Sessa, C.; Yonemori, K.; Banerjee, S.; et al. Avelumab Alone or in Combination with Chemotherapy versus Chemotherapy Alone in Platinum-Resistant or Platinum-Refractory Ovarian Cancer (JAVELIN Ovarian 200): An Open-Label, Three-Arm, Randomised, Phase 3 Study. *Lancet. Oncol.* **2021**, *22*, 1034–1046. [CrossRef] [PubMed]
157. Lee, E.K.; Xiong, N.; Cheng, S.-C.; Barry, W.T.; Penson, R.T.; Konstantinopoulos, P.A.; Hoffman, M.A.; Horowitz, N.; Dizon, D.S.; Stover, E.H.; et al. Combined Pembrolizumab and Pegylated Liposomal Doxorubicin in Platinum Resistant Ovarian Cancer: A Phase 2 Clinical Trial. *Gynecol. Oncol.* **2020**, *159*, 72–78. [CrossRef]
158. Liang, Z.D.; Long, Y.; Tsai, W.-B.; Fu, S.; Kurzrock, R.; Gagea-Iurascu, M.; Zhang, F.; Chen, H.H.W.; Hennessy, B.T.; Mills, G.B.; et al. Mechanistic Basis for Overcoming Platinum Resistance Using Copper Chelating Agents. *Mol. Cancer Ther.* **2012**, *11*, 2483–2494. [CrossRef]
159. Fu, S.; Hou, M.-M.; Wheler, J.; Hong, D.; Naing, A.; Tsimberidou, A.; Janku, F.; Zinner, R.; Piha-Paul, S.; Falchook, G.; et al. Exploratory Study of Carboplatin plus the Copper-Lowering Agent Trientine in Patients with Advanced Malignancies. *Investig. N. Drugs* **2014**, *32*, 465–472. [CrossRef]
160. Huang, Y.-F.; Kuo, M.T.; Liu, Y.-S.; Cheng, Y.-M.; Wu, P.-Y.; Chou, C.-Y. A Dose Escalation Study of Trientine Plus Carboplatin and Pegylated Liposomal Doxorubicin in Women With a First Relapse of Epithelial Ovarian, Tubal, and Peritoneal Cancer Within 12 Months After Platinum-Based Chemotherapy. *Front. Oncol.* **2019**, *9*, 437. [CrossRef]
161. Mariniello, M.; Petruzzelli, R.; Wanderlingh, L.G.; La Montagna, R.; Carissimo, A.; Pane, F.; Amoresano, A.; Ilyechova, E.Y.; Galagudza, M.M.; Catalano, F.; et al. Synthetic Lethality Screening Identifies FDA-Approved Drugs That Overcome ATP7B-Mediated Tolerance of Tumor Cells to Cisplatin. *Cancers* **2020**, *12*, 608. [CrossRef] [PubMed]
162. Pan, H.; Kim, E.; Rankin, G.O.; Rojanasakul, Y.; Tu, Y.; Chen, Y.C. Theaflavin-3,3′-Digallate Enhances the Inhibitory Effect of Cisplatin by Regulating the Copper Transporter 1 and Glutathione in Human Ovarian Cancer Cells. *Int. J. Mol. Sci.* **2018**, *19*, 117. [CrossRef] [PubMed]

medicina MDPI

Systematic Review

A Systematic Review Investigating the Difference between 1 Cycle versus 2 Cycles of Adjuvant Chemotherapy in Stage I Testicular Germ Cell Cancers

Emanuiela Florentina Rohozneanu [1,2], Ciprian Deac [2] and Călin Ioan Căinap [1,2,*]

[1] Department of Oncology, The Oncology Institute "Prof. Dr. Ion Chiricuţă" Cluj-Napoca, "Iuliu Hatieganu" University of Medicine and Pharmacy, 400015 Cluj-Napoca, Romania

[2] Department of Oncology, "Iuliu Hatieganu" University of Medicine and Pharmacy, 400012 Cluj-Napoca, Romania

* Correspondence: calincainap2015@gmail.com

Abstract: Standard care for stage I testicular germ cell cancers (seminomatous—STC or non-seminomatous—NSTC) is orchiectomy followed by active surveillance, 1 or 2 cycles of adjuvant chemotherapy, surgery or radiotherapy. The decision on the adjuvant therapeutic approach is guided by the associated risk factors of the patient and the potential related toxicity of the treatment. Currently, there is no consensus regarding the optimal number of adjuvant chemotherapy cycles. Although in terms of overall survival, there is no proven inconsistency regarding the number of cycles of adjuvant chemotherapy, and the rate of relapse may vary.

Keywords: stage I testicular germ cell cancers; seminomatous—STC; non-seminomatous—NSTC

Citation: Rohozneanu, E.F.; Deac, C.; Căinap, C.I. A Systematic Review Investigating the Difference between 1 Cycle versus 2 Cycles of Adjuvant Chemotherapy in Stage I Testicular Germ Cell Cancers. *Medicina* **2023**, *59*, 916. https://doi.org/10.3390/medicina59050916

Academic Editor: Dah-Shyong Yu

Received: 25 March 2023
Revised: 21 April 2023
Accepted: 2 May 2023
Published: 10 May 2023

1. Introduction

Background

Testicular cancer is a rare type of tumor that accounts for about 1% of all cancers in adults [1]. In 98% of cases, the cell of origin is represented by a germ cell that failed to differentiate and continued to express its pluripotency which is afterward translated into an uncontrolled malignant growth due to the accumulation of chromosomal aberrations [2]. The uncontrolled proliferation can lead to the genesis of two histological types of testicular cancer, represented by seminoma and non-seminoma, the latter being characterized within the pathology report as being a component of embryonal carcinoma, yolk sac tumor, choriocarcinoma or immature teratoma. Orchiectomy is required for diagnosis and is also the initial therapeutic approach. Further treatment depends on the histopathological features, tumor markers reference interval (α-feto protein, β-human chorionic gonadotropin, lactate dehydrogenase) and imaging-based diagnostic analysis. Stage I testicular cancer is defined by the absence of metastasis on the retroperitoneal lymph nodes and distant organs on the CT scan. The different adjuvant therapeutic approaches depicted by one or two cycles of Carboplatin, radiotherapy of the para-aortic and ipsilateral iliac lymph nodes or active surveillance for stage I seminoma and chemotherapy using one or two cycles of BEP, RPLND or active surveillance for stage I non-seminoma provide a similar outcome in terms of overall survival [3,4]. However, it can alter the risk of relapse which can fluctuate from 15–30% for stage I seminoma to 40–50% for stage I non-seminoma [5,6] Adequate management of this stage is required, as the 5-year survival rate is close to 100% [3–6].

When chemotherapy is administered, guidelines variably recommend the administration of one or two cycles, but the specific number of cycles is not well defined yet. The aim of this systematic review is to assess the difference between one versus two administered cycles of chemotherapy in stage one testicular germ cell cancers (STC and NSTC) with reference to overall and disease-free survival, short-term and long-term toxicities. Given the high rates of survival among the patient population and the associated risk of relapse, it is necessary to identify the right approach for this stage of the disease.

2. Methods

2.1. Literature Search Strategy

The articles reviewed for this paper needed a systematic search conducted in PubMed (1970–December 2022) using the following key words: 'testicular neoplasms', 'testicular cancer', 'germinal testicular cancer' and 'non-seminomatous tumors' combined with 'chemotherapy' or 'treatment'.

2.2. Selection Criteria

Papers that were reviewed for this article included the following criteria:

- Articles written and published in English
- Evidence-based clinical practice guidelines
- Randomized and non-randomized studies referring to stage I seminoma or non-seminoma treated with adjuvant chemotherapy that reported a rate of recurrence and/or survival and/or toxicity of the treatment
- If any article included other stages of seminoma or non-seminoma, outcome was reported separately for stage I disease
- If any article included other options of treatment for stage I seminoma or non-seminoma (surveillence, radiotherapy and surgery), outcome was reported separately for the patients who received chemotherapy

3. Results

3.1. Literature Search Results

We identified 30 articles that met the selection criteria of this paper and included 3 clinical practice guidelines, 2 randomized studies and 25 non-randomized studies. The results of the literature search are summarized in Table 1. The PRISMA guidelines were followed for drafting of this paper.

Table 1. Literature search results.

Study Type	Number [References]
Clinical practice guidelines	3 [7–9]
Randomized controlled studies	2 [10,11]
Non-randomized studies	25 [12–36]

3.2. Clinical Practice Guidelines

Three clinical practice guidelines were reviewed for this paper and the recommendations are summarized in Table 2.

Table 2. Guideline recommendations.

Guideline	Stage I Seminoma		Stage I Non-Seminoma	
	1 × Carboplatin AUC 7	2 × Carboplatin AUC 7	1 × BEP	2 × BEP
ESMO [7]	X	Recognizes better results, but limited data available	X	N/A
EAU [8]	X	N/A	X	X
NCCN [9]	X	X	X	Not indicated because of adverse events

BEP—Bleomycin, Cisplatin, Etoposid; ESMO—European Society for Medical Oncology; NCCN—National Comprehensive Cancer Network; EAU—European Association of Urology.

3.2.1. Clinical Stage I Seminoma

ESMO [7] and EAU [8] guidelines recognize tumor size > 4 cm and invasion of rete testis as risk factors for recurrence and recommend the administration of chemotherapy in the presence of any of these factors. Moreover, the ESMO [7] guideline highlights the potential benefit of the administration of two courses of Carboplatin; however, due to insufficient evidence, it is not recommended. In contrast, the NCCN [9] guideline does not support stratification of the patients using the aforementioned risk factors due to limited evidence that fails to prove their predictive value and, therefore, recommends the administration of one or two cycles of chemotherapy whenever active surveillance is not feasible.

3.2.2. Clinical Stage I Non-Seminoma

All three guidelines recommend active surveillance whenever it is feasible or one cycle of BEP in the presence of risk factors.

NCCN[8] does not recommend two cycles given the possible associated adverse events, but EAU[9] recognizes it as a potential option.

3.3. Randomized and Non-Randomized Studies

3.3.1. Clinical Stage I Seminoma

When deciding on the adjuvant treatment for stage I seminoma, several studies suggested that the administration of chemotherapy should be guided by the presence of one of the two risk factors, represented by tumor size and invasion of rete testis [37]. Only one of the twelve studies exposed in Tables 3 and 4 investigating the efficacy of adjuvant chemotherapy for clinical stage I seminoma used these factors to stratify the patients. Chemotherapy was mostly administered as an option of treatment chosen either by the physician or the patient.

Table 3. Studies of single cycle adjuvant chemotherapy in seminomatous testicular cancer.

Study	Eligibility	Number of Patients	Chemotherapy Regimen	Median Follow-Up (Months)	Number of Relapses
Oliver et al. [12] (1994)	-	25	1 × Carboplatin AUC 7	29	0
Dieckmann et al. [13] (2000)	-	93	1 × Carboplatin 400 mg/mp	48	8
Oliver et al. [10] (2005)	Randomized	573	1 × Carboplatin AUC 7	48	27
Dieckmann et al. [14] (2016)	-	362	1 × Carboplatin AUC 7	30	18
Tanstad et al. [15] (2011)	-	188	1 × Carboplatin AUC 7	40	7
Chau et al. [16] (2015)	-	517	1 × Carboplatin AUC 7	47.2	21
Diminutto et al. [17] (2015)	-	115	1 × Carboplatin AUC 7	22.1	6

Single-cycle adjuvant Carboplatin provided satisfactory results in multiple studies [15–17]. In the randomized prospective study published in 2005, Oliver et al. [10] not only proved the non-inferiority of Carboplatin to radiotherapy for stage I seminoma, but after a median follow-up of 4 years, the relapse rate in the single cycle chemotherapy arm was only 4.7%.

Table 4. Studies of 2 cycles of adjuvant chemotherapy in seminomatous testicular cancer.

Study	Eligibility	Number of Patients	Chemotherapy Regimen	Follow-Up (Months)	Number of Relapses
Oliver et al. [12] (1994)	-	53	2 × Carboplatin AUC 7	51	1
Dieckmann et al. [13] (2000)	-	32	2 × Carboplatin 400 mg/mp	48	0
Reiter et al. [18] (2001)	-	107	2 × Carboplatin 400 mg/mp	74	0
Steiner et al. [19] (2002)	-	108	2 × Carboplatin 400 mg/mp	59.8	2
Argirovic D [20] (2005)	-	163	2 × Carboplatin 400 mg/mp	48	3
Aparicio et al. [21] (2005)	T > 4 cm Rete testis involvement	214	2 × Carboplatin AUC 7	34	7
Koutsoukos et al. [22] (2016)	-	138	2 × Carboplatin AUC 6		5
Dieckmann et al. [14] (2016)	-	66	2 × Carboplatin AUC 7	30	1

T—tumor size.

In 2016, Dieckmann et al. [14] published a prospective non-randomized study that included a total of 725 patients with clinical stage I seminoma. Adjuvant treatment was decided by local physicians and included active surveillance, radiotherapy or chemotherapy (1 or 2 cycles of Carboplatin). After a median follow-up of 30 months, stratification of tumor size revealed that in the group of patients who received one cycle of Carboplatin, tumor dimension > 4 cm can predict a higher risk of relapse than those whose tumor is below this limit (6.8% versus 2.3%). Moreover, the risk of relapse was statistically higher for tumor sizes >5 cm (9.3%). Although the direct comparison of the number of relapses was not significantly lower for the patients treated with two courses of Carboplatin (1.5% versus 5%; $p = 0.2048$), it appears that large tumor size can predict a low efficacy of one course of Carboplatin. Similar results were obtained by Chau et al. [16] who revealed a relapse-free survival of 97.4% after one cycle of chemotherapy with Carboplatin but with a relapse rate that was nearly double for the patients with tumor size above 4 cm (5.9% versus 3.3%). Tumor size was recognized as a prognostic factor in other studies as well [22] but limited data exist regarding the role of rete testis invasion in the risk of relapse [21]. On the other hand, Oliver et al. [12] (1994) found no significant correlation between tumor size and risk of relapse which is in accordance with the results of other studies [15]. Moreover, his trial did not reveal any additional benefit in the administration of two cycles of Carboplatin when compared to one cycle, showing instead higher rates of acute and late toxicities. In contrast, the study published by Dieckmann et al. [13] (2000) revealed no relapses after two courses of Carboplatin compared to 8.6% for those treated with one cycle ($p = 0.088$). Toxicity on the reproductive system was reported, but after 20 months the median FSH levels returned to the normal range.

Reiter et al. [18] also obtained favorable results after two courses of Carboplatin without any relapses when a group of 107 patients having stage I seminoma and a toxicity profile of WHO grade I or II underwent treatment. In the study published by Steinner et al. [19], the 5-year relapse-rate after two courses of Carboplatin was 1.85% but exhibited a more hematological toxicity than previously reported: 44.4% developed thrombocytopenia, 2.8 of which was grade 4.

Aparicio et al. [21] also obtained a favorable relapse rate of 3.3% at a median follow up of 36 months and an overall 5-year survival of 100% after two cycles of Carboplatin AUC 7. Patients were stratified by tumor size (>4 cm) and invasion of rete testis, and there was a significant correlation between invasion of rete testis and relapse (DFS of 99.2% versus 91.6%; $p = 0.0108$).

3.3.2. Clinical Stage I Non-Seminoma

Tables 5 and 6 include selected trials that investigated the role of adjuvant chemotherapy in patients diagnosed with high-risk clinical stage I non-seminoma. Although the definition of high-risk disease varied, all studies had at least one risk factor in common.

Table 5. Studies of single cycle adjuvant chemotherapy in non-seminomatous testicular cancer.

Study	Eligibility	Number of Patients	Chemotherapy Regimen	Follow-Up (Months)	Number of Relapses
Oliver et al. [23] (2004)	VI LI Absence of yolk sac elements Presence of • Undifferentiated areas • Malignant teratoma • Undifferentiated or malignant trophoblastic teratoma	46	1 × BEP	33	3
Gilbert et al. [24] (2006)	VI LI Absence of yolk sac elements Presence of undifferentiated elements	22	1 × BEP	122	0
Albers et al. [11] (2008)	Randomized	191	1 × BEP	56	2
Westermann et al. [25] (2008)	VI LI ≥50% embryonal carcinoma	40	1 × BEP	96	1
Tandstad et al. [26] (2009)	VI+ VI−	157 155	1 × BEP 1 × BEP	57 49	5 2
Vidal et al. [27] (2015)	VI >50% embryonal carcinoma	40	1 × BEP	186	1
Cullen et al. [28] (2020)	VI Combined seminoma + NSGCTT	236	1 × BEP	49	4

BEP—Bleomycin, Cisplatin, Etoposid; VI—vascular inasion; LI—lymphatic invasion; NSGCTT—nonseminomatous germ cell tumors of the testis.

Only two studies conducted a direct comparison between the outcome of patients receiving one versus two cycles of BEP. Oliver et al. [23] (2004) observed that after a median follow-up of 33 months, relapses were seen in 6.5% (3/46) of patients treated with one cycle of BEP and only in 3.6% (1/28) of patients treated with two cycles of the same regimen. No significant toxicities were reported except for permanent ototoxicity in a music teacher that led to the inability to teach. Although he was treated with two cycles of BEP, many of the studies included in this paper reported ototoxicity even after one cycle of BEP [25,27] In the SWENOTECA large prospective study, Tandstad et al. [26] stratified the patients according to the LVI invasion and offered surveillance or one cycle of BEP to those without LVI and two cycles of BEP to those LVI +. Results showed that one cycle of BEP reduces the risk of relapse by 90% to both LVI + and LVI—compared to surveillance. Furthermore, after 2 cycles of BEP relapse-free survival after a median follow-up of 60 months was

100%, with no significant additional adverse events compared to one cycle of BEP except for obstipation.

Table 6. Studies of 2 cycles of BEP in non-seminomatous testicular cancer.

Study	Eligibility	Number of Patients	Chemotherapy Regimen	Follow-Up (Months)	Number of Relapses
Studer et al. [29] (1993)	VI pT > 1 Presence of embryonal carcinoma	41	2 × BEP	42	1
Pont et al. [30] (1996)	VI	29	2 × BEP	79	2
Cullen et al. [31] (1996)	Any 3 of the following criteria: • VI • LI • Absence of yolk sac elements • Presence of undifferentiated elements	114	2 × BEP	48	2
Bohlen et al. [32] (1999)	VI LI pT > 1 Presence of embryonal carcinoma	59	2 × BEP	93	1
Oliver et al. [23] (2004)	VI LI Absence of yolk sac elements Presence of • Undifferentiated areas • Malignant teratoma • Undifferentiated or malignant trophoblastic teratoma	28	2 × BEP	33	1
Chevreau et al. [33] (2004)	VI Presence of embryonal carcinoma	40	2 × BEP	113.2	0
Maroto et al. [34] (2005)	VI Local invasion of adjacent structures Presence of embryonal carcinoma	231	2 × BEP	40	2
Guney et al. [35] (2009)	VI LI pT > 1 ≥80% embryonal carcinoma Preorchiectomy AFP ≥ 80 ng/dL	71	2 × BEP	26	4
Tandstad et al. [26] (2009)	VI+ VI−	70 2	2 × BEP 2 × BEP	60 34	0 0
Bamias et al. [36] (2011)	VI LI Invasion of tunica vaginalis, spermatic cord, rete testis or scrotal wall >50% embryonal carcinoma	142	2 × BEP	79	1

BEP—Bleomycin, Cisplatin, Etoposid; VI—vascular inasion; LI—lymphatic invasion; pT—pathologycal stage of tumor according to TNM staging.

The most encouraging outcome regarding the administration of one cycle of BEP was obtained by Gilbert et al. [24]: no relapses were observed after a median follow up of 10.2 years. Similar results were published by Vidal et al. [27] and Westerman et al. [25] that obtained a relapse rate of only 2.5% and 2.7%, respectively. Results may be influenced by the small number of patients included in the studies mentioned above. Albers et al. [11] conducted a randomized phase III trial that included 191 patients comparing retroperitoneal lymph node dissection to one cycle of BEP. The authors achieved a statistically significant recurrence-free survival in favor of chemotherapy with only two relapses in the chemotherapy arm and fifteen relapses in the surgery arm after a median follow-up of 4.7 years ($p = 0.0011$). The largest and most recent prospective trial investigating the efficacy of one cycle of BEP was conducted by Cullen et al. [28] that included 246 of stage I NSGCTT. With four relapses at a median follow-up of 49 months, results showed a two year metastatic recurrence of 1.3% which is similar to the results reported for two cycles of BEP but having the advantage of low levels of serious adverse events.

Currently, there are more studies published in the literature that investigated the efficacy of two cycles of BEP than of one cycle of BEP as summarized in Table 6. In a large prospective study conducted by Cullen et al. [31], at a median follow-up of 4 years, 2 out of 114 patients had a relapse with no long-term toxicity on fertility and lung function being observed. However, the authors reported the death of a 45-year-old patient caused by a cerebrovascular event eight days after the administration of the first cycle of BEP without having hematological changes that could explain the affection. The link to the treatment was unclear. Tha data published by Maroto et al. [34] also relied on a large number of patients suffering of high-risk stage I NSGCTT (n = 231). After the administration of two cycles of BEP, only two reoccurrences have been observed. Regarding the toxicity on the reproductive system, out of the 19 patients who have fathered a child, only one needed to use cryopreserved semen. A total of 1.3% developed a tumor affecting the contralateral testicle.

In a non-randomized prospective trial, Studer et al. [29] obtained a relapse-free survival of 97.5% after the administration of two cycles of BEP at a median follow up of 42 months, with only one relapse that was mature teratoma treated surgically and without late toxicities reported. The results are consistent with the data published by Pont et al. [30] that registered 2/42 relapses after two cycles of BEP for high-risk stage I NSGCTT with no significant acute or late adverse events compared to the control group. Similar rates of relapses were reported by Guney et al. [35] (4/71).

Long-term results after the administration of two cycles of BEP were published by Bohlen et al. [32], revealing only one relapse in 59 patients followed for a median time of 93 months. One case of transient nephrotoxicity, one of neurotoxicity and one of cardiotoxicity were reported. The long-term efficacy of two cycles of BEP was also studied by Chevreau et al. [33]. At a median follow-up of 113.2 months, no relapses were observed in the 40 patients receiving two cycles of chemotherapy. Two patients developed a second cancer in the contralateral testis and no impact on fertility was observed as previously reported by other studies [30,31]. Another prospective study conducted by Bamias et al. [36] reported one relapse after a median follow-up of 79 months of 142 patients.

4. Discussion

Historically, adjuvant chemotherapy with Carboplatin in stage I testicular seminoma was used as an alternative to radiotherapy because of the growing concerns of the side effects to this treatment [13]. Currently, adjuvant Carboplatin may be administered as an option to all patients or in the presence of risk factors (tumor size > 4 cm, invasion of rete testis). However, these risk factors have been the subject of a long debate in the past years since no prospective study has been conducted in order to validate them. In this setting, a tumor size > 4 cm was correlated with an increase in the risk of relapse in contrast to rete testis invasion that lacked evidence in most of the trials [14,16,22]. The first studies that assessed the efficacy of Carboplatin in eradicating micro-metastasis used two cycles of adjuvant chemotherapy, but subsequent evidence revealed that one cycle might be equivalent. [12]. However, one course of adjuvant Carboplatin seems to be insufficient to lower this risk when tumor size is above 4 cm [14].

When comparing studies that investigated the role of adjuvant BEP in stage I non-seminoma, one should be aware that different studies used different groups of risk factors, and this could be a risk of bias. One risk factor that was common in all the studies was lymphovascular invasion. The right definition of high-risk disease for stage I non-seminoma has been investigated by many trials. One of the most significant was a meta-analysis published by Vergouwe et al. [38] that examined 23 studies and analyzed a total of 2587 patients with stage I NSTC. 29% of the patients had occult metastasis diagnosed either during follow up or at retroperitoneal lymph node dissection. Several predictors for occult metastasis were identified, but the strongest was vascular invasion defined as venous and lymphatic invasion (OR, 5.2; 95% CI, 4.0 to 6.8). The presence of embryonal carcinoma, a high pathologic stage or size of the primary tumor were also statistically significantly associated with the presence of occult metastasis but with a weaker effect.

Adjuvant chemotherapy with BEP for stage I NSTC was first explored around 1990 with the argument that using more limited chemotherapy in patients with high-risk disease will restrict exposure to a higher amount of chemotherapy in case of a relapse [39]. Although the first studies published used two cycles of chemotherapy, more recent trials proved similar results with one cycle of BEP regarding relapse-free survival. Currently, there is no agreement between the use of one or two cycles of BEP. At first glance at the number of relapses reported in Tables 5 and 6, two cycles of BEP seem to provide better results, although the difference is not obvious. The only two studies that made a direct comparison with one cycle of BEP obtained a lower number of relapses with two cycles but with no survival benefit reported [23,26]. Meanwhile, the debate is centered on the issues of dose-related toxicities from BEP. Table 7 summarizes adverse events from BEP, grouped by dose and non-dose related [39]. The acute toxicities reported in the trials mentioned in this paper were mainly hematological and gastrointestinal (nausea, vomiting). Bleomycine-induced lung injury is a well-known dose-limiting toxicity. Lung function was analyzed before and after chemotherapy describing a discreet decrease in respiratory parameters but with no symptomatic respiratory dysfunction [32] or pneumonitis reported in any of the studies selected for this paper. Perhaps one of the most concerning dose-related toxicity is infertility. Most of the studies reviewed for this article reported outcomes on fertility with the majority of the patients being able to conceive one or two years after the treatment. This was applicable not only to the patients that performed one cycle of BEP but also for those who performed two cycles of BEP with minimal toxicity on fertility [34]. This is in accordance with the results obtained by Bujan et al. [40] regarding the impact of chemotherapy and radiotherapy on spermatogenesis. The authors concluded that after two or fewer cycles of BEP sperm count returns to pretreatment levels after twelve months, but not after radiotherapy or more then two cycles of BEP.

Table 7. Toxicities of BEP chemotherapy [39].

Dose-Related	Non-Dose Related
Infertility Fatigue Pneumonitis or lung fibrosis Renal damage AnaemiaPeripheral neuropathy Ototoxicity Skin toxicity Reynaud's phenomena Avascular necrosis hip	Neutropenia Alopecia Nausea andvomiting

BEP—Bleomycin, Cisplatin, Etoposid.

5. Conclusions

Taking into account the fact that the data we currently have, we suggest that all treatment options for clinical stage I testicular cancer provide similar survival outcomes and considering the potential dose-related toxicity associated with chemotherapy, we can conclude that at the moment there is not enough evidence to support the superiority of two cycles of chemotherapy instead of one.

Funding: This research received no external funding.

Institutional Review Board Statement: Ethical review and approval were waived for this study due the format of the article.

Informed Consent Statement: Patient consent was waived due to the format of the article.

Data Availability Statement: The data analyzed in this systematic review were obtained from Pubmed electronic database.

Conflicts of Interest: The authors declare no conflict of interest.

References

1. Chung, P.; Warde, P. Testicular cancer: Germ cell tumours. *BMJ Clin. Evid.* **2016**, *2016*, 1807. [PubMed]
2. Baroni, T.; Arato, I.; Mancuso, F.; Calafiore, R.; Luca, G. On the origin of testicular germ cell tumors: From gonocytes to testicular cancer. *Front. Endocrinol.* **2019**, *10*, 343. [CrossRef] [PubMed]
3. Winter, C.; Hiester, A. Treatment of clinical stage I non-seminoma. *Asian J. Urol.* **2021**, *8*, 161–169. [CrossRef]
4. Sayed, M.M.; Nasr, A.M.; Eldin, I.M.S.; Abdelazim, Y.A. Stage I seminoma: Outcome of different treatment modalities and changes in patterns of care. A single institution experience. *Arch. Ital. Urol. Androl.* **2023**, *95*, 1–3.
5. Daugaard, G.; Gundgaard, M.G.; Mortensen, M.S.; Agerbæk, M.; Holm, N.V.; Rørth, M. Surveillance for stage I nonseminoma testicular cancer: Outcomes and long-term follow-up in a population-based cohort. *J. Clin. Oncol.* **2014**, *32*, 3817–3823. [CrossRef]
6. Mortensen, M.S.; Lauritsen, J.; Gundgaard, M.G.; Agerbæk, M.; Holm, N.V.; Christensen, I.J. A nationwide cohort study of stage I seminoma patients followed on a surveillance program. *Eur. Urol.* **2014**, *66*, 1172–1178. [CrossRef]
7. Oldenburg, J.; Berney, D.; Bokemeyer, C.; Climent, M.; Daugaard, G.; Gietema, J.; De Giorgi, U.; Haugnes, H.; Huddart, R.; Leão, R.; et al. Testicular seminoma and non-seminoma: ESMO-EURACAN Clinical Practice Guideline for diagnosis, treatment and follow-up. *Ann. Oncol.* **2022**, *33*, 362–375. [CrossRef]
8. Gilligan, T.; Lin, D.W.; Aggarwal, R.; Bagrodia, A.; Costa, D.; Davis, N.B.; Drakaki, A.; Einhorn, L.; Emamekhoo, H.; Evans, C.; et al. NCCN Clinical Practice Guidelines in Oncology (NCCN Guidelines) Testicular Cancer. *J. Natl. Compr. Cancer Netw.* **2023**, *1*, MS7–MS17.
9. Laguna, M.P.; Albers, P.; Algaba, F.; Bokemeyer, C.; Boormans, J.L.; di Nardo, D.; Fischer, S.; Fizazi, K.; Gremmels, H.; Leao, R.; et al. EAU Guidelines on Testicular Cancer. *Eur. Urol.* **2022**, 19–22. [CrossRef] [PubMed]
10. Oliver, R.T.D.; Mason, M.D.; Mead, G.M.; von der Maase, H.; Rustin, G.J.S.; Joffe, J.K.; de Wit, R.; Aass, N.; Graham, J.D.; Coleman, R.; et al. Radiotherapy versus single-dose carboplatin in adjuvant treatment of stage I seminoma: A randomised trial. *Lancet* **2005**, *366*, 293–300. [CrossRef] [PubMed]
11. Albers, P.; Siener, R.; Krege, S.; Schmelz, H.U.; Dieckmann, K.P.; Heidenreich, A. Randomized phase III trial comparing retroperitoneal lymph node dissection with one course of bleomycin and etoposide plus cisplatin chemotherapy in the adjuvant treatment of clinical stage I nonseminomatous testicular germ cell tumors: AUO trial AH 01/94 by the German Testicular Cancer Study Group. *J. Clin. Oncol.* **2008**, *26*, 2966–2972.
12. Oliver, R.; Edmonds, P.; Ong, J.; Ostrowski, M.; Jackson, A.; Baille-Johnson, H.; Williams, M.; Wiltshire, C.; Mott, T.; Pratt, W.; et al. Pilot studies of 2 and 1 course carboplatin as adjuvant for stage I seminoma: Should it be tested in a randomized trial against radiotherapy? *Int. J. Radiat. Oncol. Biol. Phys.* **1994**, *29*, 3–7. [CrossRef] [PubMed]
13. Dieckmann, K.P.; Brüggeboes, B.; Pichlmeier, U.; Küster, J.; Müllerleile, U.; Bartels, H. Adjuvant treatment of clinical stage I seminoma: Is a single course of carboplatin sufficient? *Urology* **2000**, *55*, 102–106. [CrossRef] [PubMed]
14. Dieckmann, K.-P.; Dralle-Filiz, I.; Matthies, C.; Heinzelbecker, J.; Bedke, J.; Ellinger, J.; Anheuser, P.; Souchon, R.; Pichlmeier, U. Testicular seminoma clinical stage 1: Treatment outcome on a routine care level. *J. Cancer Res. Clin. Oncol.* **2016**, *142*, 1599–1607. [CrossRef]
15. Tandstad, T.; Smaaland, R.; Solberg, A.; Bremnes, R.M.; Langberg, C.W.; Laurell, A. Management of Seminomatous Testicular Cancer: A Binational Prospective Population-Based Study from the Swedish Norwegian Testicular Cancer Study Group. *J. Clin. Oncol.* **2011**, *29*, 719–724. [CrossRef]
16. Chau, C.; Cathomas, R.; Wheater, M.; Klingbiel, D.; Fehr, M.; Bennett, J.; Markham, H.; Lee, C.; Crabb, S.J.; Geldart, T. Treatment outcome and patterns of relapse following adjuvant carboplatin for stage I testicular seminomatous germ-cell tumour: Results from a 17-year UK experience. *Ann. Oncol.* **2015**, *26*, 1865–1870. [CrossRef]
17. Diminutto, A.; Basso, U.; Maruzzo, M.; Morelli, F.; De Giorgi, U.; Perin, A. Adjuvant Carboplatin Treatment in 115 Patients with Stage I Seminoma: Retrospective Multicenter Survey. *Clin. Genitourin. Cancer* **2016**, *14*, e161–e169. [CrossRef] [PubMed]

18. Reiter, W.J.; Brodowicz, T.; Alavi, S.; Zielinski, C.C.; Kozak, W.; Maier, U.; Nöst, G.; Lipsky, H.; Marberger, M.; Kratzik, C. Twelve-year experience with two courses of adjuvant single-agent carboplatin therapy for clinical stage I seminoma. *J. Clin. Oncol.* **2001**, *19*, 101–104. [CrossRef]
19. Steiner, H.; Höltl, L.; Wirtenberger, W.; Berger, A.P.; Bartsch, G.; Hobisch, A. Long-term experience with carboplatin monotherapy for clinical stage I seminoma: A retrospective single-center study. *Urology* **2002**, *60*, 324–328. [CrossRef]
20. Argirović, D. Germ cell testicular tumors in clinical stage A and normal values of serum tumor markers post-orchiectomy: The experience in the management of 300 consecutive patients. *J. BUON* **2005**, *10*, 195–200.
21. Aparicio, J.; Germà, J.R.; del Muro, X.G.; Maroto, P.; Arranz, J.A.; Sáenz, A.; Barnadas, A.; Dorca, J.; Gumà, J.; Olmos, D.; et al. Risk-Adapted Management for Patients with Clinical Stage I Seminoma: The Second Spanish Germ Cell Cancer Cooperative Group Study. *J. Clin. Oncol.* **2005**, *23*, 8717–8723. [CrossRef] [PubMed]
22. Koutsoukos, K.; Tzannis, K.; Christodoulou, C. Two cycles of adjuvant carboplatin in stage I seminoma: 8-year experience by the Hellenic Cooperative Oncology Group (HECOG). *World J. Urol.* **2016**, *34*, 853–857. [CrossRef] [PubMed]
23. Oliver, R.; Ong, J.; Shamash, J.; Ravi, R.; Nagund, V.; Harper, P.; Ostrowski, M.; Sizer, B.; Levay, J.; Robinson, A.; et al. Long-term follow-up of Anglian germ cell cancer group surveillance versus patients with stage 1 nonseminoma treated with adjuvant chemotherapy. *Urology* **2004**, *63*, 556–561. [CrossRef] [PubMed]
24. Gilbert, D.C.; Norman, A.R.; Nicholl, J.; Dearnaley, D.P.; Horwich, A.; Huddart, R.A. Treating stage I nonseminomatous germ cell tumours with a single cycle of chemotherapy. *BJU Int.* **2006**, *98*, 67–69. [CrossRef]
25. Westermann, D.H.; Schefer, H.; Thalmann, G.N.; Karamitopoulou-Diamantis, E.; Fey, M.F.; Studer, U.E. Long-term follow up results of 1 cycle of adjuvant bleomycin, etoposide, and cisplatin chemotherapy for high-risk clinical stage I nonseminomatous germ cell tumors of the testis. *J. Urol.* **2008**, *179*, 163–166. [CrossRef]
26. Tandstad, T.; Dahl, O.; Cohn-Cedermark, G.; Cavallin-Stahl, E.; Stierner, U.; Solberg, A.; Langberg, C.; Bremnes, R.M.; Laurell, A.; Wijkstrøm, H.; et al. Risk adapted treatment in clinical stage I nonseminomatous germ cell testicular cancer: The SWENOTECA management program. *J. Clin. Oncol.* **2009**, *27*, 2122–2128. [CrossRef]
27. Vidal, A.D.; Thalmann, G.; Karamitopoulou, E.; Studer, U. Long-term outcome of patients with clinical stage I high-risk nonseminomatous germ-cell tumors 15 years after one adjuvant cycle of bleomycin, etoposide, and cisplatin chemotherapy. *Ann. Oncol.* **2015**, *26*, 374–377. [CrossRef]
28. Cullen, M.; Huddart, R.; Joffe, J.; Gardiner, D.; Maynard, L.; Hutton, P.; Mazhar, D.; Shamash, J.; Wheater, M.; White, J.; et al. The 111 Study: A Single-arm, Phase 3 Trial Evaluating One Cycle of Bleomycin, Etoposide, and Cisplatin as Adjuvant Chemotherapy in High-risk, Stage 1 Nonseminomatous or Combined Germ Cell Tumours of the Testis. *Eur. Urol.* **2020**, *77*, 344–351. [CrossRef]
29. Studer, U.E.; Fey, M.F.; Calderoni, A.; Kraft, R.; Mazzucchelli, L.; Sonntag, R.W. Adjuvant chemotherapy after orchiectomy in high-risk patients with clinical stage I non-seminomatous testicular cancer. *Eur. Urol.* **1993**, *23*, 444–449. [CrossRef]
30. Pont, J.; Albrecht, W.; Postner, G.; Sellner, F.; Angel, K.; Höltl, W. Adjuvant chemotherapy for high-risk clinical stage I nonseminomatous testicular germ cell cancer: Long-term results of a prospective trial. *J. Clin. Oncol.* **1996**, *14*, 441–448. [CrossRef]
31. Cullen, M.H.; Stenning, S.P.; Parkinson, M.C.; Fossa, S.D.; Kaye, S.B.; Horwich, A.H.; Harland, S.J.; Williams, M.V.; Jakes, R. Short-course adjuvant chemotherapy in high-risk stage I nonseminomatous germ cell tumors of the testis: A Medical Research Council report. *J. Clin. Oncol.* **1996**, *14*, 1106–1113. [CrossRef] [PubMed]
32. Bohlen, D.; Borner, M.; Sonntag, R.W.; Fey, M.F.; Studer, U.E. Long-term results following adjuvant chemotherapy in patients with clinical stage I testicular nonseminomatous germ cell tumors with high risk factors. *J. Urol.* **1999**, *161*, 1148–1152. [CrossRef] [PubMed]
33. Chevreau, C.; Mazerolles, C.; Soulié, M.; Gaspard, M.H.; Mourey, L.; Bujan, L. Long-term efficacy of two cycles of BEP regimen in high risk stage I nonseminomatous germ cell tumors with embryonal carcinoma and/or vascular invasion. *Eur. Urol.* **2004**, *46*, 209–215. [CrossRef]
34. Maroto, P.; del Muro, X.G.; Aparicio, J.; Paz-Ares, L.; Arranz, J.A.; Guma, J.; Terrassa, J.; Barnadas, J.; Dorta, J.; Germà-Lluch, J.R. Multicentre risk-adapted management for stage I nonseminomatous germ cell tumours. *Ann. Oncol.* **2005**, *16*, 1915–1920. [CrossRef] [PubMed]
35. Guney, S.; Guney, N.; Sonmez, N.C.; Ergenekon, E. Risk-adapted management for patients with clinical stage I nonseminomatous germ cell tumour of the testis. *Med. Oncol.* **2009**, *26*, 136–142. [CrossRef] [PubMed]
36. Bamias, A.; Aravantinos, G.; Kastriotis, I.; Alivizatos, G.; Anastasiou, I.; Christodoulou, C. Report of the long-term efficacy of two cycles of adjuvant bleomycin/ etoposide/cisplatin in patients with stage I testicular nonseminomatous germ-cell tumors (NSGCT): A risk adapted protocol of the Hellenic Cooperative Oncology Group. *Urol. Oncol.* **2011**, *29*, 189–193. [CrossRef]
37. Warde, P.; Specht, L.; Horwich, A.; Oliver, T.; Panzarella, T.; Gospodarowicz, M.; von der Maase, H. Prognostic factors for relapse in Prognostic factors for relapse in stage I seminoma managed by surveillance: A pooled analysis. *J. Clin. Oncol.* **2002**, *20*, 4448–4452. [CrossRef]
38. Vergouwe, Y.; Steyerberg, E.W.; Eijkemans, M.J.; Albers, P.; Habbema, J.D.F. Predictors of occult metastasis in clinical stage I nonseminoma: A systematic review. *J. Clin. Oncol.* **2003**, *21*, 4092–4099. [CrossRef]

39. Huddart, R.A.; Reid, A.M. Adjuvant Therapy for Stage IB Germ Cell Tumors: One versus Two Cycles of BEP. *Adv. Urol.* **2018**, *2018*, 8781698. [CrossRef]
40. Bujan, L.; Walschaerts, M.; Moinard, N.; Hennebicq, S.; Saias, J.; Brugnon, F.; Auger, J.; Berthaut, I.; Szerman, E.; Daudin, M.; et al. Impact of chemotherapy and radiotherapy for testicular germ cell tumors on spermatogenesis and sperm DNA: A multicenter prospective study from the CECOS network. *Fertil. Steril.* **2013**, *100*, 673–680. [CrossRef]

medicina

MDPI

Case Report

Melanoma: BRAFi Rechallenge

Christoforos S. Kosmidis [1,2], Konstantina Papadopoulou [3,*], Chrysi Maria Mystakidou [4], Evanthia Papadopoulou [5], Stylianos Mantalovas [2], Nikolaos Varsamis [1], Charilaos Koulouris [1,2], Vasiliki Theodorou [4], Konstantinos Papadopoulos [2], Christina Sevva [2], Petrina Miltiadous [4], Savvas Petanidis [6], Eleni Georgakoudi [1], Eleni Papadopoulou [1] and Sofia Baka [1]

1 European Interbalkan Medical Center, 10 Asklipiou Street, 55535 Pylaia, Greece; dr.ckosmidis@gmail.com (C.S.K.)
2 3rd Surgical Department, "AHEPA" University Hospital of Thessaloniki, School of Medicine, Aristotle University of Thessaloniki, 1st St. Kiriakidi Street, 54621 Thessaloniki, Greece
3 1st Department of Internal Medicine, G. Papanikolaou General Hospital of Thessaloniki, 57010 Thessaloniki, Greece
4 Medical School, Faculty of Health Sciences, Aristotle University of Thessaloniki, 54124 Thessaloniki, Greece
5 Shakolas Educational Centre for Clinical Medicine, University of Cyprus, Old Road Nicosia-Lemesos 215/6, 2029 Nicosia, Cyprus
6 Laboratory of Medical Biology and Genetics, Department of Medicine, Aristotle University of Thessaloniki, 54124 Thessaloniki, Greece
* Correspondence: konstantinap97@hotmail.com

Abstract: Melanoma is the most aggressive type of skin cancer. Half of melanoma cases are characterized by the mutation BRAF V600. The case presented concerns a 41-year-old patient with locally advanced melanoma, being positive in mutation BRAF V600. The patient underwent surgery and received additional targeted therapy as part of a clinical study. In subsequent disease progression, immunotherapy was used. When the disease progressed again while the patient was in a good performance status, targeted therapy was administered again, and a good response was noted, making the patient reach a statistically significant overall survival, exceeding four years. Targeted therapy has proven to be an important tool in the treatment of melanoma. The use of BRAFi targeted therapy does not exclude the option of readministration at subsequent disease progression (BRAFi rechallenge). Preclinical models suggest that the resistance mechanism of cancer cells to BRAFi therapy bends, as these cell clones lose their evolutionary advantage after stopping BRAFi. Cell clones sensitive to BRAFi may then outcompete, making the treatment effective again. Therapeutical dilemmas in the management of patients with locally advanced melanoma that progresses to metastatic cancer are discussed.

Keywords: targeted therapy; BRAFi rechallenge; metastatic melanoma; BRAF V600 mutation; BRAF inhibitor

Citation: Kosmidis, C.S.; Papadopoulou, K.; Mystakidou, C.M.; Papadopoulou, E.; Mantalovas, S.; Varsamis, N.; Koulouris, C.; Theodorou, V.; Papadopoulos, K.; Sevva, C.; et al. Melanoma: BRAFi Rechallenge. *Medicina* **2023**, *59*, 975. https://doi.org/10.3390/medicina59050975

Academic Editors: Nicolae Crisan and Călin Căinap

Received: 18 March 2023
Revised: 2 May 2023
Accepted: 11 May 2023
Published: 18 May 2023

1. Introduction

Melanoma is a very aggressive type of skin cancer. Over the past years, its incidence has increased steadily. Despite progress in therapeutic approaches, the overall survival rates have remained relatively low. The progress in the quest for better management is obvious; novel treatment options have developed and gained approval, including targeted therapy and immunotherapy.

Targeted therapy has been a major breakthrough in the last decade, addressing melanoma BRAFV600-mutant tumors. The COMBI-AD clinical trial advanced the treatment of BRAFV600E/K-mutant melanoma. In this trial, patients with Stage III melanoma, who were positive for the BRAFV600E/K mutation, received oral dabrafenib plus trametinib after the surgical resection as adjuvant therapy. The results of five-year follow up are posted, showing obvious benefit in relapse-free survival (52% vs. 36%) and distant-metastasis-free survival (65% vs. 54%) [1]. Despite the beneficial outcome that has been described, many patients receiving targeted therapy, as well as immunotherapy, experience resistance quickly, raising concerns about the next treatment step. There is intense research to forestall resistance, leading to some conclusions in clinical practice, such as the BRAFi-MEKi rechallenge [2–4].

This case report describes a patient who was diagnosed with stage IIIC melanoma. Based on the staging, it was considered a high-risk type of disease, with a five-year survival rate less than 50%. This patient was treated with targeted therapy, immunotherapy, and then targeted therapy rechallenge, reaching five-year survival mostly with a good quality of life (based on PFS [Performance Status Scale]). Rechallenge with targeted therapy is highlighted hereby; it offered survival benefit when treatment options had been eliminated, while the patient remained in good performance status.

2. Case Presentation

A 41-year-old man presented to our clinical department in February 2014 with a rapidly changing black mole on his left leg. Initially, it was a flat brownish macule, which he had for many years. The patient reported that the lesion had become slightly raised. Physical examination and review of systems showed nothing remarkable, and there was no lymphadenopathy. When asked about past medical history or family history, he did not report any relevant issues.

The excision biopsy was performed, and the histopathologic examination demonstrated obvious proliferation of atypical melanocytes in all the levels of the epidermis (pagetoid spread). Additionally, the patient demonstrated atypical melanocytes, both individually and in small nests extended into the papillary dermis. The histologic diagnosis was consistent with malignant melanoma of the superficial spreading type, and the margins were clear (0.3 cm). The Breslow thickness was 4.3 mm (Clark level IV). There was no vascular invasion or satellite lesions. Appropriate histochemistry and immunohistochemistry techniques were used to confirm the diagnosis: HMB45(+), S-100 (+). In March 2014, a wider excision was performed, and the sentinel lymph node biopsy was positive for disease. Lymph nodes clearance did not show further lymph nodes to be positive. Finally, referring to the BRAF mutation, the patient was positive for the mutation–V600E.

With the above data, in April 2014, the patient took part in the COMBI-AD clinical study. This study concerns the evaluation of adjuvant treatment with BRAF and MEK inhibitors–Dabrafenib and Trametinib, and it provided this treatment for high-risk BRAF V600 mutation-positive melanoma after surgical resection. The patient was introduced to the treatment, with no adverse events. The clinical trial has already posted some results, referring to the relapse-free survival and the distance metastasis-free survival [1].

After two years, in March 2016, during the re-examination of the patient, pelvic lymphadenopathy was detected (Figures 1 and 2), which, with the subsequent biopsy, was identified as disease progression with Vimentin(+), Melan C(+), S-100 (+), and AE1/AE3(-).

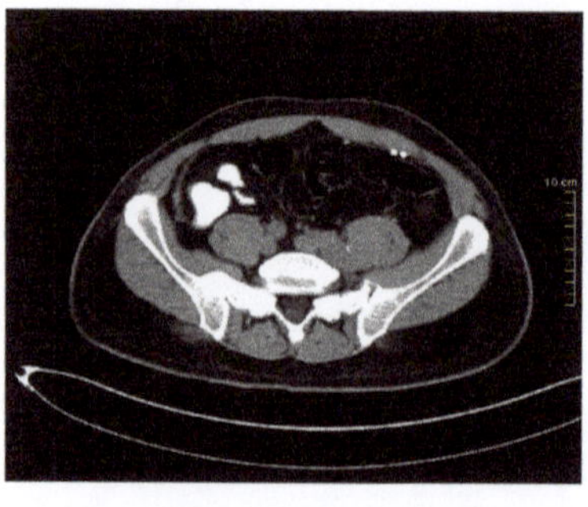

Figure 1. Pelvic lymph node metastases.

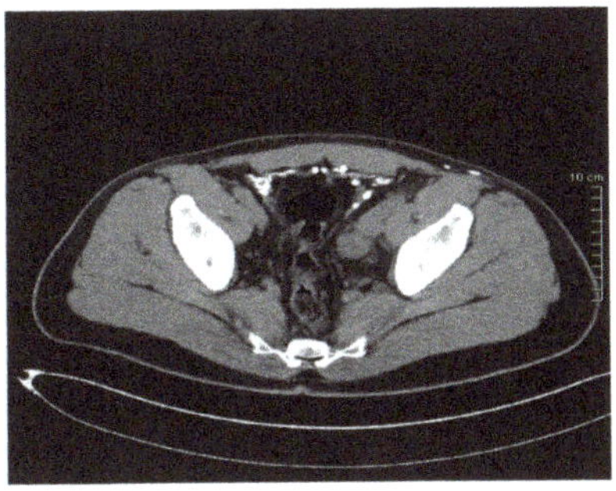

Figure 2. Pelvic lymph node metastases.

Due to disease progression, in May 2016, the patient started receiving an immunotherapy regimen as follows: Nivolumab 1 mg/kg and Ipilimumab 3 mg/kg every three weeks for four cycles, as well as maintenance immunotherapy with Nivolumab 3 mg/kg every two weeks. The patient responded to this new regimen with a significant regression of the lymph node metastasis, visible on imaging (Figure 3). During immunotherapy, the patient experienced adverse reactions. He presented grade I-II lethargy, grade I myalgia, and grade I-II diarrhea during Nivolumab + Ipilimumab combination therapy. Subsequently, with maintenance monotherapy, the patient developed grade I lethargy, grade I-II skin rash, grade I toxic ophthalmopathy, and hypothyroidism. These toxic effects were treated appropriately. Eye drops and topical corticosteroids were given for ophthalmopathy. Concurrently, the patient was treated with Levothyroxine, 50–75 µg/day. The patient remained in good condition with an excellent performance status (Karnofsky PS = 0) and continued to work during treatment.

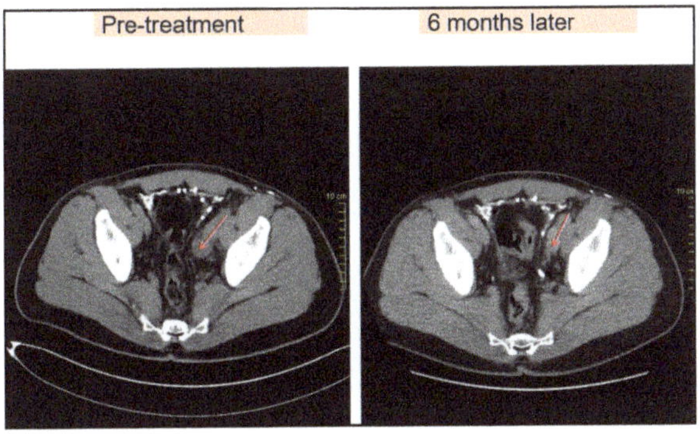

Figure 3. Pelvic lymph node metastases before and after immunotherapy, comparison showing regression (red arrow).

One year after starting immunotherapy, in May 2017, the patient experienced progression of the disease. Specifically, the CT scans showed nodular metastatic sites in the lungs, bilaterally, and metastasis in the liver. At the same time, the pelvic lymph nodes showed deterioration.

At this point, targeted therapy—BRAFi and MEKi—as well as immunotherapy, had already been used. Metastatic melanoma does not respond well to chemotherapy, as patients do not seem to benefit much in terms of overall survival. Therefore, having used the "powerful weapons" in the fight against melanoma, the management of this recurrence has raised concerns. Targeted therapy, which the patient had started years ago, was considered a good choice, as the tumor cell resistance mechanisms that may have formed may have been eliminated by stopping BRAFi and MEKi for a significant period of time [2,3,5,6].

The patient, therefore, restarted the regimen of targeted therapy with BRAF and MEK inhibitors (BRAFi and MEKi retreatment), specifically with Vemurafenib and Cobimetinib, from June 2017. The results showed a significant improvement, being that, in October 2017, the CT scan showed complete disappearance of the lung metastases, and the disease remained stable, since then, for several months. Regarding treatment toxicity, the patient developed a grade II skin rash and fever, which were treated with dose modification.

In April 2018, due to headache and dizziness, the patient underwent a neurological examination and brain imaging. Brain CT and MRI showed new brain metastases in the left occipital lobe (Figure 4). The next step was brain radiotherapy in May 2018, with 3000cGy to the whole brain and 600cGy boost. The patient subsequently continued receiving BRAFi + MEKi and showed clinical improvement. A CT scan of the brain performed after two months showed stability of the disease.

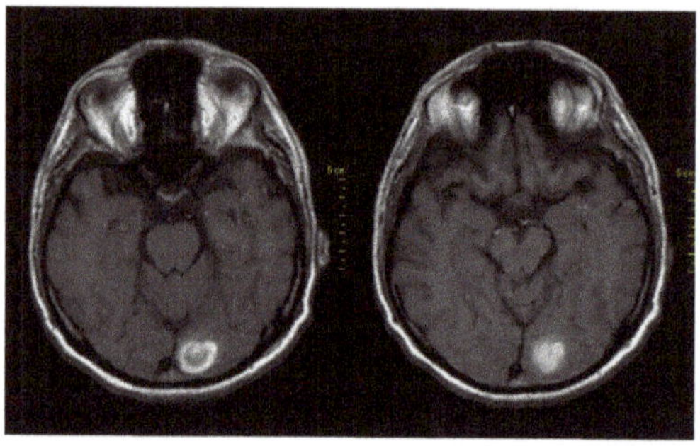

Figure 4. MRI showing brain metastases in the left occipital lobe.

The patient has since remained in good performance status—PS:0—for five months. In November 2018, due to clinical worsening of the disease, he stopped targeted therapy and continued with symptomatic treatment with corticosteroids. The overall patient survival reached five years from the initial disease onset, with good quality-of-life—Karnofsky PS:0—for a long time. The prospect of administering targeted therapy a second time to the same patient (BRAFi rechallenge/retreatment) is recommended, with optimistic results in the mentioned patient, and also in relevant preclinical and clinical studies that have been taking place in recent years.

3. Discussion

Melanoma is a type of skin cancer, which—while its incidence is constantly increasing—is better treated with the evolution in oncology therapy. The overall survival of these patients increased statistically by the emergence of immunotherapy in melanoma and the development of targeted therapies [1]. Patients with BRAF-mutated melanoma are being treated more and more effectively by employing targeted therapy and immunotherapy. Specifically, for BRAFV600-mutant patients, the development of BRAF and MEK inhibitors was highly encouraging. These advances were made in the last decade, and the patient described above derived benefit from COMBI-AD, a clinical trial that brought dabrafenib and trametinib into the adjuvant therapy setting [1]. Still, nearly all patients treated with BRAFi and MEKi eventually develop resistance. Thus, studying the results of targeted therapy rechallenge seems crucial, based on the absence of further efficient treatment regimens, other than immunotherapy. Combination treatments are also studied currently as an alternative when resistance occurs [4].

The path to precision-oriented medicine clearly includes the molecular targeting offered by BRAFi-targeted therapy, but also the recognition of the patient's prognostic profile and the time point at which each treatment regimen can act. Consequently, the decision to re-administer the aforementioned targeted therapy should be made for a selected group of patients in a selected timeframe. Studies show that prognostic factors for patients treated with combination BRAF and MEK inhibition include PS, tumor burden (number of metastatic sites), and LDH level [3]. Specifically, patients with less than three metastatic sites, low LDH, and a good score on PS scales, were the ones who seemed to benefit from a BRAFi–MEKi rechallenge–resulting in a better overall survival.

It was stated that interruption of BRAFi therapy for longer than 12 weeks—whether immunotherapy-mediated or not—appears to be able to confer an advantage to its re-administration [7,8]. The minimum time length of the break is still not well established. In a retrospective study, it has been documented that the duration of the interval between BRAFi cessation and rechallenge is not associated with survival, suggesting that a specifically long drug holiday is not required for resensitisation [3]. Further comparative studies will hopefully clarify the therapeutical outcome regarding the interval duration.

At a molecular level, many resistance mechanisms are reported to reactivate the MAPK pathway, offering a growth advantage while on inhibitor drugs. Some well established mechanisms include BRAFV600E amplification, different splicing of BRAFV600E, and other mutations in the BRAF pathway [9,10]. Adaptive resistance also develops in the absence of genomic alterations involving transcriptional changes by epigenetic mechanisms that trigger the epithelial-to-mesenchymal transition (EMT), melanocyte dedifferentiation, and neural crest stem cell-like reemergence [11,12]. Additionally, activation of autophagy is one of the known mechanisms of resistance to BRAF/MEK inhibitors.

Based on the above resistance mechanisms, studies have been made in the quest for a way to escape resistance. Hydroxychloroquine is an autophagy inhibitor, and it has been suggested in preclinical melanoma models, where it could decrease resistance to BRAF/MEK inhibitors. It has been evaluated in vivo and in safety studies, but clinical conclusions have not been noted yet [13]. Another way of escaping resistance that has been proposed is "Intermittent Treatment" with BRAF and MEK inhibitors. This is a method that has been suggested because of the phenotypic plasticity in drug resistance, but it has not been proven to show benefit. Two recent phase II trials revealed worse progression-free survival in patients with melanoma, who were treated intermittently with BRAFi and MEKi, compared with those treated with continuous therapy, with no difference in overall survival [12]. In conclusion, the considerable benefit of targeted therapy rechallenge in melanoma has triggered different approaches in the BRAFi and MEKi methods of administration in the quest for the most effective treatment in terms of delaying resistance. Still, no clinically proven alternative method for escaping resistance has been developed yet. Thus, when we are faced with progression after targeted therapy and checkpoint inhibitors, the best next step seems likely to be targeted therapy rechallenge, with that being in the context of a patient with good PFS and less than three metastatic sites.

BRAFi rechallenge in advanced-stage melanoma is supported both by cases, such as the aforementioned, and by preclinical and clinical studies looking for the mechanisms of resistance and resensitization of cancer cells to targeted therapies [5,14]. The clues are clear; more than 30% of patients who progress to BRAF or MEK inhibitors have been reported to show a second clinical response after a drug holiday period [3,7]. It is a new plausible treatment option for selected patients with BRAF-mutated melanoma who had, at first, responded to BRAFi, who have subsequently progressed, and who have completed a following treatment. The best clinical response is noted by Valpione et al. when administering combination therapy in patients with less than three metastasis sites [3]. It is supported because of the promising survival data, the known toxicity profile, and the absence of favorable treatment alternatives. The recognition of different prognostic groups offers methods for the design of clinical trials assessing intermittent dosing approaches or rechallenge therapy with BRAFi after a different regimen.

4. Conclusions

Treatment of melanoma has significantly advanced in the light of new therapies: immunotherapy and targeted therapy. However, many patients experience early resistance and progression. Patients, such as the one we described, who are young and keep a good PFS, are eligible for further treatment. Still, effective treatment options other than the aforementioned remain few in number. This case report describes the management of a patient who benefited from adjuvant targeted therapy, immunotherapy, and, when he had progressed, he was administered targeted therapy again (rechallenge), reaching a relatively long OS and a good quality-of-life for most of time. In the context of progression to both targeted therapy and checkpoint inhibitors, patients who have a good PFS, low tumor burden (less than three metastatic sites), and relatively low LDH, are noted to have the most favorable outcome with this treatment approach of rechallenge with combination BRAFi and MEKi [2,3]. Further research is still expected to shed light on this treatment option regarding best timepoint of readministration, possible combination treatment, and more detailed patient selection for targeted therapy rechallenge.

Author Contributions: C.S.K. conceptualization, formal analysis, methodology, supervision, writing—review and editing; K.P. (Konstantina Papadopoulou) project administration, visualization, conceptualization, writing—original draft; C.M.M. investigation, writing—original draft; E.P.(Evanthia Papadopoulou) investigation, writing—original draft; S.M. investigation, resources; N.V. resources, writing—original draft; C.K. investigation, resources; V.T. methodology, resources; K.P. (Konstantinos Papadopoulos) data curation, methodology; C.S. data curation, writing—original draft; P.M. methodology, resources; S.P. investigation, resources; E.G. data curation, resources; E.P. (Eleni Papadopoulou) writing—original draft; S.B. conceptualization, writing—review and editing. All authors have read and agreed to the published version of the manuscript.

Funding: This research received no external funding.

Institutional Review Board Statement: The study was conducted in accordance with the Declaration of Helsinki, and approved by the Institutional Review Board of European Interbalkan Medical Centre of Thessaloniki (protocol code 1967, 27 April 2023).

Informed Consent Statement: Informed consent was obtained from all subjects involved in the study.

Data Availability Statement: No new data were created or analyzed in this study. Data sharing is not applicable to this article.

Conflicts of Interest: The authors declare no conflict of interest.

References

1. Dummer, R.; Hauschild, A.; Santinami, M.; Atkinson, V.; Mandalà, M.; Kirkwood, J.M.; Chiarion Sileni, V.; Larkin, J.; Nyakas, M.; Dutriaux, C.; et al. Five-Year Analysis of Adjuvant Dabrafenib plus Trametinib in Stage III Melanoma. *N. Engl. J. Med.* **2020**, *383*, 1139–1148. [CrossRef] [PubMed]
2. Tietze, J.K.; Forschner, A.; Loquai, C.; Mitzel-Rink, H.; Zimmer, L.; Meiss, F.; Rafei-Shamsabadi, D.; Utikal, J.; Bergmann, M.; Meier, F.; et al. The efficacy of re-challenge with BRAF inhibitors after previous progression to BRAF inhibitors in melanoma: A retrospective multicenter study. *Oncotarget* **2018**, *9*, 34336–34346. [CrossRef] [PubMed]
3. Valpione, S.; Carlino, M.S.; Mangana, J.; Mooradian, M.J.; McArthur, G.; Schadendorf, D.; Hauschild, A.; Menzies, A.M.; Arance, A.; Ascierto, P.A.; et al. Rechallenge with BRAF-directed treatment in metastatic melanoma: A multi-institutional retrospective study. *Eur. J. Cancer* **2018**, *91*, 116–124. [CrossRef] [PubMed]
4. Manzano, J.L.; Layos, L.; Bugés, C.; de los Llanos Gil, M.; Vila, L.; Martínez-Balibrea, E.; Martínez-Cardús, A. BRAF Inhibitor Resistance in Melanoma: Mechanisms and Alternative Therapeutic Strategies. *Curr. Treat. Options Oncol.* **2022**, *23*, 1503–1521. [CrossRef]
5. Das Thakur, M.; Salangsang, F.; Landman, A.S.; Sellers, W.R.; Pryer, N.K.; Levesque, M.P.; Dummer, R.; McMahon, M.; Stuart, D.D. Modelling vemurafenib resistance in melanoma reveals a strategy to forestall drug resistance. *Nature* **2013**, *494*, 251–255. [CrossRef] [PubMed]
6. Li, Q.S.; Shen, B.N.; Xu, H.J.; Ruan, B.F. Promising Strategies for Overcoming BRAF Inhibitor Resistance Based on Known Resistance Mechanisms. *Anticancer Agents Med. Chem.* **2020**, *20*, 1415–1430. [CrossRef] [PubMed]
7. Schreuer, M.; Jansen, Y.; Planken, S.; Chevolet, I.; Seremet, T.; Kruse, V.; Neyns, B. Combination of dabrafenib plus trametinib for BRAF and MEK inhibitor pretreated patients with advanced BRAF. *Lancet Oncol.* **2017**, *18*, 464–472. [CrossRef] [PubMed]

8. Amann, V.C.; Hoffmann, D.; Mangana, J.; Dummer, R.; Goldinger, S.M. Successful retreatment with combined BRAF/MEK inhibition in metastatic BRAFV600-mutated melanoma. *J. Eur. Acad. Dermatol. Venereol.* **2017**, *31*, 1638–1640. [CrossRef] [PubMed]
9. Proietti, I.; Skroza, N.; Bernardini, N.; Tolino, E.; Balduzzi, V.; Marchesiello, A.; Michelini, S.; Volpe, S.; Mambrin, A.; Mangino, G.; et al. Mechanisms of Acquired BRAF Inhibitor Resistance in Melanoma: A Systematic Review. *Cancers* **2020**, *12*, 2801. [CrossRef] [PubMed]
10. Tangella, L.P.; Clark, M.E.; Gray, E.S. Resistance mechanisms to targeted therapy in BRAF-mutant melanoma—A mini review. *Biochim. Biophys. Acta Gen. Subj.* **2021**, *1865*, 129736. [CrossRef] [PubMed]
11. Boumahdi, S.; de Sauvage, F.J. The great escape: Tumour cell plasticity in resistance to targeted therapy. *Nat. Rev. Drug. Discov.* **2020**, *19*, 39–56. [CrossRef] [PubMed]
12. Kavran, A.J.; Stuart, S.A.; Hayashi, K.R.; Basken, J.M.; Brandhuber, B.J.; Ahn, N.G. Intermittent treatment of BRAF melanoma cells delays resistance by adaptive resensitization to drug rechallenge. *Proc. Natl. Acad. Sci. USA* **2022**, *119*, e2113535119. [CrossRef]
13. Awada, G.; Schwarze, J.K.; Tijtgat, J.; Fasolino, G.; Kruse, V.; Neyns, B. A lead-in safety study followed by a phase 2 clinical trial of dabrafenib, trametinib and hydroxychloroquine in advanced BRAFV600 mutant melanoma patients previously treated with BRAF-/MEK-inhibitors and immune checkpoint inhibitors. *Melanoma Res.* **2022**, *32*, 183–191. [CrossRef]
14. Savoia, P.; Zavattaro, E.; Cremona, O. Clinical Implications of Acquired BRAF Inhibitors Resistance in Melanoma. *Int. J. Mol. Sci.* **2020**, *21*, 9730. [CrossRef] [PubMed]

medicina MDPI

Case Report

A Rare Co-Occurrence of Maffucci Syndrome and Astrocytoma with IDH1 R132H Mutation: A Case Report

Nurali Ashirov [1], Iroda Mammadinova [1,*], Aidos Moldabekov [2], Berik Zhetpisbaev [3], Daniyar Teltayev [1], Nurzhan Ryskeldiyev [2] and Serik Akshulakov [1,2]

1 Minimal Invasive Neurosurgery Department, National Centre for Neurosurgery, Astana 010000, Kazakhstan
2 Brain Neurosurgery Department, National Centre for Neurosurgery, Astana 010000, Kazakhstan
3 Department of Pathology, National Centre for Neurosurgery, Astana 010000, Kazakhstan
* Correspondence: irodamammadinova@gmail.com

Abstract: *Background*: Maffucci syndrome is a rare genetic disorder associated with the development of multiple enchondromas and soft tissue cavernous hemangiomas, as well as an increased risk of malignant tumors. *Case Description*: Here we report a case of Maffucci syndrome in a patient who presented with a giant left frontal lobe tumor. Molecular genetic analysis of the tumor revealed an isocitrate dehydrogenase (IDH) mutation p.R132H (c.395C>A) mutation in the IDH1 gene and a heterozygous duplication of the CDKN2A genes. *Conclusions*: The presence of an IDH1 mutation is notable because this mutation is frequently seen in glial tumors and other neoplasms, and its co-occurrence with Maffucci syndrome may represent a novel risk factor for the development of gliomas. This case underscores the importance of genetic testing in patients with Maffucci syndrome who present with central nervous system tumors, as well as the need for further research to understand the relationship between IDH1 mutations and the development of gliomas in this population.

Keywords: astrocytoma; IDH-mutation; multiple enchondroma; Maffucci syndrome

Citation: Ashirov, N.; Mammadinova, I.; Moldabekov, A.; Zhetpisbaev, B.; Teltayev, D.; Ryskeldiyev, N.; Akshulakov, S. A Rare Co-Occurrence of Maffucci Syndrome and Astrocytoma with IDH1 R132H Mutation: A Case Report. *Medicina* **2023**, *59*, 1056. https://doi.org/10.3390/medicina59061056

Academic Editor: Seung-Gu Yeo

Received: 28 March 2023
Revised: 14 May 2023
Accepted: 25 May 2023
Published: 31 May 2023

1. Introduction

Enchondromatosis is a rare skeletal disorder that is characterized by the presence of multiple benign enchondromas that affect the metaphyses of the bones. These enchondromas can cause considerable deformities in the affected area, as well as multiple asymmetric edemas with unilateral prevalence. Ollier disease and Maffucci syndrome are the two most common subtypes of enchondromatosis, both of which are typically diagnosed in the first decade of life [1–3]. Apart from the presence of enchondromas, Maffucci syndrome is characterized by several vascular anomalies, such as hemangiomas of the skin and soft tissues, particularly in the limbs and abdominal wall, as well as lymphangiomas [3,4]. Enchondromas and hemangiomas in Maffucci syndrome have a higher likelihood of transforming into malignant tumors and tend to be diagnosed in adults who are over the age of 30 [2,4,5]. The exact cause of enchondromas remains unknown, but recent studies have shown that somatic mutations in metabolic enzymes, specifically isocitrate dehydrogenase 1 and 2 (IDH1/2), are a common observation in the development of enchondromas [2,6,7].

Significant advancements in cancer genetics over the past decade have revealed that genes encoding IDHs are frequently altered in various types of human malignancies, especially gliomas. Numerous studies have demonstrated that IDH mutations play a significant role in altering cellular physiology, resulting in modifications to cellular metabolism, changes to epigenomes, and abnormal regulation of redox homeostasis [7,8].

The co-occurrence of Maffucci syndrome with glial tumors, specifically astrocytomas, is an extremely rare phenomenon. The association between the two conditions is not fully understood, but some studies have suggested that the IDH mutations present in Maffucci syndrome may predispose individuals to the development of certain types of tumors, including astrocytomas.

Corvino et al. reported that the medical literature contains over thirty documented cases of glioma associated with Ollier disease, whereas the number of reported cases of astrocytoma in combination with Maffucci syndrome is significantly fewer, and the majority of those cases were described before the era of molecular genetics [9].

The available evidence suggests that gliomas in patients with enchondromatosis share similar characteristics with cartilaginous tumors and are likely caused by somatic IDH mosaicism. The timing of IDH mutation acquisition may also impact the molecular features and location of the gliomas. Patients with enchondromatosis-related gliomas tend to develop the tumors at a younger age and have a higher frequency of multicentric tumors. The molecular profile of these gliomas reveals IDH mutations and loss of ATRX expression, while no co-deletion of 1p/19q is observed, which is different from sporadic IDH-mutated gliomas [5].

The rarity of this co-occurrence poses a significant challenge for diagnosis and treatment. Due to the limited number of reported cases, there is a lack of standardized protocols for managing these patients, and the optimal treatment approach remains unclear. However, the identification of the common IDH mutations in both conditions presents a potential avenue for targeted therapy. As genetic studies on the co-occurrence of astrocytoma and Maffucci syndrome are rare, we present a confirmed case of astrocytoma associated with Maffucci syndrome through molecular genetic analysis.

2. Case Presentation

A 32-year-old female, previously diagnosed with Maffucci syndrome, presented to the hospital with a three-month history of headaches and weakness in the right upper and lower extremities. There was no birth record or family history of congenital defects, brain tumors, or bone disease. The patient had several enchondromas on her left humerus and forearm, both lower extremities. The patient had undergone ten tumor resection surgeries for multiple hemangiomas located on both sides of her abdominal wall, hand, and foot and also suffered from lymphedema affecting her right hand and foot (Figure 1). Eight years ago, she was diagnosed with a left-sided ovarian cyst and underwent a left oophorectomy, and two months before admission, she underwent pregnancy termination due to multiple fetal malformations.

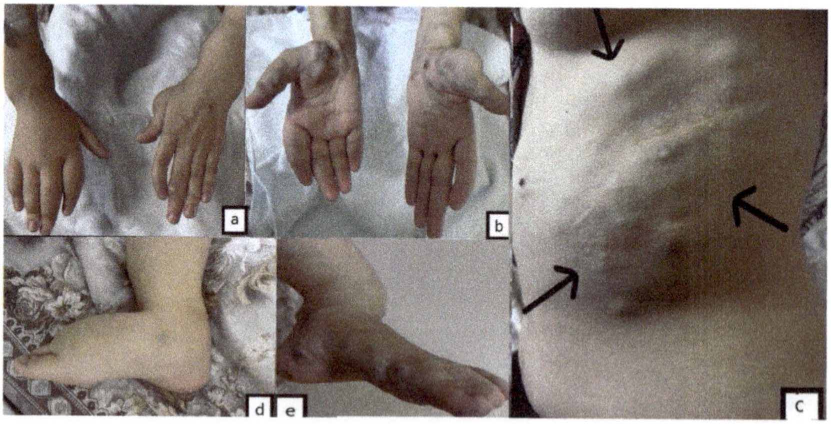

Figure 1. (**a**). Lymphedema of the right hand and cartilage lesion of the left thumb, phalanges deformity of the left hand. (**b**). Thumb and thenar hemangiomas (on both sides). (**c**). Hemangiomas of the abdominal wall and scars after hemangioma resections (arrows). (**d**). Lymphedema of the right foot. (**e**). Hemangiomas of the left foot.

Physical examination revealed multiple enchondromas affecting her left humerus, forearm, and both lower extremities, an S-shaped spinal deformity, an O-shaped deformity and lower extremities hypoplasia, as well as left upper limb hypoplasia. The patient presented with right-sided hemiparesis, and a subsequent neurological examination revealed the same.

A magnetic resonance imaging (MRI) scan showed a massive (83 × 58 × 62 mm) lesion in the left frontal lobe, invading the corpus callosum, with heterogeneous hyperintense T2 and FLAIR signals, hypointensity on T1-weighted images, and peritumoral edema. After gadolinium administration, the lesion showed heterogeneous enhancement on T1-weighted imaging. The preoperative radiological diagnosis was high-grade glioma (Figure 2).

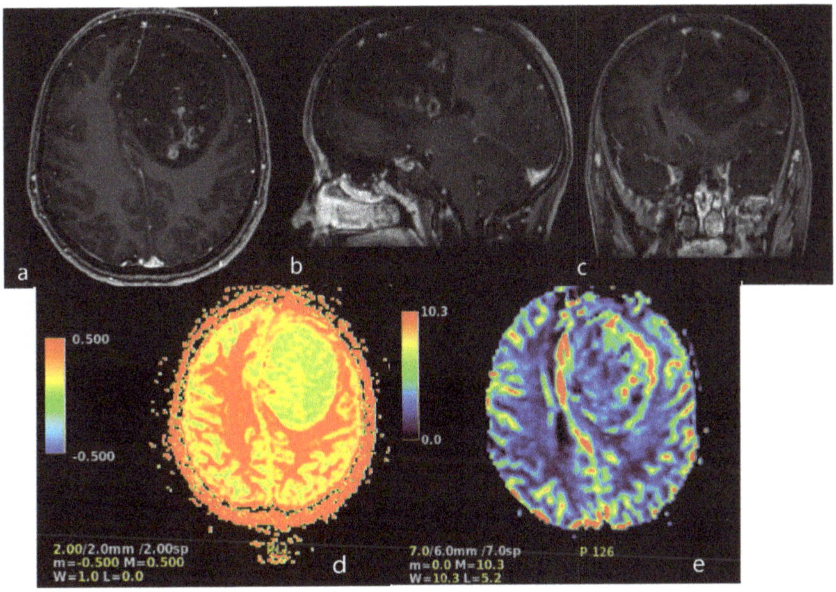

Figure 2. Brain MRI. Axial (**a**), sagittal (**b**) and frontal (**c**) contrast T1−weighted image demonstrates a large (83 × 58 × 62 mm) heterogeneously enhancing tumor in the left frontal lobe, results in a midine shift to the right side to 12 mm. Arterial Spin Labeled MRI Perfusion Imaging (**d**,**e**): elevated regional Cerebral blood perfusion (rCBF) and Cerebral blood volume (CBV) in lesion. Mean transit time (MTT) and Time to peak (TTP) is prolonged. MR spectroscopy (MRS): elevated choline/creatine peaks.

The CT scan of the chest revealed a large, well-circumscribed, destructive lytic lesion (11.7 × 13.1 × 11.2 mm) that had extended into the left lung and mediastinum, confirmed as chondroma by histological analysis. Additionally, a CT scan demonstrates numerous lesions on the sternum, left 5, 6, and 7 ribs, scapula, and humerus (on both sides) with a maximal size of 60 × 49 × 44 mm. There are numerous nodular lesions with calcifications on the anterior chest wall, the largest of which can measure up to 2.3 × 1.0 cm (Figure 3).

The patient underwent frontoparietal craniotomy for subtotal removal of the tumor. Follow-up brain MRI conducted on the second day after the surgery revealed a considerable reduction in tumor size, and neurological examinations showed improvement in the right-sided hemiparesis.

A pathological examination was carried out using an Axioskop 40 microscope by Carl Zeiss (Oberkochen, Germany) and a Pannoramic MIDI scanning microscope, with a total magnification of ×40, ×100, and ×200. The pathological examination with hematoxylin- and eosin-stained slides showed that the tissue was hypercellular, with nuclei that varied in shape and size and had hyperchromatic and polymorphic features (Figure 4). Additionally, the presence of microcysts, glomerular proliferations of vessels, and foci of coagulation necrosis was noted.

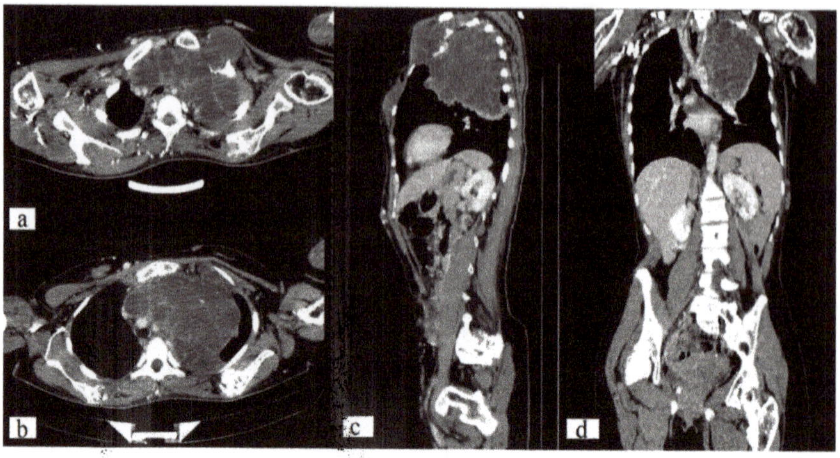

Figure 3. Axial (**a**,**b**), sagittal (**c**) and frontal (**d**) CT scan of the chest demonstrates a well-circumscribed, destructive lytic lesion of the left 1st rib (11.7 × 13.1 × 11.2 mm) and sternum (60 × 49 × 44 mm).

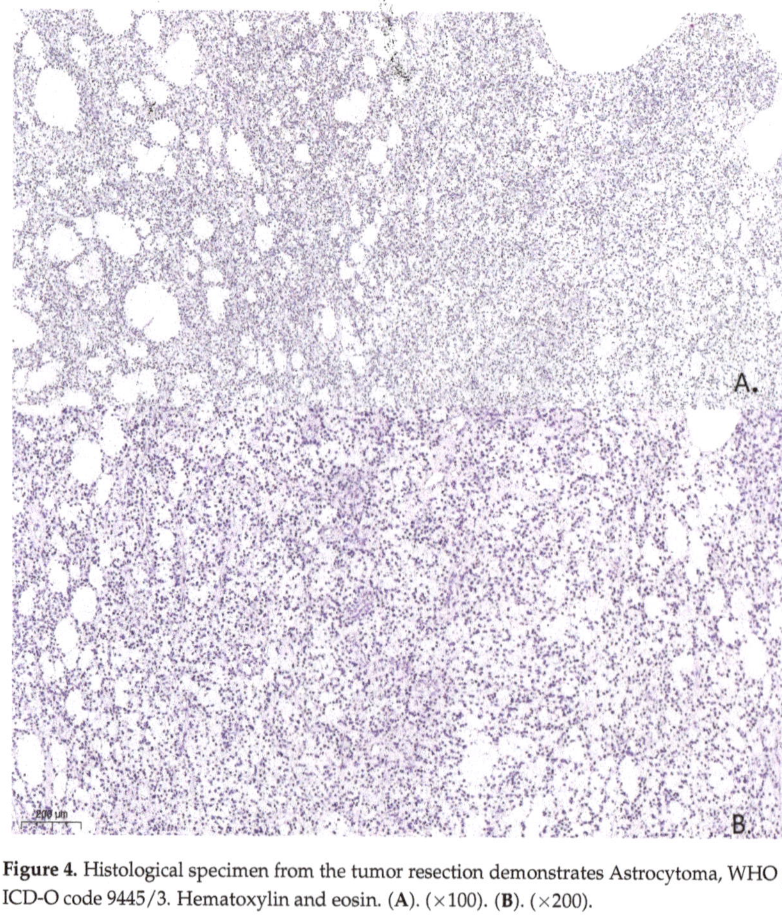

Figure 4. Histological specimen from the tumor resection demonstrates Astrocytoma, WHO grade 4, ICD-O code 9445/3. Hematoxylin and eosin. (**A**). (×100). (**B**). (×200).

The molecular genetic study identified a p.R132H (c.395C>A) mutation in the IDH1 gene and a heterozygous duplication of the CDKN2A genes, which confirmed the diagnosis of astrocytoma, IDH-mutant, WHO grade 4, ICD-O code 9445/3.

The neurological deficit regressed during the postoperative period, and the patient's symptoms gradually improved. On the tenth day following surgery, the patient was discharged for postoperative radiation and chemotherapy.

3. Discussion

This case represents a rare manifestation of Maffucci syndrome, where multiple enchondromas can be associated with the development of glial tumors in the central nervous system.

Enchondromas are benign cartilaginous tumors that arise from the medulla of bones, typically in the metaphysis. Enchondromas are the most common type of benign bone tumor and usually do not cause any symptoms.

However, when multiple enchondromas occur, they can be a sign of a more serious condition, such as Ollier disease or Maffucci syndrome. Ollier disease is a rare disorder characterized by the development of multiple enchondromas in different bones, typically unilateral in distribution with a predilection for the appendicular skeleton. The onset of Ollier disease usually occurs in the first decade of life, but it can occur in early adolescence or adulthood.

Maffucci syndrome is a rare disease identified by the development of multiple enchondromas with vascular anomalies. In addition to multiple enchondromas, individuals with Maffucci syndrome may also develop soft tissue hemangiomas, which are abnormal growths of blood vessels. The hemangiomas can be present on the skin or internal organs [4].

Both Ollier disease and Maffucci syndrome can cause significant symptoms and complications. Multiple swellings on the extremities can lead to deformity around the joints, limitations in joint mobility, scoliosis, bone shortening, leg-length discrepancy, gait disturbances, pain, loss of function, and pathological fractures. Treatment of these conditions is typically aimed at managing the symptoms and preventing complications.

The co-occurrence of Maffucci syndrome and a glial tumor is an uncommon phenomenon. In individuals with Maffucci syndrome, the occurrence of a glial tumor may present unique challenges in diagnosis and treatment due to the presence of multiple enchondromas and hemangiomas throughout the body.

It has been found that both gliomas and chondromas share a common IDH1/2 mutation mechanism. This discovery has led to a molecular analysis of brain tumor cases associated with multiple enchondromas. About 80% of individuals diagnosed with Ollier disease and Maffucci syndrome have somatic mutations in IDH1 and IDH2 detected exclusively in their tumors, which include enchondromas, chondrosarcomas, and vascular anomalies [10].

Skull base enchondromas, chondrosarcoma, and non-mesenchymal neoplasms such as glioma (less than 1/100,000 cases) represent the majority of intracranial involvement [10–13]. Additionally, a rare case of pituitary adenoma associated with systemic enchondromatosis has also been reported [14].

It is important to note that the presence of IDH1/2 mutations in glioma samples from enchondromatosis cases suggests a possible link between these two conditions [11]. The biomolecular studies performed in some reported cases of Ollier disease with brain glioma have shown a positive IDH1 mutation in a majority of cases. Corvino S et al. described 31 reported cases of Ollier disease with brain glioma, and biomolecular studies were performed only in 10 patients, where 8 patients showed positive IDH1-mutation [9]. Prokopchuk et al. described the non-skeletal neoplasms worldwide and found only 5 cases of astrocytoma associated with Maffucci syndrome [15]. According to previous studies, somatic mosaicism of IDH1/2 leads to the development of II-III grade gliomas in this disease [5,16,17]. This suggests that IDH1/2 mutations may be involved in the development of gliomas in patients with Ollier disease and other related disorders.

The presence of IDH1 mutations in astrocytomas is a common finding, and specifically, the c.395 G>A (IDH1 R132H) mutation is frequently observed in these tumors. Other genomic alterations, such as mutations in the ATRX, TP53, and CDKN2A/B genes, are also frequently observed in IDH-mutant astrocytomas [18,19]. Overall, the molecular profiling of IDH-mutant astrocytomas can provide important insights into the mechanisms underlying the development and progression of these tumors.

The hypothesis that somatic IDH mosaicism may be responsible for the development of gliomas in enchondromatosis patients is an interesting one. Bonnet et al. [5] suggested that the cartilaginous nature of these tumors may be related to the presence of IDH mutations, which are frequently observed in chondromas.

There have been a few reported cases of IDH1 R132C mutation in patients with Maffucci syndrome, and some of these cases were associated with glial tumors [5,15,16]. Additionally, IDH2 R172S and TP53 mutations have been identified in anaplastic astrocytomas in a patient with Maffucci syndrome [17]. These findings suggest that IDH mutations may play a role in the development of gliomas in patients with enchondromatosis and related disorders.

The case presented in this report is particularly interesting as it involves a grade IV astrocytoma with a c.395 G>A (IDH1 R132H) mutation. This mutation is one of the most common IDH mutations observed in gliomas [5,14,16,18], and its presence in this patient supports the hypothesis that IDH mutations may contribute to the development of gliomas in patients with enchondromatosis and related conditions.

Identification of these mutations can aid in accurately classifying these disease entities and providing targeted treatment options for patients. However, it is noteworthy that although these tumors share similar appearances, the genes responsible for their development are diverse and have distinct functions, and there is no apparent close relationship between them in terms of their signaling pathways.

The absence of specific diagnostic and surveillance guidelines for individuals with Maffucci syndrome has resulted in the delayed diagnosis of certain malignancies, particularly gliomas. This underscores the importance of having tailored medical monitoring protocols to detect any potential tumors early. The identification of the common IDH1/2 mutation mechanism in gliomas and enchondromas offers a promising avenue for further research to understand the molecular mechanisms underlying these associations. This knowledge can help in the development of effective treatments for patiets with Maffucci syndrome who may be at an increased risk of developing these tumors.

Saiji et al. reported that next-generation sequencing (NGS) analysis enables the detection of IDH mutations in enchondromas, which suggests that most, if not all, central cartilaginous tumors could be treated using novel therapeutic approaches that target IDH mutations. Despite the limited sensitivity of IDH1 R132H antibody-based immunohistochemistry due to the diverse IDH1/2 variants found in enchondromas, it confirms the existence of intratumoral mosaicism along with NGS analysis [7].

While there is still much to be learned about these genetic mutations and their roles in tumor development and progression, targeted treatments that address these specific mutations hold significant potential for improving outcomes.

The identification of common IDH mutations in both Maffucci syndrome and glial tumors, such as astrocytomas, presents a potential avenue for targeted therapy. By identifying the presence of IDH mutations in patients with Maffucci syndrome and associated glial tumors, physicians may be able to tailor treatment strategies that specifically target these mutations. This approach could potentially result in better treatment outcomes and reduced side effects compared to traditional treatments that are not specifically targeted to the genetic mutations present in these tumors.

However, it should be noted that further research is still needed to fully understand the effectiveness of targeted therapy for these patients. The rarity of the co-occurrence of

Maffucci syndrome and glial tumors means that clinical trials are limited and it is difficult to draw definitive conclusions.

It is critical for individuals with Maffucci syndrome to receive regular medical monitoring and follow-up, including imaging studies, to enable early detection of any potential tumors. Furthermore, genetic counseling may be recommended to provide individuals with a comprehensive understanding of their condition and its associated risks.

4. Conclusions

Based on the presented case of grade IV astrocytoma with a c.395 G>A (IDH1 R132H) mutation, it is important to consider IDH1/2 mutations as a predictive marker for the development of gliomas. It is also necessary to establish appropriate monitoring protocols for patients with Maffucci syndrome to detect malignant tumors at an early stage.

Further studies can be conducted to better understand the relationship between IDH1/2 mutations and the development of gliomas, as well as to investigate potential treatments that can target these mutations. It is crucial to continue advancing our knowledge in this field to improve diagnosis and treatment for patients with gliomas and related conditions.

Author Contributions: Conceptualization, N.A. and A.M.; writing—original draft preparation, I.M. and N.A.; writing—review and editing, B.Z., N.R. and D.T.; visualization, I.M.; supervision, S.A.; All authors have read and agreed to the published version of the manuscript.

Funding: This research received no external funding.

Institutional Review Board Statement: Not applicable.

Informed Consent Statement: Written informed consent has been obtained from the patient to publish this paper.

Data Availability Statement: Not acceptable.

Conflicts of Interest: The authors declare no conflict of interest.

References

1. Singh, S.; Dhingra, M.; Verma, A.; Moger, N.M.; Raja, B.S. Ollier's Disease—Rare Presentation of the Rare Disease. *J. Orthop. Case Rep.* **2021**, *11*, 31–33.
2. Zhang, H.; Alman, B.A. Enchondromatosis and Growth Plate Development. *Curr. Osteoporos. Rep.* **2021**, *19*, 40–49. [CrossRef] [PubMed]
3. Sharif, B.; Lindsay, D.; Saifuddin, A. Update on the imaging features of the enchondromatosis syndromes. *Skeletal Radiol.* **2022**, *51*, 747–762. [CrossRef]
4. El Abiad, J.M.; Robbins, S.M.; Cohen, B.; Levin, A.S.; Valle, D.L.; Morris, C.D.; de Macena Sobreira, N.L. Natural history of Ollier disease and Maffucci syndrome: Patient survey and review of clinical literature. *Am. J. Med. Genet. Part A* **2020**, *182*, 1093–1103. [CrossRef]
5. Bonnet, C.; Thomas, L.; Psimaras, D.; Bielle, F.; Vauléon, E.; Loiseau, H.; Cartalat-Carel, S.; Meyronet, D.; Dehais, C.; Honnorat, J.; et al. Characteristics of gliomas in patients with somatic IDH mosaicism. *Acta Neuropathol. Commun.* **2016**, *4*, 31. [CrossRef] [PubMed]
6. Markevičiūtė, V.; Markevičiūtė, M.Š.; Stravinskas, M. Ollier Disease: A Case Series and Literature Review. *Acta Med. Litu.* **2021**, *28*, 181–188. [CrossRef] [PubMed]
7. Saiji, E.; Pause, F.G.; Lascombes, P.; Cerato Biderbost, C.; Marq, N.L.; Berczy, M.; Merlini, L.; Rougemont, A.L. IDH1 immunohistochemistry reactivity and mosaic IDH1 or IDH2 somatic mutations in pediatric sporadic enchondroma and enchondromatosis. *Virchows Arch.* **2019**, *475*, 625–636. [CrossRef] [PubMed]
8. Han, S.; Liu, Y.; Cai, S.J.; Qian, M.; Ding, J.; Larion, M.; Gilbert, M.R.; Yang, C. IDH mutation in glioma: Molecular mechanisms and potential therapeutic targets. *Br. J. Cancer* **2020**, *122*, 1580–1589. [CrossRef]
9. Corvino, S.; Mariniello, G.; Corazzelli, G.; Franca, R.A.; Del Basso De Caro, M.; Della Monica, R.; Chiariotti, L.; Maiuri, F. Brain Gliomas and Ollier Disease: Molecular Findings as Predictive Risk Factors? *Cancers* **2022**, *14*, 3464. [CrossRef]
10. Amary, M.F.; Bacsi, K.; Maggiani, F.; Damato, S.; Halai, D.; Berisha, F.; Pollock, R.; O'Donnell, P.; Grigoriadis, A.; Diss, T.; et al. IDH1 and IDH2 mutations are frequent events in central chondrosarcoma and central and periosteal chondromas but not in other mesenchymal tumours. *J. Pathol.* **2011**, *224*, 334–343. [CrossRef] [PubMed]
11. Ding, C.; Chen, W.; Liu, F.; Xiong, M.; Chen, J. Skull Base Chondrosarcoma Caused by Ollier Disease: A Case Report and Literature Review. *World Neurosurg.* **2019**, *127*, 103–108. [CrossRef] [PubMed]

12. Nishtha, Y.; Maya, B.; Shetty, S.S.; Ganaraj, V.H.; Nupur, P.; Yasha, T.C.; Netravathi, M. A Case of Ollier's Disease with P53 Mutation Positive and IDH1 (R132H) Negative Multicentric Gliomas. *Neurol. India* **2020**, *68*, 657–659. [CrossRef] [PubMed]
13. Oushy, S.; Peris-Celda, M.; Van Gompel, J.J. Skull Base Enchondroma and Chondrosarcoma in Ollier Disease and Maffucci Syndrome. *World Neurosurg.* **2019**, *130*, 356–361. [CrossRef] [PubMed]
14. Nejo, T.; Tanaka, S.; Ikemura, M.; Nomura, M.; Takayanagi, S.; Shin, M.; Ushiku, T.; Shibahara, J.; Saito, N.; Mukasa, A. Maffucci syndrome complicated by three different central nervous system tumors sharing an IDH1 R132C mutation: Case report. *J. Neurosurg.* **2018**, *131*, 1829–1834. [CrossRef] [PubMed]
15. Prokopchuk, O.; Andres, S.; Becker, K.; Holzapfel, K.; Hartmann, D.; Friess, H. Maffucci syndrome and neoplasms: A case report and review of the literature. *BMC Res. Notes* **2016**, *9*, 126. [CrossRef] [PubMed]
16. Achiha, T.; Arita, H.; Kagawa, N.; Murase, T.; Ikeda, J.I.; Morii, E.; Kanemura, Y.; Fujimoto, Y.; Kishima, H. Enchondromatosis-associated oligodendroglioma: Case report and literature review. *Brain Tumor Pathol.* **2018**, *35*, 36–40. [CrossRef] [PubMed]
17. Moriya, K.; Kaneko, M.K.; Liu, X.; Hosaka, M.; Fujishima, F.; Sakuma, J.; Ogasawara, S.; Watanabe, M.; Sasahara, Y.; Kure, S.; et al. IDH2 and TP53 mutations are correlated with gliomagenesis in a patient with Maffucci syndrome. *Cancer Sci.* **2014**, *105*, 359–362. [CrossRef] [PubMed]
18. Louis, D.N.; Perry, A.; Wesseling, P.; Brat, D.J.; Cree, I.A.; Figarella-Branger, D.; Hawkins, C.; Ng, H.K.; Pfister, S.M.; Reifenberger, G.; et al. The 2021 WHO Classification of Tumors of the Central Nervous System: A summary. *Neuro Oncol.* **2021**, *23*, 1231–1251. [CrossRef] [PubMed]
19. Lee, S.; Kambhampati, M.; Almira-Suarez, M.I.; Ho, C.Y.; Panditharatna, E.; Berger, S.I.; Turner, J.; Van Mater, D.; Kilburn, L.; Packer, R.J.; et al. Somatic Mosaicism of IDH1 R132H Predisposes to Anaplastic Astrocytoma: A Case of Two Siblings. *Front. Oncol.* **2020**, *9*, 1507. [CrossRef] [PubMed]

medicina

MDPI

Article

BRAF V600E Mutation of Non-Small Cell Lung Cancer in Korean Patients

Hyo Yeong Ahn [1,2], Chang Hun Lee [1,3], Min Ki Lee [1,4], Jung Seop Eom [1,4], Yeon Joo Jeong [1,5], Yeong Dae Kim [1,2], Jeong Su Cho [1,2], Jonggeun Lee [2], So Jeong Lee [6], Dong Hoon Shin [1,7] and Ahrong Kim [1,3,*]

1 School of Medicine, Pusan National University, Beomeori, Mulgeum-eop, Yangsan 50612, Republic of Korea; doctorahn02@hanmail.net (H.Y.A.); ejspulm@gmail.com (J.S.E.)
2 Department of Thoracic and Cardiovascular Surgery, Biomedical Research Institute, Pusan National University Hospital, Busan 49241, Republic of Korea
3 Department of Pathology, Biomedical Research Institute, Pusan National University Hospital, Busan 49241, Republic of Korea
4 Department of Internal Medicine, Biomedical Research Institute, Pusan National University Hospital, Busan 49241, Republic of Korea
5 Department of Radiology, Biomedical Research Institute, Yangsan Pusan National University Hospital, Busan 50612, Republic of Korea
6 Department of Pathology, Seegene Medical Center, Busan 48792, Republic of Korea; gag86@naver.com
7 Department of Pathology, Biomedical Research Institute, Yangsan Pusan National University Hospital, Busan 50612, Republic of Korea
* Correspondence: ahrong2h@naver.com; Tel.: +82-51-240-7422

Citation: Ahn, H.Y.; Lee, C.H.; Lee, M.K.; Eom, J.S.; Jeong, Y.J.; Kim, Y.D.; Cho, J.S.; Lee, J.; Lee, S.J.; Shin, D.H.; et al. *BRAF* V600E Mutation of Non-Small Cell Lung Cancer in Korean Patients. *Medicina* **2023**, *59*, 1085. https://doi.org/10.3390/medicina59061085

Academic Editor: Nicolae Crisan

Received: 4 May 2023
Revised: 28 May 2023
Accepted: 30 May 2023
Published: 4 June 2023

Abstract: *Background and Objectives: BRAF* mutational status in resected non-small cell lung cancer (NSCLC) in the Korean population is poorly understood. We explored *BRAF* (particularly *BRAF* V600E) mutational status among Korean patients with NSCLC. *Materials and Methods*: This study included 378 patients with resected primary NSCLC who were enrolled from January 2015 to December 2017. The authors obtained formalin-fixed paraffin-embedded (FFPE) tissue blocks and performed peptide nucleic acid (PNA)-clamping polymerase chain reaction (PCR) for detecting BRAF V600, real-time PCR for detecting *BRAF* V600E, and immunohistochemical analyses using the mutation-specific Ventana VE1 monoclonal antibody. For positive cases in any methods mentioned above, direct Sanger sequencing was additionally performed. *Results:* The PNA-clamping method revealed the *BRAF* V600 mutation in 5 (1.3%) of the 378 patients. Among these five patients, real-time PCR, direct Sanger sequencing detected *BRAF* V600E mutations in three (0.8%) patients. Thus, two cases showed differences in their PNA-clamping and the others. Direct Sanger sequencing of PNA-clamping PCR product was performed for two cases showing negative results on direct Sanger sequencing; both contained *BRAF* mutations other than V600E. All patients harboring *BRAF* mutations had adenocarcinomas, and all patients with V600E mutation exhibited minor micropapillary components. *Conclusions*: Despite the low incidence of the *BRAF* mutation among Korean patients with NSCLC, lung adenocarcinoma patients with micropapillary components should be prioritized in terms of *BRAF* mutation testing. Immunohistochemical staining using Ventana VE1 antibody may serve as a screening examination for *BRAF* V600E.

Keywords: lung cancer; B-type Raf kinase (*BRAF*); immunohistochemistry; Ventana VE1 antibody

1. Introduction

The *BRAF* gene is responsible for encoding the V-Raf murine sarcoma viral homolog B (BRAF) kinase, which plays a crucial role in cellular signaling, survival and proliferation [1]. *BRAF* gene is located on chromosome arm 7q34 and is composed of 18 exons [2]. BRAF is associated with mitogen-activated protein kinase (MAPK) pathways including the rat sarcoma (RAS), rapidly accelerated fibrosarcoma (RAF), mitogen-activated protein/extracellular signal regulated kinase (MEK), extracellular signal-regulated kinase

(ERK), and mitogen-activated protein kinase. Mutations in the *BRAF* gene lead to sustained activation of the MAPK pathway, causing it to become a potential oncogenic driver [1]. Almost 300 different BRAF mutations were discovered in melanoma, colorectal cancer, papillary thyroid carcinoma and non-small cell lung cancers (NSCLCs) [3,4]. In addition, BRAF mutations have been classified into three classes. Class I *BRAF* mutation is RAS-independent and has higher kinase activity even in a monomer state. Class I mutation occurs in the valine residue at amino acid position 600 of exon 15; thus, it includes V600 mutations. Class II *BRAF* mutation has an intermediate kinase activity but should form homodimers to be fully activated. Finally, Class III *BRAF* mutation has an impaired kinase activity that requires RAS activation. Class II and III mutations occur either in the glycine of the G loop in exon 11 or in the activation part in exon 15 [5,6]. According to Owsley et al., Class I *BRAF* mutations represented the majority (62.1%) of all *BRAF* -mutant cases (2.4% of all cancers) in 114,662 different tumor sequencing analyses [7].

Now, dabrafenib (BRAF inhibitor) and trametinib (MEK inhibitor) combination therapy is the preferred first-line therapy for the *BRAF* V600E-mutation-positive lung cancer according to the NCCN (National Comprehensive Cancer Network) guidelines. In the French AcSe program, four patients with V600 non-E mutated lung cancer treated with vemurafenib monotherapy had outcomes comparable to the activity of vemurafenib in the *BRAF* V600E mutation [8]. Consequently, a clinical trial targeting V600 non-E mutation in lung cancer, corresponding to Class I *BRAF* mutation, is ongoing to evaluate the activity of dabrafenib and trametinib (NCT04775095). However, Class II and III *BRAF* mutations are not considered to respond to approved BRAF inhibitors [6,8]. Therefore, the evaluation of *BRAF* V600 of exon 15 mutational status, beyond V600E, could become more important.

According to the NCCN guidelines, real-time polymerase chain reaction (PCR), Sanger sequencing, and next-generation sequencing (NGS) are the most commonly recommended methods for *BRAF* mutation examination and immunohistochemistry, with an anti-BRAF p. V600E-specific monoclonal antibody recommended only after extensive validation.

In this study, *BRAF* V600 mutation, particularly the *BRAF* V600E mutational status, was explored with real-time PCR, peptide nucleic acid (PNA)-mediated clamping PCR, direct Sanger sequencing, and immunohistochemistry, which are relatively more feasible to use than NGS. The clinical and pathologic characteristics of the *BRAF* V600E mutation in non-small cell lung cancers were also investigated.

2. Materials and Methods

2.1. Patients, Tissue Specimens, and DNA

This study was performed retrospectively. Three hundred and sixty-eight patients who underwent surgical resection for primary non-small cell lung cancer between 2015 and 2017 at Pusan National University Hospital were included. Among them, five patients had synchronous primary lung cancer. The final cohort was 378 cases of primary non-small cell lung cancers. Formalin-fixed paraffin-embedded (FFPE) tissue blocks, which were made at the time of diagnosis, were used. Clinicopathological data were retrieved from the electric medical records and pathologic reports. Genomic DNA was extracted from FFPE blocks using Maxwell 16 FFPE LEV DNA Purification (Promega corp).

2.2. PNA-Mediated Clamping PCR (PNA Clamping PCR)

PNA Clamp *BRAF* mutation detection kit (Seegene, Seoul, Korea) was used. Extracted DNA was mixed with a PNA probe, primers (5′-AAACTCTTCATAATGCTTGCTCTG (forward) and 5′-GGCCAAAAATTTAATCAGTGGA (reverse)). SYBR green PCR master mix and all reactions totaled 20 µL. Real-time PCR reaction was performed according to the manufacturer's instructions using a CFX96 real-time PCR system (BioRad, Pleasanton, CA, USA). The PNA probe sequences were complementary to wild-type (V600). The PNA probe hybridizes to the wild-type *BRAF* sequence, inhibiting the amplification of the wild-type allele and enhancing preferential amplification of mutant sequences. The positive signal was detected by the intercalation of SYBR green fluorescent dye. The cycle threshold (CT) value was automatically calculated. The delta (ΔCT) value was calculated by subtracting the CT value of a test sample from the standard CT value of a control sample (ΔCT = Standard CT − Sample CT). The cutoff for the presence of mutant was ΔCT of 2. *BRAF* V600 PNA clamping PCR was performed in all 378 cases of non-small cell carcinoma.

2.3. Real-Time PCR

The real-time PCR used the Real-Q *BRAF* V600E detection kit (Real-Q; Biosewoom, Seoul, Republic of Korea). Real-time PCR was performed with CFX96 real-time PCR Detection system (Bio-Rad) according to the manufacture's instruction. The master mixture contained 12.5 μL of the 2X PCR reaction mixture and 2.5 μL of the *BRAF* probe and primer mixture. A total of 15 μL of the master mixture was dispensed into PCR tubes. Then, the extracted DNA of 10 μL (containing 50 ng of DNA) was added to each PCR tube. The sample was considered positive for V600E mutation when both the sample and the internal control were amplified and both CT value of the sample and the internal control were less than 40. If a sample showing the difference between CT value of the sample and the internal control was more than 13, the test was repeated. *BRAF* V600E real-time PCR was performed in all 378 cases of non-small cell carcinoma.

2.4. Immunohistochemistry

Immunohistochemistry was performed on the same FFPE block used for molecular testing. An automatic staining device (BenchMark XT, Ventana Medical Systems, Tucson, AZ, USA) was used for staining. All samples were cut into 3 μm thick sections and the sections were deparaffinized in an EZ prep. The slides were pretreated with CC1 (cell conditioner 1, pH8.4 buffer) for 64 min antigen retrieval and followed by pre-primary antibody peroxidase inhibition. Then, the slides were incubated with the Ventana BRAF V600E (VE1) mouse monoclonal primary antibody, and Hematoxylin II ® and Bluing Reagent was used for counterstaining. A sample known to have V600E mutation was used as a positive control. A case was considered to be positive when a signal was present in the cytoplasm [9]. Any nuclear staining was ignored.

2.5. Direct Sanger Sequencing

BRAF exon 15, which potentially contains the c.1799 T > A transversion mutation, was amplified from genomic DNA by PCR using primers 5′-AAACTCTTCATAATGCTTGCTCTG (forward) and 5′-GGCCAAAAATTTAATCAGTGGA (reverse). Amplification was performed under the following conditions: 1 cycle at 94 °C for 5 min, 40 cycles of denaturation at 94 °C for 30 s, annealing at 63 °C for 30 s, and extension at 72 °C for 30 s; then a final extension at 72°C for 5 min using BioRad C1000 (Pleasanton, CA). After purification of the PCR products, direct bidirectional sequencing was performed using the ABI 3730XL DNA Analyzer. Additionally, direct bidirectional sequencing was repeated using the BRAF PNA clamping PCR product, which is rich in mutant alleles, to detect the variants of low level. Direct Sanger sequencing using extracted DNA from FFPE blocks was performed in 5 cases of any positive results for *BRAF* V600 PNA clamping, *BRAF* V600E real-time PCR and immunohistochemistry for VE1. Particularly, direct Sanger sequencing using PNA clamping PCR product was conducted in cases of discordance in other methods.

3. Results

3.1. Clinicopathologic Characteristics of Resected Non-Small Cell Lung Cancers

A total cohort of 378 patients with resected non-small cell carcinoma was included in this study. All included patients were Korean. Basic data for included patients are shown in Table 1. Patient age ranged from 36 to 86 years (mean: 66.84 ± 8.76 years). The size of the cancer ranged from 0.9 cm to 10.0 cm (mean: 3.37 ± 1.55 cm). There were 238 males (63.0%) and 140 females (37.0%). The study cohort included 255 cases of adenocarcinoma (67.5%), 91 cases of squamous cell carcinoma (24.1%), 5 cases of adenosquamous cell carcinoma (1.3%), 9 cases of large cell neuroendocrine carcinoma (2.4%), 15 cases of sarcomatoid carcinoma (4%) and others (3 cases, 0.8%). Three hundred and five patients (80.7%) had early-stage disease (stage I and II) and the remaining 73 patients (19.3%) had advanced disease (stage III and IV). Among the patients, 168 (44.4%) patients never smoked, 112 patients (29.6%) were ex-smokers and the other 98 patients (25.9%) were current smokers. Information about an EGFR-activating mutation and ALK fluorescence in situ hybridization (FISH)/ALK D5F3 CDx Ventana immunohistochemistry was retrieved from the prior pathologic reports in electronic medical records. One hundred and twenty patients (31.7%) had EGFR-activating mutations and 11 patients (2.9%) had ALK translocation.

Table 1. Clinical characteristics of study population.

Characteristics	Number (%)
Age (years)	66.84 ± 8.76
Sex	
Male	238 (63.0)
Female	140 (37.0)
Smoking status	
Never-smoker	168 (44.4)
Current smoker	98 (25.9)
Ex-smoker	112 (29.6)
Pack-years among ever-smokers (years) *	34.68 ± 19.81
Tumor size (cm)	3.37 ± 1.55
Histologic type	
ADC	255 (67.5)
SqCC	91 (24.1)
SC	15 (4.0)
LCNEC	9 (2.4)
ADSqCC	5 (1.3)
Other	3 (0.8)
Differentiation	
WD	19 (5.0)
MD	269 (71.2)
PD	90 (23.8)
Stage	
Early (I–II)	305 (80.7)
Advanced (III–IV)	73 (19.3)
EGFR mutation	
Absent	258 (68.3)
Present	120 (31.7)
ALK translocation	
Absent	367 (97.1)
Present	11 (2.9)
Ethnicity Korean	378 (100.0)

* Among the ever-smokers, pack-year data were unavailable in six cases. ADC, Adenocarcinoma; SqCC, Squamous cell carcinoma; SC, sarcomatoid carcinoma; LCNEC, Large-cell neuroendocrine carcinoma; ADSqCC, Adenocasquamous cell carcinoma WD, Well-differentiated; MD, Moderately differentiated; PD, Poorly differentiated; EGFR, epidermal growth factor receptor; ALK, anaplastic lymphoma kinase.

3.2. BRAF V600 PNA Clamping and BRAF V600E Real-Time PCR

BRAF V600 mutation was detected in five cases (1.3%) using a PNA clamping method among 378 non-small cell carcinoma cases (Tables 2 and 3). By using BRAF real-time PCR, a BRAF V600E mutation was detected in 3 patients (0.8%) among the total 378 cohort (Tables 2 and 4), and all these positive cases for real-time PCR had positive results in PNA clamping PCR. There were two discordant cases between PNA clamping and real-time PCR.

Table 2. The results of BRAF mutation according to BRAF mutation assay.

Case Number	*BRAF* V600 PNA Clamping	*BRAF* V600E Real-Time PCR	IHC for VE1	Direct Sanger Sequencing	Direct Sanger Sequencing of the Clamping PCR Product
142	+	+	+	V600E	Not done
270	+	−	−	WT	V600K
324	+	−	−	WT	V600V, K601E
348	+	+	+	V600E	Not done
358	+	+	+	V600E	Not done

BRAF, B-type Raf kinase; IHC, immunohistochemistry; PNA, peptide nucleic acid; PCR, polymerase chain reaction; WT, Wild-type; +, positive; −, negative.

Table 3. The results of *BRAF* V600 on PNA-clamping.

Case Number	DNA Loading (ng)	Cycle Threshold (C_T)		ΔC_T-2	ΔC_T-1
		Non-PNA	V600	V600	V600
Clamping Control		**24.38**	**36.33**	**11.95**	**−1.33**
Positive Control		**24.03**	**30**	**5.97**	**5**
142	10	28.59	31.7	3.11	3.3
270	10	26.91	32.44	5.53	2.56
324	10	25.98	31.82	5.84	3.18
348	10	26.02	30.3	4.28	4.7
358	10	26.21	31.46	5.26	3.54
142	25	27.15	31.13	3.97	3.87
270	25	26.31	32.42	6.1	2.58
324	25	25.15	31.89	6.74	3.11
348	25	25.22	29.6	4.38	5.4
358	25	25.26	31.02	5.76	3.98

BRAF, B-type Raf kinase; PNA, peptide nucleic acid, PC, positive control, ΔC_T = standard C_T—sample C_T.

Table 4. The results of *BRAF* V600E on real-time PCR.

Patient Number	Internal Control C_T	Sample C_T	Result
142	25.9	31.9	+
270	24.6	NA	−
324	26.6	NA	−
348	24.7	29.6	+
358	24.4	30.6	+

BRAF, B-type Raf kinase; C_T, cycle threshold; NA, not applicable; +, positive; −, negative.

3.3. Immunohistochemistry for VE1

Immunohistochemistry for VE1 was performed in all included patients with full-face sections of FFPE blocks. Regarding the results of immunohistochemistry, three patients (0.8%) showed positive staining for tumor cytoplasm (Figure 1, Tables 2 and 5). All three cases with positive staining showed diffuse positivity for tumor cells. However, two cases had heterogenous staining intensity, though all tumor cells were positive. The other case had diffuse positivity with homogeneous intensity for tumor cytoplasm. The detailed information of staining is shown in Table 5. However, two patients, with positive results for PNA clamping, had negative immunostaining. Regarding the results of immunohistochemistry for VE1, there were two cases showing discordance with the PNA clamping method, and there was no discordant case with real-time PCR.

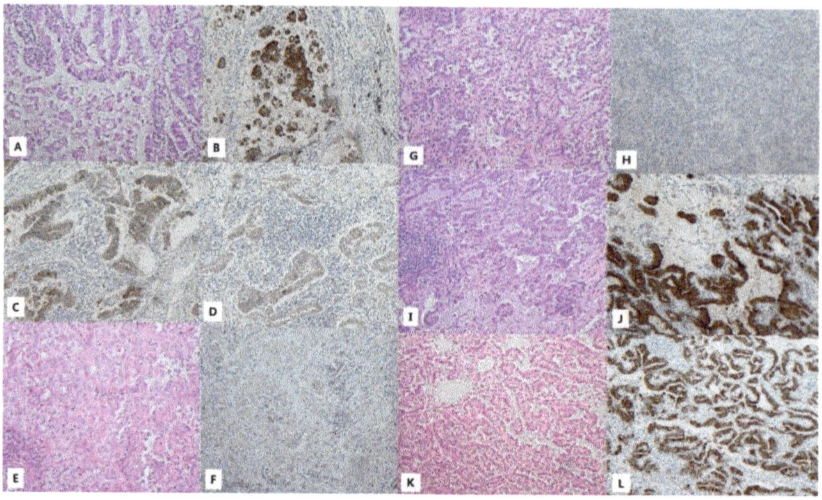

Figure 1. Histologic feature and immunohistochemistry. (**A–D**). Patient 142; all ×200. (**A**) A Hematoxylin and eosin (H&E)-stained section showing acinar and micropapillary structures. (**B**) Immunohistochemistry for VE1 (staining intensity 3). (**C**) Immunohistochemistry for VE1 (staining intensity 2). (**D**) Immunohistochemistry for VE1 (staining intensity 1). (**E,F**) Patient number 270; all ×200. (**E**) An H&E-stained section showing solid and acinar growth patterns. (**F**) Negative immunostaining for VE1. (**G,H**) Patient number 324; all ×200. (**G**) An H&E-stained section showing an acinar growth pattern. (**H**) Negative immunostaining for VE1. (**I,J**) Patient number 348; all ×200. (**I**) An H&E-stained section showing an acinar pattern and a few micropapillary structures. (**J**) Immunohistochemistry for VE1 (staining intensity 3). (**K,L**) Patient number 358; all ×200. (**K**) An H&E-stained section showing an acinar pattern and a few micropapillary structures. (**L**) Immunohistochemistry for VE1 (staining intensity 2).

Table 5. Immunohistochemistry results of patients who were V600E-positive on either or both *BRAF* PNA-clamping and *BRAF* real-time PCR.

Case Number	Proportion of Cytoplasm Positive Rate	Intensity Pattern	Intensity Scores of Tumor Cells	Result
142	100	Heterogeneous	3+: 40% 2+: 50% 1+: 10% 0: 0%	+
270	0	Homogeneous	3+: 0% 2+: 0% 1+: 0% 0: 100%	−
324	0	Homogeneous	3+: 0% 2+: 0% 1+: 0% 0: 100%	−
348	100	Homogeneous	3+: 100% 2+: 0% 1+: 0% 0: 0%	+
358	100	Heterogeneous	3+:70% 2+:30% 1+: 0% 0: 0%	+

BRAF, B-type Raf kinase; +, positive; −, negative.

3.4. Direct Sequencing

There were five patients (1.3%) who had positive results for BRAF PNA clamping, real-time PCR and immunohistochemistry. For these five patients, direct Sanger sequencing was performed. The results of Sanger sequencing were the same with those of real-time PCR and immunohistochemistry (Figure 2). Considering the PNA clamping method is a very sensitive method to detect a low allele level of mutation, direct sequencing using a clamping PCR product was performed [10]. Regarding the results of sequencing using a clamping PCR product, all cases showed mutation for the BRAF gene other than the V600E genotype (Figure 3). Finally, there were five mutated cases (1.3%) for BRAF in the total cohort. Among them, three cases (0.8%) had a V600E mutation and the other two had V600K and V600V/V601E mutations, respectively. Among the total number of BRAF mutations, V600E genotype was present in three cases, comprising 60% of the BRAF mutant.

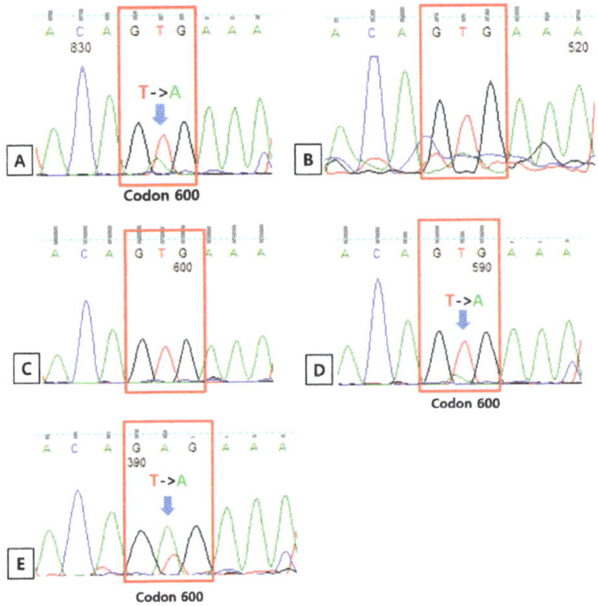

Figure 2. Results of direct Sanger sequencing. (**A**) Case number 142 carries the c.1799 T > A (p. V600E) mutation. (**B**) Case number 270 is wild-type for B-type Raf kinase (BRAF). (**C**) Case number 324 is wild-type for BRAF. (**D**) Case number 348 carries the c.1799 T > A (p. V600E) mutation. (**E**) Case number 358 carries the c.1799 T > A (p. V600E) mutation.

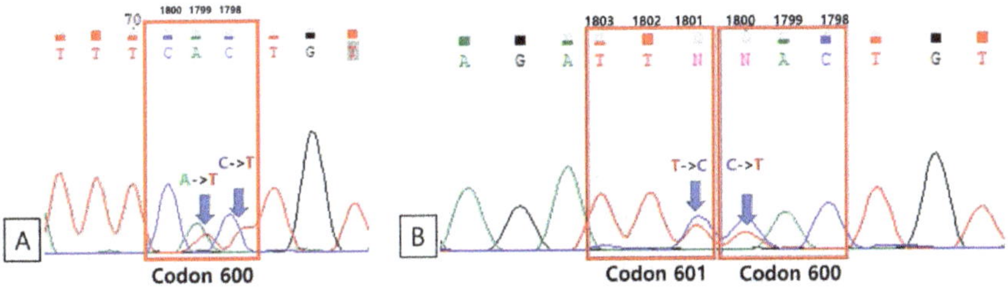

Figure 3. Direct sequencing of peptide nucleic acid (PNA) polymerase chain reaction (PCR) products reveals two p. V600E-negative cases. (**A**) Case number 270 carries the c.1798_1799 CA > TT (p. V600K) for B-type Raf kinase (BRAF) mutation. (**B**) Case number 324 carries the c.1800 C > T (p. V600V) and c.1801 T > C (p. K601E) BRAF mutations.

3.5. Clinicopathologic Aspects of BRAF Mutation in Lung Cancers

There were three cases (0.8%) of the BRAF V600E mutation among 378 non-small cell carcinomas. It was 1.2% among 255 adenocarcinoma and 246 non-small cell carcinomas without EGFR/ALK aberrations. In addition, it was 2.3% among 129 adenocarcinomas without EGFR/ALK aberrations. There were two cases of BRAF mutation other than V600E, comprising 0.5% of all non-small cell carcinoma, 0.8% of adenocarcinoma and non-small cell carcinoma without EGFR/ALK aberrations and 1.5% of adenocarcinoma without EGFR/ALK aberrations (Table 6).

Table 6. Incidence of BRAF mutations.

	BRAF Mutation	*BRAF* V600E Mutation	*BRAF* Non-V600E Mutation
	N (%)	N (%)	N (%)
NSCLC (n = 378)	5 (1.3)	3 (0.8)	2 (0.5)
Adenocarcinomas (n = 255)	5 (2.0)	3 (1.2)	2 (0.8)
NSCLC lacking *EGFR* mutation and *ALK* translocation (n = 246)	5 (2.0)	3 (1.2)	2 (0.8)
Adenocarcinomas lacking *EGFR* mutation and *ALK* translocation (n = 129)	5 (3.8)	3 (2.3)	2 (1.5)

Among the V600E mutated patients, one patient was a never-smoker and the other two were ever-smokers. There was a micropapillary component in all the V600E-mutated cases (Table 7). All patients harboring a BRAF mutation had no concomitant EGFR or ALK aberrations. Detailed clinicopathologic characteristics of individual patients with a BRAF mutation are listed in Table 7.

Table 7. Clinicopathological characteristics of individual patients with BRAF mutations.

Case No.	BRAF Mutation	Age (Years)	Sex	Smoking Status	Pack Years	Predominant Histological Subtype	Present of Micropapillary Component	pT Size (cm)	pN Stage	pM Stage	Stage
142	V600E	72	Male	Current	50	Solid	+	6.5	2	0	IIIA
270	V600K	83	Male	Current	15	Acinar	-	3.6	0	0	IB
324	V600V, K601E	67	Female	Never	0	Acinar	-	1.6	0	0	IA
348	V600E	51	Female	Never	0	Acinar	+	2.2	0	0	IA
358	V600E	52	Male	Ex-smoker	7.5	Acinar	+	1.8	0	0	IA

4. Discussion

In this study, the *BRAF* V600 mutation incidence was found in five patients (1.3%) among all cases of non-small cell carcinoma, and the *BRAF* V600E mutation was present in three patients (0.8%) with adenocarcinoma. It is relatively low when compared to most reports from the Western population [11–14]. On the other hand, the incidence is similar to that of Japanese patients [15]. In this study, there were 2.3% and 1.5% of the *BRAF* V600E mutation and *BRAF* V600 non-E mutation, respectively, among adenocarcinoma without *EGFR/ALK* alterations. According to one Korean dataset, there were four patients (1.8%) with a *BRAF* mutation among 222 Stage III/IV lung adenocarcinoma patients without *EGFR/ALK* aberrations [16]. The difference from these data probably resulted from the difference in stage distribution of the study cohort. This study included 305 cases (80.7%) of early-stage (Stage I and II) disease, contrary to their advanced stage cohort. This study included 123 cases of non-adenocarcinoma patients, and none of these patients harbored a *BRAF* mutation. However, other data reported the detection of a *BRAF* mutation in non-adenocarcinoma patients, though the incidence was very low [11,13–15,17]. Among five *BRAF* V600-mutated lung cancer patients, two patients (40%) were never-smokers and three

(60%) were ever-smokers. This is in accordance with the molecular testing guideline for the selection of patients with lung cancer for treatment, suggesting that *BRAF* mutational testing should be performed on all advanced adenocarcinoma patients, irrespective of clinical characteristics [18].

BRAF V600 non-E mutation was present in two cases. However, the result of direct sequencing showed the wild type of *BRAF* using amplified DNA extracted from FFPE blocks following sequencing using PNA clamping product detected mutation. Through PNA clamping PCR, the wild-type alleles are inhibited in the amplification process by hybridization with PNA, resulting in mutant enrichment. Though detected mutation was not the V600E genotype in this study, this result suggests that sequencing using the PNA clamping PCR product can help the detection of a mutant of low level in suspected or equivocal cases. In addition, Zengarini et al. presented some treatment effects on *BRAF* V600K mutated melanoma patients [19], and there is also an ongoing phase 2 clinical trial on the application of dabrafenib and trametinib in tumors with the *BRAF* V600E or V600K mutation including non-small cell lung cancer (ClinicalTrials.gov Identifier: NCT04439292). Additionally, *BRAF* V600E real-time PCR showed both 100% of sensitivity and specificity. All these results are in accordance with the principles of molecular and biomarker analysis for *BRAF* by NCCN guideline: "Real-time PCR, Sanger sequencing (ideally paired with tumor enrichment), and NGS are the most commonly deployed methodologies for examining *BRAF* mutation status" (NCCN Guidelines Version 3.2023).

All patients harboring the *BRAF* V600 mutation had adenocarcinoma and all patients with the V600E genotype had a micropapillary component. The result is similar to those of prior reports [9,20,21]. Theis may suggest that lung adenocarcinoma with micropapillary should be first considered to conduct *BRAF* mutation testing.

The VE1 mouse monoclonal antibody was utilized in this study. VE1 antibody is a mutation-specific antibody able to differentiate a V600E-mutated protein from wild-type and other *BRAF*-mutated proteins [22]. In this study, immunohistochemistry for VE1 showed both 100% of sensitivity and specificity. Gow et al. validated the usefulness of the Ventana VE1 antibody in lung cancer [20]. They reported that immunohistochemistry for VE1 antibody showed a 96.6% sensitivity to detect the *BRAF* V600E mutation and a 98.6% specificity to discriminate tumors without the *BRAF* V600E mutation. However, one positive case affecting the specificity value had weak positive cytoplasmic staining in 5% of tumor cells and the case had *ROS1* gene fusion. According to their criteria, the case was considered to be negative. Ilie et al. reported that VE1 immunohistochemistry is specific and sensitive to detect the *BRAF* V600E mutation [9]. Similar results were shown by Hofman et al., suggesting that VE1 staining is a rapid, specific and very sensitive method [23]. In addition, Chang et al. reported that VE1 immunohistochemistry showed almost perfect interobserver agreement, suggesting that this could be a screening test for *BRAF* mutation [24]. In present study, *BRAF* V600E mutated cases showed diffuse (100% of proportion) positivity, though the intensity was heterogeneous in two cases. Overall, these results suggest that immunohistochemistry for the Ventana VE1 antibody can be a useful screening tool in lung cancers, especially for small biopsy specimens, which must be handled with care to obtain the maximum information for treatment choice. Moreover, immunohistochemistry has many advantages over molecular diagnostics, namely because it needs much less tissue and the turn-around time is far shorter.

There are limitations in this study. The detection rate of *BRAF* mutation was only 1.3% of the study cohort, so statistical analyses could not be performed. In addition, these data are from one single institution, which makes it difficult to generalize these findings. However, this study cohort was composed of only the Korean population, and for all experiments, we only used consecutive resected samples of primary lung cancer.

5. Conclusions

BRAF V600 mutation status in resected primary non-small cell carcinoma was tested. There were five cases (1.3%) of a *BRAF* V600 mutation among 378 non-small cell carcinomas, comprising three cases of a *BRAF* V600E mutation and two cases of a *BRAF* V600 non-E mutation. All cases harboring a *BRAF* V600 mutation were adenocarcinoma without *EGFR* mutation and *ALK* translocation. All three cases of a *BRAF* V600E mutation had micropapillary component. Immunohistochemistry for Ventana VE1 antibody can be a useful screening method to detect a *BRAF* V600E mutation.

This study preliminarily suggests that the incidence of a *BRAF* V600E mutation might be low in Korean population. In addition, adenocarcinoma showing micropapillary component, especially without *EGFR/ALK* aberration, should be first considered for *BRAF* testing, including immunohistochemistry.

Author Contributions: Data curation, J.S.C., J.L., S.J.L. and D.H.S.; Formal analysis, Y.J.J. and Y.D.K.; Methodology, C.H.L., M.K.L. and J.S.E.; Writing—original draft, H.Y.A. and A.K.; Writing—review and editing, A.K. All authors have read and agreed to the published version of the manuscript.

Funding: This work was supported by a 2-Year Research Grant of Pusan National University.

Institutional Review Board Statement: The study was conducted in accordance with the Declaration of Helsinki, and approved by the Institutional Review Board of Pusan National Universitiy Hospital (IRB No 2107-011-105), approval date 28 July 2017.

Informed Consent Statement: Patient consent was waived because the research involves no more than minimal risk and waiver of informed consent will not adversely affect the rights and welfare of the subjects.

Data Availability Statement: Not applicable.

Acknowledgments: We send special thanks to Myoungsuk Jo and Jungeun Ko who provided technical support. This research was partly presented in AACR 2020 (abstract 2343).

Conflicts of Interest: The authors declare no conflict of interest.

References

1. Yan, N.; Guo, S.; Zhang, H.; Zhang, Z.; Shen, S.; Li, X. BRAF-Mutated Non-Small Cell Lung Cancer: Current Treatment Status and Future Perspective. *Front. Oncol.* **2022**, *12*, 863043. [CrossRef] [PubMed]
2. Raman, M.; Chen, W.; Cobb, M.H. Differential regulation and properties of MAPKs. *Oncogene* **2007**, *26*, 3100–3112. [CrossRef] [PubMed]
3. Forbes, S.A.; Bindal, N.; Bamford, S.; Cole, C.; Kok, C.Y.; Beare, D.; Jia, M.; Shepherd, R.; Leung, K.; Menzies, A.; et al. COSMIC: Mining complete cancer genomes in the Catalogue of Somatic Mutations in Cancer. *Nucleic Acids Res.* **2011**, *39*, D945–D950. [CrossRef]
4. Tabbò, F.; Pisano, C.; Mazieres, J.; Mezquita, L.; Nadal, E.; Planchard, D.; Pradines, A.; Santamaria, D.; Swalduz, A.; Ambrogio, C.; et al. How far we have come targeting BRAF-mutant non-small cell lung cancer (NSCLC). *Cancer Treat. Rev.* **2022**, *103*, 102335. [CrossRef] [PubMed]
5. Leonetti, A.; Facchinetti, F.; Rossi, G.; Minari, R.; Conti, A.; Friboulet, L.; Tiseo, M.; Planchard, D. BRAF in non-small cell lung cancer (NSCLC): Pickaxing another brick in the wall. *Cancer Treat. Rev.* **2018**, *66*, 82–94. [CrossRef]
6. Riudavets, M.; Cascetta, P.; Planchard, D. Targeting BRAF-mutant non-small cell lung cancer: Current status and future directions. *Lung Cancer* **2022**, *169*, 102–114. [CrossRef] [PubMed]
7. Owsley, J.; Stein, M.K.; Porter, J.; In, G.K.; Salem, M.; O'Day, S.; Elliott, A.; Poorman, K.; Gibney, G.; VanderWalde, A. Prevalence of class I-III BRAF mutations among 114,662 cancer patients in a large genomic database. *Exp. Biol. Med. (Maywood)* **2021**, *246*, 31–39. [CrossRef]
8. Mazieres, J.; Cropet, C.; Montané, L.; Barlesi, F.; Souquet, P.J.; Quantin, X.; Dubos-Arvis, C.; Otto, J.; Favier, L.; Avrillon, V.; et al. Vemurafenib in non-small-cell lung cancer patients with BRAF. *Ann. Oncol.* **2020**, *31*, 289–294. [CrossRef]
9. Ilie, M.; Long, E.; Hofman, V.; Dadone, B.; Marquette, C.H.; Mouroux, J.; Vignaud, J.M.; Begueret, H.; Merlio, J.P.; Capper, D.; et al. Diagnostic value of immunohistochemistry for the detection of the BRAFV600E mutation in primary lung adenocarcinoma Caucasian patients. *Ann. Oncol.* **2013**, *24*, 742–748. [CrossRef]
10. Kang, S.H.; Pyo, J.Y.; Yang, S.W.; Hong, S.W. Detection of BRAF V600E mutation with thyroid tissue using pyrosequencing: Comparison with PNA-clamping and real-time PCR. *Am. J. Clin. Pathol.* **2013**, *139*, 759–764. [CrossRef]
11. Luk, P.P.; Yu, B.; Ng, C.C.; Mercorella, B.; Selinger, C.; Lum, T.; Kao, S.; O'Toole, S.A.; Cooper, W.A. BRAF mutations in non-small cell lung cancer. *Transl. Lung Cancer Res.* **2015**, *4*, 142–148. [CrossRef] [PubMed]
12. Paik, P.K.; Arcila, M.E.; Fara, M.; Sima, C.S.; Miller, V.A.; Kris, M.G.; Ladanyi, M.; Riely, G.J. Clinical characteristics of patients with lung adenocarcinomas harboring BRAF mutations. *J. Clin. Oncol.* **2011**, *29*, 2046–2051. [CrossRef] [PubMed]

13. Marchetti, A.; Felicioni, L.; Malatesta, S.; Grazia Sciarrotta, M.; Guetti, L.; Chella, A.; Viola, P.; Pullara, C.; Mucilli, F.; Buttitta, F. Clinical features and outcome of patients with non-small-cell lung cancer harboring BRAF mutations. *J. Clin. Oncol.* **2011**, *29*, 3574–3579. [CrossRef] [PubMed]
14. Tissot, C.; Couraud, S.; Tanguy, R.; Bringuier, P.P.; Girard, N.; Souquet, P.J. Clinical characteristics and outcome of patients with lung cancer harboring BRAF mutations. *Lung Cancer* **2016**, *91*, 23–28. [CrossRef] [PubMed]
15. Kinno, T.; Tsuta, K.; Shiraishi, K.; Mizukami, T.; Suzuki, M.; Yoshida, A.; Suzuki, K.; Asamura, H.; Furuta, K.; Kohno, T.; et al. Clinicopathological features of nonsmall cell lung carcinomas with BRAF mutations. *Ann. Oncol.* **2014**, *25*, 138–142. [CrossRef]
16. Kim, H.C.; Kang, Y.R.; Ji, W.; Kim, Y.J.; Yoon, S.; Lee, J.C.; Choi, C.M. Frequency and clinical features of BRAF mutations among patients with stage III/IV lung adenocarcinoma without EGFR/ALK aberrations. *Onco Targets Ther.* **2019**, *12*, 6045–6052. [CrossRef]
17. Brustugun, O.T.; Khattak, A.M.; Trømborg, A.K.; Beigi, M.; Beiske, K.; Lund-Iversen, M.; Helland, Å. BRAF-mutations in non-small cell lung cancer. *Lung Cancer* **2014**, *84*, 36–38. [CrossRef]
18. Kalemkerian, G.P.; Narula, N.; Kennedy, E.B.; Biermann, W.A.; Donington, J.; Leighl, N.B.; Lew, M.; Pantelas, J.; Ramalingam, S.S.; Reck, M.; et al. Molecular Testing Guideline for the Selection of Patients With Lung Cancer for Treatment With Targeted Tyrosine Kinase Inhibitors: American Society of Clinical Oncology Endorsement of the College of American Pathologists/International Association for the Study of Lung Cancer/Association for Molecular Pathology Clinical Practice Guideline Update. *J. Clin. Oncol.* **2018**, *36*, 911–919. [CrossRef]
19. Zengarini, C.; Mussi, M.; Veronesi, G.; Alessandrini, A.; Lambertini, M.; Dika, E. BRAF V600K vs. BRAF V600E: A comparison of clinical and dermoscopic characteristics and response to immunotherapies and targeted therapies. *Clin. Exp. Dermatol.* **2022**, *47*, 1131–1136. [CrossRef]
20. Gow, C.H.; Hsieh, M.S.; Lin, Y.T.; Liu, Y.N.; Shih, J.Y. Validation of Immunohistochemistry for the Detection of. *Cancers* **2019**, *11*, 866. [CrossRef]
21. Hwang, I.; Choi, Y.L.; Lee, H.; Hwang, S.; Lee, B.; Yang, H.; Chelakkot, C.; Han, J. Selection Strategies and Practical Application of BRAF V600E-Mutated Non-Small Cell Lung Carcinoma. *Cancer Res. Treat.* **2022**, *54*, 782–792. [CrossRef] [PubMed]
22. Capper, D.; Preusser, M.; Habel, A.; Sahm, F.; Ackermann, U.; Schindler, G.; Pusch, S.; Mechtersheimer, G.; Zentgraf, H.; von Deimling, A. Assessment of BRAF V600E mutation status by immunohistochemistry with a mutation-specific monoclonal antibody. *Acta Neuropathol.* **2011**, *122*, 11–19. [CrossRef] [PubMed]
23. Hofman, V.; Benzaquen, J.; Heeke, S.; Lassalle, S.; Poudenx, M.; Long, E.; Lantéri, E.; Bordone, O.; Lespinet, V.; Tanga, V.; et al. Real-world assessment of the BRAF status in non-squamous cell lung carcinoma using VE1 immunohistochemistry: A single laboratory experience (LPCE, Nice, France). *Lung Cancer* **2020**, *145*, 58–62. [CrossRef] [PubMed]
24. Chang, S.; Choi, Y.L.; Shim, H.S.; Lee, G.K.; Ha, S.Y.; Group, K.C.P.S. Usefulness of BRAF VE1 immunohistochemistry in non-small cell lung cancers: A multi-institutional study by 15 pathologists in Korea. *J. Pathol. Transl. Med.* **2022**, *56*, 334–341. [CrossRef] [PubMed]

MDPI
St. Alban-Anlage 66
4052 Basel
Switzerland
www.mdpi.com

Medicina Editorial Office
E-mail: medicina@mdpi.com
www.mdpi.com/journal/medicina

www.ingramcontent.com/pod-product-compliance
Lightning Source LLC
Chambersburg PA
CBHW061818210926
51850CB00006B/744